D0078739

ACS SYMPOSIUM SERIES **601**

Polymers for Second-Order Nonlinear Optics

Geoffrey A. Lindsay, EDITOR
U. S. Navy

Kenneth D. Singer, EDITOR
Case Western Reserve University

Developed from a symposium co-sponsored
by the ACS Division of Polymer Chemistry, Inc.,
the Optical Society of America, and the ACS Division
of Polymeric Materials: Science and Engineering, Inc.,
at the 208th National Meeting
of the American Chemical Society,
Washington, DC,
August 21–25, 1994

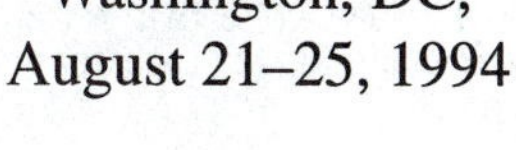

American Chemical Society, Washington, DC 1995

Library of Congress Cataloging-in-Publication Data

Polymers for second-order nonlinear optics / Geoffrey A. Lindsay, editor, Kenneth D. Singer, editor.

p. cm.—(ACS symposium series; 601)

"Developed from a symposium sponsored by the Division of Polymer Chemistry, Inc. at the 208th National Meeting of the American Chemical Society, Washington, DC, August 21–25, 1994."

Includes bibliographical references and indexes.

ISBN 0–8412–3263–6

1. Polymers—Optical properties—Congresses. 2. Nonlinear optics—Congresses. I. Lindsay, Geoffrey A., 1945– . II. Singer, Kenneth D. III. American Chemical Society. Division of Polymer Chemistry, Inc. IV. American Chemical Society. Meeting (208th: 1994: Washington, DC) V. Series.

QD381.9.066P65 1995
620.1'9204295—dc20 95–30598
CIP

This book is printed on acid-free, recycled paper. 

PRINTED IN THE UNITED STATES OF AMERICA

1995 Advisory Board

ACS Symposium Series

Foreword

THE ACS SYMPOSIUM SERIES was first published in 1974 to provide a mechanism for publishing symposia quickly in book form. The purpose of this series is to publish comprehensive books developed from symposia, which are usually "snapshots in time" of the current research being done on a topic, plus some review material on the topic. For this reason, it is necessary that the papers be published as quickly as possible.

Before a symposium-based book is put under contract, the proposed table of contents is reviewed for appropriateness to the topic and for comprehensiveness of the collection. Some papers are excluded at this point, and others are added to round out the scope of the volume. In addition, a draft of each paper is peer-reviewed prior to final acceptance or rejection. This anonymous review process is supervised by the organizer(s) of the symposium, who become the editor(s) of the book. The authors then revise their papers according to the recommendations of both the reviewers and the editors, prepare camera-ready copy, and submit the final papers to the editors, who check that all necessary revisions have been made.

As a rule, only original research papers and original review papers are included in the volumes. Verbatim reproductions of previously published papers are not accepted.

Contents

THIN FILM DEVICES

Preface

THE INVENTION OF HIGH-POWER LASERS and early experiments on nonlinear optics created a need to develop materials for nonlinear optical applications. Most of these applications remain in the tuning of laser sources. However, as video and other high-bandwidth communication emerges, applications of nonlinear optical materials for information processing, storage, and display will expand.

Research with organic and polymeric materials has been particularly promising. In the early days of research (late 1960s to late 1970s), the inherently high nonlinearity was promoted as the most desirable feature of organic nonlinear optical materials. New approaches to processing these materials, including thin-film techniques such as the monomolecular deposition and spin coating of polymer glasses, have been investigated. It is now clear that the inherent processability of organic and polymeric materials and their integrability with substrates are also key properties.

The first ACS symposium on nonlinear optical properties of organic and polymeric materials in 1982 marked the emergence of the field as a major research area. After the initial work on poled polymer glasses for electrooptic applications, the field grew even faster as a number of companies began to investigate the potential application of coatable electrooptic materials. Much effort has focused on developing new materials, and the richness of potential structures makes this effort particularly fruitful.

A hallmark of this field is its interdisciplinary nature: Chemists, physicists, and engineers have worked together to develop materials and applications. To formalize the interdisciplinary nature of the field, the Optical Society of America and the ACS Divisions of Polymer Chemistry, Inc., (POLY) and Polymeric Materials: Science and Engineering, Inc., (PMSE) formed a committee to plan joint activities of the two societies. The timing of the formation of this committee is most appropriate: Activity in the field is shifting to applications, which require close coupling between materials research and optical science and engineering.

The committee's first project was to launch a series of annual joint symposia on polymeric films for photonics applications. These symposia are held in alternating years at the annual meetings of each society. The first meeting, held at the annual meeting of the Optical Society of America in Toronto, Canada, in October 1993, attracted about 70 presentations. The second symposium was held at the ACS national meeting in

Washington, D.C., in August 1994. This meeting attracted about 120 presentations.

The symposium covered a wide range of subjects, including second- and third-order materials and effects, photorefractive and electroluminescent polymers, and passive waveguide materials. The focus of the presentations spanned the required disciplines and included the theory of optical nonlinearities; synthesis of new materials; processing of materials to activate optical nonlinearities; new methods of measurement and characterization; and device design, construction, and performance.

In this volume, we include chapters that describe in depth advances in second-order nonlinear optical polymer films produced by an orientational ordering process such as electric field poling. Shorter synopses of the entire program can be found in *Polymer Preprints*, Volume 35, Number 2.

Acknowledgments

The stimulating nature of the symposium was due to the enthusiastic participation of the presenters and attendees and the hard work of the domestic and international organizing committee; for this work we are grateful. This hard work attracted high-quality speakers from Asia, Europe, and North America and allowed for a full hearing of the wide-ranging activities in the field. We also wish to thank those organizations that provided generous support for the symposium. These include the POLY and PMSE divisions of the American Chemical Society, the Optical Society of America, and the Office of Naval Research. Participation of overseas scientists was made possible by the Tokyo and London branches of the Office of Naval Research and the London and Washington branches of the Air Force Office of Scientific Research.

We also acknowledge those whose support and assistance made the publication of this volume possible. Foremost among these are the authors who contributed the chapters. We are also indebted to the timely efforts of the reviewers for their critical evaluations of the manuscripts. We are also grateful to Anne Wilson and Barbara Pralle of the ACS Books Department, Genie Berry at China Lake, and Angie Amato at Case Western Reserve University for their contributions.

GEOFFREY A. LINDSAY
U.S. Navy, Code 474220D
Chemistry and Materials Branch
China Lake, CA 93555–6001

KENNETH D. SINGER
Case Western Reserve University
Department of Physics
10900 Euclid Avenue
Cleveland, OH 44106–7079

February 10, 1995

In Memoriam

RONALD ANDREW HENRY synthesized many new nonlinear optical chromophores late in his career. His contributions were important in the generation of 32 papers and four U.S. patents on new second-order nonlinear optical polymers (NLOP). A large number of his novel chromophores are still under investigation.

Born in Yakima, Washington, on January 26, 1916, he was the eldest son of immigrant farmers from Scotland and Canada. The University of Washington awarded him a B.S. (magna cum laude) in 1938 and a Ph.D. in chemistry in 1947. He worked for Procter and Gamble for five years. Henry then joined the U.S. Navy's laboratory at China Lake, California, as a civilian research chemist in 1947. China Lake was established during World War II as a proving ground for the rockets developed by California Institute of Technology; in 1945 a permanent Navy laboratory for research and development was created there. Henry and other chemists at China Lake were pictured in a 1948 *Life* article on "Rocket Town." Henry contributed to 175 publications and 30 U.S. patents during his career at China Lake. He was head of the organic chemistry branch for more than 16 years and was a senior research scientist later in his career.

He was among the world's top experts in high nitrogen chemistry, the synthesis of chemiluminescent compounds, and the synthesis of laser dyes. Among the many topics he worked on were morpholino compounds; substituted tetrazoles; nitroguanidine derivatives; azido, azene, imidazole, and hydrazone compounds; gas-generating compounds; burn-rate modifiers; polymeric binders; and vanadium catalysts. In 1988 he launched into the new research area of NLOP. Although he officially retired in 1979, he continued to work every day at his chemistry bench until his health deteriorated late in 1993. He died of cancer on March 13, 1994. His professional honors include the 1966 L. T. E. Thompson Award (one of the highest awards at China Lake) and the Navy Superior Civilian Service

Award in 1970. He was a member of the American Chemical Society for 56 years.

His memory of the thousands of compounds he had synthesized was legendary. When asked about making a new compound, often he would come back in a day with a vial of it in his hand because he was able to dig it out of a box of chemicals he had made 20 years ago. As a synthetic chemist, he was very skilled and productive, and his chemical intuition was extraordinary.

Henry was a walking encyclopedia, not only on organic chemistry, but also on the flora and fauna of the Sierra Nevada mountains and the Mojave Desert. He led Boy Scouts on many a trip in the wilderness and helped to build Boy Scout and Girl Scout camps in the Sierras. He and his wife, Mary Ann, counted and inventoried plant species in this area every month of the year. There are many things we will remember about Henry: his tall, lanky frame, his bright neckties that Mary Ann made, his deep booming voice that greeted everyone, "GOOD MORNING." Ron Henry loved his work and gave generously of his time and talents. This book is dedicated to his memory.

Chapter 1

Second-Order Nonlinear Optical Polymers

An Overview

Geoffrey A. Lindsay

U.S. Navy, Code 474220D, NAWCWPNS, China Lake, CA 93555–6001

The science and technology of second-order nonlinear optical polymers (NLOP) are briefly reviewed in an attempt to explain: (1) why these materials are important; (2) the origin of optical nonlinearity; (3) the strengths and weaknesses of various polymer topologies; (4) thin film processing techniques; (5) how to create and measure polar order (noncentrosymmetry); (6) stability issues; (7) fabrication of waveguides; and (8) useful devices that can be made from NLOP.

Organic polymeric thin films for photonic applications has been an exciting, rapidly evolving area of research over the last decade (*1*). This book is focused on one class of materials within this field: second-order nonlinear optical polymeric (NLOP) films (*2*). Recent developments on NLOP films portend exciting new possibilities for low cost integrated devices for the telecommunication and datacommunication industries, although NLOP devices are not yet commercial. At the heart of the new capabilities are electro-optic (EO) waveguides made from polymer films that switch optical signals from one path to another and modulate the phase or amplitude of an optical signal at greater than 40 GHz. NLOP films can also be used for sum-difference frequency generation (e.g., frequency-doubling).

Typical NLOP films are glassy polymers containing asymmetric chromophores (also called dyes), which point generally in the same direction, making the film asymmetrically polarizable. Several processes for aligning the chromophores are under development, including electric-field poling (at the softening temperature) and various self-assembly techniques. The NLOP film is dielectric and undergoes a change in index of refraction when an electric field is applied across it, which is the property of interest for optical switching and modulating.

Competing inorganic materials such as lithium niobate (*3*) and III-V materials, such as GaAs/AlGaAs multiple quantum wells (*4*), are more mature in their development but have made limited penetration in the commercial market place for high speed integrated optical devices, due in part to material property drawbacks such as fragility, high dielectric constant at microwave frequencies, and processing difficulties (*5*). NLOP overcome these particular material property deficiencies, but there are still many problems to solve, such as low cost ways to efficiently couple light into and out of the waveguides.

Obtaining a high nonlinear optical coefficient in a thermally stable polymer film has been the major quest in NLOP materials research. Chapters in this book report good progress in that regard. NLOP are now being made that have excellent stability (this is discussed in the section on stability). Today's NLOP are superior EO materials for very high speed devices due to their low dielectric constant, ε, at all frequencies (ε is less than 4 for NLOP and greater than 30 for $LiNbO_3$) (*1a*). Polymers are much easier to integrate in silicon devices and in large arrays. Today's NLOP's do not exceed $LiNbO_3$ in EO performance in low speed devices because of it's higher index of refraction, n (the figure of merits, n^3r and V_π are discussed in more detail in the section on devices); however, chromophores under development will likely change that situation in the future.

NLOP films are being developed by many worldwide companies (Akzo-Nobel, 3M, AlliedSignal, Lockheed, Hercules, IBM, Dow, NTT, Ciba-Geigy, France Telecom, and more). Several large companies (Hoechst-Celanese, AT&T, DuPont, Kodak, Sandoz) cut back their R&D programs over the last 5 years for a variety of business reasons, e.g., NLOP may not compete with direct modulation of lasers for the large volume, low speed (MHz) applications, and the potential market for very high speed devices is rather limited in the next 5 years. However, the author believes NLOP will most definitely find commercial niches in the out years as photonic systems proliferate. Most new materials need years of laboratory incubation and feedback from potential customers in order to clear all the hurdles and find business niches.

Chromophores

The molecular origin of optical nonlinearity is due to the electrical polarization of the chromophore as it interacts with electromagnetic radiation. This phenomenon is described by the Schrödinger wave equation (*6*). Second-order nonlinearity normally occurs in noncentrosymmetric materials. Therefore, asymmetric chromophores are of interest and they must be at least partially aligned in the same direction (i.e., they must have polar alignment in the NLOP film). Some symmetrical multipolar molecules can exhibit second-order nonlinearity (see reference (*1d*), but this class of chromophore will not be discussed here.

Macroscopic optical properties of NLOP films depend on the electrical polarization in the film (these relations are in the literature (*7*)). The electrical polarization in a material is simply the dipole moment per unit volume. The linear and nonlinear optical behavior of a NLOP film can be described by a series expansion of the polarization in the material, P, in powers of the applied electric fields, E:

$$P = P_o + \chi^{(1)}E + \chi^{(2)}EE + \chi^{(3)}EEE + \ldots \quad (1)$$

where P_o is the permanent (or ground state) polarization, and $\chi^{(\xi)}$ are the susceptibilities. (Note that subscripts are omitted here for brevity -- $\chi^{(2)}$ is a third-rank tensor.) The $\chi^{(1)}$ term describes ordinary linear behavior (refraction and absorption). The $\chi^{(2)}$ term (the first nonlinear term) describes the optical effects resulting from the interaction of two electrical fields, e.g., laser and radio frequency fields. The $\chi^{(2)}$ effects are sometimes referred to as "three-wave mixing," i.e., $(\omega_3;\omega_1,\omega_2)$ where ω_1 and ω_2 are frequencies of the applied fields, and ω_3 is the resulting optical frequency. The $\chi^{(3)}$ term in equation 1 describes four-wave mixing, and is important here when electric-field-induced third-order effects exist.

The same kind of mathematical treatment can be applied to an isolated chromophore's dipole moment (*8*): $\mu = \mu_o + \alpha E + \beta EE + \gamma EEE + \ldots$ where β is the chromophore's second-order susceptibility tensor (also called the quadratic hyperpolarizability, or first hyperpolarizability). Usually $\chi^{(2)}$ is related to the chromophore's β from second harmonic generation (SHG) measurements in poled NLOP films as follows:

$$\chi^{(2)} = N\beta(f^{\omega})^2 f^{2\omega} \langle \cos^3(\theta) \rangle \quad (2)$$

where N is the number of chromophores per unit volume, f^{ω} and $f^{2\omega}$ are the "local field factors" which are simple functions of the index of refraction at the fundamental and second harmonic frequencies, and θ is the average angle between the ground-state dipole moments of the chromophores and the direction in the film they would be pointing if perfectly aligned, e.g., perpendicular to the plane of the polymer film. To use equation 2, one can calculate or measure β [see Chapters 2-6 in this book for more insight on β]; and, one can estimate θ from UV-VIS absorption measurements before and after poling (*9*), or from birefringence measurements (which can be subject to error due to contributions from polymer backbone alignment during film fabrication) (*8b*).

A very successful model for calculating the value of β can be derived by considering the interaction of the ground state of the chromophore and its first excited state (the "two-state model") (*10*). The most important factors are that the chromophore have a low energy of transition between ground and first excited state, a large change in dipole moment in going between the ground state to the first excited state, and a large oscillator strength. Computer programs are available to perform the molecular orbital calculations for predicting β (e.g., MOPAC) (*11*). Calculations of β often assume the chromophore is in a vacuum. Chapter 4 in this book addresses solvent effects.

A chromophore's β can be measured from electric-field induced second harmonic generation (EFISH) measurements [see Chapter 5 in this book], and hyper-Rayleigh scattering (HRS) measurements [see Chapter 6 in this book]. If β is measured near a chromophore's charge transfer absorption band, β is resonantly enhanced. Hence, one must state the wavelength of light used to measure β. To compare different chromophores "off resonance," β measurements are often "corrected" with the two-state model to the "static case" (β_0, infinitely removed from resonance). One must be careful in comparing measured β to calculated β because different conventions are used to define hyperpolarizability (*8a*).

Operating a device near resonance, even in the small tail of a resonance, will cause absorptive losses (heating). Hence, a particular application will dictate where the chromophore's absorption peak (and band edges) must lie relative to the optical frequencies being processed in the device. This trade-off between transparency and β deserves more investigation by the NLOP community. Photothermal deflection spectroscopy offers an excellent technique for measuring very small absorption's (*12*).

In general, a good chromophore has an electron donating group connected to an electron accepting group by a large π-conjugated group. Some progress has been made in understanding more subtle structural motifs that increase the hyperpolarizability. For example, it has been shown that the degree of resonance between the charge transfer states is very important (e.g., whether the ground state is neutral, cyanine, or zwitterionic) (*13*). This degree of resonance is measured in terms of bond length alternation (BLA), defined as the difference between the length of double and single bonds of the vinylenes. It has been shown that the BLA should be about 0.05 Å (or -0.05 Å for zwitterionic chromophores) for maximum β (*13*). For some chromophores, solvatochromic effects (i.e., shifts in the absorption band due to the solvent interacting with the chromophore) have a very large effect on β (*14*). In fact β can be estimated from solvatochromic shifts (*15*). If the BLA is zero, which might actually happen if the wrong solvent is used, the β goes to zero! Hence, more attention should be paid to the solvatochromic effects [see Chapter 4 in this book], especially the effects of the polymer matrix on the $\chi^{(2)}$ properties, e.g., hydrogen bonding groups may stabilize a chromophore's polar excited state.

The actual $\chi^{(2)}$ of a NLOP film is fairly easily determined from second harmonic generation (SHG) measurements relative to a known standard [see Chapter 20 in this book]. Historically, second-order nonlinear optical susceptibilities from SHG measurements on NLOP films are referred to as "d_{ij}" coefficients, which are, by convention, 0.5 times the values of the respective $\chi_{ij}^{(2)}$ coefficients. The electro-optic (EO) coefficients of a NLOP waveguide, r_{ij}, indicate the amount of refractive index change caused by a unit increase in voltage applied across the film. Off resonance, EO coefficients are related to the d coefficients as $r \approx -4d/n^4$, where n is the index of refraction. However, for accurate predictions, resonance effects, dispersion, and local field effects must be taken into account (*16*) [and see Chapter 25 in this book]. It follows from the normal assumptions (cylindrical symmetry of the chromophores) that only two components of the second-order susceptibility tensor are important. Hence, the coefficients of interest are those describing interactions parallel to the average dipolar axis, e.g., d_{33} and r_{33}, and those perpendicular to the average dipolar axis, e.g., d_{31} and r_{31}. See reference (*7*) for details on symmetry considerations and treatment of the tensor-element subscript conventions.

Polymers

The polymer structure dictates the processability and temporal stability of the final product. Polymers of interest are typically amorphous glasses, because they are very transparent and scatter very little light. Chromophores in the polymers are usually aligned by electric field poling near the glass transition temperature, then alignment is locked in by cooling with the field on (*17*). When the field is removed, the alignment can remain frozen in the glassy matrix for many years. [Processing and stability will be covered in more detail in subsequent sections.]

Guest-host (GH) Systems. These are solid solutions: small, unattached chromophores dissolved in high molecular weight polymers. GH systems have many pitfalls and should be avoided. Generally a GH system contains about 10 to 30% by weight of the chromophore -- higher levels tend to phase separate causing light scattering. The main problem is that the glass transition temperature of the polymer host decreases dramatically due to plasticization by the chromophore guest (*18*). GH systems are also undesirable because the chromophores are labile: at elevated temperatures (at processing temperatures) they diffuse to the surface of the film and evaporate or sublime. The chromophores on the surface of a GH film can be transferred to your fingers by touching the films, and can be absorbed through the skin. They are often toxic, mutagenic, teratogenic, and carcinogenic. High molecular weight polymers are not absorbed through the skin. Hence, attaching chromophores to a high molecular weight polymer solves this problem, as well as many other problems.

Nonetheless, GH systems that meet some of the device stability requirements have been made by choosing a polymer with a very high glass transition temperature (> 250 °C) and by using a very large chromophore thereby making it less fugitive [see Chapters 2 and 7 in this book]. However, in the long term, the commercially viable NLOP will have their chromophores chemically attached to the polymer matrix.

Sidechain (SC) Polymers. SC polymers have <u>one</u> end of the chromophore chemically attached to the backbone, with most of the mass of the chromophore pendent to the backbone. They can be tethered to a backbone by a spacer group. However, it is more desirable to snug the chromophore right up to the backbone (shorten or eliminate the spacer group) -- for example, see this book's Chapters 10 - 13 for more information on the NLOP shown below:

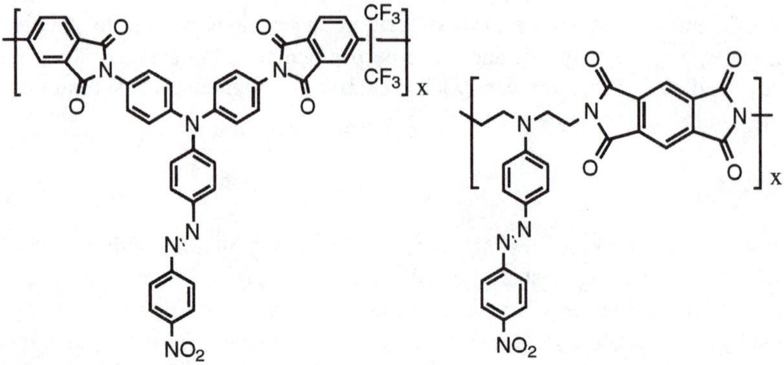

For a given glass transition temperature and a given chromophore, the SC polymers have significantly greater temporal stability than GH systems (*17,19*).

Early SC Methacrylate NLOP. Most of the earliest reported SC polymers were prepared by free radical copolymerization of methacrylic functional chromophores and methyl methacrylate (MMA) (*20*). The first generation SC methacrylic polymers are easy to process, have excellent clarity, and have excellent EO waveguide figures of merit (e.g., see reports from Hoechst-Celanese (*15a*) and Akzo-Nobel (*21*) on polymers containing the 4-dimethylamino-4'-nitrostilbene (DANS) chromophore). One of these DANS polymers was recently poled in the plane of a waveguide to give a d_{22} = 150 pm/V @ 1064 nm (resonance enhanced) and d_{22} = 26 pm/V @ 1.55 microns [see Chapter 34 in this book]. The anilino-azo-nitrobenzene chromophores (Disperse Red 1 and 19 dyes) have also been widely studied in SC NLOP (*22*). The glass transition temperature of most SC methacrylic NLOP is 120 to 140°C, but some processable acrylic polymers have a Tg as high as 205°C (*23*). The maximum $\chi^{(2)}$ properties are often obtained at about 30% to 50% chromophore concentrations (*24*). The reason for this behavior is still debated, but may be due to local field effects or to the inability of chromophores to align due to packing problems at higher concentrations. SC chromophoric acrylate polymers are for sale to the public (*25*).

Liquid Crystalline Polymers. Some of the earliest work on NLOP films (*26*), and more recent work (*27*), was based on liquid crystalline polymers. Many people found that polymeric liquid crystals (PLC's) scattered light from microdomains resulting in unacceptable optical losses in waveguide applications. However, recent reports on single domain PLC's are more promising (*28*). SC tolane-MMA copolymers were reported whose $\chi^{(2)}$ properties increased by a factor of 3 to 4 upon aging 4 months (*29*). It is possible that the high aspect ratio of the tolane groups allowed some degree of slow self-ordering (Maier-Suape thermotropic liquid crystalline interactions) during aging (*30*).

High Temperature NLOP. In the early 1990's much attention was given to increasing the thermal stability of NLOP films in order to withstand semiconductor manufacturing operations. In general, one does this by replacing aliphatic groups with aromatic groups. Most importantly, investigators at IBM-Almaden have shown that attaching aromatic groups to (eliminating aliphatic groups from) the electron donating nitrogen amine greatly improves high temperature oxidative stability, while at the same time, the aromatic groups usually increase the β of the chromophore (*31*). Aromatic groups may be necessary, but not always sufficient to ensure temporal stability at high temperatures. For example, a $\chi^{(2)}$ poly(aryl ether-oxazole) having a Tg of 242°C lost 60% of its poled order at 150°C in three hours (*32*).

Polyimides. Many investigators have turned their attention to polyimide (PI) chemistry for thermal stability (*33*). The exploration of chromophoric PI architecture's is just beginning, and very recent reports look quite encouraging (*34*), for example 100% retention of the $\chi^{(2)}$ signal for over 1000 hours at 150°C (*35*) (and see Chapters 10 - 13 in this book). Most investigators now feel that the high temperature performance of NLOP films is quite adequate to meet end-use requirements (typically > 10 years polar stability at 80 to 125°C), and a few chromophores show promise of surviving short term semiconductor processing conditions (300 to 400°C) (*31,36*).

Hydrogen bonding (*37*) and **rigid backbones** (*38*) are design features that can improve thermal stability. For example, a SC polyamide containing the DANS chromophore had a Tg of 205°C and preliminary $\chi^{(2)}$ = 40 pm/V @ 1542 nm (*39*). However, a very rigid backbone may result in poor alignment during electric-field poling, e.g., in the case of a SC polyphenylene (*40*) and a SC poly(phenylene vinylene) (*41*).

Large-β chromophores are now where more emphasis needs to be directed, especially on their attachment to thermally stable polymer backbones. These larger chromophores are difficult to solubilize and process. One of the earliest large-β chromophores in a SC MMA copolymer was reported by authors from NTT (*42*). Recently it has been shown that the incorporation of several conjugated heterocycles (thiophenes, thiazoles, and pyrroles) and several vinylene linkages, $-(CH{=}CH)_n-$, in the chromophore gives a large enhancement to β (*43*) [See Chapter 6 in this book.]

Mainchain (MC) Polymers. These polymers are comprised of chromophores which are linked in the backbone at both ends of the chromophore, and the majority of the chromophore forms part of the polymer backbone. The chromophores can be linked in a head-to-tail configuration (isoregic) (*44*), in a head-to-head configuration (syndioregic) (*45*) [and see Chapter 14 in this book], or in random head-to-tail and head-to-head configurations (aregic) (*46*).

The mainchain chromophoric topology would seem to be inherently more stable than the sidechain topology because the chromophore has fewer local degrees of freedom of motion (*47a*), but that may not be a foregone conclusion (*47b*). The MC NLOP possibilities are largely unexplored. New processing techniques to give polar orientation to isoregic mainchain NLOP are especially needed.

One comparison of 3 polymers containing the cinnamoyl chromophore, each one in a different topology, indicates that the syndioregic topology yields superior alignment and $\chi^{(2)}$ properties (see structures below):

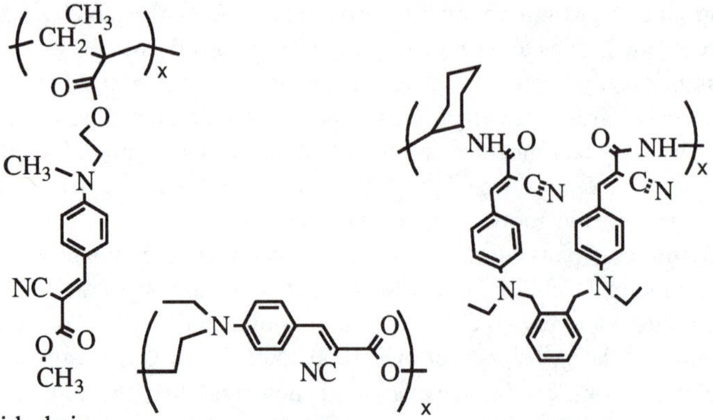

Sidechain Homopolymer	Isoregic Mainchain	Syndioregic Mainchain
$T_g = 114°C$	$T_g = 105°C$	$T_g = 205°C$
$d_{33} = 22$ pm/V	$d_{33} = 7$ pm/V	$d_{33} = 30$ pm/V
[Ref. 24(b)]	[Ref. 44(b)]	[Chapter 14 herein]

Crosslinked Polymers. Crosslinked (XL) polymers offer the promise of great long term temporal stability (below the Tg). XL polymers by their very nature must undergo the crosslinking reaction during or just after the alignment process. [See Chapter 12 in this book.] Therefore, this class of waveguide material is more difficult to manufacture, which could lead to a greater variation in the quality of the waveguide. Early reports indicated greater optical losses due to trapped density fluctuations (inhomogeneity in the crosslink density). However, more recent reports indicate that losses can approach those obtained with linear polymers (thermoplastics), for example, see Chapter 15 in this book.

Precursors to XL NLOP films can be GH, SC, and MC polymers. Polymers have been successfully crosslinked during poling using diisocyanates (*48,49*). [See Chapter 29 in this book.] Some non-hydrogen-bonded SC NLOP must be chemically crosslinked in order for the temporal stability to even approach the Tg (*50*). However, the glass transition temperatures of the urethane NLOP reported thus far have not been greater than 150°C. Other chemical reactions for crosslinking NLOP films have been reported based on 2+2 photo addition of cinnamates (*51*), epoxy cure of diamine-chromophores (*52*), and hydrosilation of allyl functional chromophores (*53*). Although some of these materials are close to the state-of-the art in stability and optical nonlinearity, it is clear that further improvements will be forthcoming. The interpenetrating polymer network (IPN) topology, first proposed by Tripathy's group for NLOP films, is a promising topology (*54*). [See Chapter 15 in this book.] In comparing thermoplastic and thermoset polymers for NLOP films, there is not yet a clear winner, but the author believes that a light degree of crosslinking is desirable for long term stability.

Polar Order

Electric-Field Poling is one technique widely used to remove centrosymmetry (i.e., to impart polar order) in the film. Thin films of the polymer are prepared for poling by spin-coating a liquid solution of the polymer onto a solid substrate. The solvent is removed by baking the film just above the glass transition temperature (Tg). An electric field is applied across the film by corona poling the film near its Tg as it sits on a grounded conductor [see Ref. (*17*)], or by charging two electrodes contacting the film. Either of these processes can create an electric field of 50 to several hundred volts per micron across the film. The electric field is applied near the Tg for 10 to 60 minutes, then the film is cooled with the field on. In situ monitoring the d or r nonlinear optical coefficients during poling is a good way to optimize poling conditions (*55*). Generally the NLO coefficients increase linearly with poling field until a condition of saturation exists (or until dielectric breakdown of the film shorts out the electrodes).

The degree of alignment is proportional to $\mu E/kT$, where μ is the ground state dipole moment of the chromophore, E is the applied electric field, k is Boltzmann's constant, and T is the poling temperature. There are several problems with electric-field poling. The polymer must be heated to high temperatures where thermal disordering of the chromophores works against the torque of the electric field. Polymers containing formal charges are difficult to pole with an electric field because at Tg the rate of migration of charges through the polymer increases many orders of magnitude and may cause dielectric breakdown. Damage to the surface of the film during poling causes defects (hence scattering losses), although protective coatings can eliminate this problem.

Optical Poling. It has been shown that the superposition of coherent fundamental and second harmonic light can be used to create oriented chromophores in a glassy polymer by a polar bleaching mechanism [see Chapter 18 in this book]. This intriguing technique might even be used for ordering octopolar chromphores (*1d*). The position of the charge transfer absorption band of the chromophore must be strategically located relative to ω and 2ω in order to achieve chromophore alignment as well as stability of the thus poled polymer. This is an area ripe for further investigation. This technique should not be confused with photo-assisted electric field poling [see Chapter 19 in this book], nor photothermal poling (*56*) in which the laser is merely used to heat the region to be poled by a d.c. electric field. And finally, photo injection of charges by illuminating a photoconductor layer on top of a grounded NLOP layer has been reported as a new way to create or enhance dipolar alignment (*57*).

Langmuir-Blodgett-Kuhn (LBK) Processing. In LBK processing (also called LB processing), the polymer molecules are designed to have hydrophilic and hydrophobic groups which cause the polymer to float in a preferred conformation on the water-air interface. These hydrophilic/hydrophobic forces are useful in orienting the chromophores normal to the plane of the film, removing the centrosymmetry. To make films by LBK processing, a solution of the polymer is deposited on water; as the solvent evaporates the NLOP film is consolidated by compression with a movable dam on a computer-controlled water trough; and, a solid substrate is dipped through the air-water interface, depositing a single molecular layer (≈ 20-30 Å thick) on the substrate (*58*). Thicker films comprised of multilayers of polymer are built up by dipping the substrate repeatedly into and/or out of the water trough, depositing one molecular layer per stroke. Turn-key multi-compartment troughs are available from several commercial suppliers, for example, NIMA (Coventry CV4 7EZ, England), NLE (Nagoya, 468, Japan), and KSV (SF-00380 Helsinki, Finland).

LBK processing has advantages over electric-field poling because it can be done at room temperature (or lower), and, one can take advantage of formal ionic charges for increasing the chromophore's β. This NLOP film processing procedure has been reduced to practice for sidechain polymers (*59*), including a liquid crystalline NLOP (*60*). The Kodak group reported making 0.5-micron waveguides for SHG with less than 1 dB optical loss (*61*). Syndioregic mainchain chromophoric polymers have been demonstrated for LBK deposition of NLOP films (*62*).

Other Self-Assembly Techniques. Investigators have attached one end of an asymmetric chromophore to a solid substrate and grown multiple layers, one layer at a time, to form NLOP films (*63*). Another exciting development in this field involves liquid crystal (LC) polymers containing chiral NLO-phore sidechains (S_C^*) that spontaneously order (with the help of a small electric field to "untwist the helical superstructure") to give polar NLOP films (*64*); and recently the same phenomenon has been demonstrated in a SC LC polymer enhanced by mechanical orientation (*65*). One isoregic mainchain chromophoric liquid crystalline polymer was reported to spontaneously form polar order, although a higher degree of order was achieved by poling (*66*). Liquid crystals are aligned by interactions with the unidirectional buffed polymer coatings on the glass cells of flat panel displays. This and other epitaxial techniques should be further explored for forming more ordered NLOP films.

Stability

Physical (or mechanical) stability refers to stability of alignment of the chromophore (related to physical aging (*67*)). It is clear that guest-host systems are much less stable than systems in which the chromophore is chemically attached to the polymer (*17, 68*). The following remarks pertain primarily to the latter. The glass transition temperature (Tg) is the primary indicator of physical stability. Amorphous polymer glasses containing polarity-aligned chromophores are said to be in a non-equilibrium state because Brownian motion (thermal energy) works to randomize the order; however, well below the Tg, a change in alignment may not be noticeable after many, many years. Furthermore, physical aging slows down with time due to the spontaneously increasing density of the glassy solid. [See Chapter 23 in this book.] The rate of randomization depends mainly on the difference between the use-temperature and the glass transition temperature (the higher the Tg, the better), the size of the chromophore relative to the amount of local free volume available, and its degree of freedom for rotational diffusion. The amount of non-equilibrium free volume present in a sample at any point in time depends upon the sample's time-temperature cooling history (the faster it is cooled, the larger the amount of free volume trapped). If NLOP films are aged in the applied electric field about 25° below the Tg for an hour before cooling them on down to the use temperature, physical aging is greatly retarded because the free volume is reduced.

Chemical stability refers to the integrity of the chemical structure of the chromophore which is under attack by heat, light, and reactions with oxygen. Examples of chemical changes are isomerization (*69*), cyclization, oxidation, and bond scission. Stability of chromophores and polymers can be screened by differential scanning calorimetry, thermal gravimetric analysis, and UV-VIS-IR spectroscopy as a function of time and temperature; but, the NLOP film must be tested in an actual waveguide device before one can state with confidence the degree of stability. Chemical changes in the NLOP waveguide will decrease the refractive index and $\chi^{(2)}$ properties [see Chapter 24 in this book]. Chemical changes can be avoided by proper design of the chromophore, processing conditions, and device packaging.

The temporal stability of a NLOP film refers to how long the physical alignment and chemical integrity of the chromophores are maintained at a given temperature. One is concerned with the temporal stability under end-use and storage conditions (typically 80 to 125°C). The processing stability refers to how well the polymer holds up under film processing and packaging conditions. These conditions may be at temperatures as high as 250 to 350°C for tens of minutes.

Optical Waveguides

A passive waveguide is simply a conduit though which light is routed, such as an optical fiber. A thin film can be a passive waveguide (called a "slab" waveguide). An active waveguide changes its index of refraction when an electric field is applied (called the Pockels effect). In an integrated device, a waveguide is typically a channel or a rib patterned in a thin transparent film. In an active waveguide, light can be switched back and forth from one channel to another, or phase-shifted.

Cladding. The waveguiding material must have a higher index of refraction than the surrounding cladding material. Light remains in the higher index material due to total internal reflection at its interface with the low index cladding material. The cladding material must be very transparent so as not to cause absorption or scattering of the evanescent energy at the surface of the guiding channel. The walls of the waveguide must be very smooth to prevent scattering of the light. For active waveguiding applications, the cladding must be thick enough to prevent the light's evanescent wave from interacting with the metal electrodes which are used to apply the electric field, because metal absorbs the light and would cause unacceptable losses. The importance and difficulty in finding suitable cladding material has not been given much attention in the literature, and is an area where more R&D is needed.

Many polymer waveguide processing techniques have already been fine tuned for passive polymer waveguides (*70*). Channels can be formed by etching grooves in the lower cladding layer (for example by UV laser ablation) and back-filling the groove with NLO polymer. Rib waveguides can be made by etching away the NLOP film on either side of a rib pattern (*71*). Etching is often done by reactive-ion etching. [See Chapter 12 in this book.] Cladding can then be back filled around the NLOP rib.

Active Waveguides. If electric field poling is used to align chromophores in the NLOP layer, any cladding between the electrodes and the NLOP layer should be slightly higher in conductivity than the NLOP layer so that the maximum voltage is applied across the NLOP layer and not across the cladding [see Chapter 33 in this book]. For high temperature poling, this can usually be accomplished by selecting a cladding material that has a slightly lower Tg than the NLO polymer. The same design rule holds for the operational waveguide. However, one cannot have a lot of free ions diffusing around in the dielectric layers or one quickly runs into problems of d.c. drift (i.e., the voltage vs. Δn relationship of the device changes with time). A sacrificial hydrophilic polymeric cladding (protective layer), such as polyacrylic acid, has been shown to give high quality corona-poled electro-optic waveguides (*72*). The hydrophilic cladding is easily removed and replaced with a different cladding polymer after corona poling.

Photobleaching can permanently lower the refractive index of NLOP material due to chemical changes in the chromophore. Thereby, light-guiding channels can be developed in the NLOP film by masking regions where you want the channels, and forming the cladding by UV-photobleaching the unmasked regions (*73*). Using polarized UV allows one to create cladding in which the transverse electric (TE) mode sees a different index profile than the transverse magnetic (TM) mode, which can be helpful in separating the two polarizations (*73a*). Specific photobleaching chemicals can also be incorporated in NLO polymers (*74*). These photo-induced bond rearrangements can be totally different for different chromophores. To be sure the chemical change is permanent and innocuous, the specific chemical changes can be identified by analytical investigation (FTIR, UV-VIS, and NMR spectroscopy) -- often model compounds are helpful.

Optical Characterization. The macroscopic optical coefficient ($\chi^{(2)}$) of NLOP films can be determined by various methods, each with a different accuracy vs. convenience trade-off (*75*). Second harmonic generation (SHG) is described in detail in Chapter 20 in this book. Useful techniques for measuring the electro-optic coefficients (r_{ij}) are the phase-modulated reflection technique (which assumes $r_{33} \approx 3r_{31}$) (*76*), and the electro-optic interferometry technique (*77*). Also see reference (*7*). When applying an electric field across a film to change its refractive index, there may be piezoelectric contributions (a change in thickness of the film with the applied electric field) (*77c*), and $\chi^{(3)}$ effects (*78*). These are usually less than 5 to 10% of the total $\chi^{(2)}$ effect.

Devices

Mach-Zehnder Interferometers. A popular optical intensity modulator design taking advantage of the Pockels effect is called a Mach-Zehnder interferometer (MZI). In a MZI, the incoming light path splits into two paths (arms) then recombines at the output. Light in one arm is phase-shifted relative to the other arm by applying an electric field across one arm and not the other (or, also across the other arm but in the opposite direction). When recombined, if light from the two arms is 180° out of phase, the intensity is extinguished (by destructive interference). If the light passes through each arm unshifted, it recombines at the same intensity minus any loss due to scattering and absorption in the waveguide. Loss reported for many NLOP at 1.3 microns (the common frequency used for long distance optical data transmission) is 0.5 to 1 dB/cm -- an acceptable loss for most devices (*1a*). In a buried NLOP waveguide (as in an integrated MZI), losses can be greater than in a slab waveguide due to scattering off of rough texture on the walls of the waveguide. NLOP having higher EO coefficients typically have red-shifted absorption peaks, and therefore higher absorption loss. The EO coefficient (r_{ij}) of the nonlinear optical material and the loss dictate how small one can make the length of the MZI (typically about 1 cm). The arms that carry (guide) the light are a few square microns in cross-section, so that many MZI's can be placed side-by-side (or stacked one on top of one another) if necessary.

Direct vs. External Modulation. If you turn a laser diode on and off by switching its power supply on and off, this is called "direct" modulation of the light intensity. If you keep the laser output fixed and turn the light on and off in the waveguide with a MZI, this is called "external" modulation. Both methods send processed optical signals down the waveguide. However, external modulation consumes less power and gives more efficient conversion at high (GHz) frequencies.

Figures of Merit. For external modulators lithium niobate is the standard for comparison because it has been under development for over 20 years. One can purchase a $LiNbO_3$ MZI which operates at about 10 GHz for about $10,000. There is not much hope of drastically lowering its cost by mass production. $LiNbO_3$ is brittle. It is difficult to grow perfect crystals every time. It is very difficult to integrate on silicon chips. It has unwanted photorefractive effects, which are not present in polymers. Its dielectric constant is ten times higher than NLOP at radar frequencies. At high frequencies, $LiNbO_3$ will require thicker, more complicated electrodes, and take more power to operate (at least twice the power required for today's NLO polymers). On the other hand, NLOP have good potential for mass production of low cost, integrated MZI's. For EO devices, all else being constant, a figure of merit commonly used is n^3r, where n is the index of refraction and r is the EO coefficient. The n for $LiNbO_3$ is about 2.2, and r is about 30 pm/V ($n^3r = 320$); whereas for polymers, n is about 1.7 and the highest r reported for polymers is about 55 pm/V [reported by S. Marder at the ACS meeting in Anaheim, CA, April 4, 1995 -- the loss was not reported], hence, $n^3r = 270$. But this figure of merit is not the whole story. For actual waveguides, a factor Γ, which takes into account the geometry of the device, must multiply n^3r [see page 299 in ref. (*1a*)]. Γ is nearly 0.9 for NLOP, but about 0.5 or less for $LiNbO_3$.

V_π, the half-wave voltage, is the voltage required to retard the phase of light 180° in one arm of a MZI. $V_\pi = t\lambda/n^3rL$, where t is the combined thickness of the guiding and cladding layers, L is the length of the waveguide over which the electric field is applied, and λ is the wavelength of light being modulated [see page 338 in ref. (*1a*).]. For today's NLOP having a 1 cm path length, 9 microns thickness (including the cladding), operating at a 1.3 micron wavelength, V_π is 4 to 5 volts. Furthermore, for polymers the frequency dispersion between electrical and optical frequencies is almost negligible: $\varepsilon \approx n^2$; whereas, for $LiNbO_3$ $\varepsilon \approx 5.8*n^2$. Newly discovered NLO chromophores have many times greater molecular nonlinearity than the best NLOP tested in MZI devices so far (but it takes research time to synthesize versions of these new chromophores which are attached to processable polymers and to get the new polymers into the hands of the device makers).

Commercial Applications. The first commercial applications for NLOP films will clearly take advantage of the polymer's low dielectric constant at high speeds (>20 GHz), the ease of integration of these films on low cost silicon microelectronics devices, and the ease of aligning optical fibers with channels in the NLOP waveguides using silicon V-groove technology (*79*). Photonic control of phased array radar may be an early application. The state-of-the-art NLOP modulator (40-GHz bandwidth) was reported by Teng in 1992 (*80*). There are several groups in the process of making 50- to 100-GHz NLOP intensity modulators, and there is one report of a 94-GHz phase modulator using a 2-methyl-4-nitroaniline organic crystal (*81*). High speed NLOP reflection modulators (for through-space interconnection) (*82*), a NLOP waveguide switch (*83*), and a NLOP optical transmitter for connecting CMOS integrated circuits in a multichip module have been reported (*84*). [Also see Chapters 30-34 in this book for waveguide devices.]

By using multiple layers, one can take advantage of the third dimension for building devices. Reports have appeared on NLOP devices in which channels are vertically stacked for optical switching (*85*), and for intensity modulation (*86*). Two vertically stacked NLOP Mach-Zehnder interferometers with less than - 17 dB crosstalk have been reported (*87*).

Using NLOP waveguides as low-cost electric field sensors should be further explored; for example, sampling the internal signals on silicon integrated chips (IC's) has been reported (*88*). RF field sensors are another application.

Frequency conversion, including second harmonic generation (SHG), is being pursued with NLOP devices. A novel waveguide device which emits SHG vertically by counter propagating the fundamental beam is described in Chapter 34 in this book (this light could be captured in an adjacent waveguide by a 45° mirror). The generation of UV femtosecond pulses by SHG in thin NLOP films has recently been demonstrated [see Chapter 35 in this book]. These ultra short pulses should find application in the analysis of fast chemical reactions. Although not optimized, fairly efficient SHG has been achieved in NLOP waveguides by various phase matching techniques. By reversing the direction of polarization inside the NLOP layer parallel to the direction of the fundamental beam propagation (*89*), a high overlap integral of the fundamental and second-harmonic modal fields can be achieved. Phase matching can also be achieved by periodically reversing polarity perpendicular to beam propagation [see Chapter 36 in this book].

Finally, large third-order effects, such as the optical Kerr effect or intensity-dependent index of refraction, are possible in second-order materials by superimposing or co-propagating a fundamental and its second harmonic. See Chapter 37 in this book which describes this effect, called "cascading." This exciting phenomenon, which is only now being explored in polymers, has great potential for building ultra fast all-optical switches.

Acknowledgments

Support from ONR Chem. & Phys. Div. and ARPA-MTO, and helpful discussions with Dr. K. Singer, Dr. W. Herman, and Dr. R. Lytel are gratefully acknowledged.

Literature Cited

1. (a) *Polymers in Lightwave and Integrated Optics, Technology and Applications*, L. A. Hornak, Ed.; Marcel Dekker: New York, 1992; (b) *1993 Technical Digest* , Series Volume 17 (OSA, Washington, D.C. 1993), and *ACS Division of Polymer Chemistry Polymer Preprints* **1994**, *35(2), 86-293*; (c) the *Proceedings of SPIE* **1988-1994** on *Nonlinear Optical Properties of Organic Materials*, volumes *971, 1147, 1337, 1560, 1775, 2025,* and *2285*; (d) *Molecular Nonlinear Optics: Materials, Physics and Devices*, J. Zyss, Ed.; Quantum Electronics Principles and Applications Series; Academic Press: New York, 1994; and, (e) G. T. Boyd and M. G. Kuzyk in *Polymers for Electronic and Photonic Applications*, C. P. Wong, Ed.; Academic Press: New York, 1993.
2. Burland, D. M; Miller, R. M; Walsh, C. A. *Chem. Rev.* **1994**, *94*, 31-75.
3. Noguchi, K.; Miyazawa, H.; Mitomi, O. *Electronic Letters Online* **1994**, 22 April, No: 19940646.
4. Cites, J. S.; Ashley, P. R. *J. Lightwave Techn.* **1992,** *12 (7)*, 1167-1173.
5. Lytel, R. *Trends in Polymer Science* **1994**, *2(4)*, 114.
6. Garito, A.; Wong, K. Y.; Zamani-Khamiri, O.; In *Nonlinear Optical and Electroactive Polymers*, P. N. Prasad and D. R. Ulrich, Eds., Plennum Press: New York, 1988.

7. (a) Prasad, P.N.; Williams, D.J. *Introduction to Nonlinear Optical Effects in Molecules and Polymers* ; John Wiley & Sons: New York, 1991; (b) Singer, K. D.; Kuzyk, M. G.; Sohn, J. E. *J. Opt. Soc. Am. B* **1987**, *4 (6)*, 968-976; (c) Chemla, D.S.; Zyss, J., Eds.; Academic: New York, 1987; Vol. 1 and 2.
8. (a) Willetts, A.; Rice, J. E.; Burland, D. M. *J. Chem. Phys.* **1992**, *97(10)*, 7590-7599; (b) Wang, C. H.; Guan, H. W. *J. Polym. Sci.:Part B: Polym. Phys.* **1993**, *31*, 1983-1988; (c) Singer, K.D.; Kuzyk, M.G.; Sohn, J.E. In *Nonlinear Optical and Electroactive Polymers*, P. N. Prasad and D. R. Ulrich, Eds., Plennum Press: New York, 1988.
9. Mortazavi, M.A.; Knoesen, A.; Kowel, S. T.; Higgins, B.G.; and Dienes, A. *J. Opt. Soc. Am. B* **1989**, *6(4)*, 733.
10. (a) Oudar, J. L.; Chemla, D. S. J. Chem. Phys. **1977**, *66*, 2664; (b) Oudar, J. L.; LePerson, H.E.; *Opt. Commun.* **1977**, *66*, 2664.
11. J. J. P. Stewart, *J. of Computational Chemistry* **1989**, *10(2)*, 221-264.
12. Skumanich, A.; Jurich, M.; Swalen, J. D. *Appl. Phys. Lett.* **1993**,*62(5)*, 446.
13. (a) Meyers, F.; Marder, S. R.; Pierce, B. M.; Bredas, J. L.; *J. Am. Chem. Soc.* **1994**, *116*, 10703; (b) Ortiz, R.; Marder, S. R.; Cheng, L.-T.; Tiemann, B. G.; Cavagnero, S.; Ziller, J. W. *J. Chem. Soc. Chem. Commun.* **1994**, 2263.
14. Marder, S. R.; Gorman, C. B.; Meyers, F.; Perry, J. W.; Bourhill, G.; Brédas, J.-L.; Pierce, B. M. *Science* **1994**, *265*, 632.
15. (a) DeMartino, R. N.; et al. In *Nonlinear Optical and Electroactive Polymers,* P. N. Prasad and D. R. Ulrich, Eds., Plennum Press: New York, 1988; and (b) Paley, M. S.; et al.; *J. Org. Chem.* **1989**, *54,* 3774-3778.
16. Moylan, C. R.; et al. *Proc. SPIE* **1993**, *2025*, 192-201.
17. Singer, K.D.; Kuzyk, M.G.; Holland, W.R.; Sohn, J.E.; Lalama, S.L.; Comizzoli, R.B.; Katz, H.E.; Schilling, M.L. *Appl. Phys. Lett.* **1988**, *53*, 1800.
18. Walsh, C. A.; et al. *Macromolecules* **1993**, *26*, 3722.
19. Bauer, S.; et al. *Appl. Phys. Lett.* **1993**, *63 (15)*, 2018.
20. (a) Leslie, T. M.; et al. *Mol.Cryst. Liq. Cryst.* **1987**, *153*, 451; (b) Esselin, S.; et al. *Proc. SPIE* **1988**, *971*, 120.
21. Möhlmann, G. R.; et al. *Proceedings of SPIE* **1990,** *1337*, 215.
22. (a) Ye, C.; Marks, T.; Yang, J.; Wong, G. K. *Macromolecules* **1987**, *20*, 2324; (b) Singer, K. D.; Sohn, J. E.; King, L. A.; Gordon, H. M.; Katz, H. E.; Dirk, C. W. *J. Opt. Soc. Am. B* **1989**, *6(7)*, 1339.
23. Herman, W. N.; Rosen, W. A.; Sperling, L. H.; Murphy, C. J.; Jain, H. *Proc. SPIE* **1991**, *1560*, 206-213.
24. (a) S'heeren, G.; Persoons, A.; Rondou, P.; Wiersma, J.; van Beylen, M.; Samyn, C. *Makromol. Chem.* **1993**, *194*, 1733; (b) Rondou, P.; Van Beylen, M.; Samyn, C.; S'heeren, G. Persoons, A. *Makromol. Chem.* **1992**, *193*, 3045.
25. (a) AdTech Systems Research, Inc, contact Dr. Sam Sinha, 1342 N. Fairfield Rd., Beavercreek, OH, 45432 (phone 513-426-3329); and (b) IBM Almaden Research Center, contact Dr. Dan Dawson, 650 Harry Road, San Jose, CA 95120-1501 (408-927-1617).
26. Meredith, G. R.; VanDusen, J. G.; Williams, D. J. *Macromolecules* **1982**, *15* , 1385.

27. (a) Yitzchaik, S.; Berkovic, G.; Krongauz, V. *Optics Lett.* **1990**, *15 (20)*, 1120; (b) Dubois, J.-C.; LeBarny, P.; Robin, P.; Lemoine, V.; Rajbenback, H. *Liq. Cryst.* **1993,** *14*, 198.
28. Paper by C. Noel on a polyether with 4,4'-oxycyanobiphenyl sidechains, presented at the Macromolecular IUPAC meeting in Akron, Ohio, July 1994.
29. Morichére, D.; Chollet, P.-A.; Fleming, W.; Jurich, M.; Smith, B. A.; Swalen, J. D. *J. Opt. Soc. Am. B* **1993**, *10*, 1894.
30. Jungbauer, D.; Teroka, I.; Yoon, D. Y.; Reck, B.; Swalen, J. D.; Twieg, R.; Willson, C. G. *J. Appl. Phys.* **1991,** *69*, 8011.
31. Moylan, C. R.; Twieg, R. J.; Lee, V. Y.; Swanson, S. A.; Betterton, K. M.; Miller, R. D. *J. Am. Chem. Soc.* **1993**, *115*, 12599.
32. Carter, K. R.; Hedrick, J. L.; Twieg, R. J.; Matray, T. J.; Walsh, C. A. *Macromolecules* **1994**, *27*, 4851.
33. (a) Lin, J. T.; Hubbard, M. A.; Marks, T. J.; Lin, W.; Wong, G. K. *Chem. Mater.* **1992**, *4*, 1148; (b) Becker,. M. W.; Sapochak, L. S.; Ghosen, R.; Xu, C.; Dalton, L. R.; Shi, Y.; Steier, W. H.; Jen, A. K.-Y. *Chem. Mater.* **1994**, *6*, 104.
34. (a) Miller, R. D.; et al. *ACS Polym. Preprints* **1994,** *35(2)*, 122; (b) Peng, Z.; Yu, L. *Macromolecules* **1994,** *27*, 2038.
35. Sotoyama, W.; Tatsuura, S.; Yoshimura, T. *Appl. Phys. Lett.* **1994**, *64(17)*, 2197.
36. Lindsay, G. A.; et al. *Proc. SPIE* **1994**, *2143*, 19.
37. Lindsay, G. A.; Henry, R. A.; Hoover, J. M.; Knoesen, A.; Mortazavi, M. A. *Macromolecules* **1992**, *25*, 4888-4894.
38. Schultz, M. *Trends in Polymer Science* **1994**, *2(4)*, 120.
39. Weder, C.; Neuenschwander, P.; Suter, U. W.; Prêtre, P.; Kaatz, P.; Günter, P. *Macromolecules* **1994**, *27*, 2181.
40. Wright, W. E.; Toplikar, E. G. *ACS Polym. Preprints* **1994**, *35(1)*, 369.
41. Hwang, D.-H.; Lee, J.-I.; Shim, H.-K.; Hwang, W.-Y.; Kim, J.-J.; Jin, J.-I. *Macromol.* **1994**, *27*, 6000.
42. (a) Amano, M.; Kaino, T. *J. Appl. Phys.* **1990**, *68(12)*, 6024; and (b) Kaino, T.; Tomaru, S. *Adv. Mater.* **1993**, *5(3)*, 172.
43. (a) Barzoukas, M.; Blanchard-Descz, M.; Josse, D.; Lehn, J.-M.; Zyss, J. *Chemical Physics*, **1989**, *133*, 323; (b) Jen, A. K-Y.; Roa, V. P.; Wong, K. Y.; Drost, K. J. *J. Chem. Soc., Chem. Commun.* **1993**, 90; (c) Roa, V. P.; et al. *ibid.*, 1118.
44. (a) Fuso, F.; Padia, A.B.; Hall, H.K., Jr. *Macromolecules* **1991**, *24*, 1710; (b) Stenger-Smith, J.; Fischer, J. W.; Henry, R. A.; Hoover, J. M.; Lindsay, G. A. *Macromol. Chem. Rapid Commun.* **1990**, *11*, 141; (c) *ibid.* **1993**, *14*, 68; and (d) Mitchell, M. A.; *Trends in Polymer Science,* **1993,** *1(5)*, 144-148.
45. (a) Lindsay, G. A.; Stenger-Smith, J. D.; Henry, R. A.; Hoover, J. M.; Nissan, R. A.; Wynne, K. J. *Macromolecules* **1992**, *25(22)*, 6075; (b) U. S. Patent 5,247,055; (c) Wright, M. E; Mullick, S. *Macromolecules* **1992**, *25*, 6045-6049; (d) Tao, X. T.; Watanabe, T.; Shimoda, S.; Zou, D. C.; Sato, H.; Miyata, S.; *Chem. Mater.* **1994**, *6*, 1961.

46. (a) Xu, C.; Wu, B; Dalton, L. R.; Ranon, P. M.; Shi, Y.; Steier, W.H., Becker, M. W. *Chemical Materials* **1993**, *5*, 1439-1444; (b) Wright, M. E.; Mullick, S.; Lackritz, H. S.; Liu, L.-Y. *Macromolecules* **1994**, *27*, 3009.
47. (a) Köhler,W.; Robello, D.R.; Dao, P. T.; Willard, C. S.; Williams, D. J.; *J. Chem. Phys.* **1990,** *93(20)*, 9157-9166; (b) Köhler, W. Robello, D. R.; Willand, C. S.; Williams, D. J.; *Macromoloecules* **1991**, *24*, 4589.
48. Shi, Y.; Ranon, P. M.; Steier, W. H.; Xu, C.; Wu, B.; Dalton, L. R. *Appl. Phys. Lett.* **1993**, *63(16)*, 2168.
49. Kitipichai, P.; LaPeruta, R.; Korenowski, G. M.; Wnek, G. E. *J. Polym. Sci.: Part A: Polym. Chem.*, **1993**, *31*, 1365.
50. Xu, C.; Wu, B.; Todorova, O.; Dalton, L. R.; Ranon, P. M.; Shi, Y.; Steier, W. H. *Macromolecules* **1993**, *26*, 5303-5309.
51. Mandal, B. K.; Kumar, J.; Huang, J. C.; Tripathy, S. *Makromol. Chem., Rapid Commun.* **1991**, *12,* 63-68.
52. (a) Page, R. H.; et al. *J. Opt. Soc. Am. B* **1990**, *7*, 1239; (b) Hubbard. M. A.; Marks, T. J.; Lin, W.; Wong, G. K. *Chem. Mater*. **1992,** *4*, 965-968.
53. Gibbon s, W. M.; Grasso, R. P.; O'Brien, M. K.; Shannon, P. J.; Sun, S. T. *Macromolecules* **1994**, *27(3)*, 771.
54. Hsiue, G. H.; et al. *Chem. Mater.* **1994**, *6*, 884-887.
55. Aramaki, S.; Okamoto, Y.; Murayama, T. *Jpn. J. Appl. Phys., Pt. 1*, **1994**, *33(10)*, 5759.
56. Yilmaz, S; Bauer, B.; Gerhard-Multhaupt, G. *Appl. Phys. Lett.* **1994**, *64(21)*, 2770.
57. Cohen, R.; Berkovic, G. *Optics Lett.* **1994**, *19(14)*, 1025.
58. Gaines, G .L. *Insoluble Monolayers at Liquid-Gas Interfaces*, Interscience: New York, 1966.
59. Hall, R. C.; Lindsay, G. A.; Anderson, B.; Kowel, S. T.; Higgins, B. G.; Stroeve, P. *Mat. Res. Soc. Symp. Proc.* **1988**, *109*, 351-356.
60. Zhou, F. O; Singer, K. D.; Lando, G; Mann, J. A. *Macromol.* **1993**, *26*, 7263.
61. (a) Mötschmann, H. R.; Penner, T. L.; Armstrong, N. J.; Ezenyilimba *J. Phys. Chem.* **1993**, *97(216)*, 3933; (b) Penner, T. L.; et al. *Nature* **1994**, *367*, 49.
62. Hoover, J. M.; Henry, R. A.; Lindsay, G. A.; Nee, S. F.; Stenger-Smith, J. D. In *Organic Materials for Non-linear Optics III*, G. J. Ashwell and D. Bloor, Eds.; the Royal Society of Chemistry: London, 1993; Special Publication 137.
63. (a) Li, D.; Ratner, M. A.; Marks, T. J., Zhang, C.; Yang, J.; Wong, G. K. *J. Am. Chem. Soc* **1990**, *112*, 7389; and (b) Yitzchaik, S.; et al. *J. Phys. Chem.* **1993**, *97*, 6958; (c) Katz, H.; et al. *Science* **1991**, *254*, 1435.
64. Wischerhoff, E.; Zentil, R. *Macromol. Chem. Phys.* **1994**, *195*, 1593-1602.
65. Benné, I.; Semmler, K.; Finkelmann, H. *Macromol. Rapid Commun.* **1994**, *15*, 295.
66. Tsibouklis, J.; et al. *Synthetic Metals* **1993**, *61*, 159-162.
67. Struik, L.C.E. *Physical Aging in Amorphous Polymers and other Materials*, Elsevier: Amsterdam, 1978.
68. Lindsay, G. A. *Trends in Polymer Science* **1993**, *1(5)*, 138.
69. Andrews, J. H.; Singer, K. D. *Applied Optics* **1993**, *32(33)*, 6703.

70. Booth, B. L. In *Polymers in Lightwave and Integrated Optics, Technology and Applications*, L. A. Hornak, Ed.; Marcel Dekker: New York, 1992.
71. Wang, W.; Chen, D.; Fetterman, H.; Shi, Y.; Steier, W. H.; Dalton, L. R. *Appl. Phys. Lett.* **1994**, *65(8)*, 929.
72. Hill, R. A.; Knoesen, A.; Mortazavi, M. A. *Appl. Phys. Lett.* **1994**, *65(14)*, 1733.
73. (a) Steier, W. H.; et al. *Proc. SPIE* **1993**, *2025*, 535-546; (b) Zyung, T.; Hwang, W.-Y.; Kim, J.-J. *Appl. Phys. Lett.* **1994**, *64(26)*, 3527.
74. Beeson, K. W.; Horn, K. A.; McFarland, M.; Nahata, A.; Wu, C. Yardley, J. T. *Proc. SPIE* **1990**, *1337*, 195-202.
75. Hayden, L. M. *ACS Polymer Preprints* **1994**, *35(2)*, 88.
76. (a) Schildkraut, et al. *Appl. Opt.* **1988**, *27*, 2839; (b) Teng, C. C.; Mann, H. T. *Appl. Phys. Lett.* **1990**, *56(18)*, 1734.
77. (a) Sigelle, M.; Hierle, R. *J. Appl. Phys.* **1981**, *52(6)*, 4199; (b) Bosshard, C.; Sutter, K.; Schlesser, R.; Günter, P. *J. Opt. Soc. Am.* **1993**, *B10*, 867; (c) Norwood, R. A.; Kuzyk, M. G.; Keosain, R. A. *J. Appl. Phys.* **1994**, *75(4)*, 1869.
78. (a) Levine, B. F.; Bethea, C. G. *J. Chem. Phys.* **1975**, *63(6)*, 2666; (b) A. Dhinojawla; Wong, G. W.; Torkelson, J. M. *J. Opt. Soc. Am. B* **1994**, *11(9)*, 1549.
79. (a) Murphy, E. J.; Rice, T. C. *IEEE J. Quant. Elec.* **1986,** *QE-22*, 928; (b) Garabedian, R.; et al. *Sensors and Actuators A* **1994**, *43*, 202-207.
80. Teng, C. C. *Appl. Phys. Lett.* **1992**, *60*, 1538.
81. Wah, C. K. L.; Lizuka, K.; Freundorfer, A. P. *Appl. Phys. Lett.* **1993**, *63(23)*, 3110.
82. Yankelevich, D.; Knoesen, A.; Eldering, C. A. *Proc. SPIE* **1991**, *1560*, 406.
83. Möhlmann, G.; et al. *Proc. SPIE* **1990**, *1337*, 215.
84. Van Eck, T. E.; Lipscomb, G. F.; Ticknor, A. J. Valley, J. F.; Lytel, R. *Applied Optics* **1992**, *31(32)*, 6823.
85. Sotoyama, W.; Tatsuura, S.; Motoyohi, K.; Yoshimura, T. *Jpn. J. Appl. Phys.* **1992**, *Pt. 2, No. 8B*, L1180.
86. Hitaka, M.; Shuto, Y.; Amano, M.; Yoshimura, R.; Tomaru, S.; Kozawaguchi, H. *Appl. Phys. Lett.* **1993**, *63(9)*, 1161.
87. Tumolillo, T. A., Jr.; Ashley, P. R. *Appl. Phys. Lett.* **1993**, *62(24)*, 3068-3070.
88. Kaino, T.; Amano, M.; Shuto, Y. In *Nonlinear Optics*, S. Miyata, Ed.; Elsevier: 1992, p. 163.
89. (a) Florsheimer, M.; Küpfer, M.; Bosshard, H.; Looser, H.; Günter, P. *Adv. Mater.* **1992**, *4*, 795; (b) Clays, K.; Schildkraut, J. S.; Williams, D. J. *J. Opt Soc. Am. B* **1994**, *11(4)*, 655; (c) Miyata, S.; Ogasawara, H.; Edel, V.; Watanabe, T. *ACS Polymer Preprints* **1994**, *35(2)*, 211.

RECEIVED April 19, 1995

Chromophores: Design, Synthesis, and Optical Measurements

Chapter 2

Design and Properties of High-Temperature Second-Order Optical Chromophores

R. F. Shi, S. Yamada, M. H. Wu, Y. M. Cai, O. Zamani-Khamiri, and A. F. Garito[1]

Department of Physics and Laboratory for Research on the Structure of Matter, University of Pennsylvania, Philadelphia, PA 19104

A new class of high temperature chromophores which are structurally similar to polyimide repeat units have successfully been developed. Both the pure chromophores and guest-host systems in polyimides have shown thermal stabilities far above 300°C. Quantum many-electron multiple configuration interaction calculations have been implemented to simulate the rich variety of structures of the new class and to understand their linear and nonlinear optical properties. A picosecond tunable laser system has been utilized to perform frequency dependent DC-induced second harmonic generation measurements on the chromophores. In the near infrared nonresonant region, values of the standard $\mu\beta$ product of 1500×10^{-48} esu have been obtained. Good agreement between experiments and calculations have been achieved.

Organic and polymeric systems are now being intensively studied because of their potential applications in optoelectronic and all optical computing devices. These systems possess many advantages over inorganic systems (*1*) (*2*). Organic systems have very large nonresonant optical nonlinearities, ultrafast and broadband electronic responses, low dielectric constants, and minimal phase mismatch between optical waves and microwaves. In addition, polymer materials are highly processable and their structures can be easily modified to meet specific needs in electrooptic (EO) waveguide applications. Multilayer waveguide devices, such as frequency converters and EO modulators (*3*), compatible with standard semiconductor device manufacturing steps, can easily be fabricated using polymer materials at relatively low costs. Compared with inorganic and semiconductor structures, polymer systems have negligible one-photon and two-photon absorptions at wavelengths designed for optical fiber telecommunications. Furthermore, with improved algorithms and computing capabilities of supercomputers, complex molecular structures can now be simulated with high

[1]Corresponding author

0097–6156/95/0601–0022$12.75/0

accuracy. Optimized structures can then be selected for synthesis and characterization, thereby considerably reducing the cost and shortening the time span in new materials development. Among different nonlinear optical processes that are being studied, a thorough understanding of the origin, the physical and chemical features has nearly been achieved for second-order optical materials. This is to be compared with third order optical materials, where a deeper understanding of the whole process and new classes of materials are required before any realistic device application is feasible (*4*).

Electric field poling processes are usually implemented to achieve the macroscopic noncentrosymmetry of the polymer system necessary for second order optical processes (*5*). In such cases, the chromophores are aligned along the applied electric field when the polymer system is heated up to above its glass transition temperature (T_g). The electric poling field is maintained when the alignment is finished and the system is cooled down to room temperatures. Finally, when the field is turned off, the chromophores are frozen in the polymer matrix, producing a net macroscopic second order optical nonlinearity. Much of the early effort was on poly(methyl methacrylate) (PMMA) for the polymer matrix and 4,-N,N-dimethylamino-4'-nitrostilbene (DANS) or its derivatives for the chromophores. Although much understanding has been achieved and various systems including guest-host, main chain, side chain, and cross-linking have been developed, the low glass transition temperature for the polymers and low decomposition temperatures for the chromophores limited, to a large extent, any device applications where stringent thermal, chemical, and photo stability conditions are required for processing and fabrication.

A new approach (*6*) to achieving high temperature stable polymer systems has recently been developed using polyimides which are well-known in microelectronics industry. These polyimides are of interest for integrated optics due to their high thermal stability (>400°C), high optical transparency, low thermal expansion comparable to that of silicon, and easy processability into planar structures. Waveguides fabricated from polyimides can be patterned by photobleaching or etching using standard semiconductor technology (*7*). In fact, new fluorinated polyimides of high optical transparency with loss less than 0.3 dB/cm at 1.3 μm have been synthesized and single mode optical waveguides using these fluorinated polyimides have been fabricated (*8*). The current research is therefore focused on finding new high temperature chromophores (*9*)(*10*) to be incorporated into polymer systems which can withstand thermal and chemical requirements in polyimide based fabrication processes and device assembly steps.

This article reports our recent development of a new class of high temperature fused-ring EO chromophores (*11*)(*12*), 1,8-naphthoylene-benzimidazoles, designed for their structural similarity to polyimide monomers and consequent easy incorporation into host polyimides. The new class was discovered prior to actual synthesis using computer aided molecular designs that implement quantum many-electron calculations. Examples of the new chromophore classes involving imides, imidazoles, and various extended aromatic fused-rings are shown in Fig. 1.

Section 2 of this article discusses thermal properties of the new systems in both their pure forms and as guest-host systems with polyimides. It is demonstrated that the side group substitution actually increases the thermal stability. In section 3, we present results of quantum many-electron calculations of these high temperature chromophore model structures, showing that the charge-correlated excited states are responsible for the relatively large second order optical responses in these chromophores. The experimental DCSHG dispersion results of these structures in different solvents, using a picosecond optical tunable laser source, will be discussed in section 4, along with their comparison with theoretical results. The EO coefficients of the chromophores will also be predicted.

THERMAL PROPERTIES

We have recently developed a new class of high temperature fused-ring EO chromophores, 1,8-naphthoylene-benzimidazoles, shown in Fig. 2. A detailed study on synthesis will be published elsewhere. Structurally, the new class is similar to polyimide monomer units, and therefore is expected to be easily incorporated into host polyimides to form guest-host systems. The benzimidazole is the main electron donating group, whose property will be enhanced when amine groups are attached. The naphthoylene backbone serves as the π bridge that provide the necessary π-electron delocalization. The twisted phenyl ring on the left side, along with the associated tert-butyl groups, improves the solubility of the chromophores in common solvents significantly. Members of the new class are highly colored, as can be seen from the linear optical absorption spectra in solution with 1,4-dioxane shown in Fig. 3. The wavelength of the first electronic transition is tunable from 430 nm to 632 nm through side group substitution.

Thermal stabilities of both the pure chromophores and the guest-host systems have been studied by thermal gravimetric analysis (TGA) and differential scanning calorimetry (DSC). The TGA data are shown in Fig. 4, demonstrating the good thermal stabilities for the whole series. The side group substitution, in our case, increases the thermal stability. For example, the parent compound, SY156, shows the 5% weight loss at 340°C, while for the amine substituted SY177, the 5% weight loss occurs at 369°C. Most importantly, the alkylated amine compound (SY215), has the best thermal stability so far, with 5% weight loss temperature reaching 398 ^{o}C. Since we are not certain at this point whether the weight loss is caused by decomposition or by sublimation, the actual decomposition temperature could possibly be higher than 398°C. Thermal stabilities of the guest-host systems were examined by studying the linear optical absorption spectra when the temperature was elevated to different stages. After spin-coating, soft vacuum baking and thermal curing, the guest-host polyimide thin films were heated to an elevated temperature and soaked for ten minutes in a nitrogen purged oven and then cooled down to room temperature before measuring the linear absorption spectrum. For SY215, the spectrum showed no change in the location or height of the absorption peak when it was dissolved in Hitachi polyimide L100 and heated up to 325°C.

THEORETICAL UNDERSTANDING

Quantum many-electron calculations are needed for the understanding of the origin, features, and behaviors of organic nonlinear optical responses (*13*). The advantages of having computation and simulation capabilities are many-fold. Calculations can provide qualitative and quantitative comparisons between theory and experiment, giving us new insights into the optical properties of the materials considered. They can guide materials scientists and synthetic chemists in the search for new materials suited for use in specific devices. Further, these quantum calculations can accurately predict new results based on available information . An example of this is the

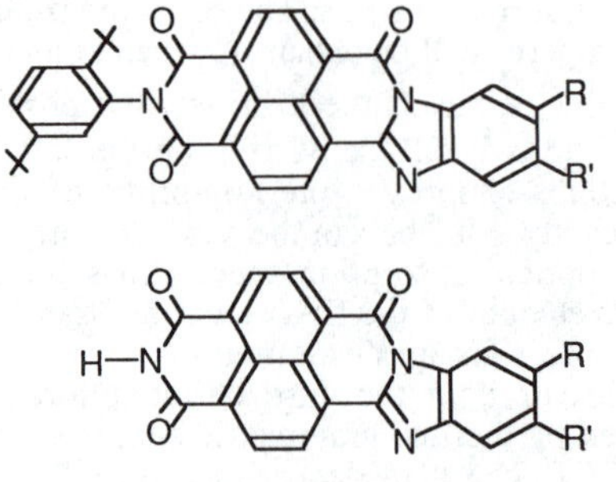

(a)

(b)

R	R'	Material
H	H	SY156
H	OCH_3	SY165
OCH_3	H	SY165A
NH_2	H	SY177
H	NH_2	SY177P
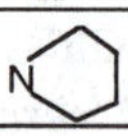 N	H	SY215

Fig. 2. Schematic molecular structures for 1,8-naphthoylene-benzimidazole chromophores. The twisted phenyl ring on the left side of the structure improves their solubilities. SY177 and SY177P are two isomers that have markedly different nonlinear optical properties, a finding which will be published elsewhere.

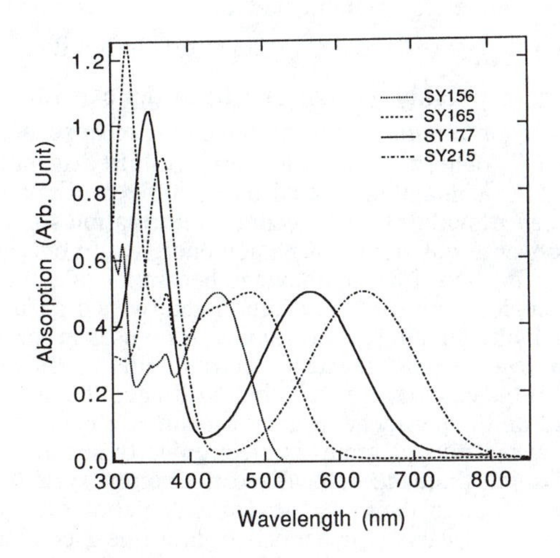

Fig. 3. Optical absorption spectra for the SY series in 1,4-dioxane solution. Tunability of the first optical transition energy, which is responsible for the color of the new class, is achieved through side group substitution.

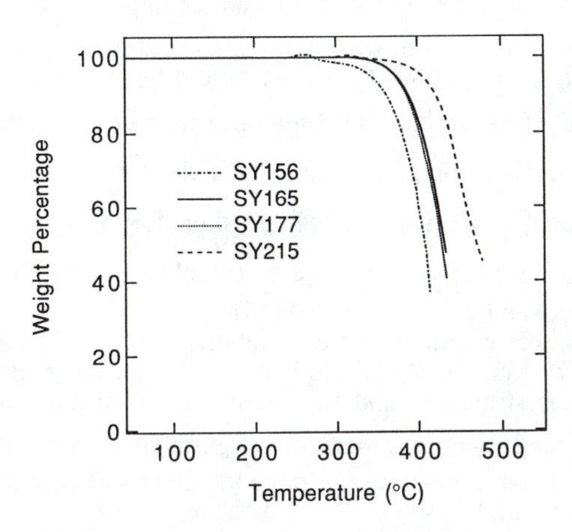

Fig. 4. Thermal gravimetric analysis (TGA) data for the pure chromophores. The 5% weight loss temperature for SY215 reaches 398°C.

prediction of r_{33} EO coefficients using $\mu\beta$ values from standard DCSHG (EFISH) experiments. Frequently, it is found that the prediction in the literature does not match the result from a direct EO measurement. One of the reasons is that in most cases, a two level model is implemented to account for different dispersion behaviors between $\beta_{ijk}(-\omega;\omega,0)$ and $\beta_{ijk}(-2\omega;\omega,\omega)$. In addition, the contribution from third order susceptibility $\gamma_{ijkl}(-2\omega;\omega,\omega,0)$ to the experimentally measured value is always difficult to remove, except in the case of hyper-Rayleigh scattering measurements (*14*), where third order susceptibility is not involved. The summation-over-states calculations based on multiple configuration interactions, on the contrary, can provide us with accurate information for various nonlinear optical properties over a wide range of photon energies. In our calculations of the model structures of the new 1,8-naphthoylene-benzimidazole class, standard bond lengths and bond angles were used. In addition, the twisted phenyl group in Fig. 2 was replaced by a hydrogen atom. Since optical responses in these compounds are determined mainly by the delocalized π-electrons, this modification will have a negligible effect on the calculated results but witll have the advantage of keeping the study focussed on the pertinent structural features. Recent NMR studies have shown that the twisted phenyl group is electronically separated from the main group, thus confirming that, the delocalized π-electron system does not extend beyond the planar portion of the molecule. Our calculation consists of three steps. First, a molecular orbital basis is generated from a linear combination of atomic orbitals (LCAO) using the self-consistent Hartree-Fock Hamiltonian with CNDO/S approximation. In the second step, electronic eigenstates, their eigen values, and transition moments between these states are calculated by implementing single and double configuration interactions (SDCI) among π molecular orbitals. In these calculations, for the SCI, the ten occupied and ten unoccupied π orbitals closest to the Fermi level are utilized. For the DCI, eight occupied and eight unoccupied π orbitals closest to the Fermi level are used. It will be shown that the second order optical response β_{ijk} $(-\omega_3;\ \omega_1,\omega_2)$ is mainly determined by the ground, S_0, and first excited, S_1, states. Therefore, although SDCI does not include all π molecular orbitals, values for β_{ijk} are accurate, while values for $\gamma_{ijkl}(-\omega_4;\omega_1,\omega_2,\omega_3)$ are still reliable. In the last step, β_{ijk} and γ_{ijkl} are calculated using the sum-over-states expression first derived by Orr and Ward (*15*).

As an example of our model calculations, we will first discuss our model structure for SY177 (see Fig. 2). Table I shows the calculated electronic excitation energies, oscillator strengths, and the x-component of the transition moments between the low-lying electronic states in the gas phase. The first excited state, S_1, is 2.44 eV above the ground state, S_0, in good agreement with the measured value of 2.21 for the compound dissolved in 1,4-dioxane.

It is worthwhile to point out that the calculated ground state dipole moment is 7.14 D, while the measured value in 1,4-dioxane is only 3.6 D. The discrepancy can be partially accounted for by the fact that in our model calculation, the twisted phenyl ring has been replaced by a hydrogen atom, which is a very weak electron donor, resulting in a smaller cancellation of the dipole moment. The discrepancy in the ground state dipole moment between calculation and measurement does not affect our calculated results for β, since, as will be discussed later, β is determined by the difference in the dipole moment between the first excited S_1 and ground S_0

states. Additionally, the calculation shows that the largest oscillator strength with respect to the ground S_0 state is from the S_1 state, while the linear absorption spectrum reveals that the S_3 state, identified from Table I, has the strongest coupling with S_0. The reason for the disagreement is not clear at the moment. But we believe that this is responsible for the slight difference in the dispersion behavior of $\mu\beta$ between theory and experiment, which will be discussed in the next section.

The tensor components of the molecular second order polarizability, β_{ijk}, are calculated using the following equation

$$\beta_{ijk}(-\omega_3;\omega_1,\omega_2) = K(-\omega_3;\omega_1,\omega_2)I_{1,2}\left(-\frac{e^3}{\hbar^2}\right) \sum_{m_1,m_2}{}' \left[\frac{r^i_{gm_1}\bar{r}^j_{m_1m_2}r^k_{m_2g}}{(\Omega_{m_1g}-\omega_3)(\Omega_{m_2g}-\omega_1)} + \frac{r^i_{gm_1}\bar{r}^k_{m_1m_2}r^j_{m_2g}}{(\Omega^*_{m_1g}+\omega_2)(\Omega^*_{m_2g}+\omega_3)} + \frac{r^k_{gm_1}\bar{r}^i_{m_1m_2}r^j_{m_2g}}{(\Omega^*_{m_1g}+\omega_2)(\Omega_{m_2g}-\omega_1)} \right]$$

where $r^i_{m_1m_2}$ is the matrix element $\langle m_1|r^i|m_2\rangle$ ($\bar{r}^i = r^i - \langle r\rangle_{gg}$). $\hbar\Omega_{m_1g}$ is the energy difference between state m_1 and the ground state g. The prime on the summations indicates a sum over all states except the ground state, and $I_{1,2}$ denotes an average over all permutations of ω_1 and ω_2 with associated indices j, k, and l. The factor $K(-\omega_3;\omega_1,\omega_2)$ is equal to $2^{-m}D$ where m is the number of nonzero frequencies in the set $\{\omega_1,\omega_2\}$ less one (unless $\omega_4=0$) and D is the number of distinguishable orderings in the set $\{\omega_1,\omega_2\}$.

Since the measurements of β are made in the solution phase, the calculated gas phase transition energies of the excited states used in the above equation were shifted so as to account for the solvatochromic shift of SY177 in 1,4-dioxane. The values for β_{ijk} calculated in this manner are $\beta_{xxx}(-2\omega;\omega;\omega)= 92.5$, $\beta_{xyy}(-2\omega;\omega,\omega)=0.2$, and $\beta_{yyx}(-2\omega;\omega,\omega)= 0.4$ in units of 10^{-30} esu and at $\hbar\omega=0.65$eV.

The experimentally measured quantity, however, is the component of β along the molecular dipole moment, or, β_μ. Table I shows that the molecular dipole moment in the model structure of SY177 is essentially along the x-axis. The expression for β_μ can then be approximated as that for β_x and is given by

$$\beta_x=\beta_{xxx} + (1/3)\sum_{i\neq x}(\beta_{xii}+2\beta_{iix})$$

We notice that β_{xxx}, having by far the largest magnitude, will essentially determine the magnitude of β_x.

Table II shows the dominant virtual excitation sequences that contribute to $\beta_{xxx}(-2\omega;\omega,\omega)$ at $\hbar\omega$=0.65 eV. The largest contribution involves the S_1 state, in the form of $S_0 \rightarrow S_1 \rightarrow S_1 \rightarrow S_0$, which is the only sequence in the standard two-level model. According to the equation for β_{ijk}, this is to be expected since S_1 has the largest transition moment, μ^x_{1g}, and dipole moment difference, $\Delta\mu^x_{1g}$, as well as the lowest transition energy with respect to the ground state. The virtual excitation sequences not included in the standard two-level model contributes about 35% to the total $\beta_{xxx}(-2\omega;\omega,\omega)$.

The importance of the S_1 state can be seen by the difference in the charge density distribution between states n and n' ($\Delta\rho_{nn'}$) defined as

$$\Delta\rho_{nn'} \equiv \rho_n - \rho_{n'}$$

where

$$\rho_m(r_1) = \int \Psi^*_m(r_1,r_2,...r_M)\Psi_m(r_1,r_2,...r_M)dr_2...dr_M.$$

The transition density matrix $\rho_{nn'}$ is defined through the expression

$$\langle\mu_{nn'}\rangle = -e\int r\rho_{nn'}(r)dr$$

with

$$\rho_{nn'}(r_1) = \int \Psi^*_n(r_1,r_2,...r_M)\Psi_{n'}(r_1,r_2,...r_M)dr_2...dr_M$$

where M is the number of valence electrons included in the molecular wavefunction. Contour diagrams of $\Delta\rho_{S_1S_0}$ and $\rho_{S_1S_0}$ are shown in Fig. 5 where solid and dashed lines correspond to increased and decreased charge density, respectively, and the contour is taken at 0.4Å above the molecular plane. Both contours clearly show the high concentration of the π electrons on the naphthoylene moiety that is transferred from the donor substituted imidazole ring upon virtual excitations. It also shows that the naphthoylene moiety serves both as an electron donor and as a bridge that provides the π electron delocalization. Equally importantly, these diagrams identify the specific atomic sites for any possible structure modifications in order to enhance the nonlinear optical effects.

Fig. 6 shows the calculated dispersion curve of $\beta_x(-2\omega;\omega,\omega)$ of SY177 as a function of input photon energy. The excitation energies have been shifted so as to take into account the solvatochromic shift, as stated above. The slope of the dispersion curve is due to the onset of the two-photon resonance at 1.11 eV resulting from the $S_0 \rightarrow S_1$ transition at 2.21 eV.

The substitution of a stronger electron donor in SY177 is expected to increase the second order susceptibility. This is reflected in the calculations for SY215 (see Fig. 2) model structure, where a dimethylamino group, which is a stronger electron donor, replaces the simple amine group in SY177. Compared with the actual SY215 compound, where the pyrrolidino is present, the model SY215 structure should have the same π-electronic properties. Table III lists the calculated

Table I. Caculated Excitation Energies, Oscillator Strengths, and Key Transition Dipole Moments for the Lowest-lying π-electron States of SY177 Model Structure in Gas Phase.

State	E(f) (eV)	$\mu^{x}_{S_0 n}$ (D)	$\mu^{x}_{S_1 n}$ (D)	$\mu^{y}_{S_0 n}$ (D)	$\mu^{y}_{S_1 n}$ (D)
S_0	0.00(0.00)	7.14	-7.15	0.72	-0.62
S_1	2.44(0.48)	-7.15	19.52	-0.62	3.27
S_2	3.32(0.02)	-1.33	-1.32	0.06	-0.37
S_3	3.48(0.08)	-2.47	-5.94	0.01	-0.23
S_4	3.58(0.02)	0.19	-1.13	-1.10	-0.29
S_5	3.87(0.17)	2.73	-0.71	2.03	1.82
S_6	3.95(0.02)	1.16	4.62	-0.12	0.23
S_7	4.44(0.02)	1.13	1.03	-0.20	1.83
S_8	4.64(0.01)	-0.03	-0.29	0.77	0.66
S_9	4.66(0.02)	-0.91	2.01	0.58	-0.72
S_{10}	4.83(0.13)	0.41	-0.26	2.27	-0.49

Table II. Largest Contributing Virtual Excitation Processes to $\beta_{xxx}(-2\omega;\omega,\omega)$ at $\hbar\omega = 0.65$ eV

m_1	m_2	Contribution (10^{-30} esu)	Cumulative (10^{-30} esu)
S_1	S_1	118.7	118.7
S_3	S_1	-11.9	106.7
S_1	S_3	-10.8	95.9
S_{13}	S_2	4.9	100.8
S_1	S_{13}	4.4	105.2
S_6	S_1	-3.8	101.4
S_1	S_6	-3.5	97.9
S_5	S_5	3.2	101.1
⋮	⋮	⋮	⋮
Total			92.5

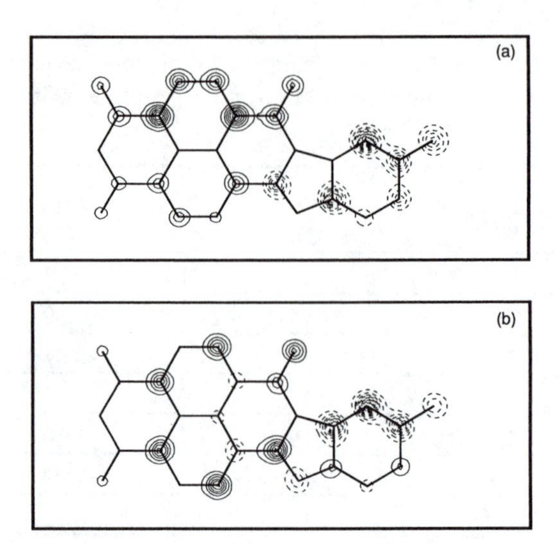

Fig. 5. Contour diagrams of the electron charge density distribution difference (a) and transition density matrix $\rho_{nn'}$ (b) between the ground state (S_0) and the first excited state (S_1) for SY177. Solid lines and dashed lines correspond to increased and decreased charge density respectively, and the contour cut is taken at 0.4Å above the molecular plane.

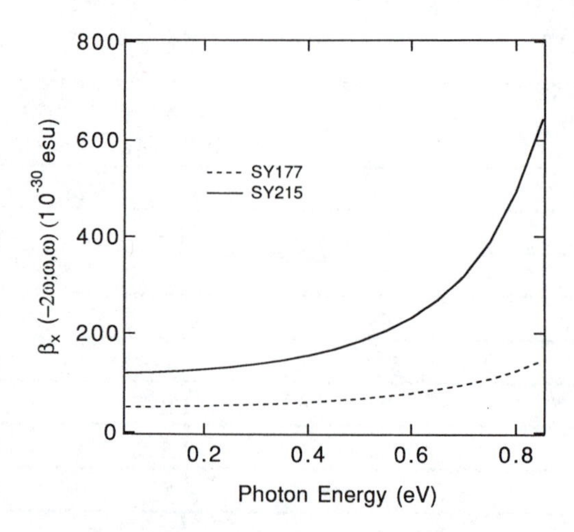

Fig. 6. Calculated $\beta_x(-2\omega;\omega,\omega)$ for SY177 in 1,4-dioxane (dashed curve) and SY215 in dichloromethane (CH_2Cl_2) (solid curve) as a function of fundamental input photon energies.

excitation energies, oscillator strengths, and key transition dipole moments for the lowest-lying π-electron states of SY215 model structure in the gas phase. Compared with SY177, the ground and first excited state dipole moments, as well as the transition moment between the two states in SY215 are enhanced. At the same time, the transition energy between the two states decreases, resulting in a red shift in the visible absorption peak. All these factors are responsible for an increased β response in SY215. This can be seen from Table IV, which shows the largest contributing virtual excitation processes to $\beta_{xxx}(-2\omega;\omega,\omega)$ for SY215 model structure in dichloromethane at $\hbar\omega$=0.65 eV. The table clearly indicates that, similar to SY177, the most dominant contributing term comes from the virtual excitation sequence $S_0 \rightarrow S_1 \rightarrow S_1 \rightarrow S_0$. The four processes next to it all contribute negatively to the overall β, thus reducing the final value for β. This shows that at least for the SY series, the standard two-level model overestimates the values for β. At $\hbar\omega$=0.65 eV, $\beta_{xxx}(-2\omega;\omega,\omega)$ has the value of 268×10^{-30} esu. Thus, more than a factor of two enhancement has been achieved compared to the value obtained for SY177. We would like to caution, however, that the value for SY215 is for the compound in dichloromethane. The corresponding value in 1,4-dioxane, which is more appropriate for the comparison, was found to be 190.7×10^{-30} esu. The reason for the difference is that the absorption peak for SY215 in dichloromethane is red shifted relative to that in 1,4-dioxane. The calculated $\beta_{xxx}(-2\omega;\omega,\omega)$ dispersion for SY215 in dichloromethane is also shown in Fig. 6. The contour diagrams for SY215 model structure, shown in Fig. 7, clearly show the enhanced charge separation responsible for the larger β values for SY215 compared to SY177.

EXPERIMENTAL RESULTS

The dispersion study of the second-order β for the high temperature chromophores was carried out using the standard DCSHG technique (*16*). Two laser sources were utilized in our measurements. One involved picosecond optical parametric generation and amplification using KTP crystals (*17*), while the other was a stimulated Raman scattering technique with hydrogen (H_2) or methane (CH_4) filled gas cells, implemented before the picosecond tunable source was available.

The standard method to obtain values for the molecular second-order β is to measure the macroscopic $\chi^{(3)}(-2\omega;\omega,\omega,0)$ of a solution containing the compound and a solvent with known $\chi^{(3)}(-2\omega;\omega,\omega,0)$. For each sample, the refractive indices and coherence lengths of the solution and BK-7 glass windows in the DCSHG cell affect the measurements. Therefore, the determination of the molecular β would require a knowledge of the $\chi^{(3)}$ values of all three components, i.e., the BK-7 glass windows, the solvent, and the solution.

For SY177, seven wavelengths were selected for the dispersion measurement using anhydrous 1,4-dioxane as the solvent. At each wavelength and its second harmonic, the refractive indices for the BK-7 glass windows were calculated using the BK-7 refractive index dispersion relation (*16*), with the results

Table III. Caculated Excitation Energies, Oscillator Strengths, and Key Transition Dipole Moments for the Lowest-lying π-electron States of SY215 Model Structure in Gas Phase.

State	E(f) (eV)	$\mu^x_{S_0n}$ (D)	$\mu^x_{S_1n}$ (D)	$\mu^y_{S_0n}$ (D)	$\mu^y_{S_1n}$ (D)
S_0	0.00(0.00)	8.50	7.64	0.30	0.32
S_1	2.02(0.45)	7.64	25.29	0.32	2.00
S_2	3.18(0.09)	-2.65	1.35	-0.34	0.72
S_3	3.34(0.14)	3.35	-6.52	0.08	0.54
S_4	3.52(0.01)	-0.30	-1.49	0.54	0.00
S_5	3.63(0.08)	-2.36	3.80	-0.08	0.75
S_6	3.68(0.09)	-1.73	-1.56	-1.88	1.83
S_7	4.19(0.05)	1.69	-1.83	-0.42	-1.73
S_8	4.36(0.06)	0.06	-0.96	1.96	0.71
S_9	4.40(0.00)	-0.26	-2.01	-0.44	1.17

Table IV. Largest Contributing Virtual Excitation Processes to $\beta_{xxx}(-2\omega;\omega,\omega)$ of SY215 in dichloromethane at $\hbar\omega = 0.65$ eV

m_1	m_2	Contribution (10^{-30} esu)	Cumulative (10^{-30} esu)
S_1	S_1	340.9	340.9
S_3	S_1	-28.2	312.7
S_1	S_3	-23.5	289.2
S_5	S_1	-10.5	278.7
S_1	S_5	-8.7	270.0
⋮	⋮	⋮	⋮
Total			268.0

listed in Table V. The coherence length, l_c, at each wavelength was then calculated using the equation $l_c = \lambda_\omega /[4(n_{2\omega}-n_\omega)]$. The $\chi_G^{(3)}(-2\omega;\omega,\omega,0)$ values for BK-7 glass were obtained through the generalized Miller rule, using the value of 3.5×10^{-14} esu at λ=1064 nm as a reference (*16*). These calculated results, listed in Table V, show that the coherence length decreases with decreasing wavelength , with the exception of λ=2148 nm, where, compared to that at λ=1907 nm, the coherence length is actually smaller. For $\chi_G^{(3)}(-2\omega;\omega,\omega,0)$, the dispersion is less than 3% over the wavelength range of interest.

The coherence lengths of 1,4-dioxane at the seven selected wavelengths were obtained through the analysis of the Maker fringes at each wavelength. The results are shown in Table VI. From l_c, $\Delta n=n_{2\omega}-n_\omega$ can be accurately calculated. The refractive indices at the fundamental wavelength have been previously determined, with the value being approximately 1.41. The values at the second harmonics were found to be $n_{2\omega}$=1.41 for Δn less than 0.005 and $n_{2\omega}$=1.42 for Δn greater than 0.005. The less than precise values for n_ω and $n_{2\omega}$ amount to less than 1% contribution to the solution $\chi^{(3)}(-2\omega;\omega,\omega,0)$. The $\chi_D^{(3)}(-2\omega;\omega,\omega,0)$ values of 1,4-dioxane were determined through a linear fit to the previously measured values (*16*) at 1907, 1543, and 1064 nm, which are 4.50, 4.93, and 5.71×10^{-14} esu, respectively. The refractive indices at both the fundamental and second harmonic, the coherence lengths and $\chi_D^{(3)}(-2\omega;\omega,\omega,0)$ for 1,4-dioxane at the seven selected wavelengths are listed in Table VI. It is noted that l_c decreases monotonically as the wavelength decreases. The dispersion of $\chi_D^{(3)}(-2\omega;\omega,\omega,0)$ throughout the range of interest is about 15%.

Stock concentrations of around 15×10^{3} mol/liter were used in the $\chi^{(3)}(-2\omega;\omega,\omega,0)$ measurements of SY177 in 1,4-dioxane. At each wavelength, five different concentrations were used, one of them being the pure 1,4-dioxane solvent. In addition, the fringe height for the pure solvent was constantly monitored to compensate for any possible long term laser power drift. Optical microscope observations (200×) showed a uniform, nonaggregated pattern for even the highest stock solutions, indicating that the solubility of SY177 in 1,4-dioxane was excellent. The $\chi^{(3)}(-2\omega;\omega,\omega,0)$ value for each concentration was calculated from the analysis of its Maker fringe. Stock solutions showed no absorption at any of the fundamental wavelengths. Corrections due to the absorption of the second harmonic were only necessary in the case of data taken at 1400 nm. In this case, the ratio of the fringe maximum to the minimum was only 1.9, compared to the other cases where, in the absence of absorption, the ratio was typically greater than 50. There was about 5 to 10% decrease in the coherence length of 1,4-dioxane relative to the stock solution, depending on the specific wavelength and the stock concentration. This is due to the change in the refractive indices of the solution, the effect of which amounts to about 1%. Therefore, one can safely assume that both n_ω and $n_{2\omega}$ for the stock solution is the same as those for the pure solvent. The decrease of the coherence length is due to the fact that $n_{2\omega}$ increases more rapidly than n_ω as a function of concentration.

In order to obtain γ', defined as $\gamma'=(\mu\beta/5KT)+\gamma$, for the chromophore, using the infinite dilution extrapolation method (*18*), we need to know the values for $\partial\chi^{(3)}/\partial C$, $\partial n^2/\partial C$, and $\partial\varepsilon/\partial C$, where $\chi^{(3)}$ is the solution susceptibility, C is the concentration, n is the refractive index of the solution, and ε is the dielectric constant. The specific volume change as a function of concentration (or weight fraction) is negligible. $\partial\chi^{(3)}/\partial C$ is obtained through the concentration dependence measurement of $\chi^{(3)}$. $\partial\varepsilon/\partial C$ is obtained in the process of ground state dipole moment measurement, where the capacitance, and thereby the dielectric constant of the solution is measured as a function of concentration. $\partial n^2/\partial C$ is normally obtained by measuring the concentration dependence of the solution refractive index. We have recently developed a new method for obtaining $\partial n^2/\partial C$ from $\partial l_c/\partial C$ shown in the following two equations (*19*):

$$\frac{\partial n_{\omega}^{2}}{\partial C}=\frac{4\lambda_{\omega}n_{\omega}}{3l_c^2}\left[4-\left(\frac{\lambda_{\omega}}{\lambda_0}\right)^2\right]\frac{\partial l_c}{\partial C},$$

$$\frac{\partial n_{2\omega}^{2}}{\partial C}=\frac{4\lambda_{\omega}n_{\omega}}{3l_c^2}\left[1-\left(\frac{\lambda_{\omega}}{\lambda_0}\right)^2\right]\frac{\partial l_c}{\partial C},$$

where λ_0 is the corresponding wavelength for the first transition energy of the chromophore. It was found that one can simply use the following equation for data analysis:

$$\frac{\mu_x\beta_x(-2\omega;\omega,\omega)}{5kT}+\langle\gamma(-2\omega;\omega,\omega,0)\rangle=\frac{1}{N_A}\left(\frac{3}{n_{\omega}^2+2}\right)^2\left(\frac{3}{n_{2\omega}^2+2}\right)\left[\frac{n_{\omega}^2+2\varepsilon}{(n_{\omega}^2+2)\varepsilon}\right]\frac{\partial\chi^{(3)}(-2\omega;\omega,\omega,0)}{\partial C},$$

where μ_x is the ground state dipole moment, β_x is the vector part of β_{ijk}, $\langle\gamma(-2\omega;\omega,\omega,0)\rangle$ is the isotropically averaged γ_{ijkl}, and N_A is the Avagadro's number.

The resultant $\mu_x\beta_x(-2\omega;\omega,\omega)+\langle\gamma(-2\omega;\omega,\omega,0)\rangle 5kT$ for SY177 in 1,4-dioxane as a function of photon energy is plotted in Fig. 8, along with the theoretical results for the model structure. In evaluating the theoretical $\mu_x\beta_x$ values, the experimental value of 3.6 D was used for μ_x. The reason for making this choice was that, as stated above, the calculation overestimates the ground state dipole moment and, since β_{ijk} values depend on $\Delta\mu$, using the calculated ground state dipole moment value would be misleading as it would introduce errors from a source which has no direct impact on the optical nonlinearity. A damping constant of $\hbar\Gamma$=0.25 eV was used in this case, derived from the linear optical absorption spectrum for the molecule in 1,4-dioxane.

Both the theoretical and experimental results, which are fairly well matched over the entire nonresonant region, show the dispersion resulting from the 2ω resonance (1.11 eV) due to the $S_0\rightarrow S_1$ transition at 2.21 eV. The error bars for each experimental data point was calculated from the linear concentration fit for $\chi^{(3)}$. The gap in the data points between 0.65 eV (1907 nm) and 0.74 eV (1670 nm) is due to

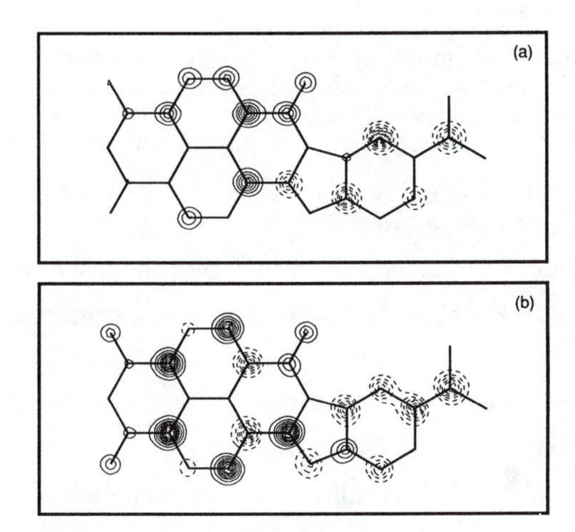

Fig. 7. Contour diagrams of the electron charge density distribution difference (a) and transition density matrix $\rho_{nn'}$ (b) between the ground state (S_0) and the first excited state (S_1) for SY215.

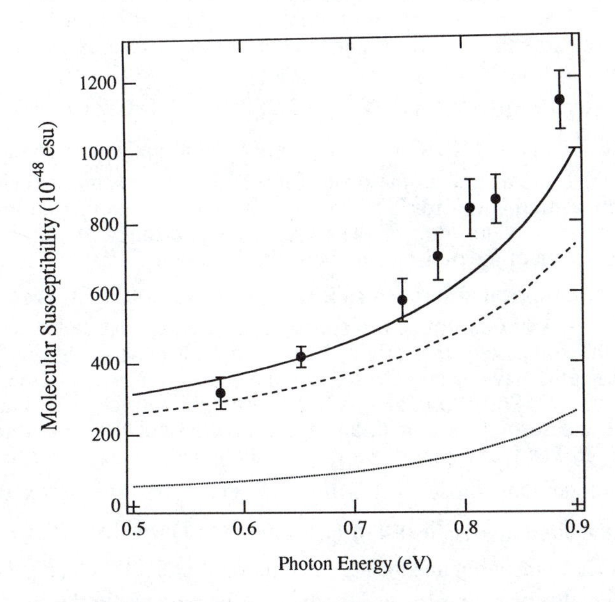

Fig. 8. Microscopic molecular susceptibilities for SY177 in 1,4-dioxane as a function of fundamental input photon energy. Points and the solid curve are experimental and theoretical values for $\mu_x\beta_x(-2\omega;\omega,\omega)+\langle\gamma(-2\omega;\omega,\omega,0)\rangle/5kT$, respectively. The dashed and dotted curves are theoretical values for $\mu_x\beta_x(-2\omega;\omega,\omega)$ and $\langle\gamma(-2\omega;\omega,\omega,0)\rangle/5kT$, respectively.

the relatively strong absorption coming from the overtone of 1,4-dioxane (*20*) peaking at around 1700 nm. The slight discrepancy between theory and experiment at high photon energies may be attributed to the inability of the theory to reproduce the oscillator strength of the second major absorption peak. The presence of the second absorption peak resulting from the $S_0 \rightarrow S_n$ transition will very likely increase the slope of the dispersion.

The theoretically calculated values for $\mu_x\beta_x(-2\omega;\omega,\omega)$ and $<\gamma(-2\omega;\omega,\omega,0)>5kT$ with T=300 K are also plotted in Fig. 8 demonstrating the sizable contribution of the third-order optical susceptibility $<\gamma(-2\omega;\omega,\omega,0)>$ to the measurement in our study. For example, the calculated value for $\mu_x\beta_x(-2\omega;\omega,\omega)$ is 333×10^{-48} esu and that for $<\gamma(-2\omega;\omega,\omega,0)>5kT$ is 80×10^{-48} esu ($<\gamma(-2\omega;\omega,\omega,0)>=386\times10^{-36}$ esu). This indicates that $<\gamma(-2\omega;\omega,\omega,0)>$ contributes 19.4% to the total measured value. The experimental determination of $<\gamma(-2\omega;\omega,\omega,0)>$ for noncentrosymmetric SY177 is difficult. We therefore obtained $\beta_x(-2\omega;\omega,\omega)$ for SY177 by subtracting the theoretical value for $<\gamma>5kT$ from the experimental data. The value for $\beta_x(-2\omega;\omega,\omega)$ obtained in this manner is 93×10^{-30} esu at $\hbar\omega$=0.65 eV.

Similar measurements have also been performed on SY215, which when dissolved in 1,4-dioxane, has a strong absorption peak at 632 nm. SY215 is the alkylated amine structure of SY177, and, as discussed before, is expected to show enhanced nonlinear optical responses compared to SY177. The experimental value for $\mu_x\beta_x(-2\omega;\omega,\omega)+<\gamma(-2\omega;\omega,\omega,0)>5kT$ of SY215 in 1,4-dioxane was determined to be 900×10^{-48} esu by DCSHG measurements at 1900 nm, more than a factor of two enhancement compared to the result for SY177 at the same wavelength. The dispersion measurements for SY215 were performed using dichloromethane (CH_2Cl_2) as the solvent due to the vibrational overtones of dioxane and the potential absorption of the molecule at the second harmonic.

The linear optical absorption spectrum of SY215 in CH_2Cl_2 is shown in Fig. 9. The visible peak at 688 nm, is red-shifted compared to that in 1,4-dioxane (632 nm) due to the solvatochromic effect. The absorption peak is also wider. Again, seven wavelengths have been selected for the dispersion measurement, ranging from 1600 nm to 1960 nm. Table VII lists the refractive indices at both the fundamental and second harmonic, and the measured coherence lengths at these wavelengths. It was found that, for a given wavelength, the coherence lengths of CH_2Cl_2 is longer than those of 1,4-dioxane. The refractive index of CH_2Cl_2 reported in the literature is n=1.424 at λ=590 nm. The values of n at different wavelengths were then obtained using the relation $\Delta n=\lambda/(4l_c)$. Once again, the less than accurate value of n contributes less than 1% uncertainty to the final molecular susceptibilities. The value of $\chi^{(3)}(-2\omega;\omega,\omega,0)$ for CH_2Cl_2 over the selected wavelength range was assumed to be 11.4×10^{14} esu,the value measured at λ=1907 nm; thus neglecting the dispersion of $\chi^{(3)}$ for CH_2Cl_2. The refractive indices, coherence lengths, and $\chi^{(3)}$ for BK-7 glass windows were calculated as before.

Table V. Fundamental and Second-Harmonic Refractive Indices, Coherence Lengths, and $\chi^{(3)}(-2\omega; \omega, \omega, 0)$ of BK-7 Glass

λ(nm)	n_{ω}	$n_{2\omega}$	l_c(μm)	$\chi^{(3)}(-2\omega;\omega,\omega,0)$ (10^{-14} esu)
2148	1.4925	1.5065	38.36	3.32
1907	1.4960	1.5082	39.08	3.32
1670	1.4992	1.5101	38.20	3.37
1600	1.5001	1.5108	37.32	3.38
1543	1.5008	1.5114	36.39	3.39
1500	1.5013	1.5119	35.58	3.40
1400	1.5025	1.5131	33.15	3.43

Table VI. Fundamental and Second-Harmonic Refractive Indices, Coherence Lengths, and $\chi^{(3)}(-2\omega; \omega, \omega, 0)$ of 1, 4-Dioxane

λ(nm)	n_{ω}	$n_{2\omega}$	l_c(μm)	$\chi^{(3)}(-2\omega;\omega,\omega,0)$ (10^{-14} esu)
2148	1.41	1.41	109.6	4.50
1907	1.41	1.41	105.7	4.50
1670	1.41	1.42	78.3	4.81
1600	1.41	1.42	73.5	4.91
1543	1.41	1.42	71.1	4.93
1500	1.41	1.42	65.4	5.05
1400	1.41	1.42	54.6	5.20

Table VII. Fundamental and Second-Harmonic Refractive Indices, Coherence Lengths of Dichloromethane

λ(nm)	n_{ω}	$n_{2\omega}$	l_c(μm)
1960	1.41	1.41	129.7
1907	1.41	1.41	124.8
1850	1.41	1.41	121.6
1800	1.41	1.41	107.1
1750	1.41	1.41	98.7
1638	1.41	1.42	72.2
1600	1.41	1.42	67.5

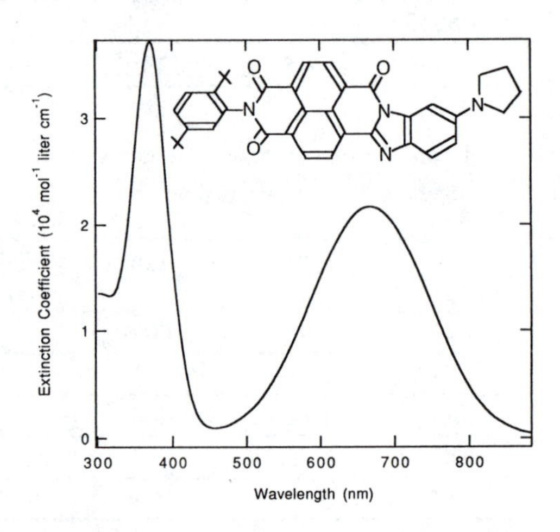

Fig. 9. Linear optical absorption spectrum for SY215 in dichloromethane (CH_2Cl_2). Compared to that in 1,4-dioxane, the peak in the visible is red-shifted and also wider in width.

Stock concentrations ranging from 6×10^{3} mol/liter (in the case where the absorption is present at the second harmonic) to 12×10^{-3} mol/liter (in the case where the absorption is absent at the second harmonic) were used in the $\chi^{(3)}(-2\omega;\omega,\omega,0)$ measurement of SY215 in CH_2Cl_2. At each wavelength, five different concentrations were used, one being the pure solvent. Each solution was checked under the optical microscope (200×) each time to ensure its quality . Similar to SY177, SY215 dissolves almost instantly in CH_2Cl_2. We were therefore able to prepare the solutions only a few hours before performing the measurement. Experimental values of $\mu_x\beta_x(-2\omega;\omega,\omega)+<\gamma(-2\omega;\omega,\omega,0)>5kT$ are plotted in Fig. 10 as a function of input photon energies.

The ground state dipole moment for SY215 was measured to be μ_x=4.4 D. Using this information and the calculated $\beta_x(-2\omega;\omega,\omega)$ and $<\gamma(-2\omega;\omega,\omega,0)>$ values, the calculated $\mu_x\beta_x(-2\omega;\omega,\omega)+<\gamma(-2\omega;\omega,\omega,0)>5kT$ is also plotted in Fig. 10. The damping constant is $\hbar\Gamma$=0.28 eV, derived from the linear absorption spectrum. Once again, both the theoretical and experimental data, which are well matched over the region of interest, show the dispersion resulting from the 2ω resonance (0.93 eV) due to the $S_0\rightarrow S_1$ transition at 1.86 eV. The feature at 1638 nm (0.76 eV) is due to the uncertainty resulting from the residual absorption at the second harmonic and not due to any possible peak. The gap in the experimental points between 1750 nm (0.71 eV) and 1638 nm (0.76 eV) is due to the strong vibrational overtone of CH_2Cl_2 in the region. Theoretically calculated $\mu_x\beta_x$ and $<\gamma>5kT$ are also plotted in Fig. 10 illustrating the contribution of $<\gamma>$ to the overall susceptibility. At $\hbar\omega$=0.65 eV, the theoretical value of 347×10^{-48} esu calculated for $<\gamma>kT$ is about 23% of the experimentally determined value of $\mu_x\beta_x+<\gamma>5kT=1468\times10^{-48}$ esu. The $\beta_x(-2\omega;\omega,\omega)$ for SY215 in CH_2Cl_2 was obtained as before by subtracting the theoretical $<\gamma>$ values from the experimental $\mu_x\beta_x+<\gamma>5kT$, leading to a value of $\beta_x(-2\omega;\omega,\omega)=268\times10^{-30}$ esu. Compared to the value of $\beta_x(-2\omega;\omega,\omega)=92\times10^{-30}$ esu for SY177 in 1,4-dioxane, there is almost a factor of three enhancement. The main factors contributing to this enhancement of β_x are the improved donor ability and the lowering of the excitation energy of the first transition in SY215 compared to SY177.

The DCSHG dispersion measurements of various structures using different solvents demonstrate the accuracy of the quantum many-electron calculations. They also reveal the origin and microscopic features of the SY series chromophores. In addition, they point to ways of improving the structures so as to enhance the optical nonlinearities. The ability to accurately describe the microscopic second order optical responses enables us to predict the electrooptic property of the chromophores incorporated into polymer matrices. Fig. 11 shows the calculated microscopic EO $\beta_x(-\omega;\omega,0)$ of SY215 as a function of photon energies, assuming a damping constant of $\hbar\Gamma$=0.20 eV. In the convention we use, $\beta_x(-\omega;\omega,0)$ is related

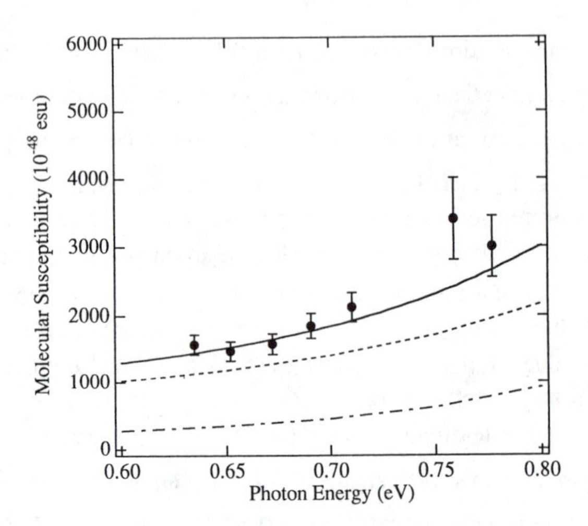

Fig. 10. Microscopic molecular susceptibilities for SY215 in dichloromethane (CH_2Cl_2) as a function of fundamental input photon energy. Points and the solid curve are experimental and theoretical values for $\mu_x\beta_x(-2\omega;\omega,\omega)+<\gamma(-2\omega;\omega,\omega,0)>5kT$, respectively. The dashed and dash-dotted curves are theoretical values for $\mu_x\beta_x(-2\omega;\omega,\omega)$ and $<\gamma(-2\omega;\omega,\omega,0)>5kT$, respectively.

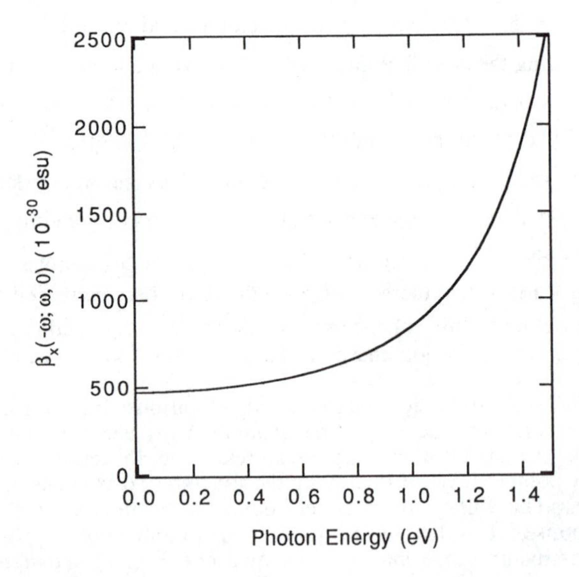

Fig. 11. Calculated $\beta_x(-\omega;\omega,0)$ for SY215, assuming 1.86 eV for the transition energy between the S_1 and S_0 states.

to $\beta_x(-2\omega;\omega,\omega)$ by $\lim_{\omega\to 0}\beta_x(-\omega;\omega,0)=4\lim_{\omega\to 0}\beta_x(-2\omega;\omega,\omega)$. As an example, we will predict the EO r_{33} response of SY215 in polymer matrices using the following formulae,

$$\chi^{(2)}_{333}(-\omega;\omega,0) = N(f^{\omega})^2 f^0 \frac{\mu_x E_{total}}{5kT}\beta_x(-\omega;\omega,0),$$

$$r_{33}=\frac{1}{n^4}\chi^{(2)}_{333}(-\omega;\omega,0),$$

where N is the number density, f^{ω} and f^0 are local field factors, E_{total} is the effective electric field which is related to the poling field (E_{poling}) by $E_{total}=f^0 E_{poling}$. The number density, N, is given by $N=N_A\rho w/M$, where $N_A=6.02\times10^{23}$ is the Avagadro's number, ρ is the density, assumed to be 1.5 gm/cm^3, w=12% is the weight fraction, and M=596.7 is the molecular weight for SY215 leading to $N=1.8\times10^{20}$ cm^{-3}. In addition, Assuming a refractive index n=1.6 and a dielectric constant $\varepsilon=3.0$, the local field factors are calculated to be $f^{\omega}=(n^2+2)/3=1.5$, and $f^0=(n^2+2)\varepsilon/(n^2+2\varepsilon)=1.6$. Using the dipole moment $\mu_x=4.4$ D, the poling field $E_{poling}=200$ V/μm, and, from the dispersion plot, $\beta_x(-\omega;\omega,0)=794\times10^{-30}$ esu at $\lambda=1.3$ μm (0.96 eV), we obtain $r_{33}=8.5$ pm/V.

CONCLUSIONS

The key to finding high speed EO waveguide applications for organic and polymer systems is that these systems must have good primary optical properties such as large optical nonlinearities, yet at the same time, maintain good secondary properties such as thermal, chemical and photo stabilities. In this paper, we have reported our recent development of high temperature fused-ring EO chromophores, 1,8-naphthoylene-benzimidazoles, which are structurally similar to polyimide monomers and are, therefore, relatively easy to incorporate in the host polyimides. These chromophores were found to possess high thermal stabilities, both in pure forms and as guest-host systems. Importantly, the TGA data showed that 1,8-naphthoylene-(3'-pyrrolidino) benzimidazole-4,5-dicarbox-N- (2,5-di-tert-butyl) phenylimide (SY215 in short), was, in its pure form, stable up to 398°C, its 5% weight loss temperature. Guest-host systems with SY215 in Hitachi polyimide L100 were stable up to 325°C.

Theoretical quantum many-electron calculations have revealed the charge-correlated nature of the second order optical responses in these chromophores. These results were confirmed quantitatively by experimental dispersion measurements using a tunable picosecond laser source on two key structures, SY177 and SY215, in two typical solvents, 1,4-dioxane and dichloromethane (CH_2Cl_2). In addition, increasing the donor strength was found to increase the dipole moments and transition moments, thereby enhancing the β responses. The calculations also showed that a two-level model is not adequate in terms of quantitative predictions; overestimating β_x by as much as 35% in some cases.

Comparison between theory and experiment demonstrated that for these chromophores, the third order contribution to the overall DCSHG signal is not negligible. The EO coefficient for SY215 in polyimide guest-host systems was also estimated. Values up to 8.5 pm/V were obtained. The study of these high temperature chromophores incorporated in various polyimides is under way and attempts to enhance both the dipole moments and the second order optical responses are being systematically carried out.

Acknowledgments

This research was generously supported by AFOSR and ARPA (grant F49620-85-C-0105). The calculations were performed on the CRAY YMP C-90 of the Pittsburgh Supercomputing Center. We are grateful for helpful discussions with Drs. Q.M. Qian and A. Panackal and Mr. W.D. Chen.

Literature Cited

1. *Polymers for Lightwave and Integrated Optics: Technology and Applications*; Hornak, L.A., Ed., Marcel Dekker: New York, NY, 1992.
2 Garito, A.F.; Shi, R.F.; Wu, M.H. *Physics Today* **1994**, 47(5), 51.
3. Teng, C.C. App. Phys. Lett. **1992**, 60, 1538.
4. Stegeman, G.I.; Torruellas, W. *Mat. Res. Soc. Sym. Proc.* **1994**. 328, 397.
5. Singer, K.D.; Kuzyk, M.G.; Sohn, J.E. *J. Opt. Am. Soc. B***1987**, 4, 968.
6. Wu, J.W.; Binkley, E.S.; Kenney, J.T.; Lytel, R.; Garito, A.F. *J. App. Phys.* **1991**, 69, 7366.
7. Binkley, E.S.; Nara, S. in *Organic Thin Films for Photonic Applications Technical Digest*, 1993, Vol 17, (Optical Society of America, Washington, DC, 1993), pp. 266.
8. Matsuura,T; Ando, S.; Matsui, S.; Hirata, H.; Sasaki, S.; Yamamoto, F. in *Organic Thin Films for Photonic Applications Technical Digest*, 1993, Vol 17, (Optical Society of America, Washington, DC, 1993), pp262.
9. Marder, S.R.; Perry, J.W. *Science* **1994**, 263, 1706.
10. Wong, K.Y.; Jen, A. K.Y. *J. App. Phys.*, **1994**, 75, 3308.
11. Shi, R.F.; Wu, M.H.; Yamada, S.; Cai, Y.M.; Garito, A.F. *App. Phys. Lett.* **1993**, 63, 1173.
12. Yamada, S.; Cai, Y.M.; Shi, R.F.; Wu, M.H. Chen, W.D.; Qian, Q.M.; Garito, A.F. *Mat. Res. Soc. Sym. Proc.* **1994**, 328, 523.
13. Heflin, J. R.; Wong, K.Y.; Zamani-Khamiri, O.; Garito, A.F. *Phys. Rev. B* **1988,** 38, 1573.
14. Clays, K.; Persoons, A. *Phys. Rev. Lett.* **1991**, 66, 2980.
15. Orr, B.J.; Ward, J.F. *Mol. Phys.* **1971**, 20, 513.
16. Heflin, J.R.; Cai, Y.M.; Garito, A.F. J. Opt. Soc. Am. B**1991**, 8, 2132.
17. Wu, M.H.; Cai, Y.M.; Garito, A.F. in *Quantum Electronics and Laser Science Conference*, 1993 OSA Technical Digest Series, Vol. 3 (Optical Society of America, Washington, DC, 1993), pp.59-60.
18. Singer, K.D.; Garito, A.F. *J. Chem. Phys.* **1981**, 75, 3572.
19. Shi, R.F.; Wu, M.H.; Yamada, S.; Cai, Y.M.; Zamani-Khamiri, O.; Garito, A.F. (To be published).
20. Goddu, R.F.; Delker, D.A. *Anal. Chem.***1960**, 32, 140.

RECEIVED January 30, 1995

Chapter 3

Hyperpolarizabilities of Push–Pull Polyenes

Molecular Orbital and Valence-Bond Charge-Transfer Models

J. W. Perry[1,2], S. R. Marder[1,2], F. Meyers[2,3], D. Lu[2], G. Chen[2], W. A. Goddard, III[2], J. L. Brédas[3], and B. M. Pierce[4]

[1]Jet Propulsion Laboratory, California Institute of Technology, Pasadena, CA 91109
[2]Beckman Institute, California Institute of Technology, Pasadena, CA 91125
[3]Center for Research on Molecular Electronics and Photonics, Université de Mons-Hainaut, Place du Parc 20, B-7000 Mons, Belgium
[4]Hughes Aircraft Company, Radar Systems, Building R2, MS, V518, P.O. Box 92426, Los Angeles, CA 90009–2426

In this paper we review two theoretical approaches that illustrate the relationships between molecular (hyper)polarizabilities and bond-length alternation (*BLA*). In one approach, we employ sum-over-states calculations at the INDO-SDCI level on $(CH_3)_2N\text{-}(CH{=}CH)_4\text{-}CHO$, including an external, static, homogeneous electric field that unes the degree of ground-state polarization and *BLA*. In the second approach, we use a simple semiclassical model, wherein the mixing of valence bond (VB) and charge-transfer (CT) states along a *BLA* coordinate is treated. Qualitatively, the two approaches give very similar structure-property trends, that are consistent with experimental hyperpolarizability data on donor-acceptor polyenes. The agreement of the calculated trends reflects the importance of VB-CT energetics in determining them.

Recent experimental work has demonstrated a correlation between the bond length alternation and the first and second hyperpolarizabilities (β and γ) of donor-acceptor (DA) substituted (push-pull) polyenes (*1, 2*). The bond length alternation, *BLA,* in DA polyenes is defined as the average of the difference in the length between adjacent carbon-carbon bonds in the polymethine, $(CH)_n$, chain. The structure of such molecules can be understood qualitatively as resulting from a superposition of valence-bond (VB) and charge-transfer (CT) limiting resonance structures (Figure 1). Experimentally, molecules have been examined where *BLA* was varied from the polyene limit of ~-0.1Å to values approaching the opposite *BLA* limit of ~0.1Å, by use of acceptors of varying strength and groups that gain aromaticity on polarization for coarse tuning. Additionally, the structures have been fine tuned by the use of solvents of varying polarity. It was shown that $\mu\beta$ (*2*) and γ (*1, 3*) exhibit positive

0097–6156/95/0601–0045$12.00/0

and negative peaks and sign changes as the ground-state polarization is increased, with *BLA* varying concomitantly from -0.12 towards 0.12Å. These experimental trends are consistent with those calculated with a finite-field AM-1 molecular orbital method, described previously (*4*). In this paper, we provide an overview of recent work, presented in part at the 1994 fall American Chemical Society meeting in Washington, DC, on detailed sum-over-states molecular orbital calculations (*5, 6*) at the intermediate neglect of differential overlap level (including configuration interaction with single and double excitations, INDO-SDCI) and a semiclassical valence-bond/charge-transfer (VB-CT) model (*7, 8*) that elucidate the relationships between *BLA*, the related π-bond order alternation, *BOA*, (*vide infra*) and α, β and γ.

Relationships between *BOA* and α, β, and γ at the INDO-SDCI level

In order to correlate α, β, and γ with geometric and electronic structure, we examined, at the semi-empirical INDO level, the molecule $(CH_3)_2N$-$(CH{=}CH)_4$-CHO (Figure 1) in the presence of a static, homogeneous electric field. Fields of high strength were used to tune the degree of ground-state polarization, *BLA* and the related parameter, *BOA* (Figure 2). The molecular polarizabilities were evaluated using a sum-over-states approach in combination with INDO-SDCI calculations (*5, 6*). These calculations allowed us to identify the electronically excited states that play an important role in the nonlinear optical response (*5, 6*). The correlation of the polarizabilities with *BOA* is shown in Figures 3-5 (note that ground-state polarization increases concomitantly with BOA). In addition, this analysis allowed us to compare the values of the polarizabilities obtained from the full sum-over-states expressions for α_{sos}, β_{sos} and γ_{sos} and simplified two or multilevel level models in which:

$$\alpha_{\mathrm{model}} \propto \frac{\mu_{\mathrm{ge}}^2}{E_{\mathrm{ge}}} \tag{1}$$

$$\beta_{\mathrm{model}} \propto \frac{\mu_{\mathrm{ge}}^2(\mu_{\mathrm{ee}}-\mu_{\mathrm{gg}})}{E_{\mathrm{ge}}^2} \tag{2}$$

$$\gamma_{\mathrm{model}} \propto \underbrace{-\left(\frac{\mu_{\mathrm{ge}}^4}{E_{\mathrm{ge}}^3}\right)}_{\mathrm{N}} + \underbrace{\sum_{\mathrm{e'}}\left(\frac{\mu_{\mathrm{ge}}^2\,\mu_{\mathrm{ee'}}^2}{E_{\mathrm{ge}}^2\,E_{\mathrm{ge'}}}\right)}_{\mathrm{T}} + \underbrace{\left(\frac{\mu_{\mathrm{ge}}^2(\mu_{\mathrm{ee}}-\mu_{\mathrm{gg}})^2}{E_{\mathrm{ge}}^3}\right)}_{\mathrm{D}} \tag{3}$$

where μ and E are the dipole matrix element and transition energy, respectively, between the subscripted states. The subscripts g, e, and e' label the ground, first excited and upper excited states, respectively.

The first-order polarizability, α, exhibits a peak at zero *BOA* (referred to here as the cyanine limit) (Figure 3). In the two-level model expression (eq. 1) α is dependent on two factors, $(1/E_{\mathrm{ge}})$ and μ_{ge}^2. As can be seen in Figure 3 there is a very good correlation between the converged sum-over-states calculation (30 states) and that of the two level model. Examination of the dependence of $(1/E_{\mathrm{ge}})$ and μ_{ge}^2 on *BOA* reveal that they both peak near the cyanine-limit (Figure 3), thus providing insight into the dependence of α on *BOA*.

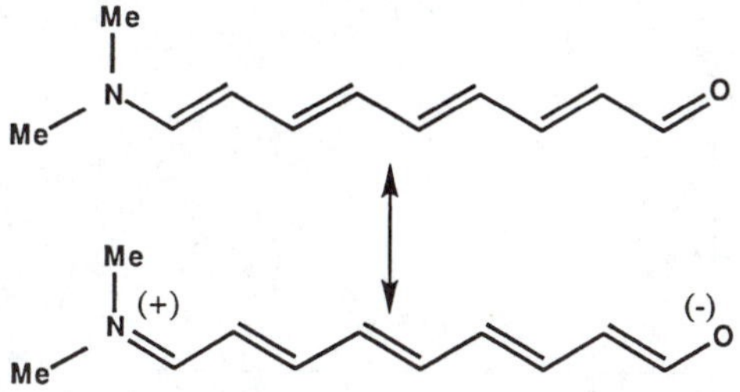

Figure 1. Valence-bond and charge-separated resonance structures for $(CH_3)_2N\text{-}(CH{=}CH)_4\text{-}CHO$.

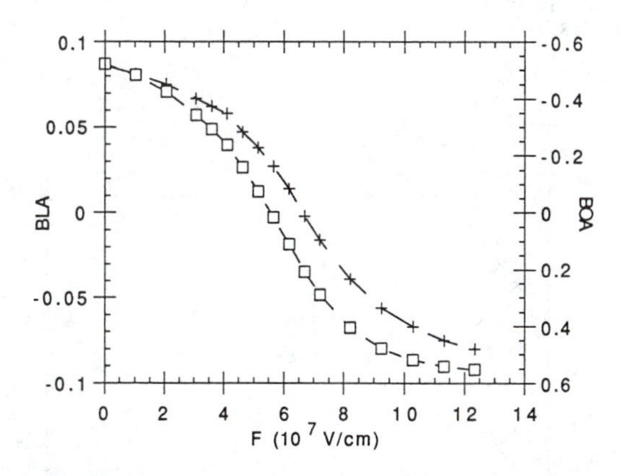

Figure 2. Evolution of *BLA* and *BOA* vs. applied field (F) for $(CH_3)_2N\text{-}(CH{=}CH)_4\text{-}CHO$.

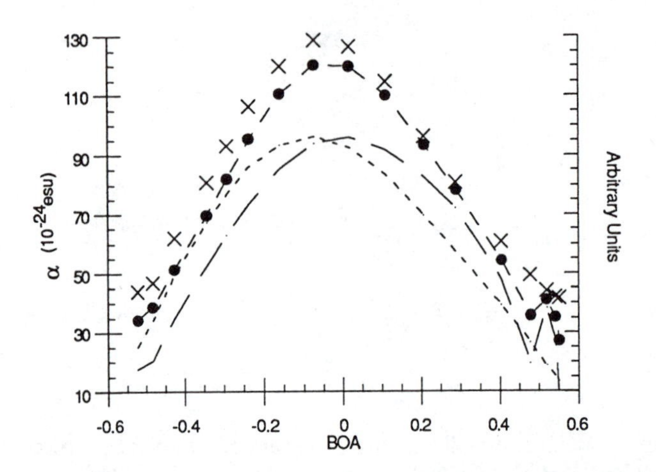

Figure 3. Evolution of: α_{sos} (crosses) and α_{model} (full circles, dashed line) in 10^{-24} esu; $1/E_{ge}$ (short dashed line) and μ_{ge} (long dashed line) in arbitrary units, plotted vs. *BOA*.

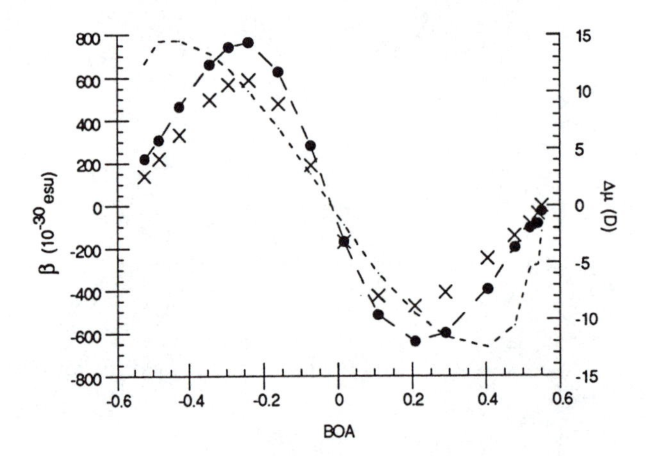

Figure 4. Evolution of: β_{sos} (crosses) and β_{model} (full circles, dashed line) in 10^{-30} esu; (μ_{ee}-μ_{gg}) (short dashed line) in Debye, plotted vs. *BOA*.

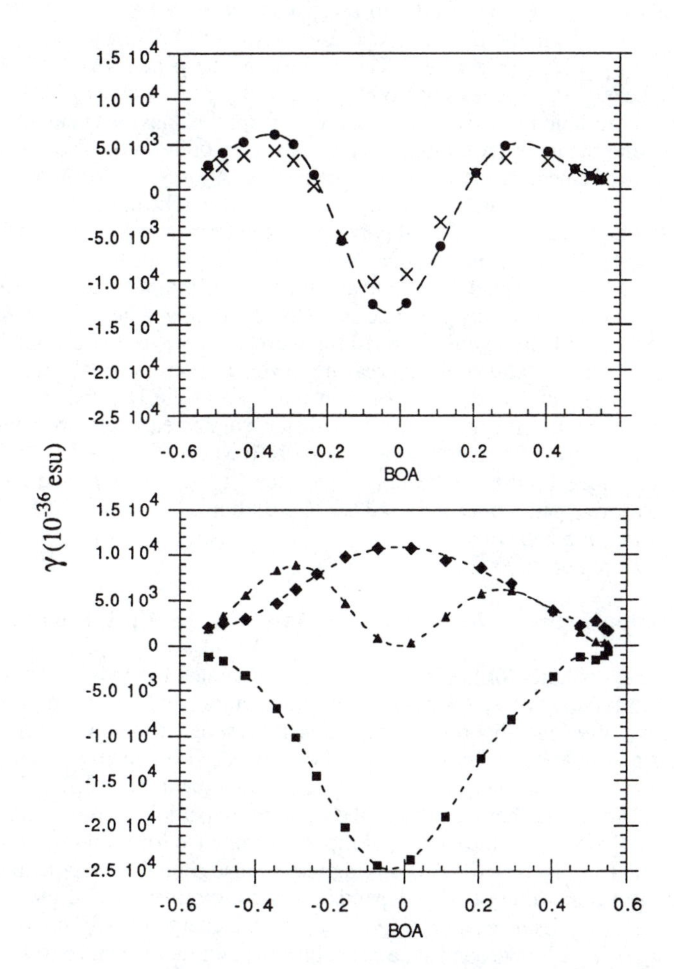

Figure 5. Evolution of: γ_{sos}(crosses) and γ_{model} (full circles, dashed line) [top], and the D-(triangles), N-(squares), and T-terms (diamonds) [bottom] in 10^{-36} esu, plotted vs. *BOA*.

The first hyperpolarizability, β, as a function of increasing ground-state polarization, exhibits a positive peak, a sign change at zero *BOA*, and a negative peak (Figure 4). Again, the two level calculation for β describes well the basic correlation of β and *BOA*, in terms of both peak position and magnitude. Accordingly, the relation between β and *BOA* can be understood in terms of the model expression (eq. 2) in which β_{model} is proportional to α_{model} scaled by a factor of $(\mu_{ee}\text{-}\mu_{gg})/E_{ge}$. Thus, the two peaks in the β curve are a result of the form of α and the sign change and peaks in $(\mu_{ee}\text{-}\mu_{gg})$ (Figure 3). The position of the β peaks are closer to zero *BOA* than the $(\mu_{ee}\text{-}\mu_{gg})$ peaks since both $(1/E_{ge}^2)$ and μ_{ge}^2 peak at zero *BOA*.

The correlation of γ with *BOA* obtained from the converged sum-over-states calculation and the three-term model (*9-12*) also agree quite well. The expression in eq. 3 consists of a negative term, N, proportional to α_{model}^2, a two-photon term, T, and a term, D, proportional to $\beta_{model}^2/\alpha_{model}$. For chromophores with large negative *BOA*, such as unsubstituted polyenes, the D term is negligible due to a small $(\mu_{ee}\text{-}\mu_{gg})$ and the T term dominates the N term resulting in positive γ (Figure 5). By increasing the ground-state polarization and thus increasing *BOA*, the D term, like β, starts to increase and hence γ increases. However, increasing the ground-state polarization toward the cyanine-limit also increases $|N|$, ultimately leading to a decrease in γ. As a result of $|N|$ increasing in this region, the peak in the γ curve occurs at a larger magnitude of *BOA* than the peak in the D term. Upon further polarization, D peaks in a positive sense and starts to decrease and $|N|$ continues to increase and thus γ decreases and goes through zero when $|N| = T + D$. At the bond-equivalent cyanine-limit the D term is zero since $(\mu_{ee}\text{-}\mu_{gg})$ is zero and the γ curve exhibits a negative peak, since both $|N|$ and T peak but $|N| > T$ (Figure 5). To a first approximation, the behavior of γ in the region of positive *BOA* mirrors that in the negative *BOA* region.

Relationships between *BLA* and α, β, and γ in the VB-CT model

The structural evolution of the DA polyenes as calculated by AM-1 or INDO-SDCI methods as a function of a strong static field can qualitatively described in terms of a superposition of valence-bond and charge-transfer resonance structures. We recently developed a simple semiclassical VB-CT model based on such a picture for DA polyenes, to illustrate the dependence of the structural evolution on VB-CT energetics and thereby provide further insight into the structure-hyperpolarizability relationships. In the VB-CT model, (*7*) the mixing of the two resonance forms along a continuous *BLA* coordinate, *q*, is treated. This mixing results in a ground-state potential surface whose equilibrium value of *BLA* depends on, among other factors, the zero-order adiabatic energy difference, $V_o = E_{CT} - E_{VB}$, and the charge transfer matrix element, *t*. When V_o is large, the ground state resembles the neutral resonance form, whereas when V_o is negative and large in magnitude, the ground state resembles the charge-transfer resonance form. For intermediate V_o, the structure evolves rapidly, with the greatest rate of change at $V_o = 0$, where the structure is cyanine-like. The value of V_o depends on the donor and acceptor strength and the difference in energy of the bridge in the two states. Additionally, since the two states have a large difference in dipole moment, $\mu_{VB} \sim 0$, $\mu_{CT} \sim e\,R_{DA}$, the application of an electric field or the screening of the charges due to the solvent dielectric response also affects the energy difference.

In order to calculate the structure-property relationships as V_o is varied, we computed the eigenstate potential surfaces (see Figure 6), obtained the equilibrium structure on the ground state potential surface, and then calculated the dipole moment, polarizability and hyperpolarizabilities, for a given value of V_o. The zero-order potential surfaces were taken to be harmonic in this simple model. The calculation was repeated over a range of V_o, from which we obtained the dependence of the polarization, polarizabilities, and *BLA* on V_o. The ground state wavefunction is written in terms of VB and CT wavefunctions as:

$$\Psi_{gr} = \sqrt{1-f}\,\psi_{VB} + \sqrt{f}\,\psi_{CT} \tag{4}$$

where f is the charge-transfer fraction, $f = 1/2 - V/2(V^2 + 4t^2)^{1/2}$. Here V represents the vertical energy difference between VB and CT harmonic surfaces at a particular q. The ground-state energy (taking E_{VB} (q = -0.12Å) = 0) for the two-state system is:

$$E_{gr} = \frac{1}{2}\left(V - \sqrt{V^2 + 4t^2}\right) \tag{5}$$

To find the equilibrium structure we solved $dE_{gr}/dq = 0$, from which one can show (7) that $q_{eq} = -0.12 + 0.24f$. Thus, *BLA* and the charge transfer fraction are linearly related in this model. The value of q_{eq} for a given value of V_o was found either by a numerical search for the minimum or by solving the equation for q_{eq} iteratively. Recognizing that $\mu = f\mu_{CT}$ and that the polarizabilities are the derivatives of μ with respect to field strength, we obtained analytic expressions for α, β, γ, and δ:

$$\alpha = \frac{2t^2\mu_{CT}^2}{E_g^3} \tag{6}$$

$$\beta = \frac{3t^2\mu_{CT}^3 V}{E_g^5} \tag{7}$$

$$\gamma = \frac{4t^2\mu_{CT}^4(V^2 - t^2)}{E_g^7} \tag{8}$$

$$\delta = \frac{5t^2\mu_{CT}^5 V(V^2 - 3t^2)}{E_g^9} \tag{9}$$

where $E_g = (V^2 + 4t^2)^{1/2} = hc/\lambda_{max}$ is the energy gap. It should be emphasized that V and E_g are the values at q_{eq} for a particular V_o. The relationships between the polarizabilities, as function of static field strength, and the structural evolution have been discussed in earlier papers (*3,7*).

To a first approximation, molecules of a given length and bridge type can be taken to have the same t and μ_{CT}, thus the polarizabilities and the structure vary primarily due to variations in V_o. Figure 7 shows the calculated dependencies for α, β, and γ as a function of V_o. The calculated dependencies exhibit very similar trends to those discussed above in the context of the molecular orbital calculations. Since E_g is at a minimum (i.e. $E_g = 2t$) when $V_o = 0$, which is where *BLA = 0*, α is peaked, β is zero, and γ is at a negative extreme. With further stabilization of the CT state, whereupon it becomes the lower energy zero-order state, V becomes negative

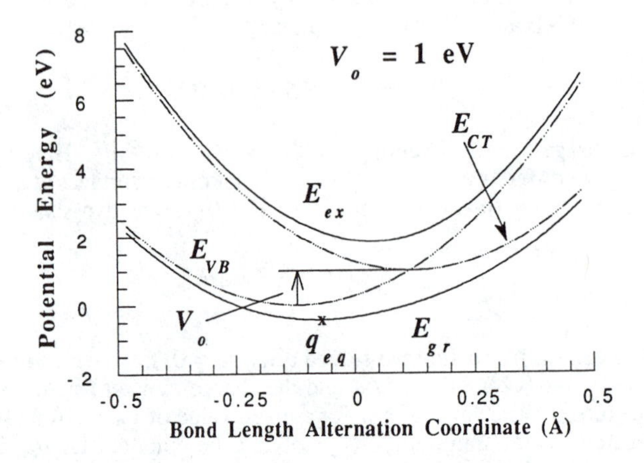

Figure 6. Potential energy surfaces for zero-order VB and CT states, and the ground and excited eigenstates, corresponding to a zero-order adiabatic energy difference of 1 eV. A harmonic force constant of 33.5 eV/Å^2 was used for the zero-order surfaces and the charge transfer matrix element, *t*, was taken as 1.18 eV. The equilibrium coordinates for VB and CT are -0.12 Å and 0.12 Å, respectively. The x marks the equilibrium coordinate (q_{eq}) on the ground state potential surface.

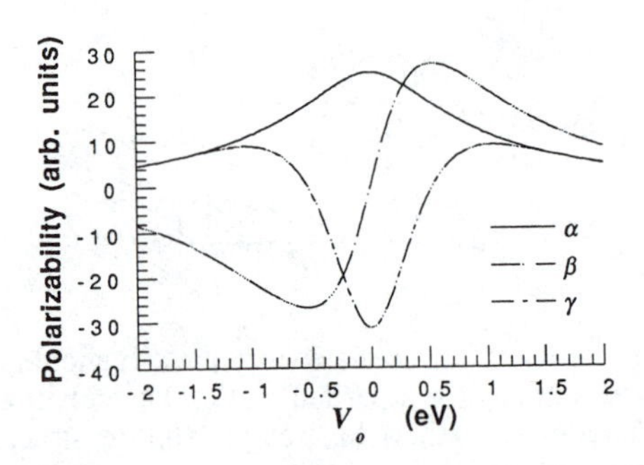

Figure 7. α, β, and γ as a function of V_o calculated using the VB-CT model for a DA polyene with nine conjugated atoms. The left extreme corresponds to zwitterionic structures and the right side to neutral polyene like structures.

and from eq. 7 so does β. Of course, this sign change is also associated with a change in sign of $(\mu_{ee}-\mu_{gg})$, which was discussed above. The value of β reaches extremes when $V = \pm t$, thus $E_g = \sqrt{5}t$, which is also where $\gamma = 0$. The value of γ reaches maxima when $V = \pm \sqrt{3}t$, whereupon $E_g = \sqrt{7}t$.

While the trends calculated using the VB-CT model agree qualitatively well with the AM-1 and INDO sum-over-states calculations, there are some small differences, particularly in the detailed shape of the γ curve (7). In the sum-over-states calculations, the positive peak value of γ (relative to that of the negative peak) is larger than is found in the valence bond model. In the VB-CT model used here only two states contribute to γ, however, as we have discussed above, there are significant positive contributions from higher lying two-photon states (the T term in eq. 3). We believe that the omission of such terms in the valence bond treatment may lead to these differences. It is possible to develop a more complete valence bond model in which various bridge localized excited states are taken into account. The treatment of such states has been described recently, (*8*) and its inclusion with the present VB-CT model is being investigated.

Figures 8 and 9 show plots of the $\mu\beta$ product, which is the microscopic quantity relevant to EFISH and poled-polymer electro-optic responses, and γ *versus* λ_{max} for a DA octatetraene calculated using the VB-CT model as V_o was varied from -2 to 2 eV. These plots illustrate the nonlinearity-transparency trade-off (as measured by λ_{max}) for DA molecules of a fixed length that vary over the domain of *BLA*. The value of $\mu\beta$, Figure 8, reaches a maximum when λ_{max} is about 490 nm (for DA polyenes with nine conjugated atoms) but decreases and goes to zero as λ_{max} reaches its maximum at ~ 520 nm (the so-called cyanine-limit). Thus, structural changes that lead to an increase in λ_{max} at fixed length do not always lead to an increase in $\mu\beta$. The lower arm of the $\mu\beta$ plot corresponds to highly polarized molecules that are dominantly zwitterionic. Since the dipole moment increases as the CT form is stabilized, the negative peak of $\mu\beta$ is substantially larger in magnitude than positive peak. The negative $\mu\beta$ peak occurs near 450 nm, substantially blue shifted relative to the positive peak. This increased nonlinearity and transparency of zwitterionic molecules near the negative extreme of $\mu\beta$ makes such molecules attractive candidates for poled polymer applications.

Because the dependence of γ on V_o is symmetric about $V_o = 0$ (Figure 7) the γ *versus* λ_{max} plot folds back on top of itself as the polarization is increased, giving the simple curve in Figure 9. Accordingly, in contrast to what was calculated for the $\mu\beta$ values, there is no intrinsic advantage to highly zwitterionic DA polyenes for γ. The positive peak of γ occurs near 400 nm but the negative peak, which is several times larger in magnitude, occurs when λ_{max} is at its maximum of 520 nm (the cyanine limit). With their large negative γ values, cyanines or DA molecules at the cyanine limit could be of interest for self-defocusing applications (like optical power limiting), or for the engineering of materials with vanishing third-order nonlinearity, through the destructive interference of the negative contribution of the cyanine and the positive nonlinearity of a host medium.

It is not uncommon to find in the literature fits of $\mu\beta$ and γ to power laws in λ_{max}. In view of the dependencies in Figures 8 and 9, it can been seen that for DA polyenes at a fixed length, a simple power law relation is not obeyed over the domain of structural variation. For DA polyenes varying in length considerable caution is needed in interpreting power law fits in length or in λ_{max}. This is because the degree of polarization and the *BLA* of the molecules could be length dependent.

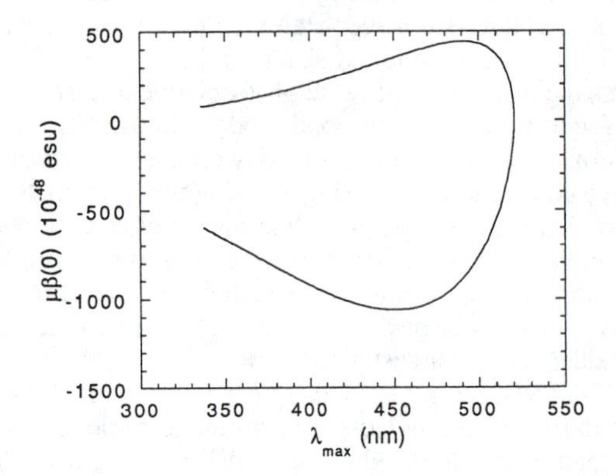

Figure 8. Dependence of $\mu\beta(0)$ on λ_{max} for a DA polyene of nine conjugated atoms calculated using the VB-CT model. The calculations were performed using the following parameters: harmonic force constant, k = 33.55 eV/Å^2, R_{DA} = 7.3 Å, t = 1.18 eV, and the charge transfer efficiency, Q = 0.5 ($\mu_{CT} = Q$ |e| R_{DA}).

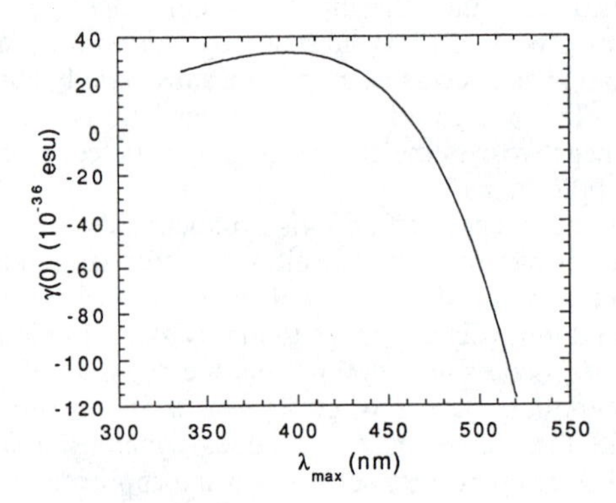

Figure 9. Dependence of $\gamma(0)$ on λ_{max} for a DA polyene of nine conjugated atoms calculated using the VB-CT model. The calculations were performed using the same parameters as for Figure 8.

Conclusions

For simple DA polyene dyes, α, β, and γ can be correlated with either *BLA* or *BOA* as demonstrated using semi-empirical INDO/SDCI calculations. The trends obtained through these calculations are in qualitative agreement with a simple model based on VB and CT mixing that involves harmonic potentials for the zero-order states. This model provides a simple physical picture for the evolution of the structure of DA polyenes as the CT state is stabilized. In this model, analytical dependencies of the polarization and polarizabilities on the energetics and interaction of VB and CT states are obtained, although the equilibrium structure must be found numerically. The agreement obtained between the different levels of theory (AM-1, INDO-SDCI, and VB-CT model) suggests that the trends obtained are robust and, therefore, provide useful guidelines for the design of optimized molecules for nonlinear optics.

Acknowledgments

The research described in this paper was performed in part by the Jet Propulsion Laboratory, (JPL) California Institute of Technology, as part of its Center for Space Microelectronics Technology and was supported by the Advanced Research Projects Agency (administered by the Air Force Office of Scientific Research) and the Ballistic Missiles Defense Initiative Organization, Innovative Science and Technology Office, through a contract with the National Aeronautics and Space Administration (NASA). Support at the Beckman Institute from the National Science Foundation, the Air Force Office of Scientific Research and the North Atlantic Treaty Organization is gratefully acknowledged. The work in Mons was carried out within the framework of the Belgium Prime Minister's Office of Science Policy "Pôle d'Attraction Interuniversitaire en Chimie Supramoléculaire et Catalyse" and "Programme d'Implulsion en Technologie de l'Information" and is partly supported by the Belgium National Fund for Scientific Research (FNRS).

Literature Cited

1. Marder, S. R.; Perry, J. W.; Bourhill, G.; Gorman, C. B.; Tiemann, B. G.; Mansour, K. *Science* **1993**, *261*, 186.
2 Bourhill, G.; Brédas, J.-L.; Cheng, L.-T.; Marder, S. R.; Meyers, F.; Perry, J. W.; Tiemann, B. G. *J. Am. Chem. Soc.* **1994**, *116*, 2619.
3. Marder, S. R.; Gorman, C. B.; Meyers, F.; Perry, J. W.; Bourhill, G.; Brédas, J.-L.; Pierce, B. M. *Science*, **1994**, *265*, 632.
4. Gorman, C. B.; Marder, S. R. *Proc. Natl. Acad. Sci. USA* **1993**, *90*, 11297.
5. Meyers, F.; Marder, S. R.; Pierce, B. M.; Brédas, J.-L. *Chem. Phys. Lett.*, **1994**, *228*, 171.
6. Meyers, F.; Marder, S. R.; Pierce, B. M.; Brédas, J.-L. *J. Am. Chem. Soc.*, **1994**, *116*, 10703.
7. Lu, D.; Chen, G.; Goddard, III, W. A.; Perry, J. W. *J. Am. Chem. Soc.*, **1994**, *116*, 10679.
8. Lu, D.; Chen, G.; Goddard, III, W. A., *J. Chem. Phys.* , **1994**, *101*, 5860.
9. Kuzyk, M. G.; Dirk, C. W., *Phys. Rev. A* **1990**, *41*, 5098.
10. Dirk, C. W.; Cheng, L. T.; Kuzyk, M. G. *Int. J. Quantum Chem.* **1992**, *43*, 27.

11. Garito, A. F.; Heflin, J. R.; Wong, K. Y.; Zamani-Khamiri, O., in *Organic Materials for Nonlinear Optics*; Hann, R. A., Bloor, D., Eds. Royal Soc. Chem. **1989**, 16.
12. Pierce, B. M., Proc. *SPIE-Int. Soc. Opt. Eng*. **1991**, *1560*, 148.

RECEIVED April 19, 1995

Chapter 4

Solvent Effects on the Molecular Quadratic Hyperpolarizabilites

I. D. L. Albert, S. di Bella, D. R. Kanis, T. J. Marks, and M. A. Ratner

Department of Chemistry and the Materials Research Center, Northwestern University, Evanston, IL 60208–3113

The Self-Consistent Reaction-Field (SCRF) theory has been employed to compute the first hyperpolarizability of a series of organic chromophores in the presence of a solvent. The solvent effect has been included in a self-consistent fashion, and hence the effect of the solvent has been included in calculating all the properties of the chromophores, namely the transition energies, the oscillator strengths of the associated transitions, the dipole moments of all relevant states, and the hyperpolarizabilities. The quadratic hyperpolarizability has been computed using the correction vector method. The results are then compared with the previously reported values of the hyperpolarizability and the experimentally observed Second-Harmonic Generation (SHG) coefficients at an excitation energy of 0.65 eV. The results show a good agreement with the experimentally observed values for many of the molecules, although there may be some overestimation of the hyperpolarizability values in cases where the ground state dipole moment of the chromophore is large. The earlier calculations, in which only the shifts in the transition energies in the presence of the solvent were used to compute the hyperpolarizabilities of the chromophores, appears to slightly underestimate the solvent shifts. This can be attributed the neglect of the effect of the solvent on the oscillator strengths and the dipole moments of the various states, which occur in the numerator of the Sum Over States (SOS) expression.

The development of organic materials for nonlinear optical (NLO) applications has been the subject of substantial current research[1-3]. In addition to having large nonlinear susceptibilities, these materials also have other optimal properties, such as high laser damage thresholds, low dielectric constants, and ultrafast response

0097–6156/95/0601–0057$12.00/0

times, which are of importance in device applications. Organic materials provide an additional advantage in that the macroscopic NLO susceptibilities of these materials are in most cases governed by the NLO characteristics of the constituent molecular chromophores, which facilitates the modelling of novel systems with optimal properties.

A number of quantum chemical methods have been directed towards the understanding and rationalization of such NLO characteristics[4]. These methods have been routinely used to estimate the NLO response of many organic molecules and to design NLO materials exhibiting optimal microscopic hyperpolarizabilities. The theoretically-estimated NLO hyperpolarizabilities are in most cases lower than the experimentally measured values. A proper comparison of the two values has been difficult owing to the difference in the environmental effects for the two determinations. While most theoretical values are calculated for an isolated molecule in the gas phase, the experimental electric-field induced second harmonic generation (EFISH) measurements are made in solution. Besides the change in the hyperpolarizability values, the solvent effects can also be seen in linear optical properties such as the energy of the optically-allowed transition (solvatochromism) and in the corresponding oscillator strengths. Thus to obtain a meaningful comparison with the experimentally observed linear and nonlinear optical properties of these systems, one must include the effect of solvent in the calculation.

There are a number of procedures for computing molecular microscopic hyperpolarizabilities in literature[4]. Of these methods, the SOS method has been used along with many semi-empirical quantum chemical model hamiltonians, such as the PPP, CNDO, INDO, and MNDO, for calculating the NLO response of both organic and metal-organic systems. The SOS method is simply a spectral representation of the response and, as such, would produce exact results if the exact eigenstates of the full hamiltonian were used. While SOS computations using semi-empirical models correlate very well with experiment, a better comparison would be possible if the environmental effects (solvent or polymer matrix) could also be included in the calculation.

A number of methods have been proposed to include solvent effects in the description of linear optical properties using continuum or semi-continuum electrostatic models based on the reaction field model[5,6]. These methods have been successfully used to predict the effect of solvent on the electronic and optical properties of many organic molecules. Recently, the reaction field model has been used to calculate the hyperpolarizability of a number of molecules including acetonitrile[7] and *p*-nitroaniline[8] in the presence of solvent using the computationally intensive *ab initio* techniques. In our earlier study[9] we estimated the shift in electronic energies of some of the low-lying states of several organic chromophores in a semi-quantitative approach, and used the shifted energies to calculate approximately the solvent-induced changes in the hyperpolarizability of these molecules.

In this paper, we use the reaction field model, originally proposed by Zerner et. al.[6] for linear optical response, to compute the effect of solvent on the quadratic hyperpolarizabilities of the high-β chromophores studied earlier. In this model, a SCRF calculation is followed by a configuration interaction calculation to generate all the excited states in the presence of a dielectric continuum. The hyperpolarizabilities are then computed using the correction vector approach.

Computational Scheme

We present here a brief outline of the SCRF model. The details of the calculation and the computational procedure for linear optical response can be found elsewhere[6]. In this procedure, the solute is placed in a spherical cavity immersed in a dielectric continuum, characterized by its dielectric constant. The electric field of the solute molecule polarizes the surrounding medium, and this new field in turn acts on the solute molecular system. Quantum mechanically, this gives an additional term, H_s, in the Hamiltonian of the isolated molecule, H_o. The total Hamiltonian can now be written as

$$H^T = H_0 + H_s \tag{1}$$

where

$$H_s = -1/2g(\varepsilon)\mu \cdot R \tag{2}$$

The total molecular energy in the presence of the solvent can now be written as

$$E^T = E_o - 1/2R\,|\langle\psi|\mu|\psi\rangle|^2 \tag{3}$$

where μ is the electric dipole moment operator and ψ is the molecular wave function (higher multipole interactions can also be included, but we do not do so here). The reaction field operator R, which is proportional to the dipole moment of the solute, μ, can be written as

$$R = g\mu \tag{4}$$

The proportionality constant g is the Onsager factor, which gives the strength of the reaction field and depends on the dielectric constant of the medium, ε, and can be written, in the case of a spherical cavity of radius, a_0, as

$$g = 2(\varepsilon - 1)/a_0^3(2\varepsilon + 1) \tag{5}$$

After variation, with the normalitzation of the wave function and eq. 3 as the constraints, the Fock operator in the presence of the solvent can now be written as

$$(F_0 - \lambda\mu g\langle\psi|\mu|\psi\rangle)|\phi_i\rangle = \varepsilon_i|\phi_i\rangle \tag{6}$$

where F_o is the effective one-electron Fock operator of the isolated solute molecule, and ϕ_i is a molecular orbital, and ε_i is the corresponding energy. λ can take the value of 1.0 or 0.5, and these values leads to two levels of including the solvent into the calculation. When the value of λ is set to 1.0, the solvent is considered as just providing an isothermal bath; in this case the solvent "cost" energy of $1/2g\mu^2$ has to be added separately. No such correction is needed if the value of λ is set to 0.5, when the solvent is included in the calculation (see ref. 6 for more details). We have used a value of 0.5 in all of the present calculations, and in this case the total energy is given by eq. 3. The total energy of any state n can now be written in terms of the dipole moment, μ_n, of the excited state n as

$$E_n^T = E_n^0 - 1/2\,g(\varepsilon)\mu_n^2 \tag{7}$$

The energy of an electronic transition in the presence of a solvent can also be written as[5]

$$\hbar\omega_{ng} = E_n^T - E_g^T - 1/2\,g(\eta^2)\mu_n(\mu_n - \mu_g) \tag{8}$$

The second term in eq. 8 corrects the energy of polarization due to the finite rate response of the solvent to the dipole of the excited state n. This correction has however not been included in our present calculation, as the NLO coefficients are calculated using the correction vector method, where the explicit values of the excitation energies of various states are not used (essentially, one simply solves the linear inhomogeneous equations). Here η is the refractive index, and $g(\eta^2)$ is given by

$$g(\eta^2) = 2(\eta^2 - 1)/a^3(2\eta^2 + 1) \tag{9}$$

In our earlier study (we call this Method A) of including the effects of solvent on the quadratic hyperpolarizability, an equation similar to eq. 8 (although the equation used appears similar, there are significant differences, for details see ref. 9 and ref. 6(a)) was used to estimate the shift in the transition energy of all the transitions, and the shifted energies were used in the SOS expression to compute the hyperpolarizability β. This procedure provides a reasonably good estimate of the shift in transition energies, but ignores any solvent-induced change in the dipole moment or the oscillator strengths of the associated transition. In the present study (hereafter called Method B) we have included the solvent effects in the SCF procedure itself, so that the shifts in the energies as well as the changes in dipole moments and the oscillator strengths are automatically included in the calculation. This is so because the solvent shift is automatically included while forming the CI matrix. Moreover, since the NLO coefficients are computed using the correction vector method, solvent effects are also automatically included in the calculation of the NLO coefficients. Details of the correction vector method for computing linear and nonlinear optical properties can be found elsewhere[10]. It should be pointed out that although we have not included the electron polarization term (second term of eq. 8) in this calculation, this effect need not be small.

Results and Discussion

In this paper we have examined two series of organic donor-acceptor chromophores (Chart I) dissolved in the same polar solvent (chloroform) for second order NLO effects. This solvent is chosen as there cannot be any additional directed bonding interactions such as hydrogen-bonding, and we expect only small changes in geometry. The calculated shifts in the lowest CT transitions and the shifts in the hyperpolarizabilities calculated at an excitation energy of 0.65 eV from the two models are given in Table 1. The values calculated from model A are taken from our earlier paper. A plot of experimental vs calculated β_{vec} values for the two calculations are shown in Figure 1. It can be seen from Table 1 and Figure 1 that, while there is a large shift in the hyperpolarizabilities of molecules **1-4** of the first series of compounds, the shifts for the molecules **5-12** are much less. This is to be expected as the solvent shift included in the SCRF procedure depends mainly on the dipole moment of the molecules and since the dipole moments of molecules **1-4** are large, we find a large shift in the hyperpolarizability values

The solvent-dependent hyperpolarizability of all these molecules has been studied using the same level of SCRF theory by Zerner et. al. (ref. 6c), however there are some differences in the two methodologies. While the geometry of the molecules used in their paper was optimized using the semi-empirical PM3 hamiltonian and the solvent-dependent hyperpolarizability was computed using the time-dependent Hartree-Fock theory, we have used idealized geometries and computed the hyperpolarizability using the correction vector method. In addition to the above differences in the methodology, we also find some differences in the conventions used in defining β. It is worth stressing that the comparison of theoretical and experimental values of β are further made difficult by the nature of conventions used in the literature[12] in defining β. If a proper comparison of our values with the values of Zerner et. al. (ref 6c) is to be made, our calculated β values must be multiplied by a factor of 2 and the experimental β values we have used must be multiplied by a factor of 6*0.58. The factor of 2 comes from the definitions of β in the two cases and factor of 6*0.58 comes because of the revised value of $d_{31}(LiIO_3)$=-4.1pm/V, used to measure the quartz reference[13] instead of the previous value of -7.1pm/V. After making all the corrections, we find that our slope of the experimental vs calculated β curve is 0.92. This is in excellent agreement with both the experiment and with Zerner's value of 1.07, the difference being attributed to the difference in molecular input geometries.

Table 1 also shows the close agreement between models A and B in estimating the CT transition energies. However the hyperpolarizability values from the two calculations show somewhat larger differences. This is so because in model A only the shift in the transition energies of the various energy levels used in the SOS procedure is calculated. However in model B, besides including the shifts in the transition energies, the shifts in the dipole moments and the oscillator strengths of the associated transitions are also included. This, in addition to decreasing the denominator of SOS expression (which is the only factor included in model A), increases the dipole moments and the oscillator strengths and hence, the numerator of the SOS expression. Thus we see a larger increase in the computed hyperpolarizability values for model B.

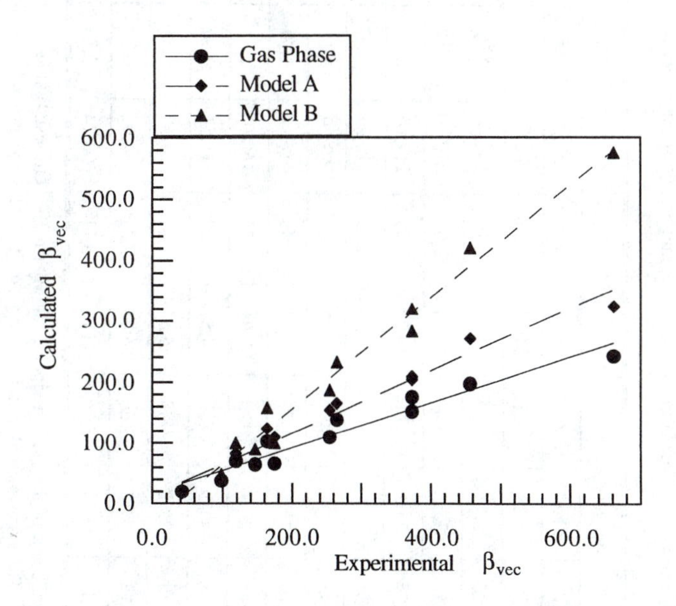

Fig. 1 Plot of computed SHG hyperpolarizability at an excitation energy of 0.65eV in the gas phase and in chloroform solution using models A and B vs experimental chloroform solution values from ref. 11. β_{vec} values are in 10^{-30} cm^5 esu^{-1}

Table 1. Comparison of the experimental[a] and the calculated linear optical properties and 1st hyperpolarizabilities of the two series of organic chromophores from models A and B in chloroform solution. The numbering of the molecules are same in chart 1.

Molecule	Gas Phase		Model A			Model B			Experimental	
	$\hbar\omega^a$	$\beta_{vec}{}^b$	$\mu_g{}^c$	$\hbar\omega^a$	$\beta_{vec}{}^b$	$\mu_g{}^c$	$\hbar\omega^a$	$\beta_{vec}{}^b$	$\hbar\omega^a$	$\beta_{vec}{}^b$
1	3.23	54.82	9.85	2.88	77.1	11.61	2.94	93.68	2.88	73.0
2	3.08	76.28	10.23	2.77	104.4	12.44	2.79	141.40	2.80	107.0
3	2.97	98.62	10.53	2.67	135.2	13.41	2.65	210.40	2.71	131.0
4	2.89	120.98	10.76	2.62	162.0	14.27	2.56	287.68	2.67	190.0
5	3.41	35.11	7.34	3.17	41.3	8.39	3.23	49.99	3.30	34.0
6	3.15	51.34	7.58	2.98	62.1	8.92	3.04	79.29	3.12	47.0
7	3.08	69.27	7.79	2.85	82.9	9.44	2.89	116.04	2.99	76.0
8	2.96	87.84	7.95	2.77	102.1	9.94	2.77	160.03	2.88	101.0
9	4.12	9.96	4.08	4.03	8.9	4.57	4.04	11.51	3.90	12.0
10	3.71	19.22	4.70	3.61	18.8	5.56	3.61	24.38	3.54	28.0
11	3.42	32.38	5.15	3.33	32.6	6.41	3.31	44.92	3.30	42.0
12	3.37	33.13	9.66	2.83	55.6	11.02	3.11	50.14	2.83	50.0

[a]CT transition energy in eV, [b]Vector component of the 1st Hyperpolarizability (in 10^{-30} cm^5 esu^{-1}) at an excitation energy of 0.65 eV. [c]Ground state dipole moment in Debye. The experimental data is taken from Ref. 11. The ground state dipole moments in the gas phase are the same as those in Model A.

To conclude, we have estimated the solvent effects on the linear optical and the first hyperpolarizabilities of two series of donor-acceptor organic chromophores using the self consistent reaction field method. We have compared these values with the values obtained using a semi-quantitative method proposed earlier. The comparison shows that the semi-quantitative method is most suitable for estimating the solvent shifts of electronic spectra and the hyperpolarizability values of molecules having relatively small dipole moments.

Acknowledgements

We thank Professor M. C. Zerner for the original ZINDO program. This research was sponsored by the NSF through the Northwestern Materials Research Center (Grant DMR-9120521) and by the AFOSR (Contract 93-1-0114).

References

1. *Molecular Nonlinear Optics: Materials, Physics, and Devices;* Zyss, J., Ed.; Academic Press: Boston, 1993.
2. Prasad, P. N.; Williams, D. J. *Introduction to Nonlinear Optical Effects in Molecules and Polymers*; Wiley: New York, 1991.
3. *Nonlinear Optical Properties of Organic Molecules and Crystals*; Chemla, D. S., Zyss, J., Eds.; Academic Press: New York, 1987 Vols. 1 and 2.
4. (a)Kanis, D. R.; Ratner, M. A.; Marks, T. J., *Chem. Rev.*, **1994**, *94*, 195,(b)Kurtz, H. A.; Stewart, J. J. P.; Dieter, K. M. *J. Comput. Chem.* , **1990**, *11*, 82, (c)Docherty, V. J.; Pugh, D.; Morley, J. O. *J. Chem. Soc. Faraday Trans. 2* **1985**, *81*, 1179, (d)Svendsen, E. N., Willand, C. S.; Albrecht, A. C. *J. Chem. Phys.* **1985**, *83*, 5760, (e)Lalama, S. J.; Garito, A. F. *Phys. Rev. A* **1979**, *20*, 1179, (f)Soos, Z. G.; Ramasesha, S. *J. Chem. Phys.*, **1989**, *89*, 1067, (g)Kanis, D. R.; Ratner, M. A.; Marks, T. J.; Zerner, M. C. *Chem. Mater.* **1991**, *3*, 19.
5. (a)Amos, A. T.; Burrows, B. L. *Adv. Quantum chem.*, **1973**, *7*, 289. (b)McRae, E. G. *J. Phys. Chem.*, **1957**, *61*, 562.
6. (a)Karelson, M. M.; Zerner, M. C. *J. Phys. Chem.*, **1992**, *96*, 6949. (b)Karelson, M. M.; Zerner, M. C., *J. Am. Chem. Soc.*, **1990**, *112*, 9405, (c)Yu, J.; Zerner, M. C. *J. Chem. Phys.*, **1994**, *100*, 7487.
7. Willets, A.; Rice, J. E. *J. Chem. Phys.*, **1993**, *99*, 426
8. (a)Mikkelsen, K. V.; Luo, Y.; Agren, H.; Jorgensen, P. *J. Chem. Phys.*, **1994**, *100*, 8240, (b)*ibid*, *Adv. Quantum Chem.* (in press)
9. Di Bella, S.; Marks, T. J.; Ratner, M. A. *J. Am. Chem. Soc.*, **1994**, *116*, 4440.
10. Albert, I. D. L.; Morley, J. O.; Pugh, D. *J. Chem. Phys.*, **1993**, *99*, 5197.
11. Cheng, L-T.; Tam, W.; Stevenson, S. H.; Meridith, G. R.; Rikken, G.; Marder, S. R. *J. Phys. Chem.*, **1991**, *95*, 10631, (b)Cheng, L-T.; Tam, W.; Marder, S. R.; Steigman, A. E.; Rikken, G.; Spangler, C. W. *ibid*, **1991**, *95*, 10643.
12. Willets, A.; Rice, J. E.; Burland, D. M.; Shelton, D. P. *J. Chem. Phys.*, **1992**, *97*, 7590.
13. Eckardt, R. C.; Masuda, H.; Fan, Y. X.; Byer, R. L. *IEEE J. Quantum Electron.*, **1990**, *26*, 922.

RECEIVED February 20, 1995

Chapter 5

Characterization of Nonlinear Optical Chromophores in Solution

Christopher R. Moylan, Robert D. Miller, Robert J. Twieg, and Victor Y. Lee

IBM Almaden Research Center, 650 Harry Road, San Jose, CA 95120–6099

The molecular hyperpolarizabilities of a variety of nonlinear optical chromophores have been measured by electric field-induced second harmonic generation (EFISH). The results, combined with thermal decomposition studies, reveal that the well-known tradeoff between thermal stability and nonlinearity has been defeated for certain compounds. Chromophores suitable for both photorefractive and electro-optic applications are identified.

Chromophore-doped polymers are being vigorously studied for use in three nonlinear optical applications. The first is for electro-optic materials,[1] which are the heart of the optoelectronic devices that convert continuous-wave diode laser beams into optical signals modulated at gigahertz frequencies. The second application is frequency doubling of diode laser light[2] so that optical storage devices can operate near 400 nm rather than 800 nm, leading to a factor of four increase in storage densities. Photorefractive materials represent the third application area,[3] and show promise for holographic storage and intensity-dependent filters, among other things. For each of these applications, nonlinear optical chromophores are required. In this paper, the standard technique for measuring the nonlinear optical properties of chromophores is briefly reviewed, and selected recent results on coumarins and thermally stable chromophores for electro-optics will be described.

Experimental Section

All hyperpolarizabilities reported in this work were obtained by electric field-induced second harmonic generation experiments (EFISH). This technique was originally developed primarily for gas-phase measurements,[4,5] and was extended to solutions by Levine and Bethea.[6,7] In an EFISH experiment, a pulsed laser is focused through a cell containing a

0097–6156/95/0601–0066$12.00/0

chromophore solution. A high voltage is applied across the cell while each laser pulse traverses the cell. Continuously maintaining an electric field of kilovolts per centimeter across the cell would induce electrochemistry, so the applied field is pulsed as well, but usually with a longer pulse length so that it is effectively a DC field during the length of each laser pulse. The applied field induces partial orientation of the polar chromophores within the solution, breaking the symmetry of the liquid and allowing $X^{(2)}$ to be nonzero.

The second harmonic generation observed from a laser pulse with circular frequency ω is proportional to the hyperpolarizability $\beta(-2\omega;\omega,\omega)$, the square of the optical electric field (E_{ω}^2), and the degree of alignment (equal to $\mu E_0/15kT$, where E_0 is the DC electric field). Unfortunately, second harmonic generation is also produced by a third-order effect, proportional to the second hyperpolarizability $\gamma(-2\omega;\omega,\omega,0)$, E_{ω}^2, and E_0. The electric field dependences of the two contributions are therefore identical. The second-order effect has an inverse temperature dependence, but in practice the experimentally accessible range of temperatures is too small to allow separation of β and γ. Since the applied electric field is not truly DC, the upconverted light due to the third-order effect appears at a slightly higher frequency than that due to the second-order effect, but this difference is too small to observe. For example, in our apparatus, the high-voltage pulse has its largest Fourier transform components at 25 and 75 kHz. The third-order signal near 954 nm thus appears at a frequency about 5×10^4 Hz higher than that of the second order signal. But since these optical frequencies are in the 10^{14} Hz regime, the difference is undetectable. In order to push the wavelength of the γ-mediated upconversion several nanometers away from the β-mediated output, the period of the applied electric field would have to be subpicosecond, which is impractical and significantly shorter than the 8 ns pulse length of the laser.

A workable, albeit complicated, method[8] of separating β from γ is to perform simultaneous EFISH and third harmonic generation experiments on solutions of symmetric compounds (whose β values are rigorously zero), thereby setting up an empirical relationship between $\gamma(-2\omega;\omega,\omega,0)$, or γ_{EFISH}, and $\gamma(-3\omega;\omega,\omega,\omega)$, or γ_{THG}. Similar EFISH and THG measurements on actual chromophore solutions allow the estimation of γ_{EFISH} from the measured γ_{THG}, and subtraction of that from the combined term, which is equal to[9]

$$\gamma + \frac{\mu\beta}{15kT} \tag{1}$$

affords an improved estimate for β. Most workers simply choose to neglect the γ term in their EFISH data analysis, recognizing that their β values are therefore typically 10-15% high. We have followed the latter procedure in this work.

The actual EFISH experiments involve measurement of second harmonic generation from solutions as a function of both path length and concentration.[10] The absolute laser intensity is not measured; therefore, the absolute $X^{(3)}$ of each solution is not measured either. The experiment yields relative values that must be referenced to a known standard. The most common standard is quartz, which in turn has been referenced to other nonlinear crystals such as KDP.[11] The recommended d_{36} for KDP is 0.39 pm/V at 1064 nm, according to Roberts;[12] the relative nonlinearities of quartz and KDP therefore imply that d_{11} for quartz is equal to 7.2 x 10^{-10} esu at that wavelength. (1 pm/V = 2.387 x 10^{-9} esu.) Meredith[13] has calculated that $d_{11}(SiO_2)$ at 1907 nm, the first Stokes line from Raman-shifted YAG output using a hydrogen cell, is 93% of the value at 1064 nm. Thus, the reference standard at 1907 nm has a nonlinearity of 6.7 x 10^{-10} esu. Those are the values that we use for experiments at those two wavelengths. Most of the published work in the field, however, has been based on higher values for the nonlinearity of KDP. Comparison of data between laboratories is therefore hazardous. A second hindrance to comparisons is that the factor of 15 in Eq. 1 has usually been given as 5, as per the original work on the subject.[6] It has been shown recently[9] that in order for β values to be compared with theoretical calculations and with measured electro-optic coefficients, β must be defined consistently for all applications, which requires the factor of 15.

A word about solvents is germane. Solvents with low dielectric constants are generally desirable, because the applied high voltage does not decrease detectably when charging a capacitative EFISH cell filled with them, because the "local field factors" which correct for the screening of the applied field by the solvent are smaller, and because small amounts of nonlinear chromophores produce large percentage changes in the second harmonic intensity. But good chromophores are strongly polar, and often require more polar solvents with higher dielectric constants. Many workers have used dioxane for their EFISH experiments, because despite having an apparently zero dipole moment and a bulk dielectric constant of only 2.209, it is capable of dissolving many polar chromophores. Unfortunately, this situation is too good to be true. It was shown more than 25 years ago[14] that dioxane is a chemical chameleon, which can flip from its lowest energy chair conformation to two different polar boat conformations in the presence of a polar solute. Therefore, dielectric data on pure dioxane do not reflect its behavior in solutions--particularly the solvent shells around polar compounds. Because the EFISH experiments yield only the product $\mu\beta$, so that separate dielectric experiments must be performed to determine μ (and from that, β), results measured in dioxane are more suspect than those measured in conformationally rigid solvents. Furthermore, dioxane out of the bottle has a substantial overtone absorption at 1907 nm, leading to absorption of 20% of the fundamental beam for each millimeter of path length. While there is apparently some evidence to suggest that the absorption is due to trace amounts of water, which can be removed upon rigorous drying,[15] the combination of the conformation problem with the absorption problem leads us to do all of our EFISH experiments in chloroform solution.

Coumarins

A good example of a class of compounds recently characterized by EFISH is the coumarins. Coumarins have been employed in electro-optic,[16] frequency doubling,[17] and photorefractive[18] applications, and are therefore of general interest in the nonlinear optics community. Yet no coumarin hyperpolarizabilities have apparently ever been measured. Accordingly, a series of fourteen amine-substituted coumarins (Figure 1) has been characterized. Our EFISH system can be operated at either 1064 or 1907 nm, as previously described.[19] Because all of the coumarins were transparent at 532 nm, they were all characterized at 1064 nm.

Molecular hyperpolarizabilities, dipole moments, and absorption maxima for the coumarins are shown in Table I. In addition to the 1064 nm hyperpolarizability, the extrapolated zero-frequency value, β_0, was determined by means of the two-level model[20] and is also included in the table. Uncertainty limits are provided for dipole moments and hyperpolarizabilities, based on the statistical uncertainties in the slopes of least-squares lines used[10] to determine μ and β. Therefore, they reflect the precision of the measurements rather than their accuracy.

It has been shown that β_0 has an explicit second-order dependence on λ_{max} and the transition dipole moment,[21] or (equivalently) an explicit third-order dependence on λ_{max} multiplied by the oscillator strength.[22] The oscillator strength itself tends to increase with λ_{max}, so the wavelength dependence of the zero-frequency hyperpolarizability is often quite strong. It has become customary to compare the two quantities for a series of similar compounds on a log-log plot, whose slope gives some indication of the amount of transparency that must be sacrificed to achieve an increase in nonlinearity within the set of chromophores. Such a plot for the coumarins has a slope of 6.1±0.7, approximately the same as for pyrazoles[23] and tolanes.[24]

Table I: Nonlinear optical data for coumarins. Wavelengths (±2) given in nm, dipole moments in debye, and hyperpolarizabilities in 10^{-30} esu.

Coumarin	λ_{max}	μ	β_{1064}	β_0
14	430	7.44±0.01	114±5	32.9±1.7
13	446	10.2±0.1	132±2	32.4±0.8
12	436	8.20±0.02	106±1	29.0±0.7
11	448	9.86±0.07	120±8	28.6±2.0
10	452	7.56±0.04	111±6	25.4±1.3
6	418	6.55±0.01	49.1±2.6	15.8±0.8
4	400	6.07±0.02	41.8±0.8	15.7±0.3
5	406	6.14±0.01	42.8±2.2	15.3±0.8
3	378	5.59±0.04	29.0±1.2	12.6±0.5
8	388	6.98±0.01	30.6±5.1	12.4±2.1
7	392	6.76±0.05	27.9±1.3	11.0±0.5
2	392	5.71±0.01	27.0±0.8	10.6±0.3
9	382	6.39±0.05	22.1±1.7	9.4±0.7
1	358	4.59±0.10	18.7±1.0	9.0±0.5

Among the 4-trifluoromethyl coumarin derivatives, the order of relative nonlinearities should reflect the order of electron donating abilities of the amine substituents. Since the amine moiety is expected to bear the positive charge resulting from intramolecular electron transfer, increasing the number of saturated carbons bonded to the nitrogen should increase the hyperpolarizability. The julolidine derivative **6** (six carbons) would therefore be expected to have the highest nonlinearity, followed by the N-methylpiperidine and dimethylamino compounds **5** and **4** (four carbons), the ethylamino and dimethylamino derivatives **3** and **2** (two carbons), and finally the simple amino derivative **1**. This is the exact order of the experimentally determined 1064 nm hyperpolarizabilities.

The relative order of zero-frequency nonlinearities for the julolidine donor coumarins is also consistent with predictions based on simple physical organic concepts. The stronger acceptors such as cyano and *m*-pyridyl had the highest nonlinearities. What is not intuitively obvious is that the coumarins with electron withdrawing groups at the 3-position were more nonlinear than those with electron withdrawing groups at the 4-position. MOPAC calculations indicate that the carbon in the 3-position bears a partial negative charge in the ground state, whereas the carbon in the 4-position has a very slight positive charge. These results suggest that intramolecular charge transfer would be better facilitated by attachment of an electron withdrawing substituent at the 3-position rather than the 4-position. Because the first excited state in nonlinear optical chromophores typically exhibits intramolecular charge transfer in the same direction but to a greater extent than in the ground state, it is reasonable that 3-substituted coumarins have larger hyperpolarizabilities than 4-substituted coumarins.

Chromophores for Photorefractive Polymer Applications

Photorefractivity is possible in photoconductive materials that also have substantial electro-optic coefficients. Optimization of nonlinear chromophores for photorefractive applications is in its infancy compared to the work that has been performed on chromophores for purely electro-optical applications (see below), but some interesting results have been obtained.

The first photorefractive polymer[25] contained the weakly nonlinear chromophore 4-nitro-1,2-phenylenediamine (or 2-amino-4-nitroaniline) crosslinked into an epoxy polymer, with a hole transport agent added to provide photoconductivity. Most subsequent photorefractive polymers have used photoconductive polymers as hosts, with NLO chromophores and photosensitizers added as dopants. Prasad and coworkers[26] used diethylaminonitrostyrene (DEANST) as a nonlinear chromophore, fullerene C_{60} as a sensitizer, and poly(vinyl carbazole), also known as PVK, as the photoconductive host to yield a photorefractive material. By using a fluorinated derivative of DEANST (known as F-DEANST) as the chromophore, and trinitrofluorenone (TNF) as the sensitizer in a PVK host, Donckers et al. prepared the first photorefractive polymer to exhibit net gain.[27] Peyghambarian and coworkers have used 2,5-dimethyl-4-(*p*-nitrophenylazo)anisole (DMNPAA) as an NLO chromophore and TNF as a sensitizer in a PVK host, with ethyl carbazole added as a plasticizer.[28] Their material currently holds the record for net gain (207 cm^{-1}) and diffraction efficiency (86%).[29] Silence et al. have come up with several chromophores[30] that provide photoconductivity as well as nonlinearity to host polymers, referring to the chromophores as "dual function dopants." Use of such chromophores allows the selection of a polymer host based on optical properties rather than photoconductivity; the best photorefractive polymer in that study used poly(methyl methacrylate) as the host, C_{60} as the sensitizer, and 1,3-dimethyl-2,2- tetramethylene-5-nitrobenzimidazoline (DTNBI) as the dual-function chromophore. It is useful to determine whether molecular nonlinearities correlate strongly or weakly with photorefractive performance. Accordingly, the chromophores DEANST, F-DEANST, DMNPAA, and DTNBI (Figure 2) have been characterized. The results are given in Table II.

In this table we introduce the reduced nonlinearity as an electro-optic figure of merit. The electro-optic coefficient for a poled polymer may be calculated[31] based on an experimentally-determined β_0:

$$r_{33}(\lambda) = \frac{2Nf_0^2 f_\lambda^2 \mu E_{pol} \beta_0 (3 - \lambda_{max}^2/\lambda^2)}{15kT_{pol} n_\lambda^4 (1 - \lambda_{max}^2/\lambda^2)^2} \tag{2}$$

In this expression, N is the molecular concentration of chromophores, f_0 and f_λ are the local field factors, E_{pol} is the electric field applied during poling, and T_{pol} is the poling temperature. Equation 2 can be rearranged as follows:

$$r_{33}(\lambda) = \frac{2w\rho N_A f_0^2 f_\lambda^2 E_{pol}}{15kT_{pol} n_\lambda^4} \left(\frac{\mu\beta_\lambda}{M}\right) \tag{3}$$

In Equation 3, w is the fraction of chromophore by weight, ρ is the density, N_A is Avogadro's number, and M is the molecular weight. Note that the molecular parameters have been collected in the parentheses; all other parameters are polymer parameters or constants. Clearly, $\mu\beta$/M, determined at the desired use wavelength, is the quantity to optimize for electro-optic polymers. For the photorefractive materials, we calculate this reduced nonlinearity at 676 nm, a common experimental wavelength.

Table II. Nonlinear optical properties of chromophores used in photorefractive polymers. f is the oscillator strength for the primary spectral transition whose peak appears at λ_{max}. Wavelengths are given in nm, dipole moments in debye, and hyperpolarizabilities in 10^{-30} esu.

Compound	λ_{max}	f	μ	β_0	$\mu\beta_{676}/M$
DEANST	282	0.26	8.04±0.34	60.2±2.5	3.04
F-DEANST	274	0.27	6.80±0.01	58.1±2.1	2.25
DTNBI	480	0.21	8.20±0.04	17.8±1.2	1.99
DMNPAA	398	0.50	5.34±0.01	34.0±4.4	1.32

It is clear from the data in Table II that the superior performance of the DMNPAA material is not due to a large electro-optic coefficient at 676 nm. Peyghambarian and coworkers agree, suggesting instead that the orientational enhancement mechanism[32] is primarily responsible for the diffraction efficiency and gain. They postulate that the birefringence of the

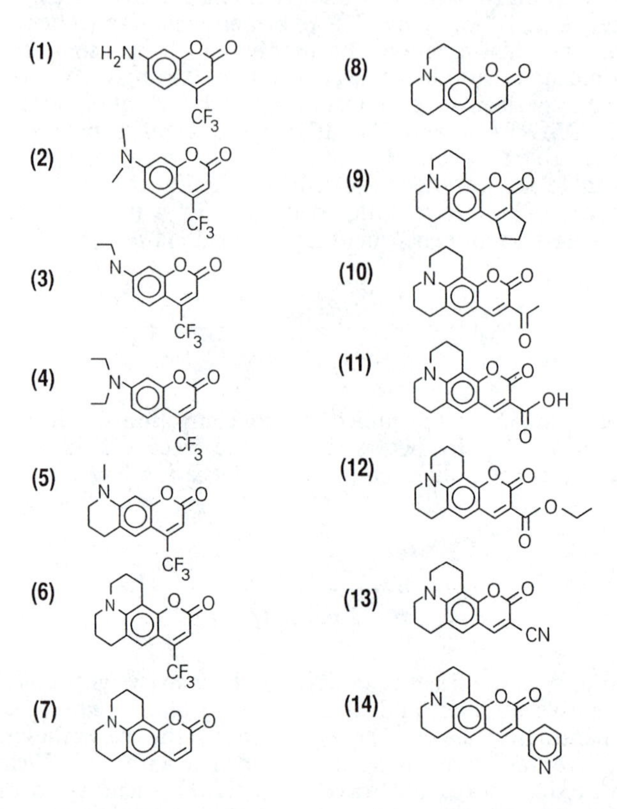

Figure 1. Structures of the coumarins studied in this work.

Figure 2. Chromophores used in photorefractive polymers.

polymer due to alternating regions of chromophore alignment and chromophore nonalignment dominates the photorefractive behavior. Table 2 shows that DMNPAA does have the largest oscillator strength among the four compounds, suggesting that its alignment would produce the greatest change in a polymer's index of refraction. 0.50 is not an unusually high oscillator strength, however. Many chromophores used for purely electro-optic applications have oscillator strengths approaching 1.0, and efforts to incorporate these compounds into photorefractive polymers are ongoing. For the remainder of this chapter, chromophores intended for use in poled electro-optic polymers will be discussed.

Diarylamino Donor Chromophores

It has been shown recently[22] that for a variety of delocalized structures containing amine donors, aryl substitution at the amine not only increases thermal stability, as anticipated, but also unexpectedly increases the molecular hyperpolarizability in most cases. This is a very unusual observation, since it evades the tradeoff that has been commonly observed between nonlinearity and thermal stability of NLO chromophores. The zero-frequency hyperpolarizability β_0 for those compounds is explicitly broken down into its component parameters below, so that the phenomenon may be better understood.

Under the assumptions of the two-level model,[20] the zero-frequency hyperpolarizability β_0 can be expressed in terms of the oscillator strength f, the difference in dipole moment between ground and excited states $\Delta\mu$, and the absorption maximum λ_{max}:

$$\beta_0 = 1.617\ f\ \Delta\mu\ \lambda^3_{max} \tag{4}$$

The factor of 1.617 is equal to $9e^2/8\pi^2hmc^3$ times the conversion constants that allow $\Delta\mu$ to be expressed in debye and $\lambda_{max}{}^3$ to be expressed in units of 10^7 nm^3. Because the oscillator strength, like λ_{max}, can be determined from a chromophore's absorption spectrum, determination of β_0 from EFISH and dielectric experiments amounts to a measurement of the excited-state dipole moment of the chromophore (to the extent that the compound is well-described by the two-level model). Furthermore, this measurement is superior to solvatochromic measurements of the same quantity, because no arbitrary estimate of the effective molecular radius is required. This benefit of EFISH experiments was recognized by Oudar and Chemla,[20] but has gone generally unexploited. Here, the relationship given above is used to determine which parameter is most responsible for the enhanced nonlinearities of aryl-substituted amine donor chromophores.

Results for chromophore structures **a-k** (Figure 3) were reported earlier.[22] For each compound, the oscillator strength of the lowest-energy electronic transition was extracted from uv/vis spectra,[33] treating each peak as a triangle for the purpose of determining its area. Using the absorption maximum from the same spectra and the hyperpolarizability from the EFISH and dielectric measurements, the values shown in Table III were obtained.

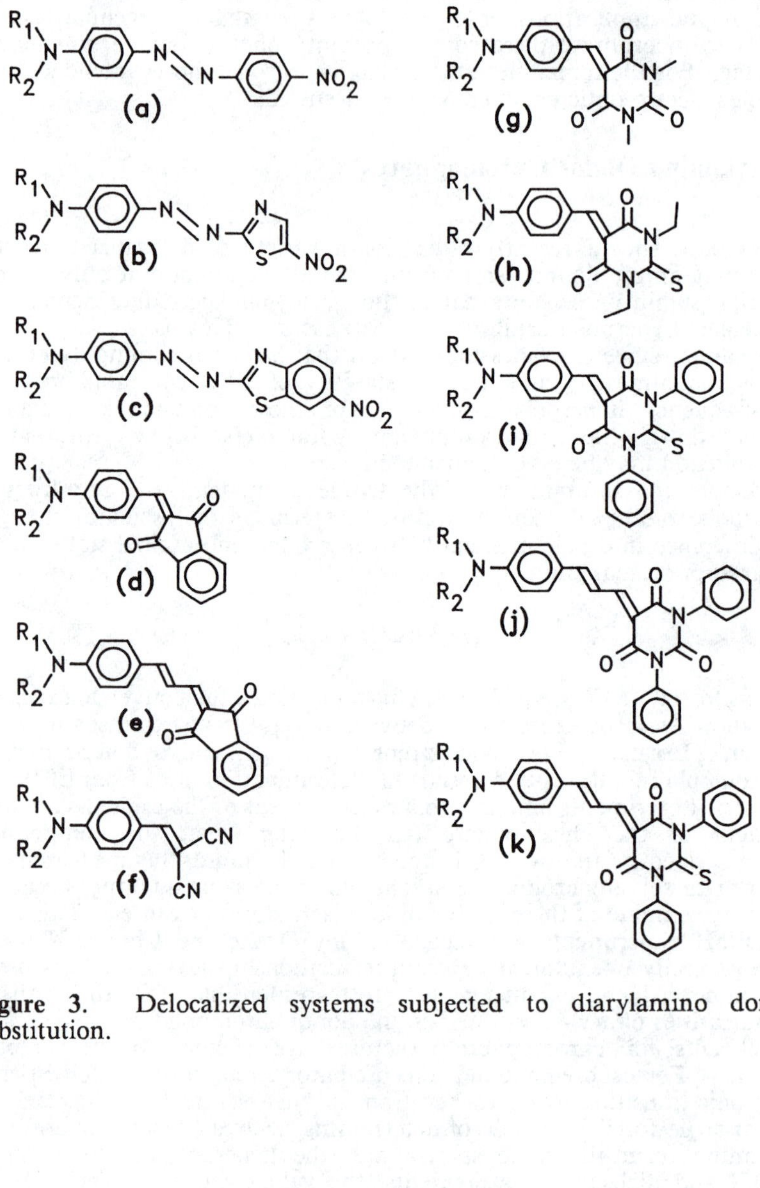

Figure 3. Delocalized systems subjected to diarylamino donor substitution.

Table III: Components of β_0 for amine donor compounds. Dipole moment differences given in debye, wavelengths in 10^7 nm^3, and β_0 in 10^{-30} esu.

Structure	R_1	R_2	f	$\Delta\mu$	λ_{max}^3	β_0
a	Me	Me	.742	5.80	11.1	76.6
	Et	Et	.685	7.16	12.1	95.2
	Ph	Ph	.619	9.12	11.5	105
b	Et	$HO(CH_2)_6$	.705	4.48	19.7	100
	Ph	Ph	.640	6.44	19.7	131
c	Et	$HO(CH_2)_6$	.881	4.94	16.5	115
	Ph	Ph	.853	6.04	16.6	138
d	Et	Et	.666	5.77	11.8	72.8
	4-MePh	4-MePh	.772	7.97	12.4	122
e	Et	Et	.809	8.04	16.8	176
	4-MePh	4-MePh	.812	8.49	16.1	179
f	$-(CH_2)_5-$		.525	4.42	8.52	31.9
	4-MePh	4-MePh	.652	4.99	9.23	48.4
g	Et	Et	.580	3.92	10.4	38.0
	4-MePh	4-MePh	.688	6.07	11.5	77.2
h	Me	Me	.631	3.12	12.1	38.2
	Et	Et	.690	2.81	12.5	39.0
	4-MePh	4-MePh	.752	5.13	13.9	86.4
i	Et	Et	.821	2.94	12.7	49.2
	4-MePh	4-MePh	.855	5.65	14.1	109
j	Et	Et	.836	6.57	20.3	180
	4-MePh	4-MePh	.925	7.32	21.0	228
k	Et	Et	.894	6.99	17.4	175
	4-MePh	4-MePh	.849	7.77	17.6	186

For all compounds in Table III, the ground state dipole moment decreases upon aryl substitution while the excited state dipole moment remains comparable. $\Delta\mu$ therefore is larger, and in most cases is the biggest factor in the observed increase in β_0. This effect is most prominent in the azobenzenes **a** and azothiazoles **b**, where aryl substitutions actually decreased λ_{max} and f but nevertheless caused β_0 to increase. For all structures except the dicyanostyrene f, $\Delta\mu$ was enhanced by aryl substitution at least as much as either of the other parameters. It would be preferable to increase $\Delta\mu$ by increasing the excited state dipole moment rather than by decreasing the ground state dipole moment, because a large ground state dipole moment allows better orientation of the chromophores within a polymer film, but the startling improvement in thermal stability due to aryl substitution greatly enhances the utility of these compounds nonetheless.

Cyano Acceptor Chromophores

Enhanced nonlinearities in cyano acceptor compounds were reported in 1987 by Katz et al.,[34] but little further work on cyano compounds was reported[35] until the thiophene studies of Rao et al.[36] in 1993. Rao and coworkers attributed the lack of follow-up work to synthetic difficulties. The assumption by other workers in the field that compounds derived from tetracyanoethylene would be chemically unstable may also have contributed. In any case, there has been a recent flurry of activity concerning these chromophores,[37] particularly with regard to incorporating them into polymers.[38-40] We were surprised at the excellent thermal stability of the dicyanostyrene chromophores we prepared as part of the study of diarylamino donor compounds.[22] We have further characterized DCM laser dye, whose use in electro-optic polymers has been demonstrated by Ermer and coworkers,[41] four symmetric DCM derivatives, and several chromophores with both diarylamino donors and cyano acceptors. The good thermal stability observed in the styrenes appears to be a general effect, according to differential scanning calorimetry (DSC) studies of these compounds. The nonlinearity-thermal stability tradeoff is not evaded upon cyano substitution as it is upon diarylamino substitution, but it is a very weak tradeoff. Cyano acceptor substitution generally produces strong enhancements in nonlinearity and modest decreases in thermal stability. Table IV shows the effect of replacing a nitro acceptor with a dicyanovinyl or a tricyanovinyl acceptor in the diphenylazo system. For calibration purposes, it is helpful to know that, according to our calculations on chromophores in polyimides, a material with a reduced nonlinearity of approximately 6.9 will exhibit an electro-optic coefficient in pm/V at 1300 nm equal to its chromophore concentration in weight percent, i.e. a polymer with 20% chromophore by weight would have an electro-optic coefficient of 20 pm/V at 1300 nm.

The parent compound in Table IV exhibits one of the highest DSC decomposition temperatures that we have observed: 393° C. Substitution of a dicvanovinyl acceptor for the nitro acceptor should, based on the data inTable IV, increase the electro-optic coefficient of an NLO polymer with the same weight loading by 59%, while decreasing the decomposition temperature by only ten degrees. Adding a third cyano group to the acceptor almost doubles the nonlinearity, while dropping the thermal stability by an additional twenty degrees.

Subjecting the experimentally-determined β_0 values to the same analysis given above for the diarylamino chromophores, we find that in this case the dominant component of the increase in β_0 is λ_{max}^3. In other words, cyano substitution dramatically decreases the energy gap between the ground and excited states, making it easier for an applied electric field to couple them. Substitution of dicyanovinyl for nitro leaves the oscillator strength and $\Delta\mu$ almost unchanged, but decreases the transition energy by 1570 cm^{-1}. Adding the third cyano moiety to the acceptor causes an even bigger red shift (2400 cm^{-1}) but also causes $\Delta\mu$ to jump from 9.9 D to 14 D.

Table IV. Effect of cyanovinyl substitution on NLO parameters of a diphenylaminonitrodiphenylazo chromophore. Wavelengths are given in nm, dipole moments in debye, hyperpolarizabilities in 10^{-30} esu, and temperatures in degrees C.

Structure	λ_{max}	f	μ	β_0	$\mu\beta_{1300}/M$	T_d
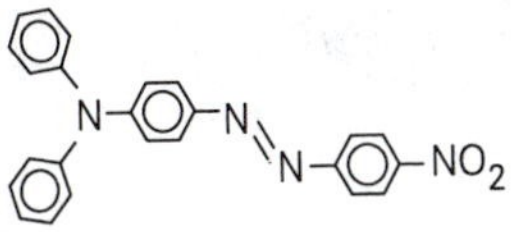 $(C_6H_5)_2N-C_6H_4-N=N-C_6H_4-NO_2$	486	0.62	5.87±0.09	105±3	2.01	393
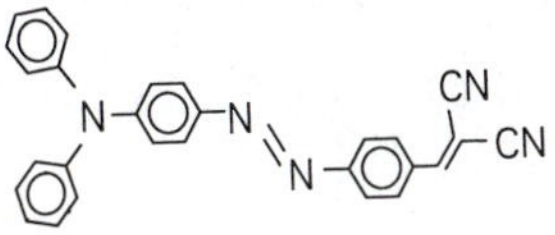 $(C_6H_5)_2N-C_6H_4-N=N-C_6H_4-CH=C(CN)_2$	526	0.64	6.74±0.03	149±5	3.20	383
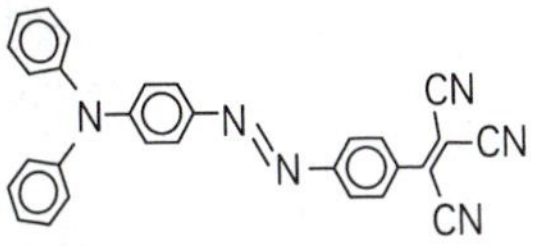 $(C_6H_5)_2N-C_6H_4-N=N-C_6H_4-C(CN)=C(CN)_2$	602	0.53	7.11±0.04	259±15	6.17	364

Defeating the Tradeoff

By judicious use of both diarylamino donors and cyano acceptors, we have synthesized a number of compounds with both excellent nonlinearity and excellent thermal stability. If we set the definition of excellence in nonlinearity at 3.0 or greater, and the definition for thermal stability at DSC decomposition temperatures greater than 300° C, seven chromophores beat the tradeoff. Those compounds are shown in Table V. For each entry, the absorption maximum is given as a rough measure of transparency, while $\mu\beta/M$ at 1300 nm corresponds to nonlinearity and T_d indicates thermal stability.

The most nonlinear compound in Table V contains a thiophene ring adjacent to the tricyanovinyl acceptor. Thiophene-containing compounds have been shown to display extremely high nonlinearities,[42] but these are achieved at the traditional cost of heavy sacrifices in thermal stability. The ditolylamino compound shown here is the only thiophene that we have found with a DSC decomposition temperature above 300° C.

Within classes of similar compounds, the nonlinearity-thermal stability tradeoff is still evident. When a nitrobenzthiazole is given a second nitro acceptor group, the nonlinearity jumps up by 70%, but the thermal stability drops by 32 degrees and the compound red shifts by 1350 cm^{-1}. When a

Table V. Chromophores with nonlinearities greater than 3.0 and thermal decomposition temperatures above 300 degrees C. Wavelengths given in nm, dipole moments in debye, and hyperpolarizabilities in 10^{-30} esu.

Structure	λ_{max}	$\mu\beta_{1300}/M$	T_d
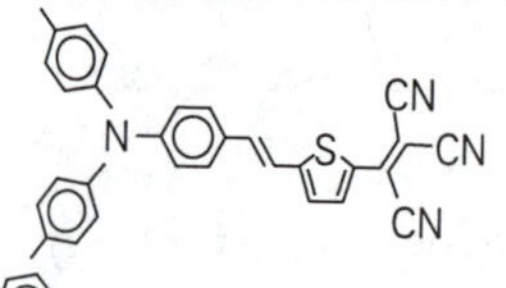	680	14.1	331
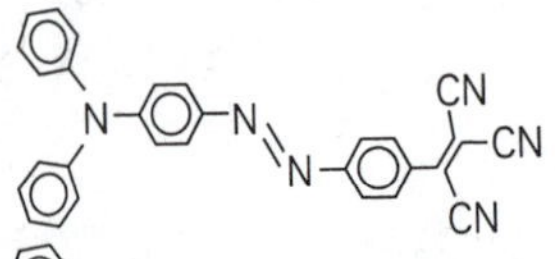	602	6.17	364
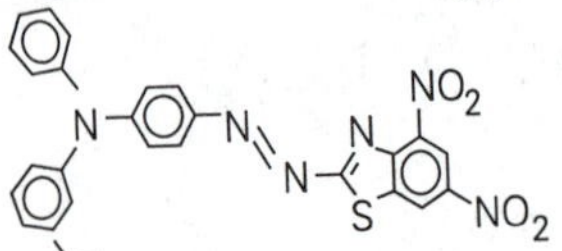	594	5.28	324
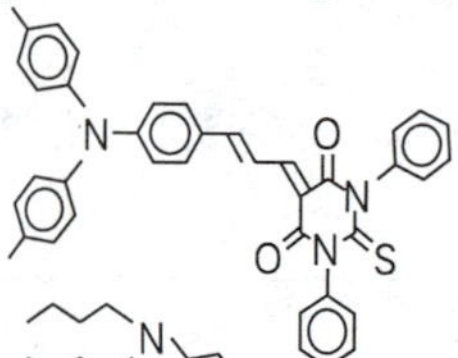	594	3.66	320
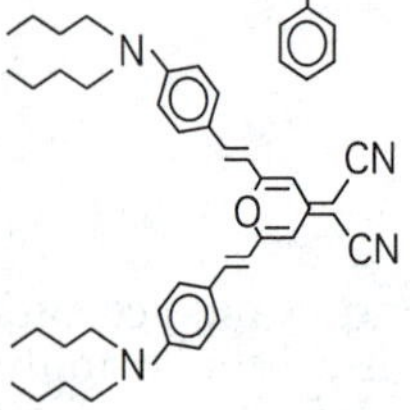	496	3.65	348
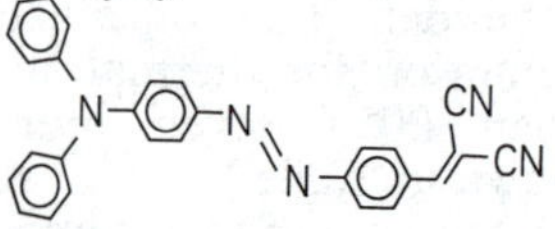	526	3.20	383
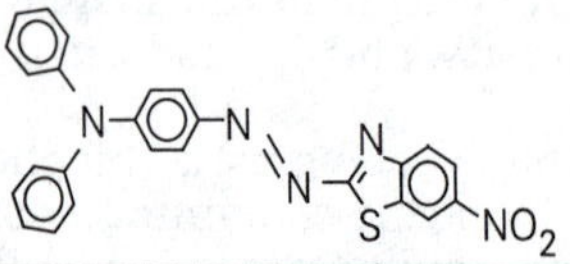	550	3.10	356

third cyano group is added to the dicyanovinyl azo compound, as discussed above, the nonlinearity increases but the thermal stability and transparency take large hits. Nevertheless, these chromophores represent the successful product of a great deal of synthesis, characterization, and calculation, and demonstrate that the tradeoff is not enforced by chemistry limitations, but only by creativity limitations. The same cannot be said, however, for the nonlinearity-transparency tradeoff, which must now be addressed more seriously than it has been. Equation 4 clearly shows that the molecular hyperpolarizability increases dramatically with the absorption maximum.

With that in mind, the dicyanopyran compound in Table 5 is of distinct interest. This compound, originally synthesized by Susan Ermer and colleagues at Lockheed and referred to as DADB, is the only compound in the table that does not possess a diarylamino donor. We have prepared the diphenylamino analogue, and a DSC scan indicates that it is 56 degrees more thermally stable than DADB. But the most interesting thing about DADB is its combination of good nonlinearity and an absorption maximum below 500 nm. We have performed some semiempirical calculations on the simple amino donor analogue of DADB, and they suggest that these compounds have two charge-transfer excited states lying close to each other in energy. Therefore, the compound is about twice as nonlinear as one would expect it to be, given its color. DADB may be the first indicator of the way around the nonlinearity-transparency tradeoff. Equation 4 is derived from the two-level model, which assumes that only one low-lying electronic state contributes to the hyperpolarizability. If that assumption does not hold, then the tradeoff indicated by Equation 4 may not hold either. If other compounds can be deliberately designed so that they have more than one charge transfer excited state, then severe red shifts may not be necessary, and the necessary transparency in the near infrared for fabricating waveguide devices will be more easily attainable.

Outperforming Lithium Niobate

Preparation of organic electro-optic materials is a worthwhile goal only if they have some advantages over the existing inorganic crystals, lithium niobate in particular. The inorganics are generally superior in thermal stability and optical loss, while the organics are cheaper, much easier to fabricate, and have infinitely variable compositions (and therefore properties). But how do they compare with regard to electro-optic performance?

There are two important electro-optic figures of merit:[1] the phase shift efficiency, proportional to n^3r, and the bandwidth per unit power,[43] proportional to n^7r^2/ε. For the latter, existing NLO polymers are already superior to lithium niobate. Its index is 2.1474 at 1300 nm, its r value is 30.8 pm/V, and its dielectric constant is 43. The figure of merit is therefore 4,700. An NLO polymer with $r=20$ pm/V, $n=1.65$, and $\varepsilon=2.7$ has a bandwidth per unit power of 4,900.

For phase shift efficiency, lithium niobate has a figure of merit equal to 305. The polymer described above would have a value of only 90. One particular ODPA-based polyimide containing the diphenylamino tricyanovinyl diphenylazo chromophore shown in Tables IV and V, if it can be prepared, is expected to exhibit an electro-optic coefficient of 52 pm/V. If its index of refraction is 1.8, its figure of merit will equal that of $LiNbO_3$.

Conclusions

The nonlinear optical properties of coumarin chromophores have been measured in solution for the first time. Chromophores used in photorefractive materials have also been characterized, and their oscillator strengths may prove to be more important than their hyperpolarizabilities. Dramatic enhancements in molecular nonlinearities without concomitant sacrifices in thermal stability have been demonstrated with the use of diarylamino donor groups and cyano acceptor groups. The nonlinearity/thermal stability tradeoff has been defeated. Cyanopyran chromophores may have the ability to defeat the nonlinearity/transparency tradeoff. We need to see if we can design other compounds with similar electronic structures (two low-lying charge transfer states). If existing functionalization and polymer synthesis methods can be extended to the best chromophores, electro-optic materials will be prepared which are not only superior to existing NLO polymers, but to lithium niobate as well.

Acknowledgments

The authors thank the National Institute of Science and Technology, the Advanced Research Projects Agency, and the Air Force Office of Scientific Research for partial support of this work.

Literature Cited

1. Singer, K.D.; Lalama, S.L.; Sohn, J.E.; Small, R.D. in *Nonlinear Optical Properties of Organic Molecules and Crystals*, D.S. Chemla and J. Zyss, eds.; Academic Press: Orlando, Florida, 1987; ch. II-8.
2. Nicoud, J.F.; Twieg, R.J. in Ref. 1, ch. II-3.
3. Moerner, W.E.; Silence, S.M. *Chem. Rev.* **1994**,*94*,127-55.
4. Hauchecorne, G.; Kerherve, F.; Mayer, G. *J. Physique* **1971**,*32*,47-62.
5. Finn, R.S.; Ward, J.F. *Phys. Rev. Lett.* **1971**,*26*,285-9.
6. Levine, B.F.; Bethea, C.G. *Appl. Phys. Lett.* **1974**,*24*,445-7.
7. Levine, B.F.; Bethea, C.G. *J. Chem. Phys.* **1975**,*63*,2666-82.
8. Cheng, L.-T.; Tam, W.; Stevenson, S.H.; Meredith, G.R.; Rikken, G.; Marder, S.R. *J. Phys. Chem.* **1991**,*95*,10631-43.
9. Willetts, A.; Rice, J.E.; Burland, D.M.; Shelton, D.P. *J. Chem. Phys.* **1992**,*97*,7590-9.
10. Singer, K.D.; Garito, A.F. *J. Chem. Phys.* **1981**,*75*,3572-80.
11. Jerphagnon, J.; Kurtz, S.K. *Phys. Rev. B* **1970**,*1*,1739-44.
12. Roberts, D.A. *IEEE J. Quantum Electron.* **1992**,*28*,2057-74.
13. Meredith, G.R. *Phys. Rev. B* **1981**,*24*,5522-32.
14. Ledger, M.B.; Suppan, P. *Spectrochim. Acta A* **1967**,*23*,3007-11.
15. Jen, A.K.-Y., personal communication, 1994.
16. Mortazavi, M.A.; Knoesen, A.; Kowel, S.T.; Henry, R.A.; Hoover, J.M.; Lindsay, G.A. *Appl. Phys. B* **1991**,*53*,287-95.

17. Yankelevich, D.R.; Dienes, A.; Knoesen, A.; Schoenlein, R.W.; Shank, C.V. *IEEE J. Quantum Electron.* **1992**,*28*,2398-403.
18. Silence, S.M.; Scott, J.C.; Hache, F.; Ginsburg, E.J.; Jenkner, P.J.; Miller, R.D.; Twieg, R.J.; Moerner, W.E. *J. Opt. Soc. Am. B* **1993**,*10*,2306-12.
19. Moylan, C.R.; Miller, R.D.; Twieg, R.J.; Betterton, K.M.; Lee, V.Y.; Matray, T.J.; Nguyen, C. *Chem. Mater.* **1993**,*5*,1499-508.
20. Oudar, J.L.; Chemla, D.S. *J. Chem. Phys.* **1977**,*66*,2664-8.
21. Moylan, C.R. *J. Chem. Phys.* **1993**,*99*,1436-7.
22. Moylan, C.R.; Twieg, R.J.; Lee, V.Y.; Swanson, S.A.; Betterton, K.M.; Miller, R.D. *J. Am. Chem. Soc.* **1993**,*115*,12599-600.
23. Miller, R.D.; Moylan, C.R.; Reiser, O.; Walsh, C.A. *Chem. Mater.* **1993**,*5*,625-32.
24. Moylan, C.R.; Walsh, C.A. *Nonlin. Opt.* **1993**,*6*,113-21.
25. Ducharme, S.; Scott, J.C.; Twieg, R.J.; Moerner, W.E. *Phys. Rev. Lett.* **1991**,*66*,1846-9.
26. Zhang, Y.; Cui, Y.; Prasad, P.N. *Phys. Rev. B* **1992**,*46*,9900-2.
27. Donckers, M.C.J.M.; Silence, S.M.; Walsh, C.A.; Hache, F.; Burland, D.M.; Moerner, W.E.; Twieg, R.J. *Opt. Lett.* **1993**,*18*,1044-6.
28. Kippelen, B.; Sandalphon; Peyghambarian, N.; Lyon, S.R.; Padias, A.B.; Hall, H.K. Jr. *Electron. Lett.* **1993**,*29*,1873-4.
29. Meerholz, K.; Volodin, B.L.; Sandalphon; Kippelen, B.; Peyghambarian, N. *Nature* **1994**,*371*,497-500.
30. Silence, S.M.; Scott, J.C.; Stankus, J.J.; Moerner, W.E.; Moylan, C.R.; Bjorklund, G.C.; Twieg, R.J. *J. Phys. Chem.*, in press.
31. Moylan, C.R.; Swanson, S.A.; Walsh, C.A.; Thackara, J.I.; Twieg, R.J.; Miller, R.D.; Lee, V.Y. *SPIE Proc.* **1993**,*2025*,192-201.
32. Moerner, W.E.; Silence, S.M.; Hache, F.; Bjorklund, G.C. *J. Opt. Soc. Am. B* **1994**,*11*,320-30.
33. Robinson, G.W.; in *Methods of Experimental Physics*; Academic Press: New York, 1962; Vol. 3, pp. 241-4.
34. Katz, H.E.; Singer, K.D.; Sohn, J.E.; Dirk, C.W.; King, L.A.; Gordon, H.M. *J. Am. Chem. Soc.* **1987**,*109*,6561-3.
35. Singer, K.D.; Sohn, J.E.; King, L.A.; Gordon, H.M.; Katz, H.E.; Dirk, C.W. *J. Opt. Soc. Am. B* **1989**,*6*,1339-49.
36. Rao, V.P.; Jen, A.K.-Y.; Wong, K.Y.; Drost, K.J. *J. Chem. Soc., Chem. Commun.* **1993**,1118-20.
37. Rao, V.P.; Cai, Y.M.; Jen, A.K.-Y. *J. Chem. Soc., Chem. Commun.* **1994**,1689-90.
38. Wong, K.Y.; Jen, A.K.-Y. *J. Appl. Phys.* **1994**,*75*,3308-10.
39. Drost, K.J.; Rao, V.P.; Jen, A.K.-Y. *J. Chem. Soc., Chem. Commun.* **1994**,369-71.
40. Jen, A.K.-Y.; Drost, K.J.; Cai, Y.; Rao, V.P.; Dalton, L.R. *J. Chem. Soc., Chem. Commun.* **1994**,965-6.
41. Ermer, S.; Valley, J.F.; Lytel, R.; Lipscomb, G.F.; Van Eck, T.E.; Girton, D.G. *Appl. Phys. Lett.* **1992**,*61*,2272-4.
42. Jen, A.K.-Y.; Rao, V.P.; Wong, K.Y.; Drost, K.J. *J. Chem. Soc., Chem. Commun.* **1993**,90-2.
43. Kaminow, I.P. *IEEE J. Quantum Electron.* **1968**,*4*,23-6.

RECEIVED February 15, 1995

Chapter 6

Experimental and Theoretical Investigation of the Second-Order Optical Properties of the Chromophore Retinal and Its Derivatives

Modeling the Bacteriorhodopsin Binding Pocket

E. Hendrickx[1], C. Dehu[2], K. Clays[1], J. L. Brédas[2], and A. Persoons[1]

[1]Center for Research on Molecular Electronics and Photonics, University of Leuven, Celestijnenlaan 200D, B-3001 Leuven, Belgium
[2]Center for Research on Molecular Electronics and Photonics, Université de Mons-Hainaut, Place du Parc 20, B-7000 Mons, Belgium

In this study we describe the second-order optical properties of retinal and related molecules. Hyper-Rayleigh scattering was used to measure the first hyperpolarizability of several retinal derivatives and to determine the solvent dependence of the first hyperpolarizability of retinal. This technique is also shown to be sensitive to fluorescence induced by multiphoton absorption, which precluded the measurement of trans retinol. By comparing the experimental results with the theoretical predictions obtained with the semi-empirical INDO/SCI/SOS method we show the validity of the two-state model for β. The two-state model is extended to explain the large solvent dependence of the first hyperpolarizability of retinal, that is also related to the degree of bond-length alternation.

Retinal is the light-absorbing chromophore of the protein Bacteriorhodopsin (bR), a constituent of the purple membrane of the bacterium *Halobacterium Halobium.* The chromophore is covalently bonded to the protein backbone by a protonated Schiff base linkage (*1*). The exact nature of the retinal binding pocket has been the subject of extensive investigations and in literature only modest agreement exists.

Photoexcitation of the chromophore starts a photocycle during which a proton is pumped from the cytoplasmic side to the extracellular side of the membrane (*2,3*). The proton gradient thus created is used by the bacterium to synthesize ATP when the concentration of oxygen in the environment is too low to generate ATP via oxidative phosphorylation. This unusual behaviour and the extraordinary thermal and photochemical stability make this protein an outstanding candidate for optical applications (*4*). The uncertainty that still exists on the exact mechanism of the photocycle has not prevented the use of the protein in devices such as light detectors (*5*), holographic media (*6*) and, recently, three-dimensional optical memories (*7*). Genetic engineering allows for modification of the protein according to the requirements of each application.

0097–6156/95/0601–0082$12.00/0

We have been able to do a direct measurement of the first hyperpolarizability β of the protein by using the hyper Rayleigh scattering technique (HRS) (*8*). The retrieved value of 2100×10^{-30} esu at 1064 nm is in good agreement with previous estimates based on two-photon absorption (*9*) and second harmonic generation from a thin poled poly(vinyl alcohol) film (*10*). This technique also was shown to be sensitive to the extent of solubilization of the protein, which resulted in a decay by a factor of ten in the apparent hyperpolarizability. A value of 2100×10^{-30} was only obtained after 100 hours in an acetate buffer (0.1 M, pH 5.0) with Triton X-100 (5% Vol.), which corresponds to monomeric bR (*11*). This value for β is to be compared with a value of 450×10^{-30} esu for dimethylaminonitrostilbene (DANS), a typical chromophore for non-linear optics. To understand this high hyperpolarizability we started a thorough experimental and theoretical study of the first hyperpolarizability of the chromophore retinal and some of its typical derivatives, which are shown in Figure 1. Only for trans retinol the first hyperpolarizability could not be determined due to fluorescence induced by multiphoton absorption.

Both the theoretical and experimental results are in excellent mutual agreement for the effects of cis-to-trans isomerization of a retinal double bond and the dependence of β on the position of the absorption band maximum. The two-state model is shown to predict the correct trends for β. By extending this two-state model and taking into account the change in molecular polarizability upon excitation, the large solvent dependence can be explained in a simple manner.

The Hyper-Rayleigh Scattering Technique

An HRS measurement is performed by measuring the intensity of the second-order scattered light on focusing an intense laser beam on an isotropic solution (*12,13*). For a detailed description of the experimental set-up the interested reader is referred to reference 14. On the molecular level, the intensity of the incoherently scattered second harmonic can be connected to the molecular first hyperpolarizability in the series expansion of the induced molecular dipole moment in the amplitude of the applied field :

$$\vec{\mu} = \alpha . \vec{E} + \beta : \vec{E} . \vec{E} + \ldots \tag{1}$$

For a liquid composed of non-centrosymmetric molecules the macroscopic polarization at frequency 2ω will be equal to :

$$\vec{P}(2\omega) = B(-2\omega;\omega,\omega) : \vec{E}(\omega)\vec{E}(\omega) \tag{2}$$

with $B(-2\omega;\omega,\omega)$ the first nonlinear macroscopic susceptibility tensor for frequency doubling. The intensity of the harmonic light is then proportional to :

$$I(2\omega) \approx B^2(-2\omega;\omega,\omega) I^2(\omega) \tag{3}$$

Neglecting dipolar orientational correlations, $B(-2\omega;\omega,\omega)$ can be decomposed in molecular β components by a transformation from laboratory to molecular coordinates and by averaging over time and molecular orientations, finally giving :

$$I(2\omega) = G I^2(\omega) \sum_i N_i [\beta^2]_i \tag{4}$$

G is a factor that includes local field factors, the scattering geometry and instrumental factors. The summation is performed over all the scattering species *i*. The expression for $[\beta^2]_i$ is a sum of molecular tensor components squared, each multiplied by a factor resulting from the orientational average over the direction cosines of the transformation from laboratory to molecular coordinates (*15,16*).

For a binary solution consisting only of solvent (S) and solute (s), equation 4 can be simplified to :

$$I(2\omega) = GI^2(\omega)(N_S[\beta^2]_S + N_s[\beta^2]_s) \qquad (5)$$

At low solute concentration, N_S can be taken as a constant and from the intercept and the slope of a plot of $I(2\omega)/I(\omega)^2$ versus N_s, we can calculate $[\beta^2]_s$ provided $[\beta^2]_S$ is known or vice-versa. This approach, called the internal reference method, effectively eliminates the need for local field correction factors at optical frequency.

Another approach is to determine the G factor from the slope of a plot of $I(2\omega)/I(\omega)^2$ versus N_S and the known EFISHG β value for a reference molecule, such as para-nitroaniline, in this solvent. From the slope of a similar plot for the solute molecule in a similar solvent and the G factor, the hyperpolarizability of this molecule can be determined. By analyzing the data in this way, it is no longer necessary to record the intensity of the second harmonic for solvents that have a low nonlinear susceptibility, such as dioxane or methanol. Furthermore, the slope of a plot of $I(2\omega)/I(\omega)^2$ versus N_S can be determined more precisely than the intercept and the $[\beta^2]_S$ that is required for applying the internal reference method is also determined by using an EFISHG β value.

Fluorescence Induced by Multiphoton Absorption. On comparing a number of results obtained with the HRS set-up to those obtained with the EFISHG set-up, we found that the HRS β values of a number of compounds were consistently higher than the EFISHG β values. The emission of visible light from a HRS sample being illuminated by an intense laser beam (1064 nm) had been noticed before. To investigate this phenomenon, the 532 nm interference filter that is always used to block out any light different from the second harmonic was replaced by a monochromator to analyse the emitted radiation.

A laser pulse (1064 nm, 10 ns, 30 mJ) was focused in a filtered 10^{-2} M solution of trans retinol (Vitamin A) in methanol. The recorded spectrum is shown in Figure 2. Trans retinol was selected because of its similarity to other retinal derivatives, which do not fluoresce under the same circumstances, and because the multiphoton absorption behaviour has been studied thoroughly (*17,18*). Since the angle between the ionone ring and the polyene chain is approximately 50° due to steric interactions between the 1,1' methyl groups and the hydrogen atom on carbon 7, all the excited states of trans retinol, which has C_1 symmetry, have the same symmetry (A). This would mean that all the excited singlet states can be investigated by both one-photon and two-photon spectroscopy. These retinyl chromophores, however, are frequently treated as belonging to the C_{2h} point group, since their excited states are somewhat similar to those of linear polyenes. Due to the weak interaction between the terminal hydroxyl group and the polyene system, this approximation applies well to trans retinol. Two-photon spectroscopy on trans retinol showed that the first excited state has ^1Ag symmetry, analogous to the linear polyenes. Due to the well-known parity rule for one and multiphoton transitions, this state can not be observed in one-photon spectroscopy. Even though the C_{2h}

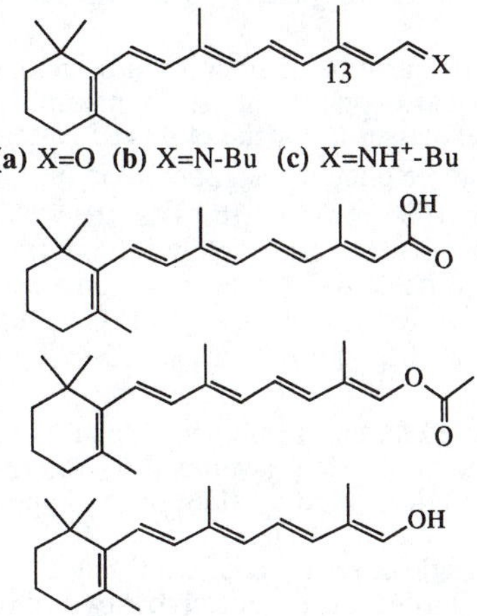

1

(a) X=O (b) X=N-Bu (c) $X=NH^{+}$-Bu

2

3

4

Figure 1. Molecular structure of the compounds studied in this work : 1(a) trans retinal; 1(b) retinal Schiff base; 1(c) retinal protonated Schiff base; 2 retinoic acid; 3 vitamin A acetate; 4 trans retinol (vitamin A).

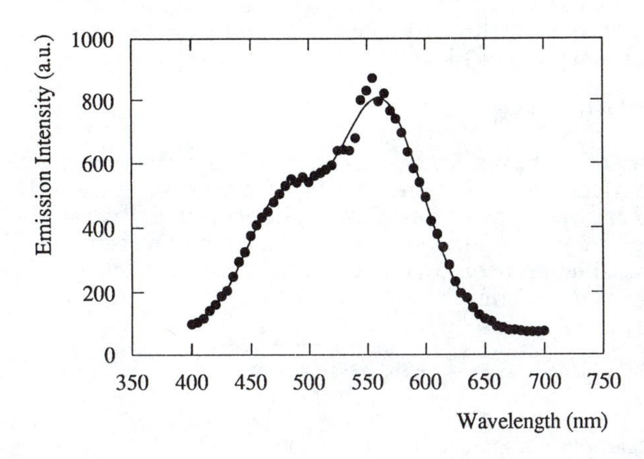

Figure 2. Emission spectrum of trans retinol on excitation with 1064 nm.

symmetry is only an approximation, the selection rules for the absorption process still hold. The maximum of the one-photon absorption to the ^{1}Bu state of trans retinol is located at 331 nm in EPA at 77 K and the two photon transition is at 350 nm (*16*). The emission maximum, under the same circumstances, is at 461 nm. This emission was postulated to originate in the ^{1}Ag state and not in the one-photon allowed ^{1}Bu state.

Two bands can clearly be distinguished in the recorded emission spectrum. Gaussian curve fitting shows that the emission maxima are located at 560 nm and at 475 nm. A possible explanation for the observed pattern is that the emission from the ^{1}Ag is pumped by three photon absorption (355 nm) into the ^{1}Bu state, as well as two photon absorption (532 nm) into the ^{1}Ag state. This two-photon absorption process thus would stimulate emission below 532 nm (the second Gauss curve). The first Gauss curve has to be caused by three-photon absorption, since its energy lies well above the energy of two-photon absorption. The maximum of this Gauss curve also coincides with that of the luminescence caused by one-photon absorption into the ^{1}Bu state.

It should be noted that no peak is visible at 532 nm. This clearly indicates that the HRS signal is completely swamped by the much stronger fluorescence, making it impossible to determine a value for the hyperpolarizability with HRS at this wavelength with the current set-up.

For other molecules, we are confident that a 3 nm bandwidth interference filter can eliminate fluorescence caused by two-photon absorption since this fluorescence is always red-shifted with respect to the two-photon excitation wavelength. A broad emission band triggered by three-photon absorption, however, will pass through the filter and artificially enhance the HRS β value. It can easily be verified whether such fluorescence is present by substituting the 532 nm interference filter for a 550 nm interference filter, or by substituting a 3 nm bandwidth interference filter at 532 nm for a 10 nm bandwidth interference filter at the same wavelength. The possibility of fluorescence stresses the importance of using an appropriate the interference filter.

Theoretical Methodology

The geometry of all the molecules was optimized using the semi-empirical Hartree-Fock Austin Model One (AM1) method (*19*). On the basis of the AM1 geometries, the norm of the vector part of the first hyperpolarizability is calculated with the Intermediate Neglect of Differential Overlap/Sum-Over-States (INDO/SOS) technique based on the perturbation expansion of the Stark energy of the molecule (*20,21*). This norm is defined as :

$$\beta_v = \sqrt{\beta_x^2 + \beta_y^2 + \beta_z^2} \tag{6}$$

where, for example, component β_x is given by :

$$\beta_x = \beta_{xxx} + \frac{\beta_{xyy} + 2\beta_{yyx} + \beta_{xxz} + 2\beta_{zzx}}{3} \tag{7}$$

The SOS method has the advantage of yielding easy access to the dynamic (i.e. frequency dependent) response in order to evaluate the dispersion in β of the present visual chromophores. In this work, we assume a damping factor of 0.1 eV for all the resonances, which is a good approximation for this kind of molecules (*9*).

If we assume that the nonlinear optical response of a molecule is dominated by the ground state and a low-lying charge-transfer excited state, a simple expression can be derived for the second harmonic generation β_{xxx} component (*22*) :

$$\beta_{xxx}(-2\omega;\omega,\omega) = \frac{3\Delta\mu_{eg} M_{eg}^2}{(\hbar\omega_{eg})^2} \frac{\omega_{eg}^2}{(1 - \frac{4\omega^2}{\omega_{eg}^2})(\omega_{eg}^2 - \omega^2)} \quad (8)$$

where $\Delta\mu_{eg}$, M_{eg} and $\hbar\omega_{eg}$ are the dipole moment difference, the electronic transition moment and the transition energy, respectively, between the ground and the first excited state. The first part of the equation is a static factor; the final part, dependent on the transition energy and the excitation energy only, can be treated as the dispersion factor.

An even simpler version of this two-state model is often employed by Marder and coworkers (*23,24*) :

$$\beta \approx \frac{\Delta\mu_{eg} M_{eg}^2}{(\Delta E)^2} \quad (9)$$

where ΔE is the energy term that takes into account the dispersion in β. We will extend the change in dipole moment upon excitation and take into account the change in linear molecular polarizability upon excitation to explain the large solvent effect on the first hyperpolarizability of trans retinal.

Solvent Dependence of the First Hyperpolarizability

The solvent dependence of the first hyperpolarizability is a topic that is currently being investigated, partly to understand the mismatch between theoretical and experimental data (*25*) and partly to allow a further optimization of NLO performance. Since it has been demonstrated that the two-state model predicts the right trends for the hyperpolarizability, it should be possible to model the solvent dependence of the first hyperpolarizability in terms of the two-state model. Work performed by Marder on donor-acceptor polyenes emphasizes the large solvent effect on the change of dipole moment on excitation (*23,24*), that is intimately linked to the degree of bond-length alternation.

Solvent Dependence of the First Hyperpolarizability of trans Retinal. We have determined the first hyperpolarizability of trans retinal in the solvents listed in Table I. For the measurements in methanol, chloroform and nitromethane, the internal reference method was applied, using the number densities and hyperpolarizabilities listed in reference 12 ($\beta_{chloroform}$=-0.49x10^{-30} esu, $\beta_{methanol}$=0.69x10^{-30} esu, $\beta_{nitromethane}$=1.82x10^{-30} esu). The measurement in dioxane was performed with para-nitroaniline (PNA) in dioxane as a reference ($\beta_{PNA,\ dioxane}$=16.9x10^{-30} esu) (*26*) and the measurement in 1,2-dichloroethane with methoxynitrostilbene (MONS) in 1,2-dichloroethane as a reference ($\beta_{MONS,\ 1,2\text{-dichloroethane}}$=119x10^{-30} esu).

The calculations performed at the Hartree-Fock ab initio level using a split-valence (3- to 21-G) basis set reproduce the experimental data. The influence of the solvent was accounted for by following Onsager's self-consistent reaction field approach as implemented in the Gaussian-92 set of programs (*27*). In this model, the solute molecule occupies a spherical cavity of radius a_o in a medium which is represented by a dielectric with a given constant (ε). If the solute molecule has a permanent dipole moment, it will induce a reflection dipole moment in the medium. The polarization of the dielectric in the electric field of the molecule itself gives rise to a reaction field. The reaction field (which is parallel to the dipole moment of the solute molecule) has the effect of enhancing the molecular dipole moment. The electrostatic effect of the solvent can be represented by an additional term (H_1) in the Hamiltonian of the isolated molecule (H_0).

$$H_{RF} = H_0 + H_1 \tag{10}$$

The perturbation term (H_1) describes the coupling between the molecular dipole operator (μ) and the reaction field (R):

$$H_1 = \mu . R \tag{11}$$

Where R is given by (*28*) :

$$\vec{R} = \frac{2(\varepsilon - 1)}{(2\varepsilon + 1)(a_0)^3}\vec{\mu} \tag{12}$$

The interaction is thus treated as a perturbation to the Hartree-Fock Hamiltonian and the calculation is continued until self-consistency is achieved in the presence of the reaction field. Note that this model exploits the reaction field theory in its most simple form, simulating all the different types of solute form by a spherical cavity and restricting the solute/solvent interaction to the dipolar term. Recent studies show that this approximation is a good starting point to simulate the solvent effect on the first hyperpolarizability of a molecule (*29,30,25*).

It was found that as the polarity increases, the average degree of bond-length alternation δr (the average value of the difference between adjacent carbon-carbon double and single bonds along the retinal conjugated backbone) decreases slightly as β increases. This relation is in line with what has been observed by Marder and co-workers in the case of donor-acceptor polyenes (*23*).

Table I. Calculated and measured solvent dependence of the first hyperpolarizability of trans retinal.

Solvent	β HRS (10^{-30} esu)	β CALC (10^{-30} esu)	δr (Å)	ε
Gas Phase		11.5	0.128	1
Dioxane	80	39.6	0.126	2.2
Chloroform	270	113	0.121	4.81
1,2-Dichloroethane	380	261	0.116	10.6
Methanol	730	598	0.109	32.6
Nitromethane	1100	641	0.108	35.9

Since there is only a small shift in the position of the absorption maximum upon changing the solvent (15 nm) (*31*), the dispersion term in β cannot account for the observed evolution. As was shown to be the case with the donor-acceptor polyenes, the change in bond-length alternation that is related to the change in dipole moment upon excitation is a crucial factor.

A similar ratio between the hyperpolarizabilities measured in methanol is found for trans retinoic acid ($\beta_{chloroform}$=110x10^{-30} esu, $\beta_{methanol}$=310x10^{-30} esu). This indicates that the main effect on the hyperpolarizability is due to the bulk influence of the solvent polarity on the molecule, and not to specific, hydrogen-bonding interactions. The first transition energy of trans retinal has been found to be sensitive to both solvent polarity and hydrogen bonding interactions, but the observed evolution is not as strong as that observed in the case of pure donor-acceptor polyenes. These specific interactions also give rise to an increase in the quantum yield of fluorescence due to state order reversal (*32*); this fluorescence was too weak, however, to affect the HRS measurements.

When a polar solute molecule is placed in a solvent cavity, it will polarize the surrounding solvent molecules, giving rise to an electric field E_{rf} that is called the reaction field. This reaction field then acts upon the polarizability of the solute molecule and enhances the molecular dipole moment (*33*):

$$\mu_{0,total} = \mu_{0,gas} + \alpha_0 E_{rf} \quad (13)$$

where $\mu_{0,gas}$ is the ground state dipole moment and α_0 is the ground state linear polarizability of the molecule in the ground state and in the solvent of choice. Since the reaction field is always in the same direction as the molecular dipole moment, which is also the direction of the main β tensor component for dipolar species such as para-nitroaniline and trans retinal, α_0 is the linear polarizability tensor component that is parallel to the dipole moment.

A similar equation can be written for the excited state dipole moment :

$$\mu_{1,total} = \mu_{1,gas} + \alpha_1 E_{rf} \quad (14)$$

where α_1 is the linear polarizability tensor component of the excited state parallel to the ground state dipole moment. For dipolar species, the ground and excited state dipole moments often lie in the same direction. Since there is insufficient time for reorientation of the surrounding solvent molecules, the orientational contribution to the reaction field will be the same as that of the ground state. On the other hand, the electrons of the solvent molecules will redistribute in response to the excited state dipole moment, thus changing the reaction field working on the molecule. For simplicity we will neglect this effect and assume that the reaction field working on the ground state and the first excited state is the same. Combining equation 13 and equation 14 gives for the change in dipole moment upon excitation :

$$\Delta\mu_{total} = \Delta\mu_{gas} + \Delta\alpha E_{rf} \quad (15)$$

Inserting this in the two-state model (equation 9) finally gives :

$$\beta \approx \beta_{gas} + \frac{M_{eg}^2 \Delta\alpha E_{rf}}{(\Delta E)^2} \tag{16}$$

Since spectrophotometric data show that the transition energy and the electronic transition moment of trans retinal do not exhibit large changes upon changing the solvent, M_{eg}^2 and $(\Delta E)^2$ can be treated as constants. A somewhat similar model was developed to explain the solvent dependence of a merocyanine dye (*34*).

This very simple model clearly demonstrates that a large change in linear polarizability upon excitation, in combination with a strong reaction field can give a significant contribution to the first hyperpolarizability. As for trans retinal, electric-field-induced changes in the optical absorption spectra (*35,36*) have shown that the increase in long-axis polarizability upon electronic excitation in the gas phase is 200 ± 20 $Å^3$. In a poly(methylmethacrylate) matrix, this increase amounts to 600 ± 100 $Å^3$. As a result, the dipole moments of the ground state and of the excited state are influenced by the reaction field in a different way. Whereas upon solvation the ground state dipole moment of trans retinal increases from 5.3 ± 0.2 D to only 6.6 ± 0.2 D, the excited state dipole moment increases from 7 ± 1 to 19.8 ± 0.7 D, which should strongly enhance the first hyperpolarizability. Even though this model is extremely simple and even does not take into account the dependence of the change in linear polarizability upon excitation on the reaction field, it clearly shows the importance of this linear polarizability factor in the case of the retinal polyene.

Even if a solvent can never simulate the field generated by the specific ionic binding site of the retinal protonated Schiff base in bacteriorhodopsin, it can generate a large reaction field through dipole-dipole and dipole-induced dipole interactions that, in combination with a large excited state polarizability, can give a contribution to β. Thus, the solvent effect can be used to investigate the relevant molecular parameter, the excited state polarizability.

First Hyperpolarizability of Retinal Derivatives.

Frequency Dependence of the First Hyperpolarizability. It is well-known that if the end group of the retinal polyene system is substituted for a more electron withdrawing group, the position of the absorption band maximum shifts to lower energies (*37*). We have determined the first hyperpolarizability of vitamin A acetate, retinoic acid, retinal Schiff base, retinal and retinal protonated Schiff base (RPSB) in methanol. To determine the calibration factor in equation 5, the internal reference method was applied with $\beta_{methanol} = 0.69 \times 10^{-30}$ esu from reference 11. Vitamin A acetate and retinal are commercially available from SIGMA-ALDRICH. Retinoic acid was a gift of BASF. These samples were used without further purification. The Schiff base of retinal was prepared by adding an 8-fold excess of N-butylamine to a 10^{-3} M solution of retinal in methanol at room temperature. N-butylamine was chosen because of its high reactivity towards retinal and its similarity to the ε-amino group of the lysine residue that links the chromophore to the Rhodopsin protein. The reaction proceeded rapidly and with high yield. The Schiff base was protonated by bubbling anhydrous HCl through the solution. Since the excess N-butylamine and HCl do not change the HRS signal significantly, no attempt was made to remove them.

Table II. Dependence of the first hyperpolarizability of retinal derivatives on the position of the main absorption band. All β values are in 10^{-30} esu.

Molecule	$\hbar\omega_{eg}$ (eV)	β 1064 HRS	β static HRS	β_v static INDO/SOS
Vitamin A acetate	3.80	140	80	2.4
Retinoic acid	3.53	310	160	38.0
Retinal Schiff base	3.40	470	220	17.9
Retinal	3.26	730	300	41.5
RPSB	2.79	3600	900	214.6

In Table II, the experimental first hyperpolarizabilities are listed, along with the position of the absorption band maxima and the static INDO/SOS results. The experimental values were corrected for dispersion using the two-level model. As can be derived from Table II, both the experimental, two-level extrapolated results and the static theoretical results show an excellent correlation between β° and $(1/\hbar\omega_{eg})^2$, which is also predicted by the two-state model.

The theoretical calculations have been carried out at the semi-empirical level and thus introduce a series of approximations with respect to the ab initio methods. Therefore, the theoretical calculations can only provide trends for the hyperpolarizability, and no absolute numbers. Furthermore, the solvent effect, which is relevant in the case of the retinal derivatives, is not accounted for. The calculations reproduce the experimental trends well. The spectroscopic properties of the molecule, required to evaluate the parameters in the two-state model, can be obtained from the INDO/SCI analysis of the first excited state. These parameters are presented in Table III.

Table III. Calculated parameters of the two-state model.

Molecule	$\lvert\Delta\mu\rvert$ (D)	M^2_{eg}	$\hbar\omega_{eg}$ (eV)
Vitamin A acetate	0.84	121.17	3.78
Retinoic acid	8.97	129.24	3.52
Retinal Schiff base	5.65	142.67	3.60
Retinal	8.92	133.46	3.50
RPSB	10.11	198.15	2.42

We observe that the evolution of the β values is governed by the evolution of the dipole moment difference between the two states and the evolution of the dipole moment difference rather than the evolution of the electronic transition dipole moment. The reversal that was observed between retinoic acid and the Schiff base emerges here again.

Effect of Trans-to-Cis Isomerization. Whereas the dependence of β on the position of the main absorption band is frequently investigated in nonlinear optics (*38*) the effect of trans-to-cis isomerization has received considerably less attention. Most of the studies are aimed at enlarging β, and from a chemically intuitive point of view a trans-to-cis isomerization would diminish β. It should be pointed out that even the trans retinal derivatives are not completely planar, due to the steric interactions between the 1,1' methyl groups and the hydrogen atom on carbon 7 which causes the 6,7 single bond to assume a 6-s-cis conformation (*7*). The AM1 optimized geometries of all the molecules indicate a torsion angle of 53° between the ring and the polyene chain. This value for the torsion angle is in excellent agreement with that found in previous studies (*39*).

We have measured the first hyperpolarizability of 9- and 13-cis retinal, their Schiff bases and protonated Schiff bases in methanol. 9- and 13-cis retinal are commercially available from SIGMA-ALDRICH and the Schiff bases and the protonated Schiff bases were prepared in a similar manner as for the trans derivatives. In Table IV, the experimental two-level extrapolated values are compared to the INDO/SOS static theoretical values. If these data are compared to the data of Table II for the trans derivatives, it is obvious that upon trans-to-cis isomerization a constant fraction of the hyperpolarizability is lost. Since this fraction is the same for the experimental and theoretical β values, theory and experiment are in excellent mutual agreement.

Table IV. Experimental (MeOH, extrapolated) and theoretical results for the 13-cis derivatives.

Molecule	β 13-cis INDO/SOS	β 13-cis HRS	β 9-cis INDO/SOS	β 9-cis HRS
Retinal Schiff base	13.7	120	13.0	100
Retinal	29.8	200	32.2	200
RPSB	151.3	600	117.3	600

The trans-to-cis evolution can be theoretically understood by analyzing the evolution of the different parameters which appear in the two-state model (see Table V). Passing from the trans to the 13-cis form hardly affects the first transition energy (the larger shift being 0.02 eV). This is in agreement with experimental data, that show that the position of the absorption band maxima of the trans derivatives is within 0.06 eV of that of their 9-cis and 13-cis counterparts. The dipole moment difference and the electronic transition moment between the two states are weaker for the cis form with respect to the trans form.

Table V. INDO/SCI calculated electronic properties of the first excited states of trans, 13-cis and 9-cis retinal derivatives. Units are in D, D^2 and eV.

Molecule	$\lvert\Delta\mu\rvert$	M^2_{eg}	$\hbar\omega_{eg}$
all-trans retinal	8.92	133.46	3.50
13-cis retinal	8.45	122.40	3.49
9-cis retinal	8.33	119.63	3.49
all-trans retinal Schiff base	5.65	142.67	3.60
13-cis retinal Schiff base	4.65	134.08	3.59
9-cis retinal Schiff base	5.09	129.94	3.59
all-trans RPSB	10.12	198.15	2.42
9-cis RPSB	9.25	162.89	2.42
13-cis RPSB	7.31	162.50	2.43

Conclusions

By comparing experimental results obtained with the hyper-Rayleigh scattering technique and theoretical INDO/SCI/SOS calculations we have performed a thorough study of the first hyperpolarizability of retinal derivatives. The two-state model was shown to predict correct trends for the hyperpolarizability and the change in dipole moment upon excitation is an important factor for understanding the effects of frequency dispersion and of *trans*-to-cis isomerization. By extending the two state model to take into account the change in linear polarizability on excitation, the large solvent effect on the hyperpolarizability of trans retinal could be explained in a simple manner. Quantum-chemical calculations also relate this effect to the solvent-induced change in bondlength alternation.

Acknowledgements

This work is partly supported by the Belgian Prime Minister Office of Science Policy "InterUniversity Attraction Pole in Supramolecular Chemistry and Catalysis" and "Impulsion Program in Information Technology", FNRS/NFWO, Flemish government and an IBM Academic joint study. We also thank BASF for the gift of all-trans retinoic acid. Eric Hendrickx is a Research Assistant of the Belgian National Fund for Scientific Research. Koen Clays is a Senior Research Associate of the Belgian National Fund for Scientific Research. The authors would like to thank Dr. S. Marder for helpful discussions.

Literature Cited

1. Birge, R.R. *Annu. Rev. Biophys. Bioeng.* **1981**, *10*, 315
2. Mathies, R.A.; Lin, S.W.; Ames, J.B.; Pollard, W.T. *Annu. Rev. Biophys. Biophys. Chem.* **1991**, *20*, 491
3. Henderson, R.; Baldwin, J.M.; Ceska, T.A.; Zemlin, F; Beckman, E; Douning, K.H. *J. Mol. Biol.* **1990**, *213*, 899
4. Oesterhelt, D.; Bräuchle, C.; Hampp, N. *Quart. Rev. Biophys.* **1991**, *24*, 425
5. Miyasaka, T; Koyama, K; Isamu, I. *Science* **1992,** *255,* 342
6. Hampp, N.; Thoma, D.; Oesterhelt, D.; Bräuchle, C. *Appl. Optics* **1992**, *31*, 1834
7. Birge, R.R. *Am. Scientist*, **1994**, *82*, 348
8. Clays, K.; Hendrickx, E.; Triest, M.; Verbiest, T.; Persoons, A.; Dehu, C.; Brédas, J.L. *Science* **1993**, *262*, 1419
9. Birge, R.R.; Zhang, C.-F. *J. Chem. Phys.* **1990**, *92*, 7178
10. Huang, J.; Chen, Z.; Lewis, A. *J . Chem. Phys.* **1989**, *93*, 3314
11. Dencher, N.A.; Heyn, M.P. *FEBS Letters* **1978**, *96*, 322
12. Clays, K.; Persoons, A. *Phys. Rev. Lett.* **1991**, *66*, 2980
13. Clays, K.; Persoons, A.; De Mayer, L. *Adv. Chem. Phys.* **1994**, *85(III)*, 455
14. Clays, K.; Persoons, A. *Rev. Sci. Instrum.* **1992**, *63*, 3285
15. Behrsohn, R.; Pao, Y.H.; Frisch, H.L. *J. Chem. Phys.* **1966**, *45*, 3484
16. Maker, P.D. *Phys. Rev. A.* **1970**, *1,* 923
17. Birge, R.R.; Bennet, J.A.; Pierce, B.M.; Thomas, T.M. *J. Am. Chem. Soc.* **1978**, *100*, 1533
18. Birge, R.R.; Pierce, B.M. *J. Chem. Phys.* **1979**, *70*, 165
19. Dewar, M.J.S.; Zoebish, E.G.; Healy, E.F.; Stewart, J.J.P. *J. Am. Chem. Soc.* **1985**, *107*, 3902
20. Orr, B.J.; Ward, J.F. *Mol. Phys.* **1971**, *20*, 513
21. Zhov, Q.L.; Heflin, J.R.; Wong, K.Y.; Zamani-Khamiri, O.; Garito, A.F. *Organic Molecules for Nonlinear Optics and Photonics:* Messier, J., Kajzar, F. Eds NATO-ARW Series, Ser. E. 239, 1991

22. Oudar, J.L. *J. Chem. Phys.* **1977**, *67*, 446
23. Marder, S.R.; Beratan, D.N.; Cheng, L.T. *Science* **1991**, *252*, 103
24. Bourhill, G.; Brédas, J.L.; Cheng, L.T.; Marder, S.R.; Meyers, F.; Perry, J.W.; Tiemann, B.G. *J. Am. Chem. Soc.* **1994,** *116*, 2619
25. Di Bella, S; Marks, T.J.; Ratner, M.A. *J. Am. Chem. Soc.* **1994**, *116*, 4440
26. Teng, C.C.; Garito, A.F. *Phys. Rev. B* **1983**, *28*, 6766
27. Gaussian-92, Frisch, M.J.; Trucks, G.W.; Head-Gordon, M.; Gill, P.M.W.; Wong, M.W.; Foresman, J.B.; Johnson, B.G.; Schlegel, H.B.; Robb, M.A.; Reploge, E.S.; Gomperts, R.; Andress, J.L.; Raghavachari, K.; Binkley, J.S.; Gonzales, C.; Martin, R.L.; Fox, D.J.; Defrees, D.J.; Baker, J.; Stewart, J.J.P.; Pople, J.A., Inc, Pitssburgh PA, **1992**
28. Onsager, L. *J. Am. Chem. Soc.* **1936**, *58*, 1486
29. Yu, J.; Zerner, M.C. *J. Chem. Phys.* **1994**, *100*, 7487
30. Mikkelsen, K.V.; Luo, Y.; Agren, H.; Jorgensen, P. *J. Chem. Phys.* **1994,** *100*, 8240
31. Das, P.K.; Becker, R.S. *J. Phys. Chem.* **1978**, *82*, 2081
32. Das, P.K.; Becker, R.S. *J. Phys. Chem.* **1978**, *82*, 2093
33. Böttcher, C.J.P. *Theory of Electric Polarization*, Elsevier, Amsterdam, 1973, 129
34. Levine, B.F.; Bethea, C.G.; Wasserman, E.; Leenders, L. *J. Chem. Phys.* **1978**, *68*, 5042.
35. Ponder, M.; Mathies, R. *J. Phys. Chem.* **1983**., *87*, 5090
36. Davidsson, Å; Johansson, L.B. *J. Phys. Chem.* **1984**, *88*, 1094
37. Rosenberg, B; Krigas, T.M. *Photochem. Photobiol.* **1967**, *6*, 769
38. Cheng, L.T.; Tam, W.; Stevenson, S.H.; Meredith, G.R.; Rikken, G.; Marder, S.R. *J. Phys. Chem.* 1991, 95, 10631
39. Nakanishi, K; Balog-Nair, V.; Arnaboldi, M.; Tsujimoto, K. *J. Am. Chem. Soc.* **1980**, *102*, 7945

RECEIVED January 30, 1995

Chapter 7

Design and Synthesis of Thermally Stable Chromophores with Low Absorption at Device Operating Wavelengths

Susan Ermer, Steven M. Lovejoy, and Doris S. Leung

Lockheed Martin Palo Alto Research Laboratory,
Research and Development Division, Organization 93–50, Building 204,
3251 Hanover Street, Palo Alto, CA 94304–1191

A series of novel, lambda-shaped chromophores combining thermal stability above 375 °C, transparency at 830 nm, and nonlinearity/MW comparable to Disperse Red 1 have been developed. These "DAD-type" chromophores are based on the dicyanomethylenepyran acceptor group and can be viewed as donor-acceptor-donor analogs of DCM, a chromophore that has already been used for demonstration devices. Thermal stability within a polyimide environment has been demonstrated using one of the chromophores, DADC, in a fully-cured polyimide waveguide device patterned by photobleaching. Functionalization for improved solubility has been demonstrated, and a readily attachable variant has been generated. The detailed syntheses and full chemical characterization of these chromophores are described.

Since the introduction of polyimide-based electro-optic (EO) materials in 1991, much progress has been made in their development (*1-3*). Our goal has been the development of a polymeric based materials system which could be generated in sufficient quantity for device prototyping. Requirements for such a system include ease of preparation, moderate electro-optic activity, transparency, and accessibility to large reproducible lots. Based on these needs, we developed a materials formulation based on commercially accessible components. These components were DCM (4-(dicyanomethylene)-2-methyl-6-(p-dimethylaminostyryl)-4*H*-pyran), a polar compound used as a laser dye, and a fluorinated polyimide, Amoco Ultradel. The Ultradel polyimides were developed for use as interlevel dielectric coatings and have been optimized for low ionic content, low dielectric constant, low moisture absorption, and high thermal stability. The main structural component of these materials is shown in Figure 1.

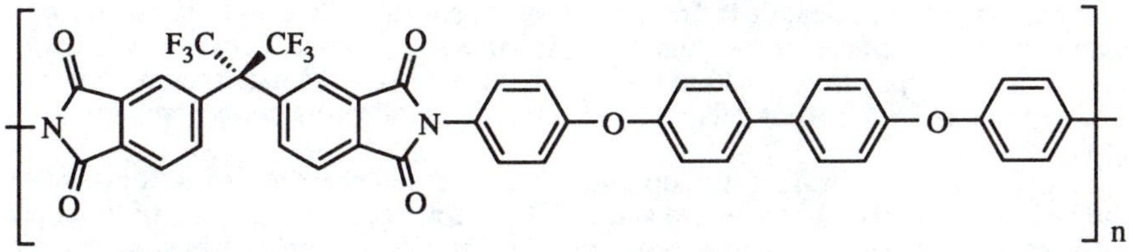

Figure 1. Main structural component of Amoco Ultradel 4212 and 3112 polyimide

0097–6156/95/0601–0095$12.00/0

These polymers are highly fluorinated. Fluorination of polymers used in optical fibers has been shown to effect properties such as dielectric constant, refractive index, water absorption and optical loss (*4*). Absorption losses in a variety of polymers such as poly(methyl)methacrylate, polycarbonate, and polysulfone, used for polymer optical fibers have been determined to result largely from higher harmonics of molecular vibrations of functional groups, particularly those of C-H, N-H, and O-H (*5*). Fluorine substituted for hydrogen in these groups shifts these absorptions, reducing optical losses in the critical waveguiding regions of the spectrum (0.8 to 1.6 microns). This explanation may be extended to the fluorinated polyimides. It is also likely that the bulky trifluoromethyl (-CF_3) groups separate the chromophoric centers and inhibit electronic inter-chain interactions.

In 1992, we described a proof-of-principle all-polyimide triple stack Mach-Zehnder (M-Z) modulator using these components as the active core layer (*6*). A cross-sectional view of the switching arm of this waveguide is shown in Figure 2. Recently, the same DCM/polyimide core layer was combined with an acrylate cladding for a high-speed device demonstration (*7*). Efforts are now needed to optimize poling in all-polyimide devices, in order to maximize poling alignment and thereby improve switching sensitivity (*8*). Studies of ionic content on electrical conduction in certain polyimide single films have been performed (*9-10*), but further work is required to fully elucidate the contribution of ionics in three layer stacks during poling.

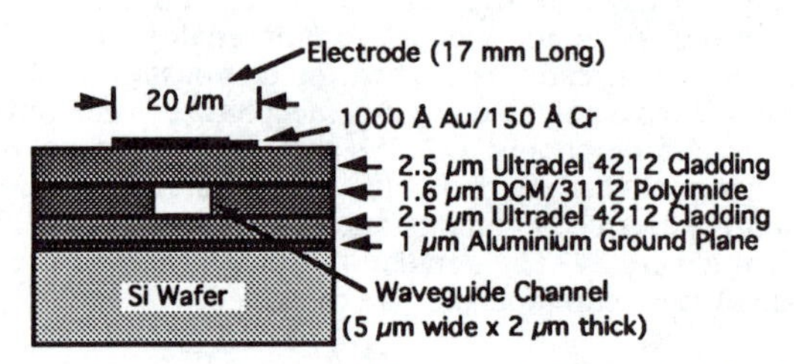

Figure 2. Diagram of DCM/Polyimide Mach-Zehnder Modulator

The guest and host components used for Mach-Zehnder prototyping were chosen for their compatibility with processing and device demonstration requirements. High thermal stability is the key advantage offered by the polyimide host materials. They are especially promising for optical interconnection devices which must be processed under the extreme conditions used for semiconductor fabrication (*11*). Advantages of DCM as guest include low absorbance at device wavelengths, photobleachability, compatibility with polyimides and their polyamic acid precursors, and commercial availability in high purity. Some of these characteristics are the result of DCM's use as a laser dye, which are designed to have sharp cut-offs on the high wavelength side of their major absorption peak, so as to avoid reabsorbing emitted light. The absorption spectrum for DCM shown in Figure 3 peaks at 480 nm in N-methylpyrrolidinone (NMP), and its optimum lasing wavelength is 661 nm. A sharp cut-off wavelength is necessary to avoid absorbance of the laser output, and this same sharp cut-off avoids excess optical loss in devices operating at 830 nm. It can be seen, also, from data presented later in this paper, that DCM offers relatively high nonlinearity for its molecular weight and transparency.

DCM is, however, less than optimum in its thermal characteristics. It plasticizes the host material and, when heated above 220 °C for significant periods of time, out-diffuses. Curing polyimides at temperatures below 220 °C compromises the superior thermal and mechanical properties that prompted their selection as host materials. This out-diffusion has been successfully controlled by physically confining the DCM using an evaporated metal layer over the waveguide, allowing cure at temperatures of up to

350 °C (*12*), but an inherently thermally-stable chromophore would be preferred. Plasticization remains a concern, since depression of the glass transition temperature, T_g , of the polyimide host is detrimental to long-term stability of the poled state (*13-15*). This concern motivated us to develop a series of chromophores which maintain the advantageous properties of DCM while improving its thermal stability. Due to our interest in 830 nm device operation, it was essential to avoid loss of transparency of the new chromophores in this region.

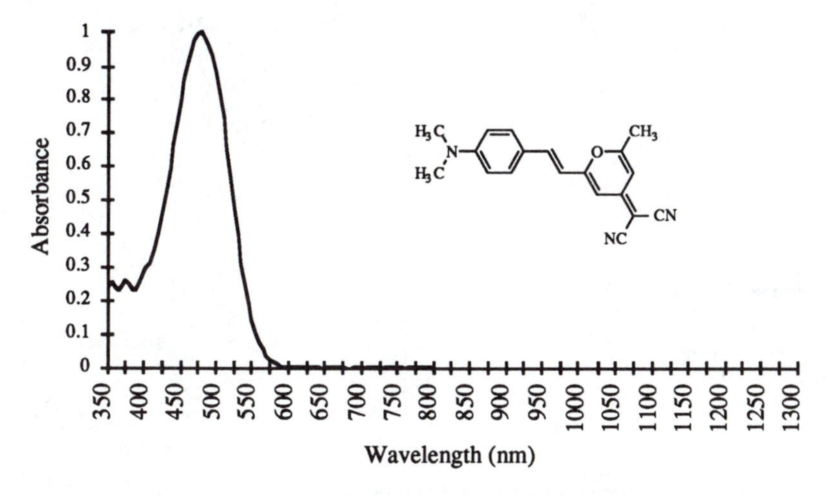

Figure 3. Absorption spectrum of DCM in NMP and its chemical structure (inset)

Design and Preparation of Thermally Stable Chromophores

Recently, several groups have published the structures of chromophores with thermal stability exceeding 300 °C. Aryl substitution on amine donors has been demonstrated to raise the thermal stability of a wide range of conventional chromophores (*16*). New chromophore classes have also been developed. Some examples are shown in Figure 4. The triarylimidazole and triaryloxazole classes of compounds have been utilized as EO chromophores by workers at IBM (*17*) and Sandia (*18*). While these two chromophores exhibit high thermal stability, their nonlinearities (β values) have been disappointing. The other two compounds shown exhibit hyperpolarizabilities comparable to standard chromophores such as DR1, but with greatly increased thermal stability. The chromophore SY177 is based on a fused-ring system which is structurally related to polyimide (*19*). It has been heated to 350 °C in a polyimide without any change in its linear absorption spectrum. Structurally related to DCM, DADC is a donor-acceptor-donor compound based on a dicyanomethylenepyran acceptor attached to two carbazole donors (*20*). The synthesis and properties of these donor-acceptor-donor analogs of DCM are the focus of this paper. Poled polymers containing these chromophores may be less susceptible to loss of EO activity due to loss of orientation as a result of their "lambda" shape (*21*). These chromophores exhibit good thermal stability at temperatures over 300 °C and maintain desirable properties such as high hyperpolarizability, transparency at device operation wavelengths, and photobleachability. The high-temperature stability and photo-bleachability of DADC has allowed the use of high processing temperatures in the fabrication of a demonstration Mach-Zehnder modulator (22). New thermally stable chromophores are continually being developed. Based on the progress seen in recent years, it is anticipated that thermal stability of chromophores will not be a significant limitation in the development of manufacturable EO materials. We have found the guest-host approach to be the most practical means of rapid chromophore development, screening, and investigation in device configuration. Once a chromophore with optimized properties for a given device application is developed, covalent attachment will be a logical next step for enhancement of loading level and long-term stability of the poled state.

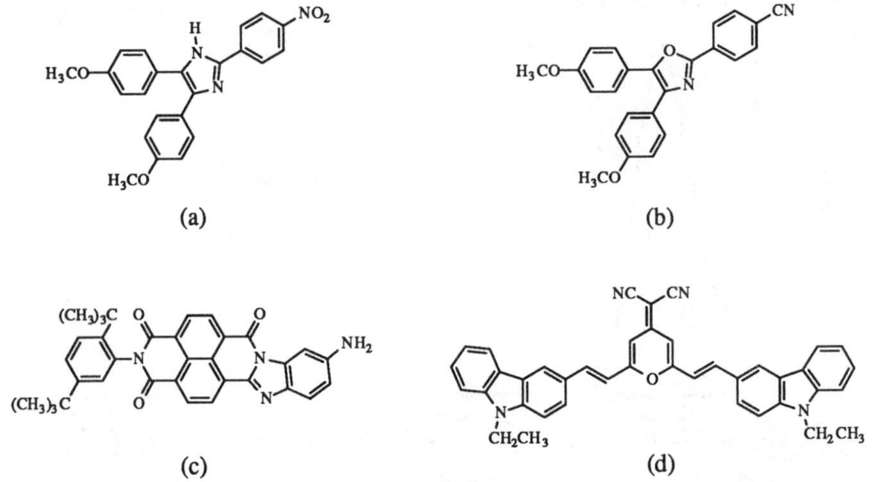

(a) (b)

(c) (d)

Figure 4. Examples of some NLO chromophores stable to >300 °C: (a) triarylimidazole, (b) triaryloxazole, (c) SY177, (d) DADC

Synthesis of Donor-Acceptor-Donor Analogs of DCM. The donor-acceptor-donor analog of DCM, (2,6-bis(2-(4-(dimethylamino)phenyl)-ethenyl)-4*H*-pyran-4-ylidene)propanedinitrile (DAD), has been previously investigated as an impurity arising from the synthesis of DCM (*23*). The DAD chemical structure may be viewed as containing two cross-conjugated DCM chromophores. This similarity in chromophores of the two compounds leads to UV-visible spectra which are very similar in appearance with nearly coincident absorption maxima. The dicyanomethylene [$=C(CN)_2$] group is a highly effective electron acceptor which has been used by a number of groups to synthesize highly active chromophores. It is mostly used as part of dicyanovinyl [$-CH=C(CN)_2$] and tricyanovinyl [$-C(CN)=C(CN)_2$] moieties (*24-26*). Substituted dicyanovinyl [$-CR=C(CN)_2$] moieties have also been used in cyclic acceptors derived from the condensation of isophorone and malononitrile (*27-29*). We began work on compounds based on the dicyanomethylenepyran moiety shortly after our first studies using DCM. This acceptor group is attractive because it allows the attachment of two different donors in a symmetrical arrangement using Knoevenagel condensation chemistry. This technique provides access to a wide range of DAD-type compounds. Recently, alternate methods of incorporation of the dicyanomethylenepyran moiety into syndioregic polymers have been described (*30*).

The synthetic scheme for the DAD-type analogs of DCM is shown in Figure 5. In general, the analogs were prepared by refluxing the appropriate aldehyde and (2,6-dimethyl-4*H*-pyran-4-ylidene)propanedinitrile together overnight in benzene or toluene using piperidine and glacial acetic acid as catalysts while removing the generated water with a Dean-Stark trap. After cooling the reaction solution, the precipitated crude analog was filtered and recrystallized (or flash chromatographed then recrystallized) to give the pure analogs. In the cases where no precipitate formed upon cooling, the reaction solvent was roto-evaporated to give the solid crude analogs then further purified. The detailed synthetic methods for the preparation of these compounds are given in the Experimental Section.

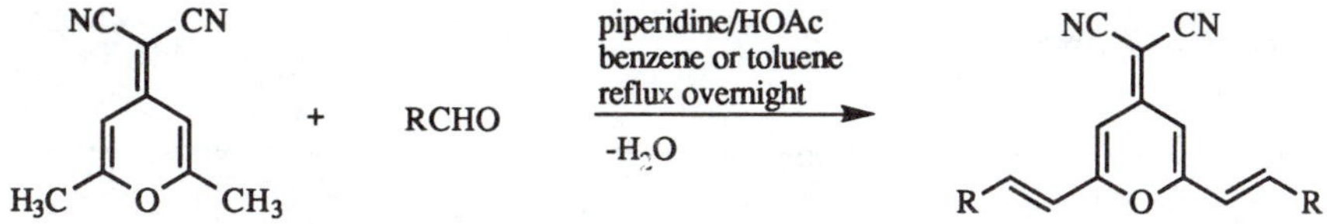

NC CN + RCHO —piperidine/HOAc, benzene or toluene, reflux overnight, $-H_2O$→ NC CN

H_3C O CH_3 R O R

Figure 5. General scheme of the synthesis of donor-acceptor-donor type analogs

Table I shows the donor-acceptor-donor analogs prepared and some of their physical properties. The visible absorption spectra were measured in N-methylpyrrolidinone (NMP), and the extinction coefficients (ε) times 10^{-4} are in parentheses. DAD's onset temperature of weight loss in air was measured in 100% O_2.

Thermal Stability. The materials in this series exhibit high thermal stabilities both in an inert atmosphere and in the presence of oxygen. We use thermogravimetric analysis (TGA) for rapid screening of the thermal stability of the neat chromophores. A sample of the compound is heated at a steady rate as weight loss is monitored. Weight loss in these compounds may occur due to either decomposition of the compound, or to an evaporative process, such as sublimation. The thermograms for a series of donor-acceptor-donor type compounds run under nitrogen atmosphere and in the presence of oxygen are shown in Figures 6 and 7, respectively. In general, the decomposition onset values (see Table I) reported by the Perkin-Elmer TGA-7 using derivative curves are generally higher than temperatures the molecules can withstand during the cure process. We believe this is due to several factors: (1) unmelted compounds are often stabilized by crystal lattice forces, (2) the host environment during cure is often highly reactive, and (3) the chromophore must withstand the higher temperature for a longer time period.

This effect is exemplified by DADT, which has TGA decomposition onset temperatures of 331 °C and 343 °C in nitrogen and air, respectively. A DADT/polyimide film was monitored during cure on a hot plate to see if the dissolved chromophore would withstand these temperatures. The film showed color loss indicative of chromophore decomposition at temperatures near 250 °C. Differential Scanning Calorimetry (DSC) run on a sample of DADT showed decomposition immediately upon melting (Moylan, C.R., IBM, personal communication, 1994). Because thermal analysis, whether by TGA or DSC, is performed on a crystalline sample, the decomposition weight loss onset temperature reported may not be predictive of the decomposition onset of a chromophore when it is solubilized within a polymer film. In summary, TGA and DSC are useful screening techniques when looking for stability trends, but do not necessarily accurately predict stability in a host environment.

Activity. Electric field-induced second harmonic (EFISH) generation experiments at 1.907 microns in chloroform on DAD, DADB, DADI, and DADC were performed by Lap-Tak Cheng (du Pont) and have been reported previously (*31*). These values are summarized in Table II, together with reference values for Disperse Red 1 (DR1) and DCM run in the same laboratory. The values for β_0 were calculated using the standard two-level model approximation. The units for both β and β_0 are 10^{-30} esu. The μ and β values for DADI are imprecise due to the insolubility of DADI in the EFISH solvent and are shown in parentheses.

Table I. Donor-acceptor-donor type analog physical properties

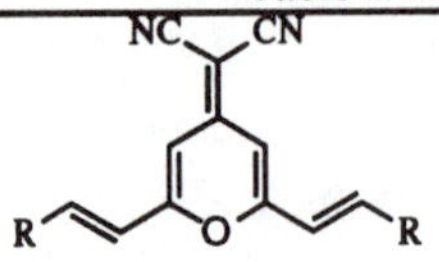 R= (structure: NC, CN, R, O, R)	Acronym	mp (°C)	λ_{max} (nm)	Onset temperature of weight loss in air (°C)	Onset temperature of weight loss in N_2 (°C)
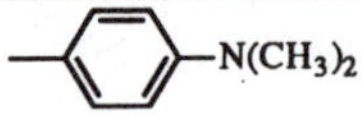 $N(CH_3)_2$	DAD	262.7-263	491 (6.83)	330	392
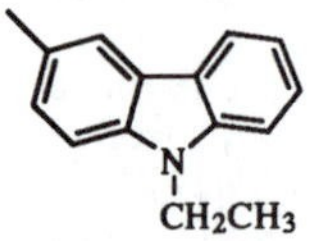 N, CH_2CH_3	DADC	337-338	459 (7.15)	375	389
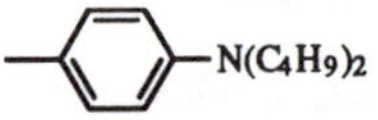 $N(C_4H_9)_2$	DADB	249.6-250.4	500 (6.63)	371	387
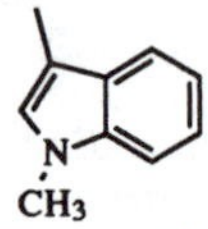 N, CH_3	DADI	315-325	465 (6.27)	374	392
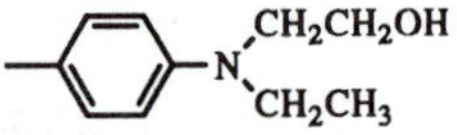 N, CH_2CH_2OH, CH_2CH_3	DADOH	222.5-224.5	499 (7.05)	363	372
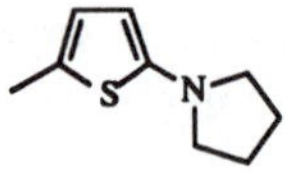 S, N	DADT	>350	529 (6.33)	331	343
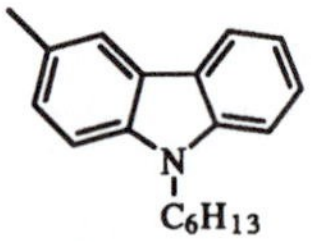 N, C_6H_{13}	DADCH	226.4-226.9	460 (6.98)	413	432
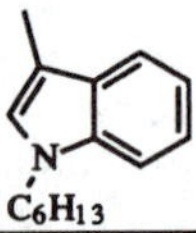 N, C_6H_{13}	DADIH	194.3-195	466 (6.22)	364	411

Table II. EFISH results for DAD-type analogs and standards DR1 and DCM

Acronym	Formula	MW	μ(D)	β	$\mu\beta$	β_0	$\mu\beta_0$	$\mu\beta_0$/MW
DR1	$C_{16}H_{18}N_4O_3$	314.4	7.5	70	525	50	360	1.15
DCM	$C_{19}H_{17}N_3O$	303.4	8.6	82	705.2	58	498	1.64
DAD	$C_{28}H_{26}N_4O$	434.3	7.9	102	805.8	70	556	1.28
DADB	$C_{40}H_{50}N_4O$	602.9	9.7	160	1552	108	1048	1.74
DADI	$C_{30}H_{22}N_4O$	454.5	(10.8)	(250)				
DADC	$C_{40}H_{30}N_4O$	582.7	8.5	95	807.5	69	586	1.01

SOURCE: Adapted from reference *31*.

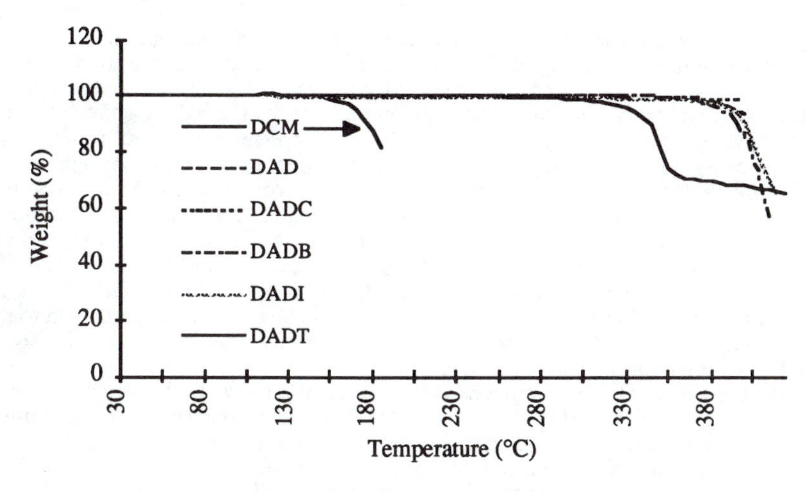

Figure 6. TGA of DAD-type analogs and DCM in N_2

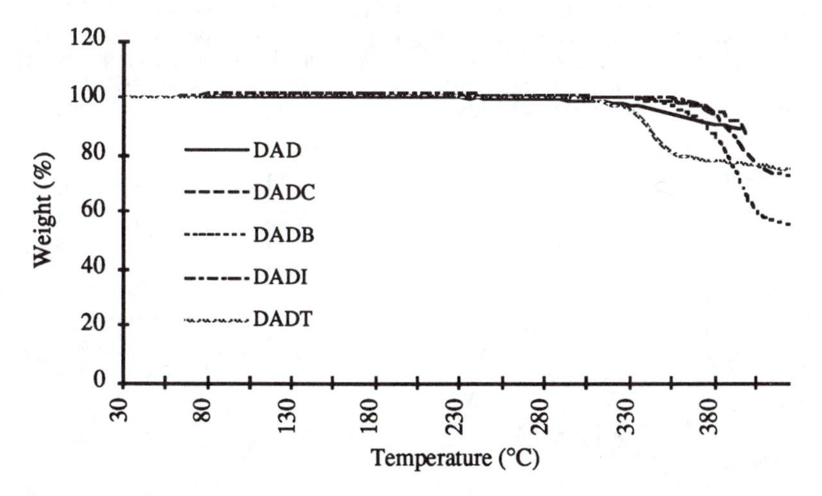

Figure 7. TGA of DAD-type analogs in the presence of O_2

Transparency. Transparency (very low absorption) at device operation wavelength is critical. The design of highly nonlinear chromophores for use in devices operating a shorter wavelengths (830 nm, for example) is particularly challenging. High nonlinear optical activity in chromophores is generally correlated with long-wavelength absorptions in the visible spectrum. This correlation is not surprising given the electronic basis of nonlinear phenomena and the strategies for improving it in a given class of compounds. Designing for enhanced nonlinear activity commonly involves a combination of structural modifications to the chromophore including: (1) lengthening the conjugated electron backbone, (2) using highly polarizable π-electron systems in the backbone and, (3) adding more/stronger electron donor and acceptor groups at opposite ends of the molecule. Marder and co-workers have shown that a balanced approach must be taken when using these methods to avoid bypassing the state of maximum bond length alternation (*32-33*). Bond length alternation and nonlinearity are reduced if either the ground (neutral) or charge-separated state are excessively favored in a given solution or polymer environment. Molecules with highly delocalized π-electron systems tend to have low-energy transitions, particularly when heteroatoms are present, and these low energy transitions are excited by longer wavelengths of light. As a result, in more highly delocalized systems with conventional electron-donors and acceptors, the major absorption peaks are red-shifted, often substantially. These shifted absorptions can extend into device operating regions. We have seen the effects of residual absorption on the high-wavelength side of a peak extend at least 200 nm beyond the position of the absorption maxima, even when the peak appears to be relatively sharp. This absorption "tail" is sufficient to raise optical loss in a loaded polymer film to an unacceptable level. Designing for improved nonlinearity, therefore, often involves accepting lowered transparency as a trade-off.

Most of the DAD-type compounds perform remarkably well in the transparency-activity trade-off. DAD, DADC, DADB, and DADI all have nonresonant nonlinearities higher than those of common chromophores such as DR1. When the activity per molecular weight is taken into consideration, their nonlinearities are still comparable. Absorption cut-off values render these compounds potentially useful for 830 nm device operation. An exception is DADT, which exhibits the red-shift commonly seen in thiophene-containing chromophores. Based on the work of others who have developed chromophores containing a heteroaromatic group in place of the phenyl group (*34*), we expect that the nonlinearity of DADT is quite high. Unfortunately, the extreme insolubility of DADT has precluded measurements of its μ or β values. The visible absorption spectra of some representative DAD-type analogs are shown in Figure 8.

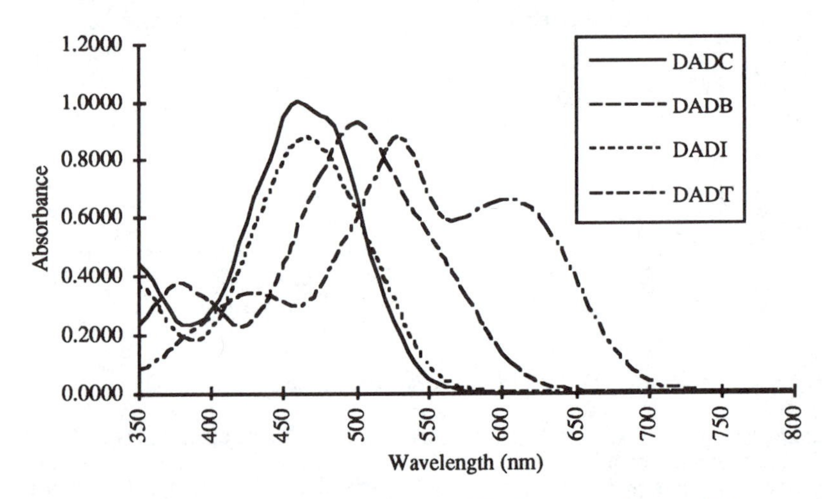

Figure 8. Visible absorption spectra of DAD-type analogs in NMP

Improvements of Solubility and Loading Levels in Guest-Host Systems

Overall, the solubilities of the DAD-type compounds exceeded our expectations. However, in some cases, notably DADI and DADT, insolubility rendered EFISH measurements impossible or unreliable. In other instances, measurements in solvents could be performed but reasonable loading levels (>5% by weight) in polyimides and their polyamic acid precursors were not possible, or else could only be achieved using a single solvent (DADC, for example). This led us to solubilize these compounds by attachment of aliphatic and polar groups. The nitrogen atom present in the donor moiety is the most accessible point for functionalization. Hexyl substitution to make DADIH and DADCH improved solubility significantly. Decomposition onset temperatures of the neat chromophores were determined by TGA in both nitrogen and air (see Figures 9 and 10). These temperatures remain well above 400 °C in nitrogen atmosphere. The effect of attaching longer aliphatic and polar substituents to the amino group on the decomposition onset temperature values is not well understood at this time.

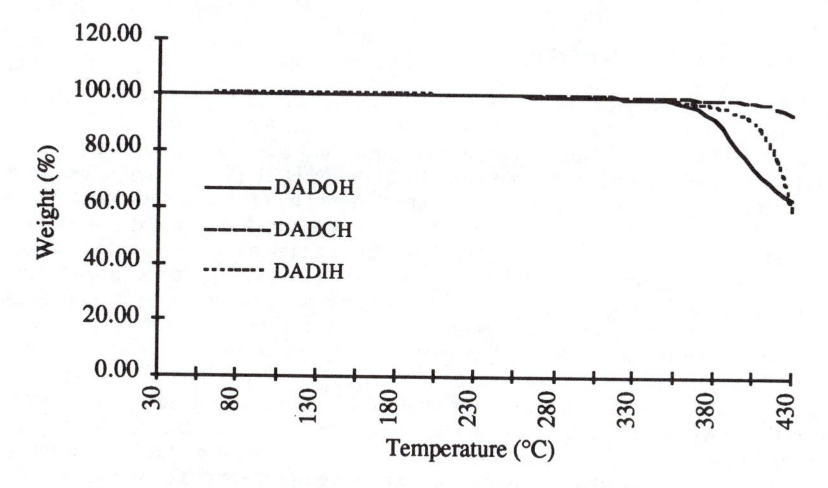

Figure 9. TGA of hexyl- and hydroxyl-substituted DAD-type analogs in N_2

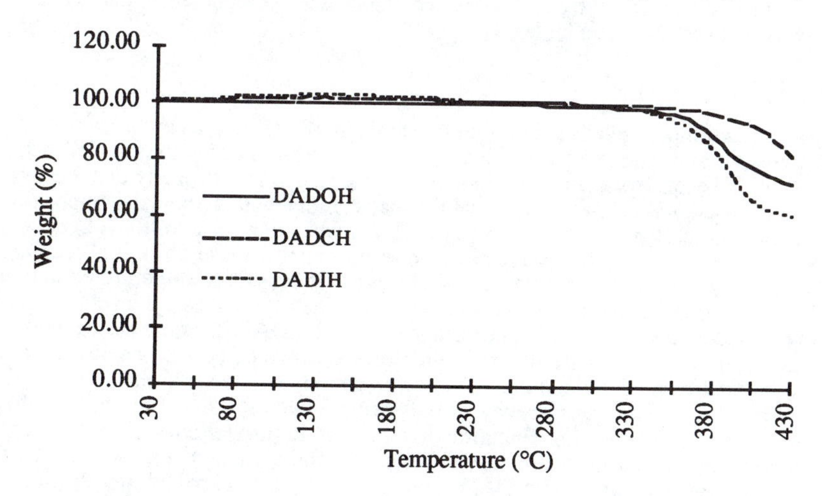

Figure 10. TGA of hexyl- and hydroxyl-substituted DAD-type analogs in air

Addition of the polar hydroxyethyl group to the nitrogen donor brought further increases in solubility. The chromophore with two such groups, DADOH, can be loaded into some polyamic acid spin dopes at 30% by weight. Thermal stability is slightly lowered, with onset of decomposition occurring at 363 °C and 372 °C in air and nitrogen, respectively. This level of stability is adequate for complete polyimide cure in a guest-host system. The utility of DADOH may go beyond guest-host systems, however. Because two functionalized donor ends of the molecule are available for reaction, this compound is a logical candidate for covalent attachment to a polymer system.

Beyond Guest-Host Systems

Many researchers have recently pursued covalent attachment of standard donor-acceptor chromophores, such as azo dyes, to polyimides in order to improve EO polymer stability (*35-41*). This attachment approach has the dual advantages of circumventing out-diffusion and improving chromophore loading levels. In these systems the upper limit of processing temperatures is still primarily limited by the inherent decomposition temperature of the chromophore. Attachment is tedious and removes the advantage afforded by use of semiconductor grade polyimide host materials, which are invariably lower in ionic contaminants than polymers produced under standard research laboratory conditions. Guest-host systems provided a way to rapidly screen a large variety of chromophores with a modest expenditure of effort. Once a chromophore with optimized properties for a given device application is developed, covalent attachment will be a logical next step for enhancement of loading level and long-term stability of the poled state. Covalent incorporation of an attachable, thermally-stable chromophore such as DADOH within a polyimide backbone could generate a material stable to short-duration, high-temperature excursions during processing. Such a material would be suitable for on-chip processing applications which require tolerance of semiconductor assembly conditions. Short-term stability could be improved also, because the polyimide could be cured at the recommended time and temperature, resulting in a more fully densified system with a higher T_g.

Summary

A series of chromophores based on the dicyanomethylenepyran acceptor group have been prepared and characterized. These donor-acceptor-donor analogs of DCM are most exciting because of their combination of properties, rather than any one. Most of these new compounds exhibit exceptional thermal stability, with thermal decomposition onset temperatures above 375 °C in nitrogen. Even when adjusting for the longer heat exposure times and reactive polyamic acid environment, these chromophores are expected to withstand polyimide cure and device processing temperatures above 300 °C, optimizing the thermal and mechanical properties of the host polymer. This has already been demonstrated for one of the chromophores, DADC, which has been used in a demonstration Mach-Zehnder modulator. Nonlinearities for the DAD-type chromophores are on the same order as for lower-stability common chromophores such as Disperse Red 1. The transparency of these compounds renders them suitable for devices operating at 830 nm. Functionalization for improved solubility has been demonstrated without significant loss in thermal stability, and a readily attachable variant has been generated. The DAD-type chromophores described are all photo-bleachable, enabling waveguide patterning in films with standard lithographic techniques. These new chromophores are promising and are expected to help further the development of fully cured polyimide electro-optic materials.

Experimental

All organic chemicals were purchased from Aldrich Chemical Company except as noted and were used as received. (2-(2-(4-(Dimethylamino)phenyl)ethenyl)-6-methyl-4*H*-pyran-4-ylidene)propanedinitrile (DCM) was obtained from Exciton (Dayton, OH). The preparation of (2,6-Bis(2-(4-(dimethylamino)phenyl)ethenyl)-4*H*-pyran-4-ylidene)-propanedinitrile (DAD) has been described previously (*42*). 4-(Ethyl(2-hydroxyethyl)-amino)benzaldehyde was synthesized by the method of Stenger-Smith and co-workers (*43*), and (2,6-dimethyl-4*H*-pyran-4-ylidene)propanedinitrile was prepared according to Woods (*44*). Melting points (mp) were determined in capillary tubes on a Mel-Temp II capillary melting point apparatus utilizing a digital thermometer and are uncorrected. Proton nuclear magnetic resonance (^{1}H NMR) spectra were obtained at room temperature (RT) on a Varian XL-300 NMR spectrometer using deuteriochloroform ($CDCl_3$) as solvent and tetramethylsilane (TMS) as an internal reference unless otherwise indicated. Chemical shifts are reported in parts per million (ppm). Fourier transform infrared (FT-IR) spectra were measured on a KVB/Analect RFX300 FT-IR spectrometer. Ultraviolet-visible (UV-vis) spectra were obtained on a Varian Cary 5E spectrometer. Elemental microanalyses were performed by Desert Analytics (Tucson, AZ).

9-Hexylcarbazole.
Sodium hydride (1.22 g, 30.5 mmol, 60% dispersion in mineral oil) was slowly added to a stirring solution of carbazole (5.00 g, 30 mmol) in dry, freshly distilled N,N-dimethylformamide (20 mL) at 0 °C. The reaction mixture was allowed to warm to RT, and 1-bromohexane (4.95 g, 30 mmol) was added dropwise. After stirring overnight, the reaction solvent was roto-evaporated, and the resulting crude material was redissolved in ethyl acetate (300 mL) and extracted with water (2 X 150 mL) and brine (1 X 100 mL). The organic phase was dried with anhydrous magnesium sulfate, filtered, and roto-evaporated. The crude solid was recrystallized from methanol to give the product as colorless needles, 4.51 g (67%): mp 63.3-64.2 °C (lit. (*45*) mp 66.8-67.5 °C).

9-Hexyl-3-carbazolecarboxaldehyde.
A stirring solution of 9-hexylcarbazole (4.43 g, 17.6 mmol) in dry, freshly distilled N,N-dimethylformamide (12 mL) at 0 °C was treated with phosphorus oxychloride (1.8 mL, 19.4 mmol, Mallinckrodt). The reaction solution was stirred at 90 °C overnight, then cooled to RT, poured into ice water (100 mL), neutralized with 5% aqueous sodium hydroxide, and extracted with dichloromethane (2 X 200 mL). The combined organic extracts were washed with brine (1 X 100 mL), dried with anhydrous magnesium sulfate, filtered, and roto-evaporated. Flash chromatography (7/3 dichloromethane/hexane) gave the product as a yellow oil, 3.91 g (79%).

1-Hexylindole.
Sodium hydride (1.71 g, 71.4 mmol, 60% dispersion in mineral oil) was slowly added to a stirring solution of indole (5.00 g, 42.7 mmol, Sigma) in dry, freshly distilled N,N-dimethylformamide (20 mL) at 0 °C. The mixture was stirred for one hour at 0 °C, then 1-bromohexane (7.05 g, 42.7 mmol) was added dropwise, and the reaction mixture was stirred overnight at RT. The reaction solvent was roto-evaporated, and water (100 mL) was added. The reaction mixture was extracted with ethyl acetate (2 X 150 mL) after neutralization with 5% aqueous sodium hydroxide. The combined organic extracts were washed with brine, dried with anhydrous magnesium sulfate, filtered, and roto-evaporated. Vacuum distillation gave the product as a water-white oil, 5.82 g (68%): bp 95-97 °C (0.06 mmHg); ^{1}H NMR ($CDCl_3$) δ 0.92 (t, $J = 6$ Hz, 3H, CH_3), 1.26-1.40 (m, 6H, aliphatic CH_2), 1.79-1.91 (m, 2H, $NCH_2C\underline{H_2}$), 4.12 (t, $J =$

7.4 Hz, 2H, NCH_2), 6.51-7.70 (m, 6H, aromatic H); Anal. Calcd for $C_{14}H_{19}N$: C, 83.53; H, 9.51; N, 6.96. Found: C, 83.60; H, 9.54; N, 6.65.

1-Hexyl-3-indolecarboxaldehyde.
A stirring solution of 1-hexylindole (5.69 g, 28.3 mmol) in dry, freshly distilled N,N-dimethylformamide (14 mL) was cooled to 0 °C, and phosphorus oxychloride (2.9 mL, 31.1 mmol, Mallinckrodt) was added dropwise. The reaction solution was stirred overnight at 90 °C, cooled to RT, added to ice water, and hydrolyzed with 5% aqueous sodium hydroxide. The aqueous mixture was extracted with dichloromethane (2 X 200 mL). The combined organic extracts were washed with water (1 X 100 mL) and brine (1 X 100 mL), dried with anhydrous magnesium sulfate, filtered, and roto-evaporated. Flash chromatography (dichloromethane) gave the product as a yellow oil, 0.80 g (12%).

5-(1-Pyrrolidinyl)-2-thiophenecarboxaldehyde.
A mixture of 5-bromo-2-thiophenecarboxaldehyde (25.00 g, 124 mmol), pyrrolidine (10.24 g, 144 mmol), anhydrous potassium carbonate (19.89 g, 144 mmol), and Aliquat 336 (0.5 mL) in dimethylsulfoxide (125 mL) was stirred mechanically under argon at 110 °C for 2 days. After cooling to RT, the reaction mixture was added to water (700 mL) and extracted with dichloromethane (3 X 125 mL). The combined organic extracts were washed with water (1 X 200 mL) and brine (1 X 100 mL), then dried with anhydrous magnesium sulfate, filtered, and roto-evaporated. The resulting crude solid was flash chromatographed (9/1 dichloromethane/ethyl acetate) and recrystallized from benzene/hexane to give the product as amber flakes, 8.97 g (40%): mp 104-106 °C (lit. (*46*) mp 102-106 °C).

(2,6-Bis(2-(9-ethyl-9*H*-carbazol-3-yl)ethenyl)-4*H*-pyran-4-ylidene)propanedinitrile (DADC). A solution of 9-ethyl-3-carbazolecarboxaldehyde (10.00 g, 44.8 mmol), (2,6-dimethyl-4*H*-pyran-4-ylidene)propanedinitrile (3.86 g, 22.4 mmol), and piperidine (2 mL) in toluene (100 mL) was refluxed overnight using a Dean-Stark trap. The reaction solution was cooled to RT, and the solvent was roto-evaporated. The crude solid was recrystallized twice from N,N-dimethylformamide/ethyl acetate to give the product as an orange solid, 1.39 g (11%): mp 337.6-338.4 °C; ^1H NMR ($CDCl_3$) δ 1.49 (t, J = 7.2 Hz, 6H, two $CH_2C\underline{H}_3$), 4.42 (q, J = 7.2 Hz, 4H, two $C\underline{H}_2CH_3$), 6.64 (s, 2H, two pyran ring CH), 6.79 and 7.77 (two d, J = 15.8 Hz, 2H each, trans double bond CH), 7.30-8.34 (m, 14H, aromatic H); FT-IR (KBr) 2203 and 2188 cm^{-1} (CN stretch); UV-vis λ_{max} (NMP) 459 nm ($\varepsilon = 7.15 \times 10^4$); Anal. Calcd for $C_{40}H_{30}N_4O$: C, 82.45; H, 5.19; N, 9.61. Found: C, 82.34; N, 4.98; N, 9.34.

(2,6-Bis(2-(4-(dibutylamino)phenyl)ethenyl)-4*H*-pyran-4-ylidene)propanedinitrile (DADB). Prepared as above from 4-(dibutylamino)benzaldehyde in 37% yield. Two recrystallizations from toluene gave the product as a red-orange solid: mp 249.6-250.4 °C; FT-IR (KBr) 2207 cm^{-1} (CN stretch); UV-vis λ_{max} (NMP) 500 nm ($\varepsilon = 6.63 \times 10^4$); Anal. Calcd for $C_{40}H_{50}N_4O$: C, 79.69; H, 8.36; N, 9.29. Found: C, 79.91; H, 8.28; N, 9.37.

(2,6-Bis(2-(1-methylindol-3-yl)ethenyl)-4*H*-pyran-4-ylidene)propanedinitrile (DADI). Prepared as above from 1-methylindole-3-carboxaldehyde in 84% yield. Recrystallization from N,N-dimethylacetamide/water gave the product as red needles: mp >300 °C; ^1H NMR (DMSO-d_6) δ 3.88 (s, 6H, two NCH_3), 6.91 (s, 2H, pyran ring CH), 7.19 and 7.97 (two d, J = 16 Hz, 2H each, trans double bond CH), 7.99 (s, 1H, indole ring CH), 7.22-8.28 (m, 8H, aromatic H); FT-IR (KBr) 2202 and 2187 cm^{-1} (CN stretch); UV-vis λ_{max} (NMP) 465 nm ($\varepsilon = 6.27 \times 10^4$); Anal. Calcd for $C_{30}H_{22}N_4O$: C, 79.28; H, 4.88; N, 12.33. Found: C, 79.38; H, 4.75; N, 12.30.

(2,6-Bis(2-(4-(ethyl(2-hydroxyethyl)amino)phenyl)ethenyl)-4*H*-pyran-4-ylidene)propanedinitrile (DADOH). A solution of 4-(ethyl(2-hydroxyethyl)amino)benzaldehyde (3.43 g, 17.8 mmol), (2,6-dimethyl-4*H*-pyran-4-ylidene)propanedinitrile (1.53 g, 8.89 mmol), piperidine (2 mL), and glacial acetic acid (1 mL) in toluene (100 mL) was refluxed overnight using a Dean-Stark trap. The reaction solution was cooled to RT, and the solvent was roto-evaporated. Flash chromatography (9/1 dichloromethane/ethyl acetate) and recrystallization from toluene gave the product as a dark brown powder, 0.75 g (16%): mp 222.5-224.5 °C; ^{1}H NMR ($CDCl_3$) δ 1.20 (t, *J* = 6.8 Hz, 6H, two $NCH_2C\underline{H}_3$), 1.74 (t, *J* = 5.5 Hz, 2H, two OH), 3.49 (q, *J* = 6.8 Hz, 4H, two $NC\underline{H}_2CH_3$), 3.54 (t, *J* = 5.9 Hz, 4H, two $NC\underline{H}_2CH_2OH$), 3.83 (dt, *J* = 5.5 and 5.9 Hz, 4H, two $NCH_2C\underline{H}_2OH$), 6.43 and 7.41 (two d, *J* = 15.5 Hz, 2H each, trans double bond CH), 6.49 (s, 2H, two cyclohexene ring CH), 6.74 and 7.41 (two d, *J* = 8.7 Hz, 4H each, aromatic H); FT-IR (KBr) 2204 and 2189 cm^{-1} (CN stretch); UV-vis λ_{max} (NMP) 499 nm ($\varepsilon = 7.05 \times 10^4$); Anal. Calcd for $C_{32}H_{34}N_4O_3$: C, 73.54; H, 6.56; N, 10.72. Found: C, 73.08; H, 6.42; N, 10.54.

(2,6-Bis(2-(5-(1-pyrrolidinyl)thien-2-yl)ethenyl)-4*H*-pyran-4-ylidene)propanedinitrile (DADT). A solution of 5-(1-pyrrolidinyl)-2-thiophenecarboxaldehyde (1.45 g, 8.00 mmol), (2,6-dimethyl-4*H*-pyran-4-ylidene)propanedinitrile (0.69 g, 4.00 mmol), piperidine (2 mL), and glacial acetic acid (1 mL) in benzene (100 mL) was refluxed overnight under argon using a Dean-Stark trap. The reaction mixture was cooled to RT and filtered. The crude product was washed with benzene, then recrystallized from N,N-dimethylformamide to give the product as brown crystals, 1.14 g (57%): mp >350 °C; FT-IR (KBr) 2180 and 2196 cm^{-1} (CN stretch); UV-vis λ_{max} (NMP) 603 nm ($\varepsilon = 4.75 \times 10^4$), 529 nm ($\varepsilon = 6.33 \times 10^4$), 432 nm ($\varepsilon = 2.46 \times 10^4$); Anal. Calcd for $C_{28}H_{26}N_4OS_2$: C, 67.44; H, 5.26; N, 11.24; S, 12.86. Found: C, 67.34; H, 5.38; N, 11.42; S, 12.62.

(2,6-Bis(2-(9-hexyl-9*H*-carbazol-3-yl)ethenyl)-4*H*-pyran-4-ylidene)propanedinitrile (DADCH). A solution of 9-hexyl-3-carbazolecarboxaldehyde (3.76 g, 13.4 mmol), (2,6-dimethyl-4*H*-pyran-4-ylidene)propanedinitrile (1.15 g, 6.7 mmol), and piperidine (2 mL) in toluene (25 mL) was refluxed overnight using a Dean-Stark trap. The reaction solution was cooled to RT, and the precipitated crude solid was filtered and recrystallized from toluene to give the product as dark orange crystals, 1.63 g (35%): mp 226.4-226.9 °C; ^{1}H NMR ($CDCl_3$) δ 0.88 (t, *J* = 6.5 Hz, 6H, two CH_3), 1.23-1.46 (m, 12H, aliphatic CH_2), 1.82-1.94 (m, 4H, aliphatic CH_2), 4.31 (t, *J* = 7.2 Hz, 4H, two NCH_2), 6.49 (s, 2H, two pyran ring CH), 6.69 and 7.67 (two d, *J* = 16 Hz, 2H each, four trans double bond CH), 7.28-8.28 (m, 14H, aromatic H); FT-IR (KBr) 2206 cm^{-1} (CN stretch); UV-vis λ_{max} (NMP) 460 nm ($\varepsilon = 6.98 \times 10^4$); Anal. Calcd for $C_{48}H_{46}N_4$: C, 82.97; H, 6.67; N, 8.06. Found: C, 82.58; H, 6.52; N, 8.29.

(2,6-Bis(2-(1-hexylindol-3-yl)ethenyl)-4*H*-pyran-4-ylidene)propanedinitrile (DADIH). A solution of 1-hexyl-3-indolecarboxaldehyde (0.80 g, 3.47 mmol), (2,6-dimethyl-4*H*-pyran-4-ylidene)propanedinitrile (0.30 g, 1.74 mmol), and piperidine (1 mL) in toluene (25 mL) was refluxed for 5 hours using a Dean-Stark trap. The reaction solution was cooled to RT, roto-evaporated, redissolved in dichloromethane, passed through a short bed of silica, and roto-evaporated again. The resulting solid was recrystallized from toluene to give the product as a red crystalline solid, 0.64 g (62%): mp 194.3-195 °C; ^{1}H NMR ($CDCl_3$) δ 0.91 (t, *J* = 6 Hz, 6H, two CH_3), 1.26-1.44 (m, 12H, aliphatic CH_2), 1.85-1.98 (m, 4H, two $NCH_2C\underline{H}_2$), 4.20 (t, *J* = 7 Hz, 4H, two NCH_2), 6.57 (s, 2H, two cyclohexene ring CH), 6.72 (d, *J* = 16 Hz, 2H, two trans double bond CH), 7.30-7.47 (m, 6H, aromatic H), 7.52 (s, 2H, two indole ring C2 H), 7.76 (d, *J* = 16 Hz, 2H, two trans double bond CH), 7.98 (dd, *J* = 6.3 and 2 Hz, 2H, aromatic H); FT-IR (KBr) 2203 cm^{-1} (CN stretch); UV-vis λ_{max} (NMP) 466 nm (ε = 6.22 X

10^4); Anal. Calcd for $C_{40}H_{42}N_4O$: C, 80.78; H, 7.11; N, 9.42. Found: C, 80.47; H, 6.90; N, 9.25.

Acknowledgments

Portions of this work have been funded by Lockheed Independent Research and Development and by John Zetts of Air Force Wright Laboratory (WL/MLPO), whose support and interest in this technology is greatly appreciated. We thank S. Marder and S. Gilmour (both of CIT/JPL) for advice related to the preparation of the thiophene-containing donor aldehyde, M.M. Steiner and J. Zegarski of the Lockheed Chemistry Department for thermal analyses and FT-IR characterization of the materials, and L.-T. Cheng (du Pont) for EFISH measurements. Finally, we acknowledge members of the Lockheed Photonics group, including D.G. Girton, W. Anderson, T.E. Van Eck, and J. Marley, for continuing discussion and guidance which have been invaluable to the development of chromophores compatible with device requirements.

Literature Cited

1. Wu, J.W.; Valley, J.F.; Ermer, S.; Binkley, E.S.; Kenney, J.T.; Lipscomb, G.F.; Lytel, R. *Appl. Phys. Lett.* **1991**, *58*, 225-227.
2. Wu, J.W.; Binkley, E.S.; Kenney, J.T.; Lytel, R.; Garito, A.F. *J. Appl. Phys.* **1991**, *69,* 7366-7368.
3. Ermer, S.; Kenney, J.T.; Wu, J.W.; Valley, J.F.; Lytel, R.; Garito, A.F. *ACS Polymer Preprints* **1991**, *32(3)*, 92-93.
4. Matsuura, T.; Hasuda, Y.; Nishi, S.; Yamamoto, F. *Macromolecules* **1991**, *24*, 5001-5005.
5. Takezawa, Y.; Taketani, N.; Tanno, S.; Ohara, S. *J. Polym. Sci., Part B: Polym. Phys.* **1992**, *30*, 879-885.
6. Ermer, S.; Valley, J.F.; Lytel, R.; Lipscomb, G.F.; Van Eck, T.E.; Girton, D.G. *Appl. Phys. Lett.* **1992**, *61(19)*, 2272-2274.
7. Girton, D.G.; Anderson, W.W.; Valley, J.F.; Van Eck, T.E.; Dries, L.J.; Marley, J.A.; Ermer, S., this volume.
8. Girton, D.G.; Ermer, S.; Valley, J.F.; Van Eck, T.E.; Lovejoy, S.M.; Leung, D.S.; Marley, J. *ACS Polymer Preprints* **1994**, *35(2)*, 219-220.
9. Neuhaus, H.J.; Day, D.R.; Senturia, S.D. *J. Electron. Mat.* **1985**, *14*, 379-404.
10. Smith, F.W.; Neuhaus, H.J.; Senturia, S.D.; Feit, Z.; Day, D.R.; Lewis, T.J. *J. Electron. Mat.* **1987**, *16*, 93-106.
11. Lytel, R.; Lipscomb, G.F.; Binkley, E.S.; Kenney, J.T.; Ticknor, A.J. In *Materials for Nonlinear Optics: Chemical Perspectives*, Marder, S.R., Sohn, J.E., and Stucky, G.D., Eds.; ACS Symposium Series 455; American Chemical Society: Washington, DC, 1991; pp 103-112.
12. Fujimoto, H.H.; Das, S.; Valley, J.F.; Stiller, M.; Dries, L.; Girton, D.; Van Eck, T.; Ermer, S.; Binkley, E.S.; Nurse, J.C.; Kenney, J.T. In *Electrical, Optical, and Magnetic Properties of Organic Solid State Materials*, Garito, A.F., Jen, A.K.-Y., Lee, C.Y.-C., and Dalton, L.R., Eds.; Materials Research Society Symposium Proceedings 328; Materials Research Society: Pittsburgh, PA, 1993; pp 553-564.
13. Liu, L.-Y.; Ramkrishna, D.; Lackritz, H.S. *Macromolecules* **1994**, *27*, 5987-5999.
14. Valley, J.F.; Wu, J.W.; Ermer, S.; Stiller, M.; Binkley, E.S.; Kenney, J.T.; Lipscomb, G.F.; Lytel, R. *Appl. Phys. Lett.* **1992**, *60*, 160-162.
15. Stähelin, M.; Burland, D.M.; Ebert, M.; Miller, R.D.; Smith, B.A.; Twieg, R.J.; Volksen, W.; Walsh, C.A. *Appl. Phys. Lett.* **1992**, *61*, 1626-1628.
16. Twieg, R.J.; Burland, D.M.; Hedrick, J.; Lee, V.Y.; Miller, R.D.; Moylan, C.R.; Seymour, C.M.; Volksen, W.; Walsh, C.A. In *Organic, Metallo-organic, and*

Polymeric Materials for Nonlinear Optical Applications, Marder, S.R. and Perry, J.W., Eds.; Proc. SPIE 2143; Society of Photo-Optical Instrumentation Engineers: Bellingham, WA, 1994; pp 2-13.
17. Moylan, C.R.; Miller, R.D.; Twieg, R.J.; Betterton, K.M.; Lee, V.Y.; Matray, T.J.; Nguyen, C. *Chem. Mater.* **1993**, *5*, 1499-1508.
18. Cahill, P.A.; Seager, C.H.; Meinhardt, M.B.; Beuhler, A.J.; Wargowski, D.A.; Singer, K.D.; Kowalczyk, T.C.; Kosc, T.Z. In *Nonlinear Optical Properties of Organic Materials VI*, Möhlmann, G.R., Ed.; Proc. SPIE 2025; Society of Photo-Optical Instrumentation Engineers: Bellingham, WA, 1993; pp 48-55.
19. Shi, R.F.; Wu, M.H.; Yamada, S.; Cai, Y.M.; Garito, A.F. *Appl. Phys. Lett.* **1993**, *63*, 1173-1175.
20. Ermer, S.; Leung, D.; Lovejoy, S.; Valley, J.; Stiller, M. In *Organic Thin Films for Photonic Applications Technical Digest, Volume 17*; Optical Society of America: Washington, DC, 1993; pp 50-53.
21. Yamamoto, H.; Katogi, S.; Watanabe, T.; Sato, H.; Miyata, S.; Hosomi, T. *Appl. Phys. Lett.* **1992**, *60*, 935-937.
22. Girton, D.; Ermer, S.; Valley, J.F.; Van Eck, T.E. In *Organic Thin Films for Photonic Applications Technical Digest, Volume 17*; Optical Society of America: Washington, DC, 1993; pp 70-72.
23. Hammond, P.R. *Optics Comm.* **1979**, *29(3)*, 331-333.
24. Katz, H.E.; Dirk, C.W.; Singer, K.D.; Sohn, J.E. In *Advances in Nonlinear Polymers and Inorganic Crystals, Liquid Crystals, and Laser Media*, Musikant, S., Ed.; Proc. SPIE 824; Society of Photo-Optical Instrumentation Engineers: Bellingham, WA, 1987; pp 86-92.
25. Jen, A.K.-Y.; Rao, V.P.; Wong, K.Y.; Drost, K.J. *J. Chem. Soc., Chem. Commun.* **1993**, 90-92.
26. Rao, V.P.; Jen, A.K.-Y.; Wong, K.Y.; Drost, K.J. *J. Chem. Soc., Chem. Commun.* **1993**, 1118-1120.
27. Lemke, R. *Chem. Ber.* **1970**, *103*, 1894-1899.
28. Mignani, G; Soula, G.; Meyrueix, R. Fr. Patent 2 636 441, 1990; *Chem. Abstr.* **1990**, *113*, 181123n.
29. Gadret, G; Kajzar, F.; Raimond, P. In *Nonlinear Optical Properties of Organic Materials IV*, Singer, K.D., Ed.; Proc. SPIE 1560; Society of Photo-Optical Instrumentation Engineers: Bellingham, WA, 1991; pp 226-237.
30. Stenger-Smith, J.D.; Henry, R.A.; Chafin, A.P.; Lindsay, G.A. *ACS Polymer Preprints* **1994**, *35(2)*, 140-141.
31. Ermer, S.; Girton, D.G.; Leung, D.S.; Lovejoy, S.M.; Valley, J.F.; Van Eck, T.E.; Cheng, L.-T. *Nonlinear Optics* **1994**, in press.
32. Marder, S.R.; Beratan, D.N.; Cheng, L.-T. *Science* **1991**, *252*, 103-106.
33. Marder, S.R.; Perry, J.W.; Tiemann, B.G.; Gorman, C.B.; Gilmour, S.; Biddle, S.; Bourhill, G. *J. Am. Chem. Soc.* **1993**, *115*, 2524-2526.
34. Dirk, C.W.; Katz, H.E.; Schilling, M.L.; King, L.A. *Chem. Mater.* **1990**, *2*, 700-705.
35. Lin, J.T.; Hubbard, M.A.; Marks, T.J.; Lin, W.; Wong, G.K. *Chem. Mater.* **1992**, *4*, 1148-1150.
36. Zysset, B.; Ahlheim,. M.; Stähelin, M.; Lehr, F.; Prêtre, P.; Kaatz, P.; Günter, P. In *Nonlinear Optical Properties of Organic Materials VI*, Möhlmann, G.R., Ed.; Proc. SPIE 2025; Society of Photo-Optical Instrumentation Engineers: Bellingham, WA, 1993; pp 70-77.
37. Meyrueix, R.; Tapolsky, G.; Dickens, M.; Lecomte, J.-P. In *Nonlinear Optical Properties of Organic Materials VI*, Möhlmann, G.R., Ed.; Proc. SPIE 2025; Society of Photo-Optical Instrumentation Engineers: Bellingham, WA, 1993; pp 117-128.

38. Becker, M.W.; Sapochak, L.S.; Ghosen, R.; Xu, C.; Dalton, L.R.; Shi, Y.; Steier, W.H.; Jen, A.K.-Y. *Chem. Mater.* **1994**, *6*, 104-106.
39. Sotoyama, W.; Tatsuura, S.; Yoshimura, T. *Appl. Phys. Lett.* **1994**, *64*, 2197-2199.
40. Jen, A.K.-Y.; Drost, K.J.; Rao, V.P.; Cai, Y.; Liu, Y.-J.; Mininni, R.M.; Kenney, J.T.; Binkley, E.S.; Marder, S.R.; Dalton, L.R.; Xu, C. *ACS Polymer Preprints* **1994**, *35(2)*, 130-131.
41. Yu, D.; Yu, L. *ACS Polymer Preprints* **1994**, *35(2)*, 132-133.
42. Ermer, S.; Valley, J.F.; Lytel, R.; Lipscomb, G.F.; Van Eck, T.E.; Girton, D.G.; Leung, D.S.; Lovejoy, S.M. In *Organic and Biological Optoelectronics*, Rentzepis, P.M., Ed.; Proc. SPIE 1853; Society of Photo-Optical Instrumentation Engineers: Bellingham, WA, 1993; pp 183-192.
43. Stenger-Smith, J.D.; Fischer, J.W.; Henry, R.A.; Hoover, J.M.; Nadler, M.P.; Nissan, R.A.; Lindsay, G.A. *J. Polym. Sci., Part A: Polym. Chem.* **1991**, *29*, 1623-1631.
44. Woods, L.L. *J. Am. Chem. Soc.* **1958**, *80*, 1440-1442.
45. Hopfinger, A.; Cislo, J.; Zielinska, E. *Rocz. Chem.* **1968**, *42(5)*, 823-828; *Chem. Abstr.* **1969**, *70*, 3729a.
46. Hartmann, H.; Scheithauer, S. *J. Prakt. Chem.* **1969**, *311*, 827-843.

RECEIVED January 30, 1995

Chapter 8

Thermally Stable Nonlinear Optical Moiety-Doped Polyimides for Photonic Devices

H. K. Kim[1], H. J. Lee[2], M. H. Lee[2], S. K. Han[2], H. Y. Kim[2], K. H. Kang[2], Y. H. Min[2], and Y. H. Won[2]

[1]Department of Macromolecular Science, Han Nam University, 133 Ojung-Dong Daeduck-Gu, Taejon, Korea 300–791
[2]Photonic Switching Section, ETRI, P.O. Box 106, Yusung, Taejon, Korea 305-600

New NLO moiety-doped polyimides as host-guest systems were developed. NLO moieties based on both dialkyl amino alkyl sulfone stilbenes (DASS) and dialkyl amino nitro stilbenes (DANS) were synthesized as guest chromophores. Owing to the newly introduced flexible alkyl chains to the NLO chromophore terminal groups, the compatibility between guest chromophores and host polyamic acid in NMP solvent was improved and the solubility of the alkylated-NLO moieties to the host polyimide was increased up to the 40 wt %. These polyimide-based guest-host systems exhibited a significant improvement in the thermal stability at high temperatures exceeding 250 °C by the TGA. The electro-optic coefficient at 632.8 nm is 13 pm/V for the 40 wt % DASS-doped polyimide system poled at the 135 V/mm. These new materials are promising candidates for photonic switching devices with low operating voltage.

Recently, poled NLO (or EO) polymeric materials have been considered potential candidates for high speed integrated electro-optic devices such as Mach-Zehnder modulators and directional couplers[1,2]. Compared to inorganic materials, these poled EO polymers offer a number of advantages, such as the subpicosecond response time, the lower power dissipation, ease of processability as thin film with semiconductor technologies[1]. It is strongly noted that poled EO polymeric materials must retain thermal stability and EO thermal stability at both manufacturing and end-use environments, which require long-term stability up to 125 °C and short excursions to the temperature of 300 °C or higher. In many efforts to obtain such materials, two types of poled EO polymers such as side chain EO polymers and guest-host systems were intensively studied. The side chain EO polymer systems are composed of the organic molecule units which are covalently bound to polymer backbones[3,4]. In the guest-host system, the EO organic molecules (guest) are dissolved into a polymer matrix. Much attention has been paid on the development of polyimide host-guest systems[5,6]. Our main material efforts have been focussed on improving the electro-

0097–6156/95/0601–0111$12.00/0

optic coefficient of materials by increasing the solubility of EO moieties in a polyimide as well as improving EO thermal stability with high glass transition temperature and no sublimation during curing or poling process. For this purpose, we have chemically modified the EO moieties based on both dialkylamino alkylsulfone stilbenes (DASS) and dialkylamino nitro stilbenes (DANS), yielding new alkylated-EO moieties as guest chromophores. Furthermore, new NLO moiety-doped polyimides as host-guest systems are being demonstrated in making multilayer-stack poled photonic devices.

II. EXPERIMENTAL

1. Synthesis of 4-(diethylamino)-4'-(hexyl sulfonyl) stilbene (DASS 62)

(1) Preparation of 4-Hexyl-sulfonyl toluene

A two-phase system composed of 1-bromohexane (130 g, 0.79 mol) in toluene (200 mL) and p-toluene sulfonic acid sodium salts (100 g, 0.56 mol) and 10 g of tetrabutylammonium bromide in water (200 mL) was stirred overnight at 80 ^{o}C. After the reaction system was cooled to room temperature, water (400 mL) was added, and the organic layer was extracted three times with toluene. The extractor was concentrated after dried with anhydrous $MgSO_4$. The product was distilled in vacuo to produce 106 g (79 %, yield) of a viscous liquid. bp 130 ^{o}C/0.1 mmHg: ^{1}H-NMR ($CDCl_3$) d 0.80 (t, 3H), 1.23 (m, 6H), 1.66 (m, 2H), 2.41 (s, 3H), 3.03 (t, 2H), 7.34 (d, 2H), 7.76 (d, 2H).

(2) Preparation of 4-Hexyl-sulfone benzyl bromide.

4-Hexylsulfonyl toluene (34.8 g, 0.145 mol) was dissolved in 200 mL of CCl_4 and refluxed under nitrogen atmosphere. NBS (25.81 g, 0.145 mol) and benzoyl peroxide (0.4 g) were added to the solution at reflux for 12 h. The reaction mixture was cooled and filtered off the succimide. The filterate washed with water and dried with $MgSO_4$. The product mixture was concentrated after the solvent was removed, the residue was a mixture of 30 % unreacted starting material and 70 % brominated product and then used without further purification for later reaction. ^{1}H-NMR ($CDCl_3$) d 0.75 (t, 3H), 1.24 (m, 6H), 1.64 (m, 2H), 3.04 (t, 2H), 4.42 (s, 2H), 7.51 (d, 2H), 7.80 (d, 2H).

(3) Preparation of 4-Hexyl-sulfone benzyl phosphonate.

The distilled triethyl phosphite (8.3 g, 0.05 mol) was heated at reflux, and 13 g (0. 04 mol) of 4-hexyl-sulfone benzyl bromide was added dropwise while stirring, at such a rate that a gentle reflux was maintained. When the addition was completed, the reaction was refluxed for an additional 2 h. The mixture was cooled to 25 ^{o}C, and the product mixture was distilled off the volatile materials. The crude viscous liquid was isolated by cromatography (eluent ethyl acetate : hexane, 5 : 1). 68 % yield. ^{1}H-NMR ($CDCl_3$) d 0.82 (t, 3H), 1.26 (m, 12H), 1.66 (m, 2H), 3.04 (t, 2H), 3.26 (d, 2H), 4.08 (q, 4H), 7.49 (d, 2H), 7.83 (d, 2H).

(4) Preparation of 4-(diethylamino)-4'-(hexyl sulfonyl) stilbene (DASS 62).

In a dry 250 mL three-necked, round-bottomed flask equipped with a sealed stirrer, a nitrogen inlet, and a reflux condenser topped with a Drierite tube were placed, under nitrogen, 90 mL of 1,2-dimethoxy ethane (DME) and sodium hydride (2.4 g, 0.06 mol,

60 % dispersion in mineral oil). The suspension was stirred for 1 min. To the suspension was added 7.08 g (0.04 mol) of N,N-diethylamino benzaldehyde, and the mixture was stirred for 5 min to ensure complete dissolution. To this solution was added 15 g (0.04 mol) of 4-hexyl-sulfone benzyl phosphonate in 30 mL DME, and the reaction mixture was refluxed with vigorous stirring for 5 h. The bright yellow solution was poured over 200 g of crushed ice. And the solid thus formed was washed into the ice mixture. The yellow solid was collected by filtration, washed with cold water, and air dried. Recrystallization from ethanol / methylene chloride yielded 7.8 g (49 %) of yellow crystals: mp 96 ^{o}C ; ^{1}H-NMR ($CDCl_3$) d 0.80 (t, 3H), 1.17 (t, 6H), 1.23 (m, 6H), 1.67 (m, 2H), 3.05 (t, 2H), 3.36 (q, 4H), 6.66 (d, 2H), 6.90 (d, 1H), 7.20 (d, 1H), 7.41 (d, 2H), 7.59 (d, 2H), 7.82 (d, 2H). ^{13}C-NMR ($CDCl_3$) d 12.6, 13.9, 22. 3, 22.7, 27.9, 31.2, 44.4, 56.5, 111.5, 121.2 123.5, 126.1, 128.4, 128.5, 132.9, 136.0, 143.9, 149.1.

2. Synthesis of 4-(Dihexylamino)-4'-nitrostilbene (DANS 6)

(1) Preparation of 4-(dihexylamino)benzaldehyde.

A mixture of 25 g (0.2 mol), 37 g (0.2 mol) of dihexylamine, 21.3 g (0.2 mol) of sodium carbonate, 150 mL of dimethylsulfoxide, and 0.5 g of distearyldimethyl-ammonium chloride was heated under nitrogen while stirring at 110 ^{o}C for 113 h. The reaction mixture was cooled and poured into 1.5 L of water, and the resulting solution was extracted with ethyl ether (4 x 300 mL). The combined extracts were washed with water and dried with anhydrous $MgSO_4$, and the solvent was removed at reduced pressure. The residue was fractionally distilled in vacuo to provide 32.4 g (56 %) of a pale yellow oil : bp 162 - 185 ^{o}C/0.1mmHg ; ^{1}H-NMR ($CDCl_3$) d 0.89 (t, 6H), 1.23 (m, 12H), 1.60 (m, 4H), 3.31 (t, 4H), 6.63 (d, 2H), 7.69 (d, 2H), 9.67 (s, 1H).

(2) Preparation of 4-(Dihexylamino)-4'-nitrostilbene (DANS 6)

A mixture of 12.1 g (0.042 mol) of 4-(dihexylamino)benzaldehyde, 8.2 g (0.045 mol) of 4-nitrophenylacetic acid, 3.57 g (0.042 mol) of piperidine and 100 mL of xylene was heated while stirring at reflux or 20 h with continuous removal of water by using a Dean-Stark apparatus. Approximately half of the xylene was distilled, 30 mL of heptane was added, and the residue was cooled to -30 ^{o}C. Red crystals separated which were collected and recrystallized from toluene / heptane to yield 4.8 g (28 %) of a red solid: mp 82 ^{o}C ; ^{1}H-NMR ($CDCl_3$) d 0.89 (t, 6H), 1.30 (m, 12H), 1. 58 (m, 4H), 3.28 (t, 4H), 6.62 (d, 2H), 6.90 (d, 1H), 7.21 (d, 1H), 7.40 (d, 2H), 7.54 (d, 2H), 8.17 (d, 2H). ^{13}C-NMR ($CDCl_3$) d 14.0, 22.7, 26.8, 27.3, 31.7, 51.0, 111.5, 120. 8, 123.4, 124.1, 125.9, 128.6, 133.8, 145.2, 145.7, 148.9.

3. Thin film preparation

Guest chromophores (DASS-62 & DANS-6) with the content of 20 wt % and 40 wt % were mixed in Amoco Ultradel 3112/4212 polyimides (U-PI 3112/4212) or Hitachi isoindoroquinazorin-dione polyimide (PIQ-2200) using a magnetic stirrer. The solutions in all cases were filtered using a 0.45 μm teflon membrane filter. Thin films were prepared by spin-coating the solution on various substrates such as a silicon wafer, indiumtinoxide (ITO) coated glass, quartz and NaCl discs. The thin films were soft-baked for 5 minutes at 120 °C and imidized (cured) for several hours at various temperatures. The thin films were analyzed by IR (Bomem MB-100) and UV spectroscopy (Shimadzu UV-3100S). The film thickness was measured by a surface profilometer (Tencor instruments, Alpha Step 300).

4. Thermal stability

The thermal stability of guest chromophore-doped polyimide host-guest systems were studied by thermal gravimetric analysis (TGA) (Dupont TGA 9900). The weight loss as a function of temperature was followed by TGA at 10 °C/minute under a nitrogen purge of 50 cc/minute.

5. Optical property measurements

The refractive index of guest chromophore-doped polyimide host-guest systems at the wavelength of 632.8 nm or 1300 nm was measured with the prism coupling method. The He-Ne laser beam (or the laser diode) was coupled into and out of the polymer films spin-coated on Si-wafers with a Gadolinium Gallium Garnet (GGG) prism. The electro-optic coefficients of samples were measured using the simple reflection technique reported by Teng and Man[7].

III. RESULTS AND DISCUSSION

Host-guest system design:

The EO moiety such as dimethylaminonitrostilbene (DANS-1), Disperse Red (DR-1), etc., was iniatially mixed into poly(methylmethacrylate) (PMMA) as a host polymer[8,9]. These host-guest systems had limited thermal stability and limited miscibility. The concentration of the EO moiety was kept below 15 %. Recently, it was reported that polyimide based-host guest systems were suitable for multilayer-stack poled polymer electro-optic devices[5,6]. The host polymers in these systems used Amoco Ultradel 3112/4212 polyimides (U-PI 3112/4212) and Hitachi PIQ-2200 polyimide isoindoroquinazorindione (PIQ-2200). These polyimides exhibited thermal stability to withstand electronic processing conditions without structural deterioration and optical transparency at the operating wavelength of the devices. However, the thermal stability of the EO moiety and the compatibilty of the EO moiety with polyimides are still needed to be improved.

We have synthesized long alkylated diethylamino hexylsulfone stilbene (DASS-62) and dihexylaminonitro stilbene (DANS-6) as guest chromophores to increase the solubility of EO moieties in a polyimide and to enhance the electro-optic coefficient of the materials as well as improve the EO termal stability with neither sublimation nor decomposition during curing or poling process. The main synthetic routes of these long alkylated guest chromophores are shown in Scheme 1.

The relative EO coefficient for EO moieties and the maximum solubility of various DANS and DASS moieties mixed into either an U-PI 3112/4212 or a PIQ-220 polyimide are summarized in Table 1. The similar values of $\beta\mu_g$ were theoretically obtained for DANS and DASS moieties, where β is the hyperpolarizability (λ= 1907nm) and μ_g is the ground state dipole moment[10,11]. When long alkyl groups were obtained for DANS and DASS moieties, where β is the hyperpolarizability (λ= 1907nm) and μ_g is the ground state dipole moment[10,11]. When long alkyl groups were introduced into these EO moieties, the solubility of long alkylated EO moieties was experimentally enhanced up to the 50 weight percent of the active EO moieties. By increasing the solubility of the active EO moieties in the polyimide host, these long alkylated EO moiety-doped polyimide systems exhibited the higher overall-EO coefficient, as estimated in relative r*. The UV-Vis spectroscopy shows that the π–π* absorption bands of DASS-62 and DANS-6 are centered around 370 nm and around 450 nm, respectively.

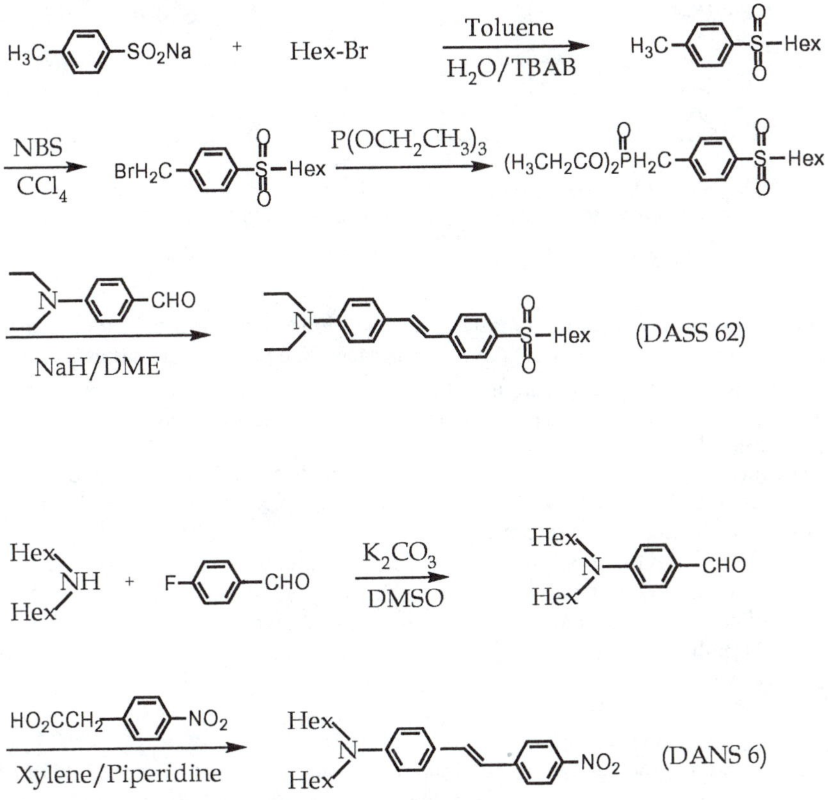
H3C–⟨⟩–SO2Na + Hex-Br
Toluene
H2O/TBAB
H3C–⟨⟩–S(O2)–Hex
NBS
CCl4
BrH2C–⟨⟩–S(O2)–Hex
P(OCH2CH3)3
(H3CH2CO)2PH2C–⟨⟩–S(O2)–Hex
N–⟨⟩–CHO
NaH/DME
S–Hex
(DASS 62)
Hex
NH
F–⟨⟩–CHO
K2CO3
DMSO
CHO
HO2CCH2–⟨⟩–NO2
Xylene/Piperidine
NO2
(DANS 6)

Scheme 1. Synthetic Routes of DASS-62 and DANS-6 as Guest Chromophores.

The curing condition and thermal stability of guest chromophore-doped polyimide systems were studied by TGA. Polyimide PIQ-2200 has two distinct stages in the curing process with different temperature ranges. The first stage is the imidization stage in which the polyamic acid condenses to form the imide rings at the temperature of lower than 200 °C and the second is a densification stage in which the polyimide iminizes to form a bicyclic analogue at the temperature of 210 °C (for at least 4 hrs).

Table 1. Relative EO Coefficient (r*) for NLO Moieties.

NLO moiety (M.W.)	$\beta \mu g$ (λ = 1907nm)	max. sol. (wt %)	#relative r* (r/r_{DANS-1})
DANS-1 (268)	135	15	1
DANS-6 (408)	135	30	1.30
DASS-11 (301)	113	15	0.84
DASS-62 (399)	113	50	1.87

#The relative EO coefficient (r*) was estimated as a standard of r_{DANS-1} which is calculated from the following equation:

$$r \propto \beta \mu g \ / \ (M.W.) \ \times \ \text{maximum solubility (wt \%)} \ / \ 100.$$

However, U-PI 3112/4212 has one-step process of imidization at 200 °C for at least 3 hours. IR spectroscopy indicates that the characteristic peaks of the carbonyl groups in polyamic acid disappeared around 1660 cm^{-1} and two new peaks corresponding to the characteristic peaks of the carbonyl group in polyimide appeared around 1775 cm^{-1} and around 1723 cm^{-1}.

The thermal stability of these host-guest systems is summarized in Table 2. TGA traces show that 40 wt% DASS-62 and 40 wt% DANS-6 in polyimides are all thermally stable at high temperature exceeding 250 °C. Practically, however, 40 wt% DASS-62 in a PIQ-220 polyimide is thermally stable up to 210 °C and 40 wt% DANS-6 in a PIQ-220 polyimide was thermally stable up to 190 °C with no sublimation. Experimentally, the glass transition temperature (T_g) was estimated during the poling process. The T_g of 20 wt% DASS-62 is ~ 170°C and that of 40 wt% DASS-62 is ~ 160 °C.

Refractive index measurement

The refracrive indices for the polyimide host-guest systems at both 632.8 nm and 1300 nm were measured with the prism coupler. The laser light was coupled into and out of the polymer films spin-coated on Si-wafers with a Gadolinium Gallium Garnet (GGG) prism. In this topology, the Si wafer acts as the substrate with a higher refractive index than that of the polymer film. As a result, the polymer film forms a leaky quasi-waveguide, since the light is guided by the film-air interface while the reflection at the film-substrate interface is leaky[12,13]. The beam is totally reflected except for the coupling conditions to waveguiding modes of both TM and TE modes which determine the guiding mode angles. A calculation based on the theory given by Tien et. al.[14] is used to determine the film refractive index and thickness from the measured guiding mode angles. The measured refractive indices of polyimide host-guest system cured at 200 °C for 4 hours are presented in Table 3.

Table 2. Thermal Stability of NLO Moiety-doped Polyimides for NLO Applications

Guest	Host	subl. temp. (°C)*	Product of Guest
DANS-6	PIQ-2200	190	this work
DASS-62	PIQ-2200	210#	this work
DCM	PIQ-2200	190	exciton
Lophine 1	PIQ-2200	200	IBM

* No sublimation of the guest chromophores was observed at the given temperature.

The decomposition temperature was detected at 250 °C by TGA.

Table 3. Refractive index measurement of polyimide host-guest systems cured at 200 °C for 2 hours on a hot plate at the wavelength of 1300 nm.

Host	Guest	n_{TE}	n_{TM}
PIQ	-	1.6460 (1.6860)*	1.6431 (1.6822)
PIQ	DASS-62 (40 wt %)	1.6469 (1.6904)	1.6439 (1.6871)
UPI-3112	-	1.5858	1.5784
UPI-3112	DASS-62 (40 wt%)	1.6018	1.5959

* indices of refraction were measured at the wavelength of 632.8 nm.

Determination of the electro-optic coefficients:

The electro-optic coefficients of a DASS-62/PIQ-2200 system were presently measured using the simple reflection technique reported by Teng and Man[8]. A thin film was spun on ITO coated glass adjusting the spin speed to produce the film thickness of ~ 2 μm. For the poling and the reflection, another electrode was fabricated by thermal evaporation to form an over 0.1 μm - thick film of aluminum on top of the polymer layer. For poling, the sample was heated to about 160 ~ 170 °C in a convection oven and a voltage was applied between Al and ITO electrodes measuring the current. The results of the measurements are plotted in Figure 1. The filled symbols present the electro-optic coefficients which are calculated with the maximum modulated signals. However, the signal is complicated by the interference effects between the reflected beams from the two sides of the polymer film. To reduce the spurious modulating effects by the interference, the magnitudes of the modulated signals biased at the +90° and -90° phase retardation between the s- and p-waves are averaged. These results are also plotted with open symbols in Figure 1. As expected, the electro-optic coefficient (r_{33}) increases with increasing the amount of guest chromophores and the intensity of the poling field. The maximum electro-optic coefficient in the present study is ~ 13 pm/V for the 40 wt% DASS-62 poled at the 135 V/μm. Similarly, for the sample of 40 wt% DASS-62 in a PIQ-2200 polyimide, the measured value of the electro-optic coefficient (r_{33}) with a poling field of 40 V/μm at the wavelength of 1300 nm was about 5.2 pm/V. At the present, the electro-optic coefficient (r_{33}) as a function of the poling field is being measured.

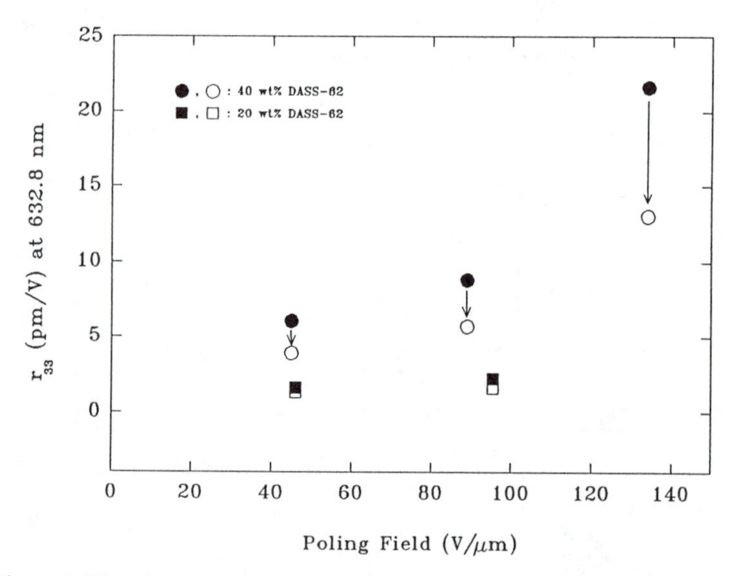

Figure 1. The electro-optic coefficients (r_{33}) at the wavelength of 632.8 nm as a function of poling field.

IV. SUMMARY

New NLO moiety-doped polyimides as host-guest systems are being demonstrated in making photonic devices. Long alkylated-NLO moieties based on both dialkylamino alkyl sulfone stilbenes (DASS) and dialkylamino nitro stilbenes (DANS) exhibited a significant improvement in the thermal stability at high temperatures exceeding 250°C. They retain remarkably high solubility in polyimides, improving the electro-optic coefficients(r_{33}) of these polyimide-based guest-host systems. The r_{33} value in the present study has been increased to 13 pm/V for the 40 wt% DASS-62-doped polymer system poled at the 135 V/μm. However, further increase up to 25 pm/V may easily be achieved by increasing the amount of guest moieties and/or the intensity of the poling field. Currently, various DASS analogues were systematically designed and synthesized, and were chemically attached to the polymer backbone of PMMA, providing side chain NLO polymers. These NLO polymeric materials are being characterized.

V. REFERENCES

1. E. van Tomme, P. P. van Daele, R.G. Baets, and P. E. Lagasse, IEEE Journal of Quantum Electronics, 27, 778-787 (1991).
2. G. H. Cross, A. Donaldson, R. W. Gymer, S. Mann, N. J. Parsons, D. K. Haas, H. T. Man, and H. N. Yoon, SPIE, 1177, 79 (1989).
3. A. K. M. Rahman, B. K. Mandal, X. F. Zhu, J. Kumar, and S. K. Tripathy, Mat. Res. Soc. Symp. Proc., 214, 67 (1991).
4. C. P. J. M. van der Vorst, W. H. G. Horsthuis, and G. R. Mohlman, "Polymers for Lightwave and Integrated Optics: Technology and Applications", edited by L. A. Hornak, pp365-395, Marcel Dekker, Inc. New York, 1992.
5. J. W. Wu, E. S. Binkley, J. T. Kenny, R. Lytel, and A. F. Garito, J. Appl. Phys., 69, 7366 (1991).
6. S. Ermer, J. F. Valley, R. Lytel, G. F. Lipscomb, T. E. Eck, and D. G. Girton, Appl. Phys. Lett., 61, 2272 (1992).
7. C. C. Teng and H. T. Man, Appl. Phys. Lett., 56, 1734 (1990).
8. (a) K. D. Singer, W. R. Holland, M. G. Kuzyk, G. L. Wolk, H. E. Katz, M. L. Schilling, and P. A. Cahill, SPIE, 1147, 233 (1989); (b) P. Pantelis and J. R. Hill, "Polymers for Lightwave and Integrated Optics : Technology and Applications", edited by L. A. Hornak, pp. 343-363, Marcel Dekker, Inc. New York, 1992.
9. H. L. Hampsch, J. Yang, G. K. Wong, and J. M. Torkelson, Macromolecules, 21, 526 (1988).
10. A. Ulman, C. S. Willand, W. Kohler, D. R. Robello, D. J. Williams, and L. Handley, J. Am. Chem. Soc., 112, 7083 (1990).
11. S. Stahelin, D. M. Burland, M. Ebert, R. D. Miller, B. A. Smith, R. J. Twieg, W. Volksen, and C. A. Walsh, Appl. Phys. Lett., 61, 1626 (1992).
12. T. N. Ding and E. Garmire, Appl. Opt., 22, 3177 (1983).
13. Y. Shuto, M. Amano, and T. Kaino, Jpn. J. Appl. Phys., 30, 320 (1991).
14. P. K. Tien, R. Ulrich and R. J. Martin, Appl. Phys. Lett., 14, 291 (1969).

RECEIVED January 30, 1995

Chapter 9

Design and Synthesis of a Polar Dipyrromethene Dye

M. B. Meinhardt[1], P. A. Cahill[1,3], T. C. Kowalczyk[2], and K. D. Singer[2]

[1]Chemistry of Materials Department, Mail Stop 0368, Sandia National Laboratories, Albuquerque, NM 87185–0368
[2]Department of Physics, Case Western Reserve University, Cleveland, OH 44106–7079

Semiempirical methods were applied to the design of a new second order nonlinear optical (NLO) dye through polar (noncentrosymmetric) modifications to the symmetric dipyrromethene boron difluoride chromophore. Computational evaluations of candidate structures suggested that a synthetically accessible methoxyindole modification would have second order NLO properties. This new dye consists of 4 fused rings, is soluble in polar organic solvents and has a large molar extinction coefficient (86×10^3). Its measured hyperpolarizability, β, is -44×10^{-30} esu at 1367 nm. The methoxyindole therefore induces moderate asymmetry to the chromophore.

The large potential market for electrooptic materials and devices for high speed data transfer, either as part of telecommunications or CATV networks or within computer backplanes, has prompted efforts towards the synthesis, processing and evaluation of new organic nonlinear optical (NLO) materials. Such devices would operate through the linear electrooptic effect that requires noncentrosymmetry on both molecular and macroscopic scales. Stability of the poled state is one of many requirements placed on these materials; other requirements include suitable and stable refractive indices for core and cladding, thermal stability as high as 350 °C, electrooptic coefficients greater than 30 pm/V, suitable electrical resistivity in both the core and cladding for efficient poling, and excellent optical transparency (losses < 0.3 dB/cm) at operating wavelengths. Simultaneous attainment of all these parameters has proved extremely difficult. Because organic NLO dyes are often the limiting factor in the ultimate thermal stability of a NLO material, a promising approach to improved materials is through the synthesis of new dyes.

Dye Design

Our approach is based on Marder et al.'s observations that a maximum in β, the first hyperpolarizability, occurs near the zero-bond alternation, or cyanine limit, in polar (noncentrosymmetric) chromophores (*1*). Therefore the nonlinearity of organic dyes

[3]Corresponding author

0097–6156/95/0601–0120$12.00/0

may be maximized through asymmetric modifications to known, centrosymmetric cyanine or cyanine-like dyes (Figure 1). This approach might also lead to dyes with narrow electronic absorption spectra and correspondingly lower absorption losses to the red of the principal charge-transfer absorption which gives rise to the nonlinear optical effect. Poling of cationic dyes such as the cyanines is problematic; therefore, charge neutralization must first be addressed.

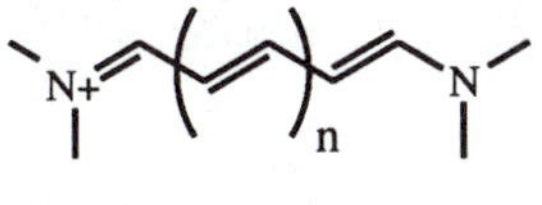

Figure 1. Cyanine dye structure.

An internally charge compensated class of cyanine dyes are the dipyrromethene difluoroborates shown in Figure 2. (*2*) Highly fluorescent symmetric dipyrromethenes are commercially available as biological probes (*3*) and laser dyes (*4*). Derivatives that are soluble in either organic (R_2 = alkyl) or aqueous media (R_2 = sulfonate) are known. In addition, the dipyrromethenes are among the most photochemically stable dyes. (*5*) High fluorescence quantum yields have been reported over a wide spectral range. (*3*) Therefore, this chromophore is a good starting point for the synthesis of second order NLO dyes by asymmetric modification.

Figure 2. Dipyrromethene difluoroborate dye structure.

The type and location of donor and acceptor groups on this chromophore that will maximize second order NLO properties is not obvious, however, because the ends of the cyanine chromophore are coordinated to the boron atom and therefore not available for direct modification. A polar substitution pattern must provide for both a charge transfer transition that is related to the strong absorption in the symmetric molecule and a ground state dipole moment which is substantially parallel to this transition. Candidate structures were therefore evaluated computationally with MOPAC using the AM1 basis for geometry optimization. Spectroscopic INDO/S methods with configuration interaction (ZINDO) were used for electronic spectra estimation. (*6*)

Direct calculation of the first hyperpolarizability of the candidate molecules with these semiempirical methods was considered, but a lack of closely related model compounds that could be used to verify such methods were not available. The two-state model, however, provides a means of relating molecular hyperpolarizability to information readily obtainable from spectral calculations and was used to estimate the magnitudes of the hyperpolarizability of candidate molecules.(*7*)

$$\beta(-2\omega;\omega,\omega) \cong \frac{3e^2}{2hm} \frac{\omega_{eg}^2 f \Delta\mu}{(\omega_{eg}{}^2 - \omega^2)(\omega_{eg}{}^2 - 4\omega^2)} \quad (1)$$

In this expression ω_{eg} is the frequency of the electronic transition, f is the oscillator strength, and $\Delta\mu$ is the change in the dipole moment. A figure of merit roughly proportional to $\mu\beta$ was used (2) to rank candidate structures ($\mu\beta$ is the important factor in poled films; only β might be used for ranking crystals). Its factors are the ground state dipole moment, oscillator strengths (f), cos Φ, the projection of the transition dipole moment onto μ_g, and λ^2 (because β trends as $1/\omega^2$). A factor of dm could also be included, but was considered separately (see Table 2). This figure of merit was then used to order and rationally select a synthetic target molecule. Such a

$$FOM = f \bullet \mu_g \bullet \cos\phi \bullet \lambda^2 \tag{2}$$

figure of merit tacitly assumes that the lowest energy charge transfer absorption gives rise to the majority of the NLO effect, or alternately, that the two-state molecule is a good approximation for this class of dyes. This is probably a valid assumption because the lowest energy electronic transition in dipyrromethenes (and cyanines in general) is well separated from other transitions.

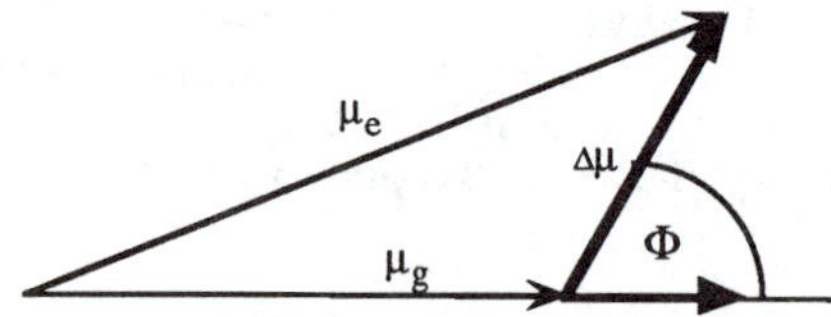

Figure 3. Projection of the change in dipole moment onto the ground state dipole.

Computational Results and Discussion

Direct substitution of the dipyrromethene framework (Figure 2) was attempted first, but substitutions of donor/acceptor pairs at R_1 or R_2 did not couple strongly with the symmetric chromophore's electronic structure, i.e., the HOMO and LUMO coefficients at these carbons are small. After other similar substitution patterns led to the same weak coupling, a variation based on benzannelation at R_1 and R_2 to give the indole-pyrrole-methene chromophore shown in Figure 4 suggested stronger coupling.

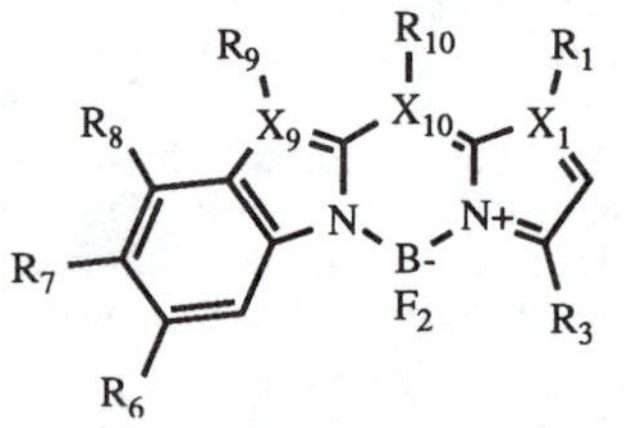

Figure 4. Structure of an indole-pyrrole-methene difluoroborate chromophore.

Ground and excited state dipoles (μ), oscillator strengths (f) and absorption maxima (λ) were extracted from INDO/S calculations for the series of molecules listed in Table 1. The projection angle Φ was calculated from the dot product of the ground and excited state dipoles as depicted in Figure 3. Results of ZINDO and figure of merit calculations from these systematic structural variation of the dipyrromethene framework are given in Table 2.

Table 1. Candidate Indole-Pyrrole-Difluoroborates

Compound	X1	X9	X10	R9	R3	R6	R7	R1	R10
A	C	C	C	Me	Me	MeO	H	Me	H
B	C	C	C	Ph	Me	MeO	H	Me	H
C	C	C	C	Ph	Ph	MeO	H	Me	H
D	C	C	C	Me	Me	MeO	H	Ph	H
E	C	C	C	Me	Me	Me2N	H	Me	H
F	C	C	C	Me	Me	MeO	MeO	Me	H
G	C	C	C	Me	Me	MeO	MeO	Ph	H
H	N	C	C	-	H	MeO	H	H	H
I	N	C	C	-	H	MeO	H	Me	H
J	N	C	C	-	H	MeO	H	Ph	H
K	C	N	C	Me	Me	MeO	H	-	H
L	C	N	C	Me	Me	MeO	MeO	-	H
M	C	N	C	Me	Me	MeO	MeO	-	Me
N	C	N	C	Ph	Ph	MeO	MeO	-	H
O	C	C	N	Me	Me	MeO	H	H	-
P	C	C	N	Me	Me	MeO	H	Me	-
Q	C	C	N	Me	Me	Me2N	H	Me	-
R	C	C	N	Me	Me	MeO	H	MeO	-
S	N	C	N	-	H	MeO	H	Me	-
T	C	N	N	Me	Me	MeO	H	-	-
U	C	N	N	Me	Me	MeO	H	-	-
V	C	N	N	Me	Me	MeO	MeO	-	-
W	N	N	N	-	H	MeO	H	-	-

Table 2. Figure of Merit (FOM) Computations Summary. A C.I. level of 14 was used in these INDO/S calculations.

Compound	μ_g	$\Delta\mu_e$	$\Delta\mu$	λ	f	cos ϕ	ϕ	FOM
A	7.73	10.6	16.7	502	1.06	-0.643	130	1.33e+06
B	7.44	11.2	8.21	515	1.14	0.677	47.4	1.52e+06
C	7.37	11.5	8.96	532	1.16	0.623	51.5	1.51e+06
D	7.48	10.9	9.10	509	1.09	0.560	55.9	1.18e+06
E	5.83	11.0	10.3	522	1.10	0.380	67.7	6.61e+05
F	8.96	10.7	9.09	510	1.05	0.583	54.3	1.43e+06
G	8.83	10.8	9.74	513	1.08	0.525	58.3	1.32e+06
H	2.63	11.2	10.5	537	1.10	0.375	68.0	3.15e+05
I	3.24	11.3	12.1	535	1.12	-0.118	96.7	1.22e+05
J	3.34	11.5	12.6	541	1.16	-0.187	101	2.12e+05
K	9.37	9.57	18.9	516	0.835	-0.985	170	2.05e+06
L	10.0	9.75	19.6	534	0.838	-0.959	164	2.29e+06
M	9.97	9.50	3.77	519	0.818	0.926	22.2	2.04e+06
N	10.3	10.5	3.08	552	0.940	0.956	17.0	2.81e+06
O	6.81	11.2	5.91	548	1.07	0.894	26.6	1.95e+06
P	6.56	11.2	17.3	550	1.07	-0.887	152	1.88e+06
Q	3.81	11.7	14.8	570	1.14	-0.739	138	1.04e+06
R	3.71	11.6	14.6	544	1.16	-0.778	141	9.91e+05
S	2.27	12.1	14.2	590	1.16	-0.913	156	8.39e+05
T	10.7	9.46	20.1	578	0.730	-0.990	172	2.57e+06
U	10.9	9.32	2.09	575	0.712	0.990	7.92	2.54e+06
V	10.9	9.57	20.5	600	0.720	-0.999	178	2.83e+06
W	8.50	10.6	18.4	641	0.830	-0.842	147	2.44e+06

The calculations summarized above suggest that benzannelation induces noncentrosymmetry into the chromophore and that the methoxy substituent additionally polarizes the ground state. Phenyl substituents lead to greater bathochromic shifts than methyl, but the effects are small. Such substitutions may separately impart synthetic advantages in directing electrophilic additions or in blocking sites otherwise available for side reactions.

Further examination of the data in Table 2 (Figure 5 contains a graph of this data) permits several generalizations to be made concerning structure-property relationships within various substitution patterns and the degree and location of heteroatom (nitrogen) incorporation into the dipyrromethene chromophore. Compounds A-G are modifications of an "all-carbon" frame. These compounds have very similar μ_g (7.3-8.8 D), λ_{max} (500-530 nm), and good oscillator strengths. As a group these compounds have less than optimal projections of the electronic transitions as reflected by Φ in the range of 47-67°. All these compounds include a MeO donor, except for E, which includes a Me_2N donor. None of these chromophores include an electron acceptor substituent such as nitro or nitrile because the calculations indicated that the bridging methene (X_{10}) acts as a strong acceptor in this unusual chromophore.

Compounds H-J, where X_1 = N, have the lowest FOM for the series. The resulting structure is an indole-imidazole combination in which electron donor are present at both ends of the chromophore. This is consistent with the low calculated values for μ_g (2.6-3.3 D) and Φ's of 67-100° (transition dipole nearly perpendicular to the ground state dipole).

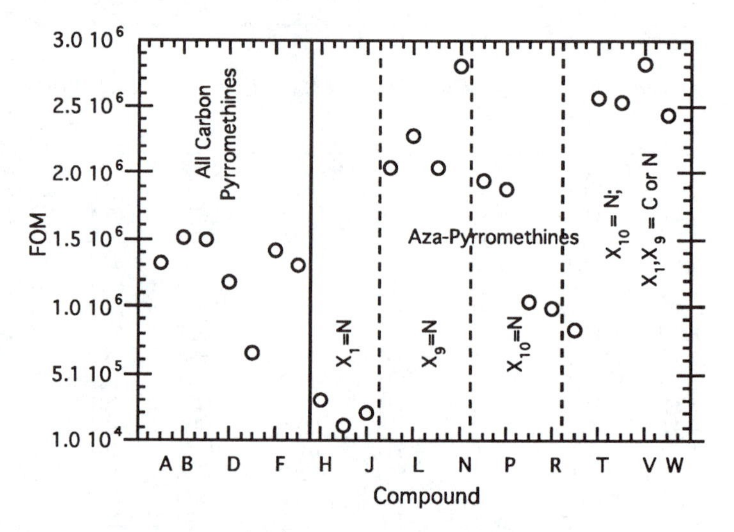

Figure 5. Graph of computational results. The highest Figures of Merit are calculated for benzimidazole (X_9 = N) donors.

Series K-N, X_9 = N, are composed of a benzimidazole-pyrrole framework and show the largest figures of merit of the chromophores linked by a carbon at X_{10}. This group has large μ_g (9.3-10.2 D) and Φ's within 20° of parallel with μ_g Absorption maxima are on the order of 530 nm. Strong oscillator strengths are also observed. Series O-R, X_{10} = N, demonstrated low to moderate μ_g, excellent f and λ in the range of 550 nm. These compounds have variable FOM performance and have moderate μ_g and mid-range Φ values.

Compounds S-W represent a series of di- and tri-aza frameworks with widely varying FOMs. For compound S, μ_g was exceptionally low (2.2 D) in contrast to diaza compounds T-V with μ_g values 10.7-10.9 D. This latter group showed the longest absorption maxima, 575-600 nm, with modest values for f (0.71). Excellent projections were noted for these compounds, particularly compound V with a projection within 2° of parallel. The synthesis of these compounds has not yet been examined.

Experimental Results and Discussion

Compound A (Scheme I) was chosen for synthesis on the basis of the calculations of the asymmetry combined with synthetic accessibility. The experimental program consisted of a multistep synthesis and a hyperpolarizability measurement of the final chromophores. Detailed experimental procedures will be available separately, but all compounds were fully characterized by standard spectroscopic methods. Purity was additionally established on the final chromophore by HPLC. Observed absorption maxima were shifted by about 10% to the red of the values predicted from INDO/S computations. Specifically for compound **A** the calculated absorption maximum is 502 nm (see above) and the observed maximum was 55 nm (see below). This is consistent with the accuracy of the ZINDO methods and the wavelengths shifts associated with moving from vacuum (ZINDO) to a condensed phase.

Syntheses Compound **A** was chosen as the "all carbon" chromophore for synthesis, in part due to synthetic accessibility as shown in Scheme I, below. The methoxyindole itself is prepared is several steps by literature methods. (*8-10*) Condensation of this indole with a pyrrole aldehyde leads to the desired product which is purified by chromatography in low to moderate yields. Interestingly, symmetric products are also formed and must result by transfer of the aldehyde from the pyrrole to the indole and subsequent reaction between identical ring systems.

Scheme I.

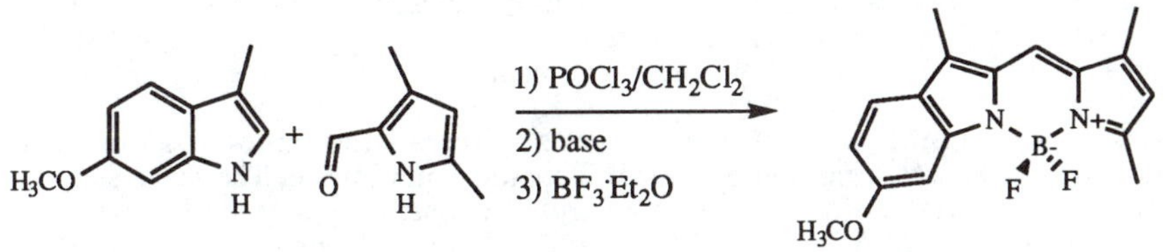

H_3CO ... N H + O N H — 1) $POCl_3/CH_2Cl_2$ 2) base 3) $BF_3{\cdot}Et_2O$ → H_3CO ... N N+ B- F F

A

Scheme II.

H_3CO ... N H — $POCl_3/DMF$ → H_3CO ... N H O

H_3CO ... N H O + Ph ... N H Ph — 1) $POCl_3/CH_2Cl_2$ 2) base 3) $BF_3{\cdot}Et_2O$ → H_3CO ... N N+ B- F F Ph Ph

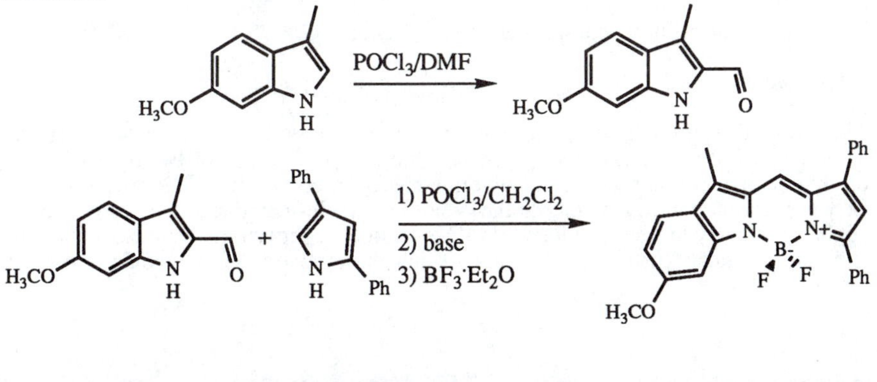

The diphenyl analog to **A** was also obtained using the 2-formylindole and 2,4-diphenylpyrrole as shown in Scheme II. Symmetrical products were again observed and indicate transfer of the aldehyde functional group between the indole and pyrrole. Reaction product mixtures were generally complex. Yields ranged from 15-50% in the final condensation step.

Compound **A**'s absorbance maximum (λ_{max}) is 555 nm in benzene with a molar absorptivity (ε) of 86 x 10^3 L/mol-cm (Figure 6). The large molar absorptivity indicates that the oscillator strength of the initial cyanine chromophore is not lost in the asymmetric system. The absorption maxima are only slightly solvatochromic.

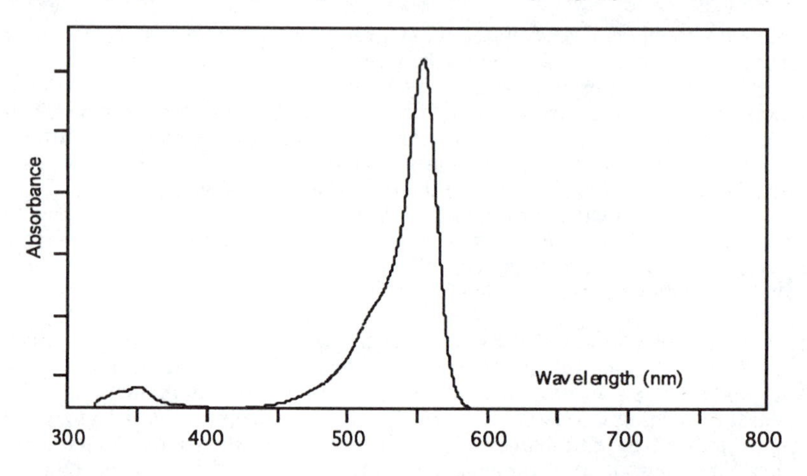

Figure 6. Absorption spectrum of methoxyindole-pyrrole-methene, **A**.

Thermal Stability Thin polyimide films containing chromophore **A** showed an onset of thermal decomposition between 200° and 225°C, much lower than the approximately 300° C onset for a symmetrical tetraphenyldipyrromethene, but similar to that of other symmetrical alkylated dipyrromethenes.

Hyperpolarizability The hyperpolarizability of the indolepyrromethene chromophore **A** was measured by EFISH techniques in dioxane. The value suggests only a slight deviation from centrosymmetry. The data is summarized below. The hyperpolarizability of **A** is about 10% that of DANS (dimethylaminonitrostilbene).

Polarizability	(α)	4.9 x 10^{-23} esu
Dipole Moment	(μ)	7.3 x 10^{-18}
Hyperpolarizability, 1367 nm	(β_{1367})	-44 x 10^{-30}
Hyperpolarizability, extrapolated to zero frequency	(β_0)	-12 x 10^{-30}

Conclusions

The rational approach to chromophore design detailed in this work has been demonstrated to facilitate chromophore selection. The synthesis of two dyes selected on the basis of semiempirical calculations has been completed. The magnitude of first hyperpolarizability of the methoxyindole-pyrrole-methene indicates that only slight asymmetry has been induced into the symmetric dipyrromethene chromophore and that significant further modification of this chromophore will be required to meet the goals outlined above. The most direct route to such new chromophores may be through dimethoxybenzimidazole precursors that would lead to compounds similar to compound **T, V,** and **W**. These compounds have much larger figures of merit and large transition dipoles that may lead to large hyperpolarizabilities.

Acknowledgments

This research was performed, in part, at Sandia National Laboratories and was supported by the U.S. Department of Energy under contract DE-AC04-94AL85000. Helpful discussions with D. R. Wheeler and C. C. Henderson (SNL) and A. J. Beuhler (Amoco Chemical Co., present address: Motorola) are acknowledged. This project could not have been completed without the help of R. Steppel et al. (Exciton) who scaled up the synthesis of the indole-pyrrole dye.

References

1. Marder, S. R.; Gorman, C. B., Meyers, F.; Perry, J. W.; Bourhill, G.; Bredas, J.-L.; Pierce, B. M. *Science* **1994,** *265,* 632.
2. For a recent review see Boyer, J. H.; Haag, A. M.; Sathyamoorthi, G.; Soong, M.-L.; Thangaraj, K. *Heteroatom Chem.* **1993,** *4,* 39.
3. Molecular Probes, Eugene, Oregon.
4. Exciton, Dayton, Ohio.
5. Hermes, R. E.; Allik, T. H.; Chandra, S.; Hutchinson, J. A. *Appl. Phys. Lett.* **1993,** *63,* 877.
6. LaChapelle, M.; Belletête, M., Poulin, M.; Godbout, N.; LeGrand, F.; Héroux, A.; Brisse, F.; Durocher, G. *J. Phys. Chem.* **1991,** *95,* 9764.
7. Oudar, J. L. and Chemla, D. S. *J. Chem .Phys.* **1977,** *66*, 2664.
8. Fleming, I., Woolias, M. *J. C. S. Perkins Trans. I* **1979,** 827.
9. Silverstein, R. M., Ryskiewicz, E. E. and Willard, C. *Org. Syn. Coll. Vol. III*, 831.
10. Deady, L. W. *Tetrahedron,* **1967,** *23*, 3505.

RECEIVED May 3, 1995

POLYMERS: SYNTHESIS AND CHARACTERIZATION

Chapter 10

High-Temperature Nonlinear Polyimides for $\chi^{(2)}$ Applications

R. D. Miller, D. M. Burland, M. Jurich, V. Y. Lee, C. R. Moylan, R. J. Twieg, J. Thackara, T. Verbiest, W. Volksen, and C. A. Walsh

IBM Almaden Research Center, 650 Harry Road, San Jose, CA 95120–6099

A variety of thermally stable NLO chromophores have been attached via a short tethers to 3,5-diaminophenol to produce NLO substituted aromatic diamine monomers suitable for polyimide formation. This may be accomplished either via nucleophilic displacement or by Mitsunobu coupling for base-sensitive materials. The functionalized diamines have been incorporated into a variety of polyimide materials yielding thermally stable polymers with Tg's ranging from 210 – 230°C. These materials may be poled at or near the polymer glass temperature and the polar alignment thus induced was stable for long periods at 100°C. The improved thermal stability of the chromophores containing diarylamino donor substitution was also manifested in the cured polymers and is also apparently not significantly jeopardized by the inclusion of an alkyl tether.

Organic nonlinear optical (NLO) materials are of interest because of the large and fast intrinsic nonlinearities, high damage threshold limits and ease of processing (*1,2*). They have been studied in various forms, including: crystals, organic glasses, Langmuir-Blodgett films, vapor deposited films, poled polymers, etc. For organic modulator and switch applications, poled polymers are the materials of choice.

It is now apparent that fully integrated polymeric modulators and switches will operate continuously at elevated temperatures (80 – 100°C) and could experience brief excursions to 250°C or more for short periods during the various integration and packaging steps (*3,4*). This places additional demands on the device materials beyond just high nonlinearity. The elevated operating temperatures will require thermal stability for the NLO polymer; both orientational stability of the induced polar order and intrinsic chemical and thermal stability of the chromophore and polymer. Although much experimental effort has been expended to improve the orientational stability, somewhat less has been devoted to the actual development of thermally stable chromophores.

0097–6156/95/0601–0130$12.00/0

Initially, orientational stability was sought using crosslinking techniques to restrict the chromophore rotational mobility (*5*). This approach has resulted in materials with large nonlinearities and impressive polar order stability, however, these systems are often difficult to process into operating devices due to embrittlement, cracking, adhesive breakdown, dimensional changes upon curing, etc.

Although factors such as chromophore size, poling procedures, polymer aging, chromophore-polymer interactions, etc., can influence the relaxation rate of poled polymer systems, it is clear that the glass transition temperature (Tg) of the system plays a primary role (*5,6*). For this reason, thermoplastic materials with high glass temperatures have recently attracted considerable interest. This has included both host-guest composites (*7 – 13*) as well as polymers with chemically bonded chromophores (*5,14 – 20*). Furthermore, it has recently been suggested that a Tg which is 100°C or more above the highest continuous use temperature is necessary to maintain the polar order for long periods of time (*21*). This has driven the current quest for very high Tg NLO polymers. The high temperature poling process, coupled with the need for low optical losses and dimensional stability in a polymeric waveguide configuration, create a new set of difficulties to be addressed for high temperature NLO systems.

Thermally Stable NLO Chromophores

Recent studies on NLO chromophore thermal stability have shown that the more usual and unfortunate relationship between chromophore nonlinearity and thermal lability is not inviolable (*22 – 24*). In fact, we have recently (*17,21,22,25,26*) demonstrated that the thermal stabilities of a variety of stilbene, tolane and azo derivatives can often be greatly improved simply by the substitution of diarylamino donor substituents for the more common dialkylamino derivatives without seriously compromising chromophore nonlinearity. This characteristic is apparent for a wide variety of electron accepting and electron transmitting functionality. The pertinent thermal and nonlinear data for a variety of nitro substituted chromophores is shown in Table I. The column headed $\mu\beta_{1.3}(-\omega;\omega,0)$ provides a comparison of the calculated hyperpolarizability for the electro-optic effect derived from $\beta(2\omega;\omega,\omega)$ measured by electric field induced second harmonic generation (EFISH, 1.907 μm) extrapolated to 1.3 μm times the measured molecular dipole moment in Debye (*27*). This is a particularly pertinent quantity for electro-optic poled polymer applications in the infrared. A related quantity ($\mu\beta_{1.3}(-\omega;\omega,0)$/MW) represents a crude attempt to scale the poled molecular nonlinearities to molecular size (approximated here by molecular weight). This quantity, which we define as the reduced molecular nonlinearity, will be relevant to the macroscopic bulk nonlinearity which depends on the number density (molecules/cm^3) of the NLO chromophore (*1,2*). The onset decomposition temperature, designated as T_d (*21*) in Table I, is derived from differential scanning calorimetry (DSC) data run at a heating rate of 20°/minute. This number, when compared with use temperatures derived from a complete three-part thermal analysis (*21*), which we have developed for the assessment of the thermal stabilities of potential NLO chromophores, provides only an upper limit. In practice, temperatures often as much as 60 – 80°C lower than T_d, as determined by DSC, may be more appropriate for estimating the

Table I. The thermal, linear and nonlinear properties of a variety of nitro substituted NLO chromophores: a. absorption maximum in chloroform; b. quadratic molecular hyperpolarizability extrapolated to zero frequency assuming the validity of the two-state model (*5*); c. values of the molecular nonlinearity for the electro-optic effect extrapolated to 1.3 μm $\times$ the dipole moment (D) (*25*); d. calculated values for the reduced nonlinearities extrapolated to 1.3 μm.

Structure	T_d (°C)	λ_{max} (nm) [a]	μ(D)	β_o (esu) $\times 10^{30}$ [b]	$\mu\beta_{1.3}(-\omega,\omega,0)$ $\times 10^{30}$ cm^5· D/esu [c]	$\frac{\mu\beta_{1.3}}{MW} \times 10^{30}$ [d]
Ph_2N–C₆H₄–N=N–(6-nitrobenzothiazol-2-yl) **1**	356	550	7.21	71.8	1400	3.2
Ph_2N–C₆H₄–N=N–(5-nitrothiazol-2-yl) **2**	298	582	6.89	68.2	1300	3.2
Ph_2N–C₆H₄–CH=CH–(6-nitrobenzothiazol-2-yl) **3**	367	458	5.48	57.2	757	1.7
Ph_2N–C₆H₄–CH=CH–(5-nitrothiazol-2-yl) **4**	325	492	5.18	56.8	738	1.9
Ph_2N–C₆H₄–N=N–C₆H₄–NO_2 **5**	393	486	5.87	54.3	794	2.0
Ph_2N–C₆H₄–CH=CH–C₆H₄–NO_2 **6**	358	436	4.75	37.3	419	1.1
Ph_2N–C₆H₄–≡–C₆H₄–NO_2 **7**	336	418	4.84	28.2	317	0.9
H_2N–C₆H₄–N=N–C₆H₄–NO_2 **8 (DO3)**	303	410	6.57	27.1	345	1.4

thermal robustness of NLO chromophores for poled polymer applications. For a variety of substituted chromophores, the data in Table I and reference 22 show the efficacy of diarylamino donor substitution. We have somewhat arbitrarily targeted chromophores with reduced nonlinearities $\geq 2.0 \times 10^{-30}$ $cm^5{\cdot}D/esu$ and T_d temperatures $\geq 350°C$ for further study. The thermal analysis data recorded at scan rates of 20°/minute, while not predictive in an absolute sense, are useful in a relative sense for comparing prospective chromophores. In general, we find that the thermal stability of the nitro substituted azobenzene derivatives and stilbenes are often comparable, although exceptions have been observed (*22*). These, in turn, are both more stable than the corresponding tolane derivatives. Similarly, the molecular nonlinearities for the azo and stilbene derivatives are also often comparable. For these reasons, as well as for synthetic efficacy, the azo derivatives were selected over the related stilbenes for detailed study.

Synthesis

In principle, the most straightforward route to N,N-diarylaminophenylazobenzene derivatives would be the direct azo coupling of a substituted benzene diazonium salt with an appropriately substituted triphenylamine derivative (*28*). A phase transfer variant of this procedure, in fact, works for the preparation of the parent 4-N,N-diphenylamino-4′-nitroazobenzene **5**, albeit in moderate yield (47%) (*29*). However, this technique fails completely for substituted triarylamino derivatives containing strongly electron donating substituents. The problem is illustrated below for N-p-methoxyphenyldiphenylamine **9**. In this particular case, a complex mixture results from attempted azo coupling, from which the dearylated azo dye DO1 [**11**] is isolated as the major reaction product even when run under phase transfer conditions. Similar problems resulted with triarylamine derivatives substituted with other electron donating substituents such as amino, acetamido, trifluoroacetamido, etc. This is unfortunate, since such substituents are useful for the attachment of tether functionality for ultimate incorporation into a polymer (vide infra).

We have found, however, that a variety of diarylamino substituted azobenzene derivatives may be prepared directly from preformed, commercially available azo dyes by a variation of the Ullmann coupling procedure using cationic phase transfer polyethers (*30*). The reaction itself and some pertinent examples are shown in Table II. In the case of 4-N,N-diphenylamino-4′-nitroazobenzene (**5**, entry 1), the isolated yield of the product by Ullmann coupling was almost twice that obtained by the diazonium coupling route (81 vs. 47%). A wide variety of aryl substituted azo dyes can be generated in this manner, as shown in Table II. We have routinely employed two synthetic variants of this procedure. Method A uses a solid-liquid phase transfer procedure (*31*) utilizing 18-crown-6 and anhydrous potassium carbonate in a high boiling solvent such as o-dichlorobenzene. Reaction temperatures of 170−180°C and run times of 5−14 hours are typical. After completion of the reaction, the solvent is removed under vacuum and the residue purified by flash column chromatography (*32*). A second procedure (Method B) utilizing triethylene glycol dimethyl ether as both the solvent and phase transfer medium and higher reaction temperatures (205−210°C) for 2−6 hours has been used in a few cases. Using this procedure, the reaction

Table II. Preparation of aryl substituted azo dyes by modified Ullmann coupling. Procedure A: solid-liquid phase transfer (based on 1 mmol of the azo starting material); 4 mmol Cu, 0.3 mmol 18-crown-6, 2.5 mmol K_2CO_3, o-dichlorobenzene, 185°C, 12 – 18h; Procedure B: 4 mmol Cu, 2.5 mmol K_2CO_3, triethylene glycol dimethylether, 210°C, 2 – 5h.

$$R^1-N(H)-C_6H_4-N{=}N-C_6H_4-NO_2 + ArI \xrightarrow[\Delta]{Cu} R^1-N(Ar)-C_6H_4-N{=}N-C_6H_4-NO_2$$

Entry	R^1	Aryl Halide	Procedure [a]	Product	Yield (%)	mp(°C)	λ_{max} (nm)
1	Ph D01(11)	C_6H_5-I	B	$Ph_2N-C_6H_4-N{=}N-C_6H_4-NO_2$ **5**	81	144–6	486
2	Ph	$MeO-C_6H_4-I$	A	$MeO-C_6H_4-N(Ph)-C_6H_4-N{=}N-C_6H_4-NO_2$ **12**	68	154–5	498
3	Ph	$I-C_6H_4-I$	B	$I-C_6H_4-N(Ph)-C_6H_4-N{=}N-C_6H_4-NO_2$ **13**	74	171–3	486
4	Ph	t-BuPh$_2$SiO~~O$-C_6H_4-I$	A	t-BuPh$_2$SiO~~O$-C_6H_4-N(Ph)-C_6H_4-N{=}N-C_6H_4-NO_2$ **14**	50	- - - [b]	496
5	Ph	$(CH_3)_2CHC(O)NH-C_6H_4-I$	A	$(CH_3)_2CHC(O)NH-C_6H_4-N(Ph)-C_6H_4-N{=}N-C_6H_4-NO_2$ **15**	56	169	496
6	Ph	$H_2N-C_6H_4-I$	A	$H_2N-C_6H_4-N(Ph)-C_6H_4-N{=}N-C_6H_4-NO_2$ **16**	61	155–7	498
7	Ph	3-MeO$-C_6H_4-I$	B	3-MeO$-C_6H_4-N(Ph)-C_6H_4-N{=}N-C_6H_4-NO_2$ **17**	55	116	492
8	H D03(8)	$MeO-C_6H_4-I$	B	$(MeO-C_6H_4)_2N-C_6H_4-N{=}N-C_6H_4-NO_2$ **18**	40	168	510
9	H	$(CH_3)_2CHC(O)NH-C_6H_4-I$	A	$(CH_3)_2CHC(O)NH-C_6H_4-NH-C_6H_4-N{=}N-C_6H_4-NO_2$ **19a**	19	210–12	448
				$((CH_3)_2CHC(O)NH-C_6H_4)_2N-C_6H_4-N{=}N-C_6H_4-NO_2$ **19b**	40	255	502

is monitored by TLC, and when complete, the mixture is filtered, diluted with water and extracted with ethyl acetate. Examination of Table II shows that the synthetic technique is quite general. By controlling reagent stoichiometry, one can achieve monoarylation with bifunctional reagents (entry 3), arylation with reactively functionalized iodides (entries 5, 6 and 9) and one pot bis-arylation of primary amines (entries 8 and 9). The direct production of 4-N,N-diarylamino-4′-nitroazobenzenes containing a flexible tether group (entry 4) is particularly pertinent, as the silicon protecting group is easily removed with 1M tetrabutylammonium fluoride in THF and the resulting alcohol functionality can be used to generate polymerizable NLO monomers (vide infra). Attempts to utilize aryl iodides containing unprotected alcohol functionality directly in the Ullmann coupling were unsuccessful.

Since the ultimate goal of this work was the preparation of polyimide derivatives containing thermally stable chromophores, a simple route to NLO functionalized aromatic diamines was developed which is shown in Scheme 1. The alcohol **20a** is the commercially available azo dye DR1. The diphenylamino substituted derivative **20b** was prepared as described in Reference 33. The functionalized aromatic diamines **M1** and **M2** were prepared by nucleophilic displacement from the corresponding NLO functionalized tosylates **21a** and **21b** and subsequent hydrolysis. Under strongly basic conditions, the alkylation reaction takes place chemoselectively, primarily on oxygen. The bifunctional monomer **M3** could also be prepared in 53% overall yield using the same procedure starting from 3,5′-dihydroxy-4,4′-diaminobiphenyl (**23**).

Although this procedure works well in some cases, it fails completely for a number of important examples. For example, attempted displacement from the tosylate derivative of **24** led to immediate decoloration of the solution. The starting 6-nitrobenzothiazole substituted alcohol starting material **24** was prepared by azo coupling of 6-nitrobenzothiazole-2-diazonium tetrafluoroborate **25** to 4-N-(4′-hydroxyethylphenyl) diphenylamine **26**. The latter was prepared by the Ullmann coupling of diphenylamine with t-butyldiphenylsilyl protected p-(β-hydroxyethyl)iodobenzene followed by treatment with 1M Bu_4NF in THF. Similarly, the reaction also failed for a variety of other chromophores containing base-sensitive tricyanovinyl acceptor groups. For these thermally stable, but reagent-sensitive chromophores, another procedure was developed utilizing the Mitsunobu (*34*) coupling reaction shown in Scheme 2 (*17,35*). Since the reaction failed using the free amine, the amino substituents were first selectively protected as trifluoroacetamides. The protecting groups could be easily removed after coupling by treatment with weak base. Using this procedure, the overall yield of **M4** from **24** was ~50%. The yield of **M2** could also be improved from 27% (Scheme 1a) to 60% using this procedure. The N-t-butoxycarbonyl protecting group can also be utilized when the functionality is sensitive to basic hydrolysis (as is observed for chromophores containing tricyanovinyl substituents) (*35*). In these cases, deprotection is accomplished using trifluoroacetic acid in methylene chloride at room temperature.

In summary, using the procedures described, the NLO functionalized aromatic diamine monomers **M1**–**M4** could be prepared and purified.

NLO Functionalized Polyimide Derivatives

The diamines **M1** – **M4** could be converted into a variety of polyimide derivatives using standard condensation polymerization procedures. In a dry box, equimolar quantities of the diamines and a suitable bis-anhydride (for example, either oxydiphthalic anhydride (ODPA) **30** or hexafluoroisopropylidenediphthalic anhydride (6FDA) **31** were mixed in dry N-methylpyrrolidinone (NMP) and stirred at room temperature overnight to produce homogeneous solutions of the amic acids **PA1**, **PA2**, **PA4** (Scheme 3). The corresponding amic acids derived from **M3** could be prepared in a similar fashion. Amic acid films were prepared by dilution of the NMP solutions with 15 – 20% o-xylene and spin casting. The spun films could be thermally imidized by heating slowly from 150 – 250°C. Alternatively, the amic acids in NMP could be chemically imidized by heating with acetic anhydride-pyridine. The polyimides thus formed were isolated by precipitation into water or water-methanol followed by filtration, washing and vacuum drying. The polyimide derivatives derived from ODPA (**PI-2a**, **PI-3**) are insoluble in common solvents and films were prepared by casting the amic acids followed by thermal curing. The hexafluoroisopropylidene substituted polymers (**PI-1**, **PI-2b**, **PI-4**), however, were readily soluble in solvents such as cyclohexanone, 1,1,2,2-tetrachloroethane, cyclopentanone, etc., and films were prepared by spin casting. The solvent-spun polyimide films were processed by heating in a stepwise fashion at 150, 200, 250°C for 1/2 h under N_2 and finally to 275°C for 15 minutes. The Tg values of the polymers **PI-1** to **PI-4** ranged from 210 – 230°C, as measured by DSC analysis at a heating rate of 20°/minute. The long wavelength visible absorption maxima of the NLO chromophores in **PI-2a,b**, **PI-3** appeared around 500 nm while that of **PI-1**, which contains a DR1 subunit, absorbed at shorter wavelengths (~ 474 nm). The benzothiazole derivative **PI-4** was more purple in color and absorbed at 548 nm. The TGA scans of the cured polyimides **PI-1** – **PI-4** showed no significant weight loss below 350°C. Comparative DSC analysis of **PI-1** and **PI-2b** show clearly the improved thermal stability of the latter. For these particular examples, the corresponding onset decomposition temperatures (T_d) were 338°C and 376°C respectively (see Figure 1). It is interesting to note that the incorporation of an alkyl tether group does not seem to effect the thermal stability of the chromophore in the polymer very much (*27*).

The thermal stability of the cast and processed polyimide derivatives was also studied by variable temperature UV-visible spectroscopy. For these studies, relatively thin films (0.5 – 0.7 μm) were utilized and the change in the optical density at the long wavelength maximum was monitored upon heating. Figure 2 shows the direct comparison between two 6F-polyimides **PI-1** and **PI-2b**. **PI-1** contains a dialkylamino substituted chromophore while the **PI-2b** has diarylamino donor substitution. The former is reasonably stable to heating under N_2 at 250°C, but decomposes rapidly at 275°C. **PI-2b** is completely stable at 275°C and is still reasonably stable at 300°C. Above 300°C, all of the polyimides described here show some signs of decomposition, as evidenced by changes in their UV-visible spectra. The 6-nitrobenzothiazole substituted polymer (**PI-4**) was only slightly less stable than the other diarylamino substituted derivatives and lost ~ 8% of the original absorbance at 548 nm upon heating at 275°C for 95 minutes. The fact

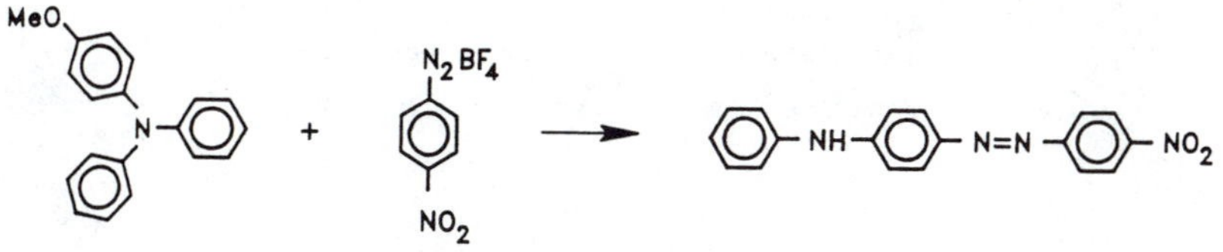

MeO
N_2BF_4
N
+
→
NH
N=N
NO_2
NO_2

9 10 11 (D01)

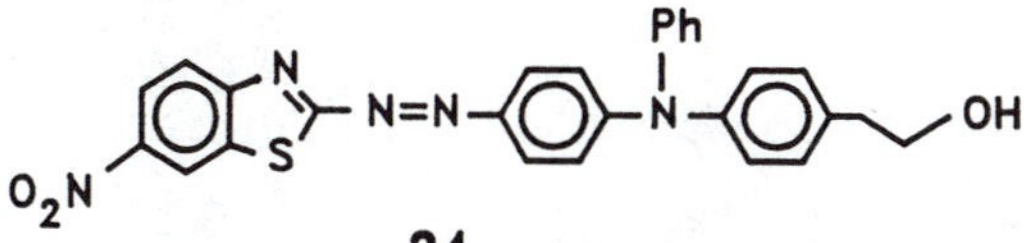

Ph
N
N=N
N
OH
S
O_2N

24

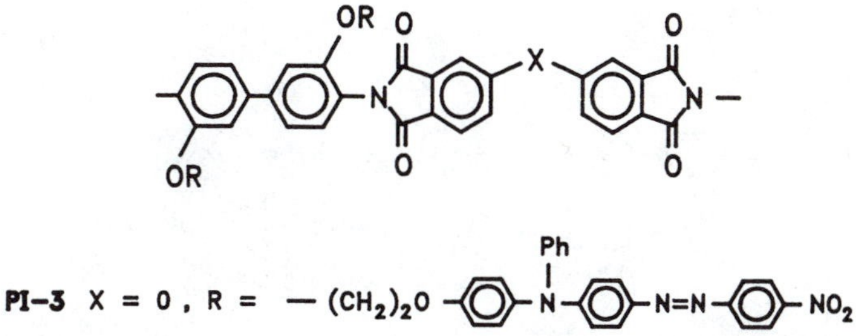

OR O O
X
N
N —
OR O O

Ph

PI-3 X = O , R = — $(CH_2)_2O$ — N — N=N — NO_2

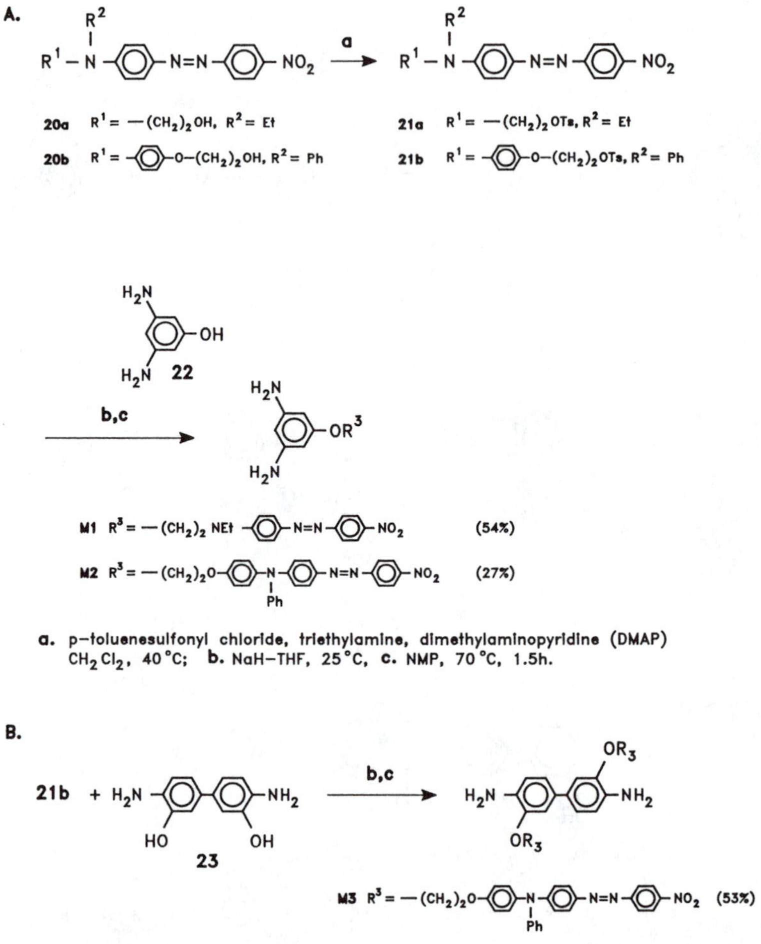

A.

20a $R^1 = -(CH_2)_2OH$, $R^2 = Et$

20b $R^1 = -C_6H_4-O-(CH_2)_2OH$, $R^2 = Ph$

21a $R^1 = -(CH_2)_2OTs$, $R^2 = Et$

21b $R^1 = -C_6H_4-O-(CH_2)_2OTs$, $R^2 = Ph$

22

M1 $R^3 = -(CH_2)_2NEt-C_6H_4-N{=}N-C_6H_4-NO_2$ (54%)

M2 $R^3 = -(CH_2)_2O-C_6H_4-N(Ph)-C_6H_4-N{=}N-C_6H_4-NO_2$ (27%)

a. p-toluenesulfonyl chloride, triethylamine, dimethylaminopyridine (DMAP) CH_2Cl_2, 40 °C; b. NaH-THF, 25 °C, c. NMP, 70 °C, 1.5h.

B.

21b + **23** $\xrightarrow{b,c}$

M3 $R^3 = -(CH_2)_2O-C_6H_4-N(Ph)-C_6H_4-N{=}N-C_6H_4-NO_2$ (53%)

Scheme 1. The preparation of NLO functionalized aromatic diamines by nucleophilic substitution.

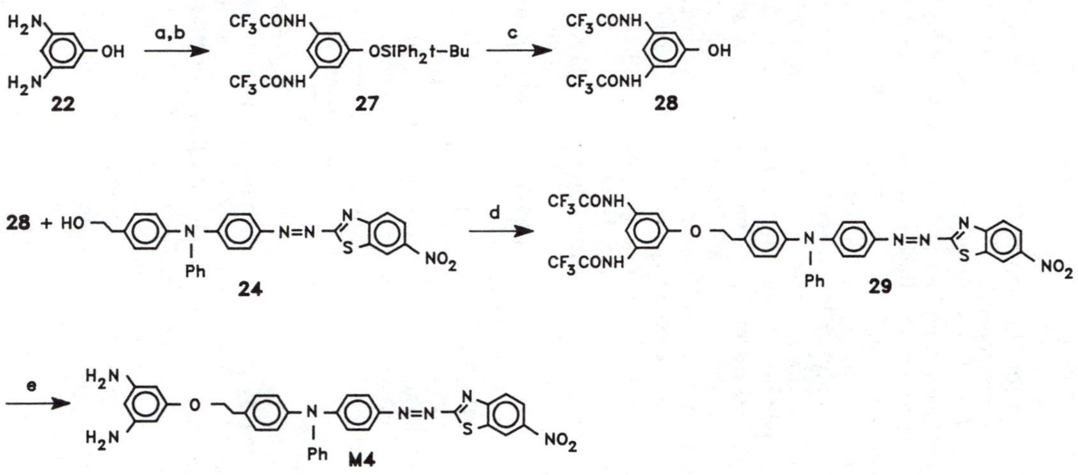

a. t–$BuPh_2SiCl$, triethylamine, CH_3CN, reflux, 2h; b. $(CF_3CO)_2O$, pyridine, Δ reflux, 2h;
c. Bu_4NF–THF(1M), 25°C, 1h; d. Diisopropylazodicarboxylate, THF, Ph_3P; e. K_2CO_3, MeOH–H_2O (3:2).

Scheme 2. The preparation of NLO functionalized aromatic diamines by the Mitsunobu reaction coupled with selective deprotection.

that chromophore decomposition as determined by variable temperature UV-Visible studies occurs at lower temperatures than predicted by DSC may be attributed to the heating rate dependence of the data obtained by the latter technique (*21*).

Nonlinearity Measurements

The electro-optic coefficients (r_{33}) for the polymers **PI-1** through **PI-4** were measured by one of three procedures depending on the poling technique, and the results are shown in Table III. For samples that were poled by corona discharge, an attenuated total reflection (ATR) technique was employed (*36*). When single films were electrode poled, an ellipsometry technique was utilized (*37,38*). A heterodyne technique (*39,40*) was used to determine r_{33} of polymers poled in modulator geometries. With the second method, rather thin films (NLO polymer layer only) were utilized (1.5 – 2.5 μm) and the poling field was maintained at a relatively low and constant value (~ 75 V/μm) to avoid any possibility of shorting and damage to the film. The values of the electro-optic coefficient (r_{33}) obtained in these electrode poling experiments would be expected to increase linearly with field (*5*). In the case of **PI-2b**, a triple-stack phase modulator composed of two crosslinked acrylate buffer layers sandwiching the NLO polymer layer was constructed and larger poling fields were applied (~ 250 V/μm). The r_{33} value measured in this modulator was similar to those obtained both by corona poling and by single polymer layer ellipsometry upon extrapolating the latter to the modulator poling voltage. In the case of two related NLO polyimides, the r_{33} values measured either at 250 V/μm and/or extrapolated to this poling field compared quite favorably with those predicted by calculation (*27*) (i.e., 8.8 pm/V for **PI-2a** and 8.3 pm/V for **PI-2b**, respectively). However, the measured value for the 6-nitrobenzothiazole polymer **PI-4** seems anomalously low based on the substantially larger hyperpolarizability of the chromophore (*22*). Rough calculations of the r_{33} value predicted for this polymer indicated an expected value of around 20 pm/V at 250 V/μm; considerably larger than the initial experimental value. It is possible that the poling is less efficient for this more extended chromophore.

Greatly improved thermal orientational stability for the polyimides **PI-1** – **PI-4** would be anticipated by virtue of their relatively high glass transition temperatures (see Table III). Relaxation data obtained by monitoring the intensity of the second harmonic signal of films of **PI-2a** and **PI-2b** poled in a corona field (~ 220 V/μm, poling temperatures 210 – 220°C) are shown in Figure 3. After a small decay in the signal of ~ 5%, each sample was stable for over 1000h at 100°C. In addition, preliminary results suggest that annealing the sample of **PI-2b** at 175°C for ~ 3h in the poling field increases the characteristic relaxation time τ (derived from a stretched exponential fit (*5*) of the decay data) by ~ 50%.

Conclusions

In summary, we have described the preparation of a number of thermally stable diarylamino substituted nitroazobenzene derivatives by a modified Ullmann coupling procedure using commercial azo dyes. The procedure appears quite general and is applicable to materials containing potentially reactive tether functionality for formation of NLO functionalized aromatic

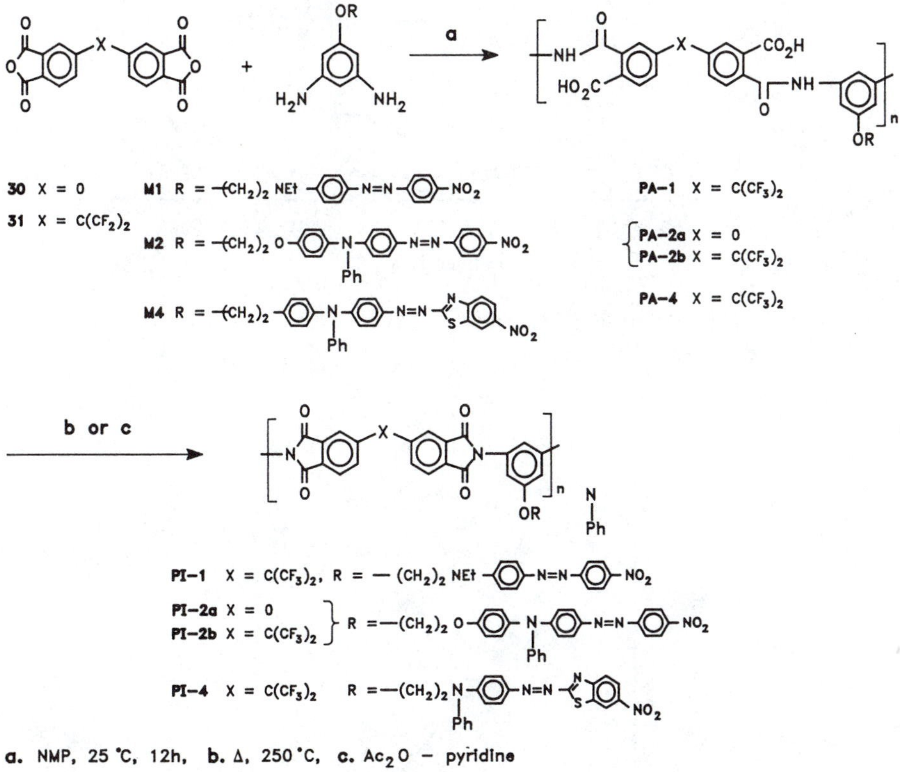

30 X = O
31 X = $C(CF_2)_2$

M1 R = $-(CH_2)_2$ NEt–C_6H_4–N=N–C_6H_4–NO_2

M2 R = $-(CH_2)_2$ O–C_6H_4–N(Ph)–C_6H_4–N=N–C_6H_4–NO_2

M4 R = $-(CH_2)_2$–C_6H_4–N(Ph)–C_6H_4–N=N–(benzothiazole)–NO_2

PA-1 X = $C(CF_3)_2$
PA-2a X = O
PA-2b X = $C(CF_3)_2$
PA-4 X = $C(CF_3)_2$

PI-1 X = $C(CF_3)_2$, R = $-(CH_2)_2$ NEt–C_6H_4–N=N–C_6H_4–NO_2

PI-2a X = O
PI-2b X = $C(CF_3)_2$
R = $-(CH_2)_2$ O–C_6H_4–N(Ph)–C_6H_4–N=N–C_6H_4–NO_2

PI-4 X = $C(CF_3)_2$ R = $-(CH_2)_2$ N(Ph)–C_6H_4–N=N–(benzothiazole)–NO_2

a. NMP, 25 °C, 12h, b. Δ, 250 °C, c. Ac_2O – pyridine

Scheme 3. The preparation of NLO functionalized side chain polyimides by condensation polymerization.

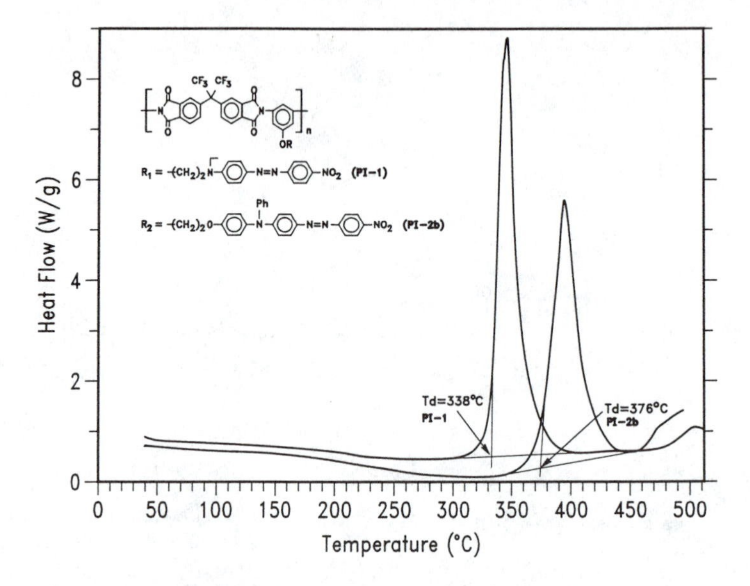

Figure 1. DSC analysis of the polyimides PI-1 and PI-2b measured at a heating rate of 20°/minute.

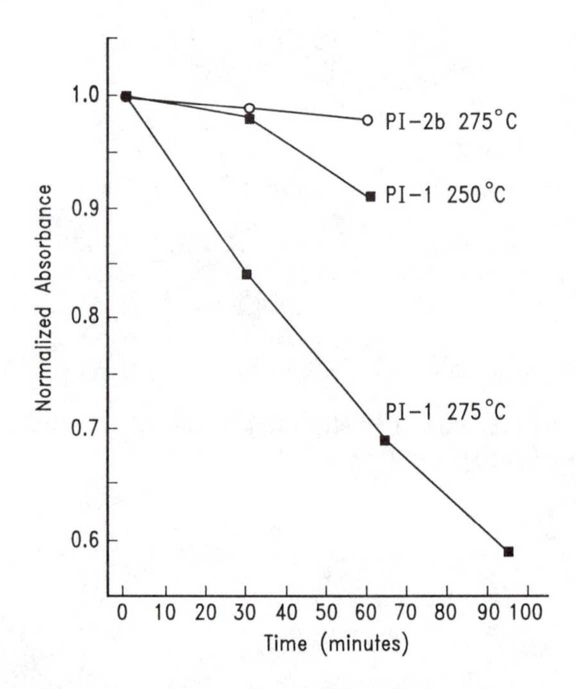

Figure 2. Variable temperature UV-Visible studies of films of PI-1 and PI-2b. Absorbance measurements at the λ_{max} of the long wavelength visible transition (474 nm for PI-1 and 498 nm for PI-2b).

Table III. The thermal, linear and nonlinear optical properties of a variety of NLO functionalized side chain polyimides: a. poling temperature; b. polymer glass transition temperature as measured by DSC analysis at a heating rate of 20°/minute; c. electrode poling (75 V/μm), r_{33} value measured by ellipsometry; d. corona poling (~230 V/μm), r_{33} measured at 1.3 μm by attenuated total reflection (ATR); e. corona poling (200 V/μm); f. measured in a stacked three-level waveguide phase modulator; g. corona poling (230 V/μm), r_{33} measured by ATR at 633 nm and extrapolated to 1.3 μm assuming the validity of the two-state model (5).

Polyimide	T_{pol}[a] (°C)	T_g[b] (°C)	λ_{max} (nm)	Chromophore Wt. %	r_{33} (1.3μm) (pm/V)
PI-1	205	213	474	36	3.75[c]
PI-2a	210	223	500	47	8.2-10.0[d]
PI-2b	220	228	498	45	{7.0[e], 8.1[f], 3.1[c]}
PI-3	210	210	496	58	13.0[g]
PI-4	220	225	548	53	2.45[c]

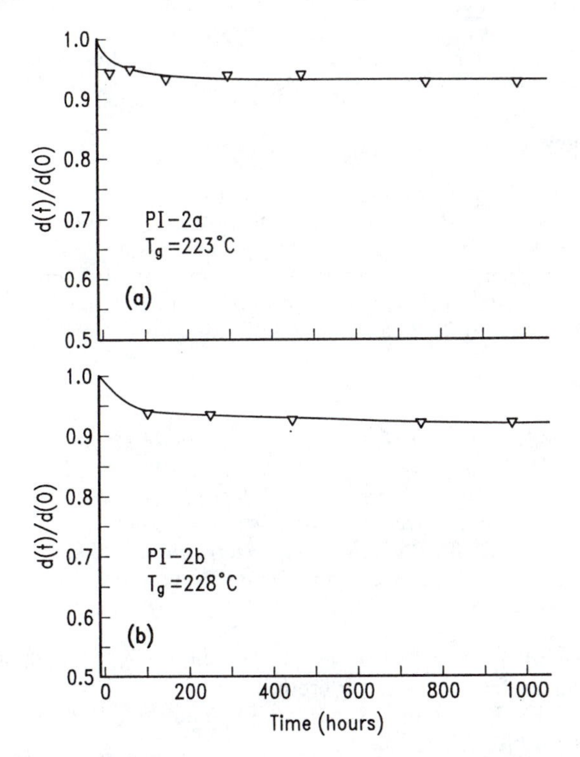

Figure 3. Decay in the second harmonic coefficient at 100°C, laser fundamental 1.047 μm: a. PI-2a, Tg = 223°C; b. PI-2b, Tg = 228.

diamines. At the same time, a chromophore containing the intrinsically more nonlinear 6-nitrobenzothiazole unit with a functionalized tether was prepared by direct azo coupling. The chromophores containing hydroxyalkyl or hydroalkoxy tethers could be connected to 3,5-diaminophenol via an ether linkage either by direct nucleophilic displacement or by the Mitsunobu coupling of the alcohol. The latter procedure is quite general and allows the preparation of thermally stable, but reagent sensitive, monomers which cannot be coupled by nucleophilic displacement under basic conditions. The NLO functionalized diamines can then be incorporated into polyimides by standard condensation polymerization techniques. Polyimide films can be prepared either via the functionalized polyamic acids by thermal curing or from the preformed polyimides themselves, when soluble. These techniques produced a variety of functionalized polyimide derivatives with glass transition temperatures ranging from 210 – 230°C. The improved thermal stability of chromophores containing diarylamino relative to dialkylamino substitution was demonstrated both by DSC analysis and by variable temperature UV-visible studies. The polymers could be poled as the polyamic acid precursors and as polyimides using either corona or electrode poling. In all but one case, the measured electro-optic coefficients were consistent with the approximate values calculated based on EFISH hyperpolarizability measurements. The thermal stability of the polar order of typical corona poled samples was excellent, showing a decrease in the second harmonic intensity of less than 10% over 1000h at 100°C. The stability could be further improved by aging at elevated temperatures. These preliminary results suggest that by varying the NLO chromophore and/or the dianhydride, one could anticipate the preparation of NLO polymers with both improved nonlinearities and elevated glass temperatures.

Acknowledgments

The authors gratefully acknowledge partial funding support from the AFOSR (F49620-92-C-0025) and the NIST-ATP program (Cooperative Agreement No. 70NANB2H1246) for this project. One of us (T.V.) would like to acknowledge support from the Commission for Educational Exchange between the USA, Belgium and Luxembourg, The Belgian National Fund for Scientific Research (NFWO) and from IBM Belgium.

Literature Cited

1. *Nonlinear Optical Properties of Organic Molecules and Crystals*; Chemla, D. S., Zyss, J., Eds.; Academic Press: New York, 1987.
2. Prasad, P. N.; Williams, D. J. *Introduction to Nonlinear Effects in Monomers and Polymers*; John Wiley & Sons: New York, 1991.
3. Boyd, G. T. In *Polymers for Electronic and Photonic Applications*; Wong, C. P., Ed.; Academic Press: New York, 1993; p. 467.
4. Lipscomb, G. F.; Lytel, R. *Mol. Cryst. Liq. Cryst. Sci. Technol. B, Nonlinear Optics* **1992**, *3*, 41.
5. Burland, D. M.; Miller, R. D.; Walsh, C. A. *Chem. Rev.* **1994**, *94*, 31 and references cited therein.
6. Man, H. T.; Chiang, K.; Hass, D.; Teng, C. C.; Yoon, H. N. *Proc. SPIE* **1990**, *1213*, 77.

7. Wu, J.; Valley, J.; Ermer, S.; Binkley, E.; Kenney, J.; Lipscomb, G.; Lytel, R. *Appl. Phys. Lett.* **1991**, *58*, 225.
8. Wu, J. W.; Valley, J. F.; Stiller, M.; Ermer, S.; Binkley, E. S.; Kenney, J. T.; Lipscomb, G. F.; Lytel, R. *Proc. SPIE* **1991**, *1560*, 196.
9. Wu, J. W.; Binkley, E. S.; Kenney, J. T.; Lytel, R.; Garito, A. F. *J. Appl. Phys.* **1991**, *69*, 7366.
10. Stähelin, M.; Burland, D. M.; Ebert, M.; Miller, R. D.; Smith, B. A.; Twieg, R. J.; Volksen, W.; Walsh, C. A. *Appl. Phys. Lett.* **1992**, *61*, 1626.
11. Walsh, C. A.; Burland, D. M.; Lee, V. Y.; Miller, R. D.; Smith, B. A.; Twieg, R. J.; Volksen, W. *Macromolecules* **1993**, *26*, 3720.
12. Stähelin, M.; Walsh, C. A.; Burland, D. M.; Miller, R. D.; Twieg, R. J.; Volksen, W. *J. Appl. Phys.* **1993**, *73*, 8471.
13. Wong, K. Y.; Jen, A. K.-Y. *J. Appl. Phys.* **1994**, *75*, 3308.
14. Bales, S. E.; Brennan, D. J.; Gulotty, R. J.; Haag, A. P.; Inbasekanan, M. N. U.S. Patent 5,208,299 (1993).
15. Lindsay, G. A.; Stenger-Smith, J. D.; Henry, R. A.; Hoover, J. M.; Nissan, R. A.; Wynne, K. J. *Macromolecules* **1992**, *25*, 6075.
16. Yang, S.; Peng, Z.; Yu, L. *Macromolecules* **1994**, *27*, 5858.
17. Moylan, C. R.; Twieg, R. J.; Lee, V. Y.; Miller, R. D.; Volksen, W.; Thackara, J. I.; Walsh, C. A. *Proc. SPIE* **1994**, *2285*, 17.
18. Zysset, B.; Ahlheim, M.; Stähelin, M.; Lehr, F.; Prêtre, P.; Karitz, P.; Günter, P. *Proc. SPIE* **1993**, *2025*, 70.
19. Becker, M. W.; Spochak, L. S.; Ghosen, R.; Xu, C.; Dalton, L. R.; Shi, Y.; Steier, W. H.; Jen, A. K.-Y. *Chem. Mater.* **1994**, *6*, 104.
20. Lon, J. T.; Hubbard, M. A.; Marks, T. J.; Lin, W.; Wong, G. K. *Chem. Mater.* **1992**, *4*, 1148.
21. Miller, R. D.; Betterton, K. M.; Burland, D. M.; Lee, V. Y.; Moylan, C. R.; Twieg, R. J.; Walsh, C. A.; Volksen, W. *Proc. SPIE* **1994**, *2042*, 354.
22. Moylan, C. R.; Twieg, R. J.; Lee, V. Y.; Swanson, S. A.; Betterton, K. M.; Miller, R. D. *J. Am. Chem. Soc.* **1993**, *115*, 12,599.
23. Ermer, S.; Leung, D. S.; Lovejoy, S. M.; Valley, J. F.; Stiller, M. *Proceedings Organic Films for Photonic Applications Technical Digest*, American Chemical Society and Optical Society of America, 1993, Vol. 17, p. 50.
24. Shi, R. F.; Wu, M. H.; Yamada, S.; Cai, Y. M.; Garito, A. F. *Appl. Phys. Lett.* **1993**, *63*, 1173.
25. Twieg, R. J.; Betterton, K. M.; Burland, D. M.; Lee, V. Y.; Miller, R. D.; Moylan, C. R.; Volksen, W.; Walsh, C. A. *Proc. SPIE* **1993**, *2025*, 94.
26. Twieg, R. J.; Burland, D. M.; Hedrick, J.; Lee, V. Y.; Miller, R. D.; Moylan, C. R.; Seymour, C. M.; Volksen, W.; Walsh, C. A. *Proc. SPIE* **1994**, *2143*, 1.
27. Moylan, C. R.; Swanson, S. A.; Walsh, C. A.; Thackara, J. I.; Twieg, R. J.; Miller, R. D.; Lee, V. Y. *Proc. SPIE* **1993**, *2025*, 192.
28. *The Chemistry of Diazonium and Diazo Groups, Parts 1 and 2*; Patai, S., Ed.; J. Wiley & Sons: New York, 1973.
29. Ellwood, M.; Griffiths, J. *J. Chem. Soc. Chem. Comm.* **1980**, *18*.
30. Miller, R. D.; Lee, V. Y.; Twieg, R. J. (submitted for publication).

31. Gauthier, S.; Fréchet, J. M. J. *Synthesis* **1987**, 383.
32. Still, C.; Kahn, M.; Mitra, A. *J. Org. Chem.* **1978**, *43*, 2973.
33. Miller, R. D.; Burland, D. M.; Dawson, D.; Hedrick, J.; Lee, V. Y.; Moylan, C. R.; Twieg, R. J.; Volksen, W.; Walsh, C. A. *Polym. Prepr.* **1994**, *35(2)*, 122.
34. Hughes, D. L. In *Organic Reactions*; Paquette, L. A., Ed.; John Wiley & Sons: New York, 1992, Vol. 42, Chap. 2.
35. Miller, R. D.; Hawker, C.; Lee, V. Y. (submitted for publication).
36. Moricheré, D.; Dentan, V.; Kazjar, F.; Robin, P.; Levy, Y.; Dumont, M. *Optics. Commun.* **1989**, *74*, 69.
37. Teng, C.; Man, H. *Appl. Phys. Lett.* **1991**, *58*, 435.
38. Chollet, P.-A.; Gadret, G.; Kajzar, F.; Raimond, P. *Thin Solid Films* **1994**, *242*, 132.
39. Valley, J. F.; Wu, J. W.; Valencia, C. L. *Appl. Phys. Lett.* **1990**, *57*, 1084.
40. Thackara, J. I.; Jurich, M.; Swalen, J. D. *J. Opt. Soc. Am. B* **1994**, *11*, 835.

RECEIVED March 30, 1995

Chapter 11

Highly Efficient and Thermally Stable Second-Order Nonlinear Optical Chromophores and Electrooptic Polymers

Alex K.-Y. Jen[1], Varanasi Pushkara Rao[1], and Jayaraman Chandrasekhar[2]

[1]EniChem America, Inc., Research and Development Center, 2000 Cornwall Road, Monmouth Junction, NJ 08852
[2]Department of Organic Chemistry, Indian Institute of Science, Bangalore 560012, India

Organic polymeric electro-optic (E-O) materials have attracted significant attention because of their potential use as fast and efficient components of integrated photonic devices *(1,2)*. However, the practical application of these materials in optical devices is somewhat limited by the stringent material requirements imposed by the device design, fabrication processes and operating environments. Among the various material requirements, the most notable ones are large electro-optic coefficients (r33) and high thermal stability *(3)*. The design of poled polymeric materials with high electro-optic activity (r33) involves the optimization of the percent incorporation of efficient (large $\beta\mu$) second order nonlinear optical (NLO) chromophores into the polymer matrices and the effective creation of poling-induced non-centrosymmetric structures. The factors that affect the material stability are a) the inherent thermal stability of the NLO chromophores, b) the chemical stability of the NLO chromophores during the polymer processing conditions, and c) the long-term dipolar alignment stability at high temperatures. Although considerable progress has been made in achieving these properties *(4)*, organic polymeric materials suitable for practical E-O device applications are yet to be developed. This chapter highlights some of our approaches in the optimization of molecular and material nonlinear optical and thermal properties.

Push-Pull Heteroaromatic Stilbenes: Experimental and Theoretical Studies

For some time, the EniChem group has been studying thiophene incorporated donor-acceptor (D-A) compounds for second order nonlinear optical applications *(5-9)*. Our initial studies focused on the replacement of the benzene rings of the well known N,N-diethylamino-4-nitrostilbene (DANS, 1) with thiophene moieties *(5)*. The rationale was that the lower resonance energy of thiophene relative to benzene should reduce the aromatic delocalization and thereby increase the electronic transmission between donor and acceptor substituents resulting in an increase of β. Indeed, the experimental EFISH studies (1.907 μm, dioxane solvent) revealed that the $\beta\mu$ value of 2 was significantly higher than that of 1 (Table I). This result in conjunction with the comparable dipole moments of 1 (7.3 debye) and 2 (7.1 debye) demonstrated that the increase in $\beta\mu$ upon thiophene substitution is solely due to the increase in first-order hyperpolarizability, β.

0097–6156/95/0601–0147$12.00/0

Table I: Experimental Linear and Nonlinear Optical Properties (EFISH data, measured at 1.907 μm) of a Donor-Acceptor Substituted Stilbene and its Thiophene Analog.

Compound	λ_{max} (nm)	μ (debye)	$\beta\mu$ 10^{48} (esu)	β 10^{30} (esu)	β_0 10^{30} (esu)
1 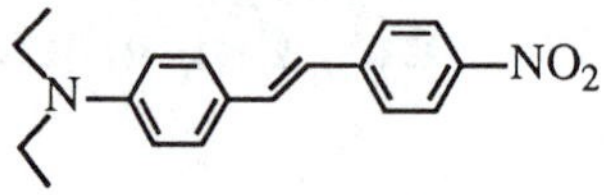 (diethylamino–phenyl–CH=CH–phenyl–NO_2)	424	7.3	580	79.5	60.6
2 (pyrrolidino–thienyl–CH=CH–thienyl–NO_2) 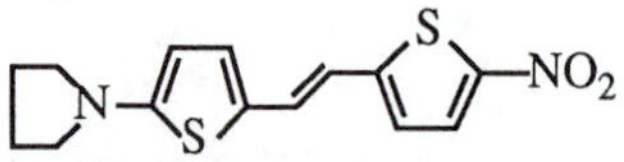	516	7.1	1040	146.5	95.6

In view of the results obtained with thiophene stilbene derivatives such as 2, we were interested in understanding the role of other common five-membered heteroaromatic rings such as furan and pyrrole in influencing the molecular NLO properties. While there are some scattered reports on furan substituted NLO compounds *(9,10)*, pyrrole substituted NLO compounds have never been reported. In order to gain some insight into how pyrrole and furan affect the molecular nonlinearity with respect to thiophene, we initiated theoretical investigations of the second order NLO responses of two classes (3-5 and 6-8) of nitro-amino substituted heteroaromatic stilbenes (Table II). These structures differ only in the nature of heteroaromatic rings and their positions in the molecular frame-work. In stilbenes 3-5, the ring attached to the donor moiety is the heteroaromatic ring, whereas in structures 6-8, the acceptor substituted ring is the heteroaromatic ring.

First, geometry optimizations were performed at the MNDO level *(11,12)* assuming planar geometries for all the structures considered. Then, complete geometry optimizations (which also include dihedral angles) were carried out using AM1 method *(11,12)*. Hyperpolarizability calculations were performed on the planar as well as the fully optimized geometries. We employed the commonly used finite-field and sum-over-states methods for the computation of the first-hyperpolarizability *(13)*. In the finite-field method, the molecular hyperpolarizabilities are obtained from the energy derivatives of the ground state energy. The alternative sum-over-states approach involves the use of a perturbation expression in terms of the excited states of the molecule. Both methods have been extensively used for the calculation of β in conjunction with semi-empirical as well as ab initio molecular orbital methods. We performed calculations both at the ab initio (4-31G) and semi-empirical (AM1/CI) levels *(8, 14,15)*. In the latter, single and pair excitations over a space of 18 orbitals were considered. Contributions from about 150 singlet states were included. However, as the first excited state is believed to be the dominant excited state that contributes to the first-hyperpolarizability in many molecular systems *(10,16)*, we have also highlighted some of the relevant excited state parameters such as the transition wave length (λ) and the change in dipole moments ($\Delta\mu$) between the ground and the first excited states from sum-over-states calculations.

Three sets of calculated hyperpolarizability values are reported here. Using planar geometries, the first two sets of β values were computed at ab initio and AM1/CI levels. Additional AM1/CI calculations were performed on fully optimized geometries. In all the structures examined, the β values obtained from AM1 calculations are higher (by a factor of 2-3) than those obtained from ab initio calculations. However, the trends in β values are virtually identical in all the three sets of calculations. The β values obtained for fully optimized geometries are lower than those obtained for planar geometries and this emphasizes the relationship between molecular planarity and effective charge-transfer in donor-acceptor substituted stilbenes *(17)*. Comparison of the computed β values obtained for 3-5 indicates that pyrrole and furan placed on the donor end are more effective in enhancing the molecular nonlinear response than thiophene (3 < 4 < 5). On the other hand, β values obtained for structures 6-8 suggest that the converse is true (6 > 7 > 8) when the substituted heteroaromatic rings act as acceptors.

Table II: Calculated Hyperpolarizabilities (β_0), Transition Energies (λ) and Change in Dipole Moments (Δμ) for Several Amino-Nitro Substituted Heteroaromatic Stilbenes

Compound	*ab initio*	*AM1*	*AM1(fully opt. Geom.)*		
	$\beta_0 10^{30}$ (esu)	$\beta_0 10^{30}$ (esu)	$\beta_0 10^{30}$ (esu)	λ (nm)	Δμ (debye)
3 H_2N–(S)–NO_2	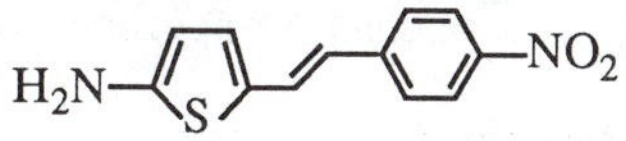 62.5	161.8	142.9	376.6	28.6
4 H_2N–(O)–NO_2	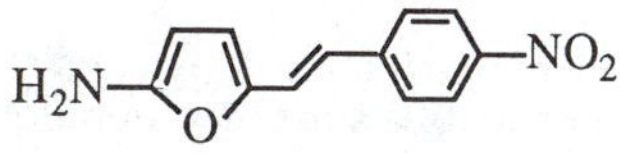 69.4	174.1	153.1	379.5	32.1
5 H_2N–(NH)–NO_2	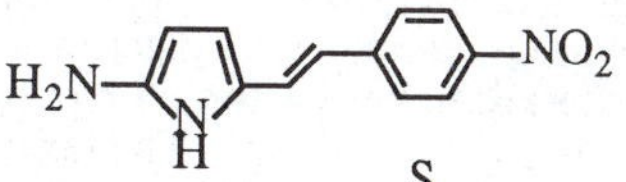 90.4	212.6	182.9	384.2	35.5
6 H_2N–(S)–NO_2	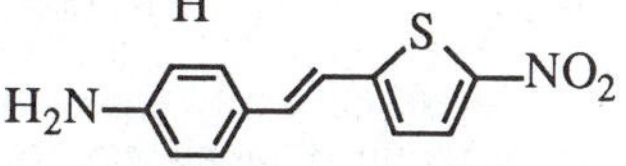 58.7	183.1	165.1	380.5	30.6
7 H_2N–(O)–NO_2	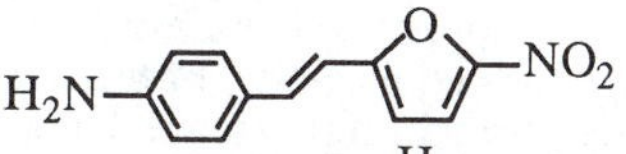 51.5	153.9	138.4	373.4	27.4
8 H_2N–(NH)–NO_2	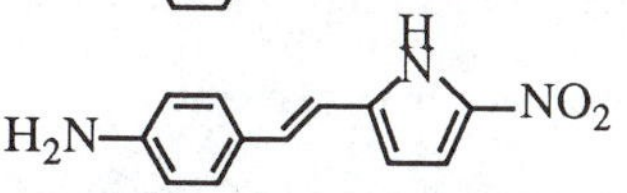 43.2	132.7	105.1	365.0	22.2

These trends suggest that heteroaromatics play a subtle role in influencing the molecular nonlinear optical properties of donor-acceptor compounds. Because the second order nonlinear optical responses depend on the extent of intramolecular charge-transfer in the donor-acceptor systems, we tend to correlate the observed trends in β to a few relevant parameters of heteroaromatic rings. These parameters include the resonance energy as well as the electron-rich or electron-deficient characteristics of heteroaromatic rings. While the resonance energy of heteroaromatic rings affects the electronic transmission between donor and acceptor substituents, their electron-rich or deficient characteristics affect the overall electron-donating and electron-accepting effects. For the heteroaromatic rings considered here, the experimental resonance energies decrease in the order: thiophene (28 Kcal/mol) > pyrrole (21 Kcal/mol) > furan (16 Kcal/mol); their electron-rich character (assessed from the chemical reactivity) increases in the order: thiophene << furan < pyrrole *(18)*. If the aromatic delocalization is a dominant factor in determining β, one would expect much higher calculated β values for furan derivatives 4 and 7, with respect to the corresponding pyrrole and thiophene derivatives (5 and 8, 3 and 6, respectively). The calculated β values indicate that it may not be the case; instead they suggest that furan and pyrrole either enhance or decrease the molecular nonlinearity relative to thiophene depending on where they are located in the molecular frame-work. The highly electron-rich nature of pyrrole is due to the ring-nitrogen which serves as an additional electron-donor and assists the amino group to produce a more powerful electron-donating effect in compound 5, whereas in compound 8, the opposite effect occurs because the ring nitrogen counteracts the electron-withdrawing effect of the nitro group. The resulting electron-donating effect in 5 may account for its higher β value, whereas the lower β value of 8 can be attributed to its weaker electron-withdrawing effect. One would expect a similar trend with furan containing compounds 4 and 7, as furan is also very electron-rich in nature.

Another important piece of information which can be discerned from our calculations is the relationship between β and the first excited state properties (e.g., the change in dipole moment between the ground and excited states [$\Delta\mu$] and the transition wave length [λ]) in this class of compounds. Based on the perturbation expression, larger values of λ, $\Delta\mu$, and transition dipole matrix element r would lead to an increased β. Within each of the two series of heteroaromatic compounds, r values are roughly constant, while transition wavelengths show a small variation consistent with the trends in β. The dominant factor determining the variations in β appears to be $\Delta\mu$.

Highly Efficient Nonlinear Optical Chromophores

An approach which was successfully implemented in the design and synthesis of highly efficient nonlinear optical chromophores involves the incorporation of the highly electron-defficient tricyanovinyl group in electron-donor substituted thiophene systems *(19,20)*. Although the role of the tricyanovinyl group as an electron-acceptor in the design of nonlinear optical chromophores has been known for a long time, very limited number of its benzenoid derivatives have been reported *(21)*. The rarity of this class of compounds can be attributed to the difficulty in functionalizing the benzenoid systems (e.g. donor substituted stilbenes, azobenzenes and schiff bases) with the tricyanovinyl group. We found that electron-donor substituted thiophenes are several orders of magnitude more reactive than the corresponding benzenoid systems in reacting with tetracyanoethylene. Utilizing this enhanced reactivity of thiophenes, we synthesized several tricyanovinyl derivatives containing extended conjugated thiophene systems. Table III contains a few of the highly efficient tricyanovinylthiophene NLO chromophores synthesized in our laboratories.

Table III: Optical Properties of Some Highly Efficient NLO Chromophores

Compound	λ_{max} (nm)	$\beta\mu$ 10^{48} (esu)	$\beta_0\mu$ 10^{48} (esu)
9	640	6200	3023
10	653	7400	3469
11	662	9100	4146

EFISH studies (measured at 1.907 μm in dioxane solvent) revealed that 9 had one of the largest βμ values so far reported for stilbene type donor-acceptor compounds. We have since demonstrated and observed a much enhanced nonlinearity by increasing the conjugation length as seen in compounds 10 and 11. The relative magnitude of the βμ values of 10 and 11 indicate that olefinic moieties are more effective in increasing the molecular nonlinearity than thienylvinyl units in designing extended conjugated systems. To translate these impressive molecular nonlinear optical properties into material properties, we have incorporated compounds 10 and 11 into acrylate and polyimide matrices and studied their poling behavior and the electro-optic response. These material properties are discussed in a latter section.

Tradeoffs Between Thermal Stability and Molecular Nonlinearity

For the organic NLO chromophores to be useful for electro-optic device applications, the chromophores must posses large nonlinear optical coefficients as well as high thermal stability *(3)*. Although the tricyanovinylthiophene derivatives 9-11 possess dramatically enhanced molecular nonlinearities, their stability at high temperatures (>250°C) is very much limited by the olefinic conjugating linkages present. At high temperatures, the olefinic linkages undergo cis-trans isomerization which leads to a significant reduction of molecular nonlinearity. Furthermore, this cis-trans isomerization results in a local concentration of π-electron density, which in turn favors a reaction with electrophiles such as singlet oxygen. To eliminate the thermal instability associated with olefinic linkages, we designed a new class of materials *(22)* lacking olefinic bonds, such as 12, and also a set of compounds based on bithiophenes and fused-thiophenes (13 and 14) in Table IV.

Table IV: Optical Properties of Thermally Stable and Efficient NLO Chromophores

Compound	λ_{max} (nm)	$\beta\mu$ 10^{48} (esu)	T_m (°C)	T_d (°C)
12	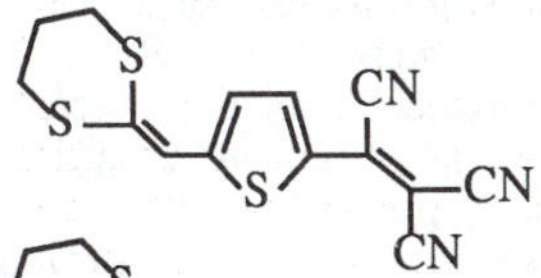 556	940	230	290
13	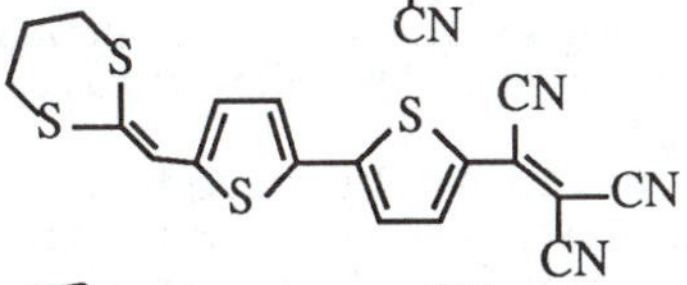 594	1500	215	305
14	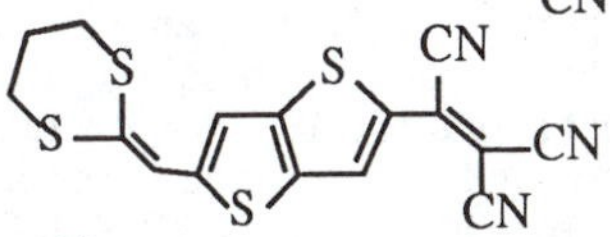 570	2200	270	310
15	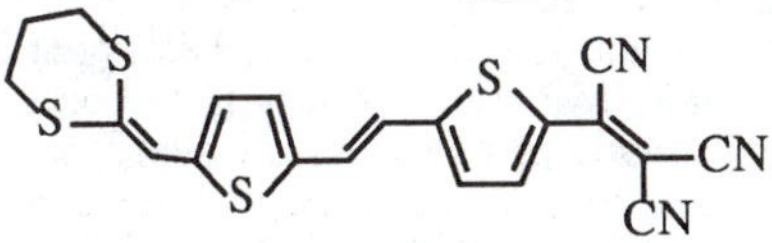 612	3420	252	250

In these compounds, the ketendithioacetal group *(9)* serves as the electron-donor and the tricyanovinyl group as the electron-acceptor. The molecular nonlinear optical properties obtained from EFISH experiments (measured at 1.907 μm in dioxane solvent) and the thermal stability data obtained from DSC technique (20°C per minute) are presented in Table IV. The experimental βμ values indicate that the thiophene-stilbene 15 possesses higher molecular nonlinearity than the fused-thiophene 14 which in turn possesses higher activity than the bithiophene 13 derivative. The molecular nonlinearity of 15 can be attributed to its longer and effective π-conjugation offered by the thiophene-stilbene unit. The higher βμ value obtained for 14 in comparison with 13 can be attributed to the better planarity of the fused-thiophene compared to the bithiophene.

Inherent thermal stability of these compounds were obtained by differential scanning calorimetry (DSC). All the samples were heated in a sealed DSC pan at the heating rate of 20°C per minute, and the decomposition temperatures (T_d) were estimated from the intercept of the leading edge of the decomposition exotherm with the base line. All the four compounds (11-15) possess high thermal stability (>250°C). The bithiophene and fused-thiophene derivatives 13 and 14, respectively, were more stable than the thiophene-stilbene derivative 15. Since the T_d's vary with the heating rate, care must be taken when interpreting the results obtained from this study. For this reason, we have also evaluated their stabilities with isothermal heating process at two different temperatures, 250°C and 275°C, respectively, for half an hour. At 250°C all the four compounds studied were stable, whereas only compounds 13 & 14 were stable at 275°C. These thermal stabilities may be further improved by functionalizing the bithiophene and the fused-thiophene with donor and acceptor groups possessing much higher thermal stabilities.

Electro-optic Polymers: Polyimide Based Guest-Host Materials

In designing nonlinear optical polymers for practical electro-optic device applications, the choice of the polymer system plays a major role in preserving the stability of poling induced noncentrosymmetric structures at high temperatures over extended periods of time. Among the many polymeric systems, polyimides appear to be ideal candidates because of their high glass transition temperatures, low dielectric constants and compatibility with semiconductor processes *(23)*. Recently, several reports on organic thin-film based electro-optic polymers utilized polyimides as host matrices *(24-27)*. Although the reported results with polyimide host matrices are very encouraging in view of optimizing the long-term poling stability, the chemical degradation of NLO chromophores within the polymer environment during the high temperature imidization and poling processes imposes limitations on the selection of chromophores.

Using polyimides as host matrices, we have developed a new class of electro-optic materials containing some of our highly efficient heteroaromatic chromophores *(28)*. One such chromophore used for this purpose is the tricyanovinylthiophene derivative 9 (Table III). Guest-host systems were prepared by dissolving 10-15 wt. % of 9 in Hitachi PIQ-2200 polyamic acid (solvent system: NMP). Polymeric thin-films were formed by spin-coating the combined dye/polymer solution on an indium tin oxide coated glass substrate. The films were kept in a vacuum oven at 120°C for 10 h to remove the solvent. At this stage, the films were imidized by heating at 200°C for 30 minutes. During this process, residual NMP complexed with the polyamic acid and the by products of imidization were, to a large extent, eliminated from the films. A thin layer of gold was vacuum evaporated onto the imidized polymer films to serve as the top electrode for poling. The samples were poled at 220°C for 10 minutes with an applied dc electric field (1.0 MV/cm).

Comparison of the absorption spectra of the films before and after curing, revealed that the tricyanovinylthiophene derivative 9 was intact during the high temperature imidization process. The electro-optic activity of the poled polymer films was measured at 1.3 μm by an experimental setup similar to that described by Teng and Mann *(29)*. The measured electro-optic coefficients were in the range of 15-20 pm/V for polymer films containing 10-15 wt. % of 9. These values can be further improved by optimizing the poling parameters such as poling field and poling temperatures. The stability of electro-optic activity was examined at 120°C, before and after physical aging. Non-physically aged sample showed an initial drop of 20 % and an additional drop of 20% over next 600 h. On the other hand, the same sample when physically aged showed an initial drop of only 10% and a gradual drop of another 10% over 600 h. Thus, through the use of high glass transition temperature polyimide and physical aging, we achieved a significant improvement of the thermal stability of the poled polymer films.

Electro-optic Polymers: Side-Chain Polyimides

Despite the improvement in the long-term stability of the electro-optic activity, the polyimide based guest-host materials approach has certain disadvantages. These include: a) sublimation of NLO chromophore from the polymeric films at device fabrication temperatures (>275°C), b) higher loading levels of chromophores leading to plasticization effects, thus resulting in lower Tg's and lower long-term thermal stabilities. To overcome these deficiencies associated with guest-host materials, various approaches have been developed to covalently bond the NLO chromophores to the polymer matrices *(30)*. However, very few reports on NLO functionalized polyimides have appeared, perhaps due to synthetic difficulties *(31-37)*. Dalton's group *(31)* and Yu's group *(32)* have developed synthetic schemes for aliphatic polyimides containing Disperse Red and DANS as side chains, respectively. In this connection, we have developed *(33)* a novel two stage process to obtain side-chain aliphatic polyimides containing tricyanovinyl substituted NLO chromophores. Although these aliphatic polyimides provide fairly promising results, there is a strong need to further improve their mechanical properties (for multilayer integration) and thermal stability in order to accommodate the processing requirements of electro-optic devices. Aromatic polyimides possess much higher thermal stability (higher glass-transition temperatures) and provide almost unlimited variations of monomers for fine-tuning the structural and electrical properties *(36-37)*. For this reason, we developed a synthetic methodology for the incorporation of tricyanovinyl chromophores into aromatic polyimides (Scheme I).

Compound 18 was obtained in 92% yield by condensing 2,5-dinitrophenol 16 with 2-(N-ethylanilino)ethanol 17 in the presence of triphenylphosphine and diethylazodicarboxylate (DEAD) in tetrahydrofuran [Mitsunobu reaction,*(38)*]. The dinitro compound 18 was then reduced by hydrogenation in dimethylformamide (DMF) with a catalytic amount of palladium/carbon to give the diamine 19 as the viscous liquid (80% yield). The diamine 19 was reacted with 2,2'-bis-(3,4-dicarboxyphenyl)hexafluoropropane dianhydride (6FDA) 20 (1 equiv.) in NMP at 0°C (N_2). The concentration of this reaction mixture was adjusted to obtain a solution with 15% by mass solid content. After approximately 12 h, 1,2-dichlorobenzene (same amount as original NMP) was added to the solution as a co-solvent to aid in the removal of water formed during the imidization process. This process was performed by refluxing the resulting solution for 3 h. The hot solution was added drop-wise to methanol and the precipitate 21 was collected. Spectroscopic evidence indicate that 21 was completely imidized under these conditions. The post tricyanovinylation of polyimide 21 was carried out by dissolving 21 and tetracyanoethylene (TCNE) (1.1 equiv.) in DMF and heating at 70°C (N_2) for 24 h

(92% yield). The structures of 21 and 22 were confirmed by conventional spectroscopic techniques (IR, UV and ^{1}H NMR) and elemental analysis.

Scheme 1

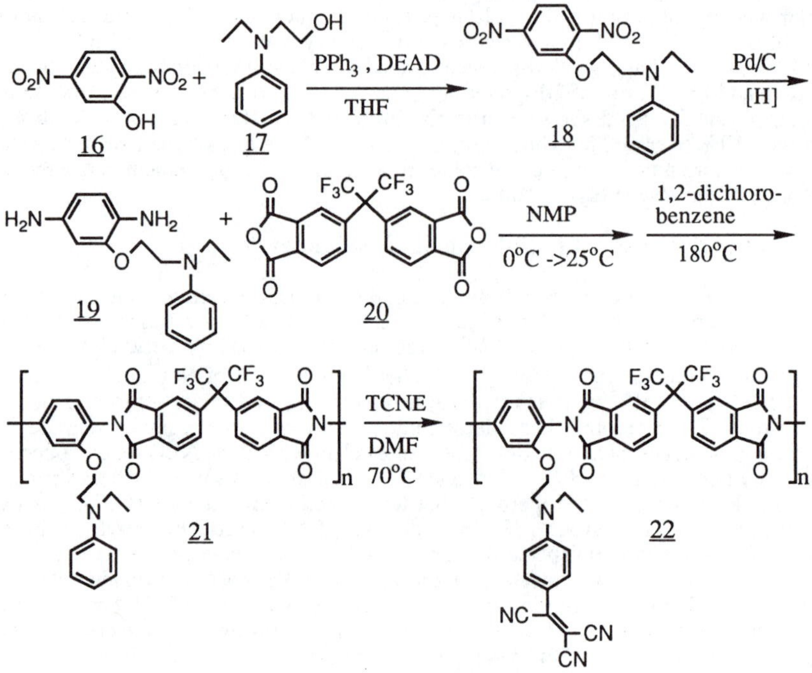

The polyimide 22 (32 % m/m of the NLO chromophore N,N-diethylamino-4-tricyanovinyl-benzene) was very soluble in polar solvents such as cyclohexanone, dimethyl sulfoxide and dimethylacetamide. The Tg (224°C) of this aromatic polyimide was 350°C higher than the aliphatic polyimide containing the same chromophore and the loading level *(33)*. This could be attributed to the hindrance of the polymer chain motion by the more rigid aromatic structure of this polyimide. From the TGA studies, the polyimide 22 was thermally stable up to 310°C under nitrogen atmosphere. These studies were based on dynamic heating at a relatively faster rate (20°C/min) and cause some ambiguity in determining the inherent thermal stability of the polymer. For this reason, we heated a thin film sample of the polymer on a hot stage isothermally at 250°C (0.5 h) and 275°C (0.5 h). The π-π^* charge-transfer absorption band of the chromophore was used to monitor the decomposition temperature. At 275°C, there was less than 5% change in the intensity of the absorption indicating the high thermal stability of this NLO polymer. Thin films (1-2 μm) of 22 were spin-coated onto an indium tin oxide (ITO) glass substrate using a 17 % m/m solution (filtered through 0.2 μm syringe filter) of the resin in dimethylacetamide. The films were kept in a vacuum oven at 120°C for 10 days and then briefly heated in a hot stage at 215°C (N_2) for 15 minutes to ensure the removal of the residual solvent. A thin layer of gold was vacuum evaporated on the polyimide films to serve as the top electrode for poling. The samples were poled at 215°C for 5 minutes with an applied DC electric field of

0.5 $MVcm^{-1}$, cooled to room temperature and the poling field was subsequently removed. The E-O activity (r_{33}) of the polymer film was measured at a wavelength of 0.83 μm and was 15 pm/V.

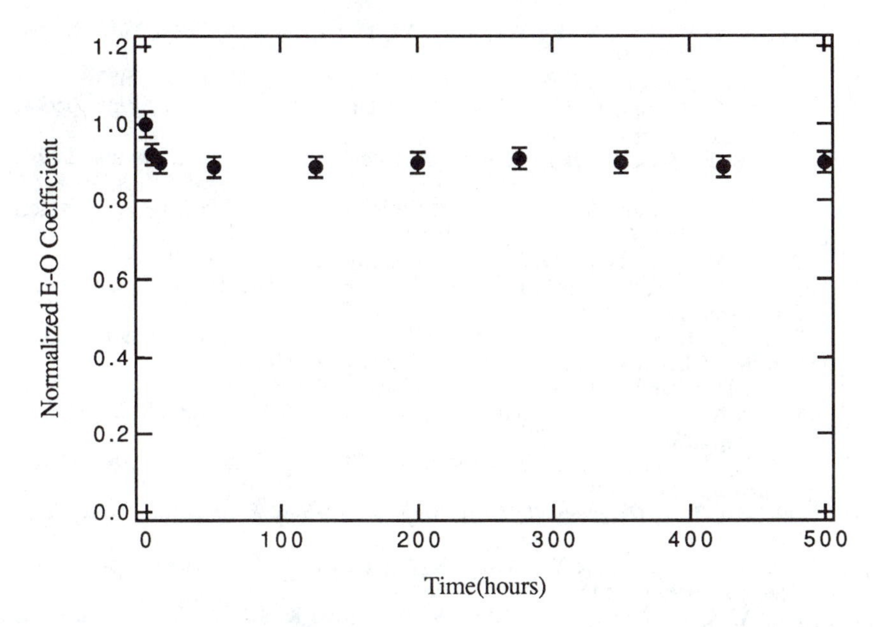

Figure I: Temporal Stability of the Poled Side-chain Polyimide 22 at 100°C in air (Normalized r_{33} is plotted as a function of baking time).

The thermal stability of the poled side-chain polyimide was demonstrated by heating the poled sample in an oven at 100°C for over 1000 h. The E-O activity of the sample shows an initial drop to 90% of its original value within 10 h then remains unchanged (Figure I). Thus, through the use of side-chain aromatic polyimides we have demonstrated the improved thermal stability of the poled NLO films. We are currently involved in developing electro-optic polyimides that contain highly active and thermally stable heteroaromatic NLO chromophores as side-chains.

Acknowledgements

We thank Dr. K. J. Drost, Y. Cai, K. Y. Wong and Y. Liu for their contributions in the synthesis and characterization of some of the compounds and Dr. N. Namboothiri for the computational work presented in this chapter. We are grateful to Dr. R. M. Mininni for his continuing support and encouragement. Work at IISc was supported by a grant from ISRO-IISc Space Technology Cell.

Literature Cited

1. *Polymers for Lightwave and Integrated Optics;* Hornak, L. A., Ed.; Marcel Dekker, New York, 1992.
2. *Nonlinear Optical Properties of Organic Molecules and Crystals;* Chemla, D. S.; Zyss, J., Eds.; Academic Press, New York, 1987.

3. Lytel, R.; Lipscomb, G. S. *Mater. Res. Soc. Symp. Proc.* **1992**, *247*, 17.
4. *Materials for Nonlinear Optics: Chemical Perspectives;* Marder, S. R.; Sohn, J. E.; Stucky, G. D., Eds.; ACS Symposium Series 455, Washington, 1991.
5. Rao, V. P.; Jen, A. K-Y.; Wong, K. Y.; Drost, K. J. *Tetrahedron Lett.* **1993**, *34*, 1747.
6. Jen, A. K-Y.; Rao, V. P.; Wong, K. Y.; Drost, K. J. *J. Chem. Soc. Chem. Commun.* **1993**, 90.
7. Wong, K. Y.; Jen, A. K-Y.; Rao, V. P. *Phys. Rev. A* **1994**, *49*, 3077.
8. Wong, K. Y.; Jen, A. K-Y.; Rao, V. P.; Drost, K. J. *J. Chem. Phys.* **1994**, *100*, 6818.
9. Rao, V. P.; Cai, Y.; Jen, A. K-Y. *J. Chem. Soc. Chem. Commun.* **1994**, 1689.
10. Cheng, L. T.; Tam, W.; Marder, S. R.; Stiegman, A. E.; Rikken, G.; Spangler, C. W.; *J. Phys. Chem.* **1991**, *95*, 10631.
11. Dewar, M. J. S.; Thiel, W. *J. Am. Chem. Soc.* **1977**, *99*, 4899.
12. Dewar, M. J. S.; Zoebisch, E. G.; Healy, E. F.; Stewart, J. J. P. *J. Am. Chem. Soc.* **1985**, *107*, 3902.
13. Kanis, D. R.; Ratner, M. A.; Marks, T. J. *Chem. Rev.* **1994**, *94*, 195.
14. Jain, M.; Chandrasekhar, J. *J. Phys. Chem.* **1993**, *97*, 4044.
15. Clark, T.; Chandrasekhar, J. *Israel J. Chem.* **1993**, *33*, 435.
16. Singer, K. D.; Sohn, J. E.; King, L. A.; Gordon, H. M.; Katz, H. E.; Dirk, C. W. *J. Opt. Soc. Am. B* **1989**, *6*, 1339.
17. Kodaka, M.; Fukaya, T.; Yonemoto, K.; Shibuya, I. *J. Chem. Soc. Chem. Commun.* **1990**, 1609.
18. Gilchrist, T. L. *Heterocyclic Chemistry*, John Wiley & Sons Inc., New York, 1985.
19. Rao, V. P.; Jen, A. K-Y.; Wong, K. Y.; Drost, K. J. *J. Chem. Soc. Chem. Commun.* **1993**, 1118.
20. Jen, A. K-Y.; Wong, K. Y.; Rao, V. P.; Drost, K. J.; Cai, Y. *J. Electronic Mater.* **1994**, *23*, 653.
21. Katz, H. E.; Singer, K. D.; Sohn, J. E.; Dirk, C. W.; King, L. A.; Gordon, H. M. *J. Am. Chem. Soc.* **1987**, *109*, 6561.
22. Rao, V. P.; Wong, K. Y.; Jen, A. K-Y.; Drost, K. J. *Chem. Mater.* **1994**, *6*, 2210.
23. *Polyimides;* Wilson, D.; Stenzenberger, H. D.; Hergenrother, P. M., Eds.; Chapman and Hall, New York, 1990.
24. Wu, J.; Valley, J. F.; Ermer, S.; Binkley, E. S.; Kenney, J. T.; Lipscomb, G. F.; Lytel, R. *Appl. Phys. Lett.* **1991**, *58*, 225.
25. Wu, J.; Valley, J. F.; Ermer, S.; Binkley, E. S.; Kenney, J. T. *Appl. Phys. Lett.* **1991**, *59*, 2214.
26. Valley, J. F.; Wu, J.; Ermer, S.; Stiller, M.; Binkley, E. S.; Kenney, J. T.; Lipscomb, G. F.; Lytel, R. *Appl. Phys. Lett.* **1992**, *60*, 160.
27. Jeng, R. J.; Chen, Y. M.; Jain, A. K.; Kumar, J.; Tripathy, S. K.; *Chem. Mater.* **1992**, *4*, 1141.
28. Wong, K. Y.; Jen, A. K-Y. *J. Appl. Phys.* **1994**, *75*, 3308.
29. Teng, C. C.; Mann, H. A. *Appl. Phys. Lett.* **1990**, *56*, 1734.
30. Burland, D. M.; Miller, R. D.; Walsh, C. A. *Chem. Rev.* **1994**, *94*, 31.
31. Becker, M.; Sapochak, L.; Ghosen, R.; Xu, C.; Dalton, L. R.; Shi, Y.; Steier, W. H.; Jen, A. K-Y. *Chem. Mater.* **1994**, *6*, 104.
32. Peng, Z.; Yu, L. *Macromolecules* **1994**, *27*, 2638.
33. Jen, A. K-Y.; Drost, K. J.; Cai, Y.; Rao, V. P.; Dalton, L. R. *J. Chem. Soc. Chem. Commun.* **1994**, 965.
34. Zysset, B.; Ahlheim, M.; Stahelin, M.; Lehr, F.; Pretre, P.; Kaatz, P.; Gunter, P. *Proc. SPIE* **1993**, *2025*, 70.

35. Park, J.; Marks, T.; Yang, J.; Wong, G. K. *Chem. Mater.* **1990**, *2*, 229.
36. Jen, A. K-Y.; Liu, Y.; Cai, Y.; Rao, V. P.; Dalton, L. R. *J. Chem. Soc. Chem. Commun.* **1994**, 2711.
37. Yu, D.; Yu, L. *Macromolecules* **1994**, *27*, 6718.
38. O. Mitsunobu, *Synthesis* **1981**, 1.

RECEIVED April 25, 1995

Chapter 12

Techniques of Ultrastructure Synthesis Relevant to the Fabrication of Electrooptic Modulators

L. R. Dalton[1], B. Wu[1], A. W. Harper[1], R. Ghosn[1], Y. Ra[1], Z. Liang[1], R. Montgomery[1], S. Kalluri[2], Y. Shi[3], W. H. Steier[2], and Alex K.-Y. Jen[4]

[1]Loker Hydrocarbon Research Institute, University of Southern California, Los Angeles, CA 90089–1661
[2]Department of Electrical Engineering, University of Southern California, Los Angeles, CA 90089–0483
[3]Tacan Corporation, 2330 Faraday Avenue, Carlsbad, CA 93111
[4]EniChem America, Inc., Research and Development Center, 2000 Cornwall Road, Monmouth Junction, NJ 08852

The production of thermally stable second order nonlinear optical lattices and the processing of these lattices into buried channel waveguides which can be integrated with fiber optic transmission lines and electronic drive circuitry is discussed. Three techniques for the production of hardened NLO lattices are reviewed; namely, (1) the use of DEC chromophores, with reactive functionalities at both the donor and acceptor ends, to created a hardened lattice by two-step processing with the final step an intermolecular crosslinking reaction, (2) the generation of polyimides containing covalently incorporated NLO chromophores, and (3) the exploitation of thermosetting schemes including sol-gel processing where the NLO chromophore is covalently attached to the reacting species. The adaptation of these three schemes to incorporate high $\mu\beta$ chromophores is discussed. Also reviewed is the production of buried channel waveguides in hardened NLO materials by (1) photochemical processing, (2) reactive ion etching and electron cyclotron resonance etching, and (3) spatially selective poling.

Rapid developments in telecommunications, including the increased use of high bandwidth fiber optic transmission cable, are creating a need for improved electro-optic modulators and directional couplers. However, before commercial utilization of polymeric modulators can be contemplated a variety of criteria must be satisfied including reasonable optical nonlinearity; low optical loss; adequate thermal stability to withstand the temperature variations encountered in device fabrication, integration, and operation; processibility to permit coupling to existing fiber optic lines and electronic drive components; and low manufacturing cost. Indeed, commercial viability will likely require that the performance/cost ratio of polymeric modulators significantly exceed that of the competition (modulated lasers and inorganic modulators).

0097–6156/95/0601–0158$12.00/0

Although octopolar chromophores have been proposed (*1*), modulator development to date has focused upon the use of dipolar chromophores. Unfortunately, dipole-dipole repulsion opposes the straight forward assembly of such chromophores into noncentrosymmetric lattices; thus, production of stable lattices by self-assembly techniques has yet to prove useful for the general fabrication of electro-optic modulators (*2,3*). Sequential synthesis techniques, exploiting Langmuir-Blodgett techniques, Merrifield-type covalent coupling reactions, ionic coupling, and molecular beam epitaxy methods, have produced interesting structures but have not yet produced materials useful in fabricating electro-optic modulators. The method of choice in the fabrication of early prototype modulators has been electric field poling near the glass transition temperature of polymers containing NLO chromophores either as dopants or a covalently bonded components of the polymer structure (*4,5,6,7*). Of course, this approach suffers from the dilemma that electric field poling does not produce a thermodynamically stable lattice and relaxation of poling-induced order following removal of the poling field is a problem which must be addressed. A simple solution would seem to be to find a polymer with a glass transition temperature sufficiently greater than operational and processing temperatures that no significant relaxation of poling-induced order would occur. However, although such polymers are known to exist, unrealistic thermal stability requirements are placed upon NLO chromophores if they are to withstand the temperatures used in poling such materials. A more practical approach appears to be to find polymer materials, which incorporate NLO chromophores, that can be spin cast into optical quality films (that is, exhibit reasonable solubility in processing solvents), can be poled at modest temperatures, and then chemically hardened into a high glass transition lattice capable of withstanding subsequent processing and in-field use. Exploration of this latter approach has been a major objective of our laboratory and is the primary focus of this article.

In pursuing hardened NLO lattices, two quite philosophically different approaches have become popular; the first involves doping of chromophores into polymeric or oligomeric materials to form composite materials while the second focuses upon covalent coupling of the chromophore so that the chromophore is attached by one or more points of attachment to a final hardened polymer lattice. The obvious advantage of the former approach is that of ease and cost of producing NLO materials; frequently, commercially available chromophores can simply be blended with commercially available polymers, e.g., the isotropic poly(ether imide) Ultem (General Electric). Disadvantages of such composite materials include (1) poor chromophore loading (number density) due to finite solubility of the chromophore guest in the polymer host, (2) phase separation and aggregation of chromophores, (3) sublimation of chromophores at high processing temperatures, (4) a plasticizing effect on the host lattice, and (5) dissolution of chromophores with application of cladding layers. In practice, the above limitations of guest/host composites translate into inadequate optical nonlinearity (because of poor loading and problems encountered in poling) and unacceptable limitations placed upon processing for the fabrication of devices. Consequently, we have chosen to focus upon schemes which involve covalent incorporation of chromophores even though such schemes require initially greater synthetic effort.

Lattice Hardening Reactions Utilizing Covalently Incorporated Chromophores.

We proceed to describe the production of hardened NLO materials employing the three schemes alluded to above. A fourth approach, namely the production of interpenetrating networks (*8*), will not be discussed here but is discussed elsewhere in this volume.

Two-Step Processing of Hardened Lattices Using Double-End Crosslinkable (DEC) Chromophores. The general concept of employing DEC chromophores has been introduced elsewhere (*9*). A representative example, employing an azobenzene chromophore, together with representative data on the thermal and temporal stability of the second order optical nonlinearity (measured by second harmonic generation) is given in figure 1. The advantage of the DEC scheme is that spin casting and initial poling of the NLO prepolymer can be carried out without interference from the lattice hardening (crosslinking) reaction. Indeed, AdTech has taken advantage of this feature to make the prepolymer shown in figure 1 commercially available. Recent work has focused upon adaptation of high $\mu\beta$ chromophores, e.g., those incorporating heteroaromatic groups and acceptor groups such as thiobarbituric acid and 3-phenyl-5-isoxazolone, to the DEC format. A representative example is shown in synthesis **scheme 1.**

Scheme 1.

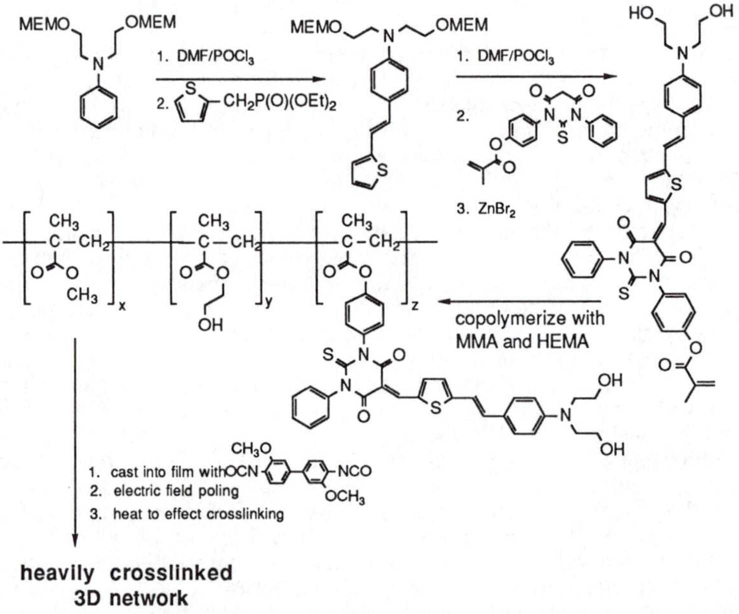

1. $DMF/POCl_3$
2. $CH_2P(O)(OEt)_2$

1. $DMF/POCl_3$
2.
3. $ZnBr_2$

copolymerize with MMA and HEMA

1. cast into film with OCN–[…]–NCO
2. electric field poling
3. heat to effect crosslinking

heavily crosslinked 3D network

Although a number of such materials have been successfully synthesized and chemically characterized, NLO characterization has not been completed at this time. Thus, it is not possible at this time to quantitatively specify the improvement in optical nonlinearity that is to be realized with high $\mu\beta$ chromophores incorporated into hardened DEC lattices or to specify the thermal stability of such lattices.

DEC Monomer

Polymerized with MMA and HEMA

Polymer 2

Crosslinker (XL)

1) Casting film with XL
2) Electric poling
3) Thermal cross-linking

Crosslinked polymer

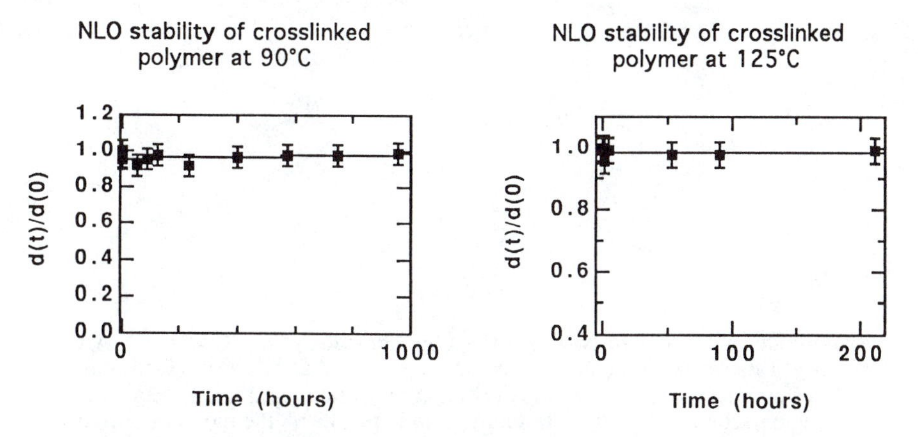

Figure 1. The production of a hardened NLO lattice from a DEC chromophore is shown. Such a chromophore has reactive functionalities at both donor and acceptor ends of the NLO chromophore. Also shown is the thermal stability of the optical nonlinearity for the chromophore shown. The temporal stability of second harmonic generation (relative second harmonic generation efficiency, d(t)/d(0)) is shown at 90 and 125°C.

Two Step Processing of Hardened Polyimide Lattices Covalently Incorporating NLO Chromophores. In **schemes 2 and 3**, we illustrate two general schemes for producing hardened polyimide lattices containing covalently incorporated NLO chromophores. Again, a two step processing protocol is pursued. Spin casting and initial poling is accomplished using a polyamic acid precursor polymer which is then hardened by thermally or chemically effecting an intramolecular condensation. As with intermolecular crosslinking of the DEC approach discussed above, imidization results in a dramatic improvement in glass transition temperature and in the thermal stability of NLO activity (*10*). Representative thermal stability data is given in figure 2. Again, more recent research efforts have focused upon incorporating chromophores with larger $\mu\beta$ values than the azobenzenes of schemes 2 and 3. An example of our efforts in this regard are discussed in detail elsewhere in this volume (*11*).

Scheme 2.

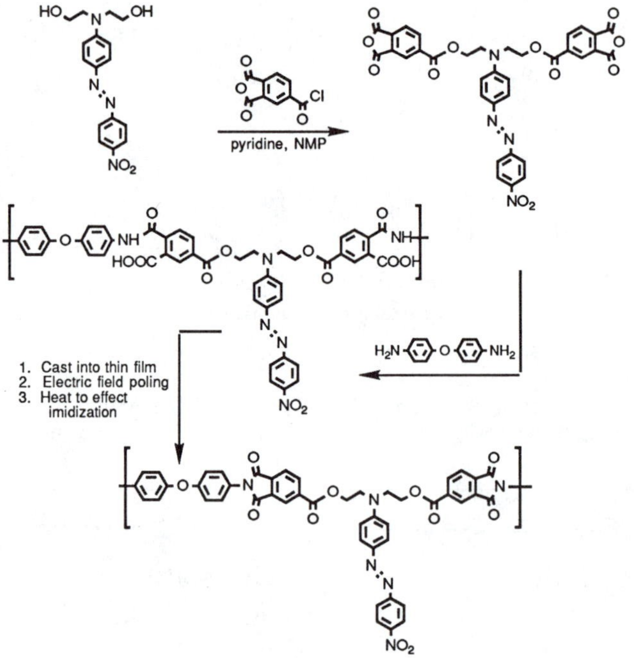

Production of Hardened NLO Lattices by Thermosetting Reactions. A large number of thermosetting reactions have been explored as has been noted in recent reviews by Burland and coworkers (*12*) and by Dalton, et al. (*13*). Here we will focus upon **schemes 4 and 5**. In thermosetting reactions, the monomers involved in the thermally-induced condensation reactions are reacted to the point of producing materials with sufficient viscosity to permit spin casting of films and poling of films without problems from ionic conductivity. Care must be exercised that the hardening reactions are not permitted to proceed to the point of interfering with reorientation of the chromophores under the influence of the poling field.

Scheme 3.

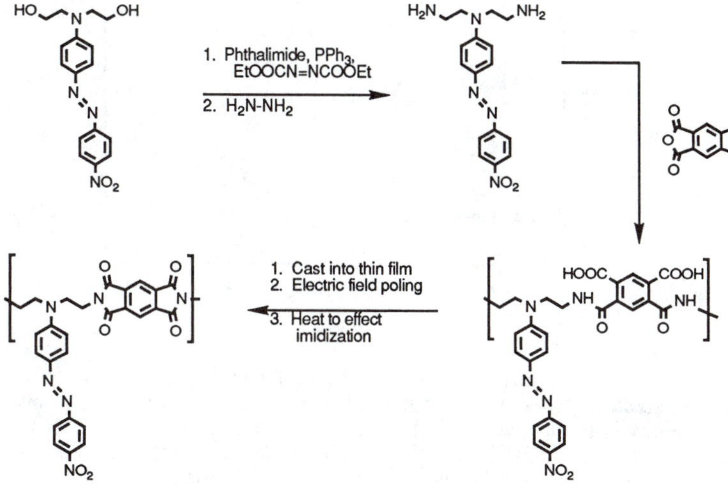

HO
N
OH
N
N
NO2
1. Phthalimide, PPh3,
EtOOCN=NCOOEt
2. H2N-NH2
H2N
N
NH2
N
N
NO2
1. Cast into thin film
2. Electric field poling
3. Heat to effect
imidization
HOOC
COOH
NH
NH

The importance of poling protocol is illustrated in figure 3. When following scheme 5, a poling protocol which involves stepped increases in temperature results in a factor of 1.5 increase in optical nonlinearity compared to simply ramping the temperature to the final temperature (180°C). An optical nonlinearity (second harmonic coefficient, d_{33}) of 27 pm/V (at 1064 nm) was obtained by the stepped protocol. The thermal stability of the optical nonlinearity obtained for the sol-gel material of scheme 5 is shown in figure 4. For the material of scheme 4, optical nonlinearities (d_{33} values) of 59 pm/V and 95 pm/V (at 1064 nm) were obtained for the tri and tetralink thermosetting chromophores, respectively. Representative thermal stability data is shown in figure 5.

Scheme 4.

HO
N
OH
N
N
O S:O
(CH2)2
OH
+
CH3
NCO
NCO
1. Spin cast onto substrate
2. Electric field poling
3. Heat to effect crosslinking
Trilink
3D stabilized
NLO matrix
HO
N
(CH2)6
N
OH
O S O
(CH2)2
OH
1. Spin cast onto substrate
2. Electric field poling
3. Heat to effect crosslinking
Tetralink
3D stabilized
NLO matrix

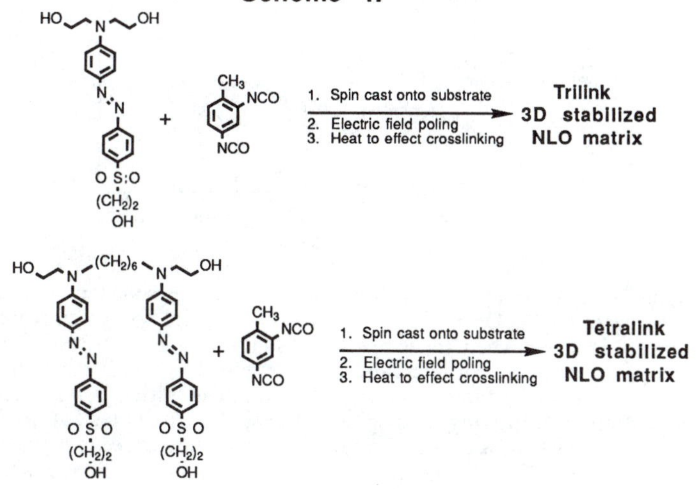

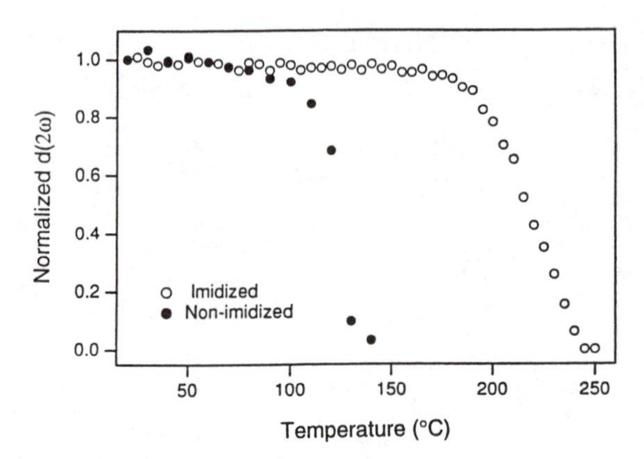

Figure 2. A dynamic assay of the thermal stability of the material synthesized in scheme 2 is shown. The second harmonic generation efficiency is monitored while increasing the temperature of the NLO material at a rate of 10°C per minute. This assay of the thermal stability of poling-induced order is analogous to the assay of the thermal stability of chemical structure effected by thermal gravimetric analysis.

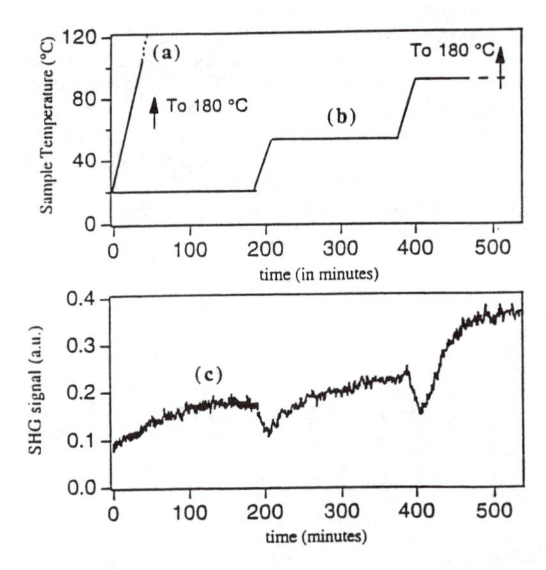

Figure 3. (a) A normal poling protocol is shown. Here the temperature is increased to 180°C while the poling field is applied. (b) A step poling protocol is shown where the temperature is increased in 20 to 30°C steps in approximately 3 hour intervals. (c) The second harmonic generation signal is shown for the step poling protocol. The initial drop in SHG immediately following a step can likely be attributed to ionic conductivity which occurs as the sample temperature exceeds the glass transition temperature. Toward the end of the 3 hour period a leveling off of SHG is observed which reflects the fact that the lattice has hardened to the point of inhibiting further poling-induced alignment of chromophores.

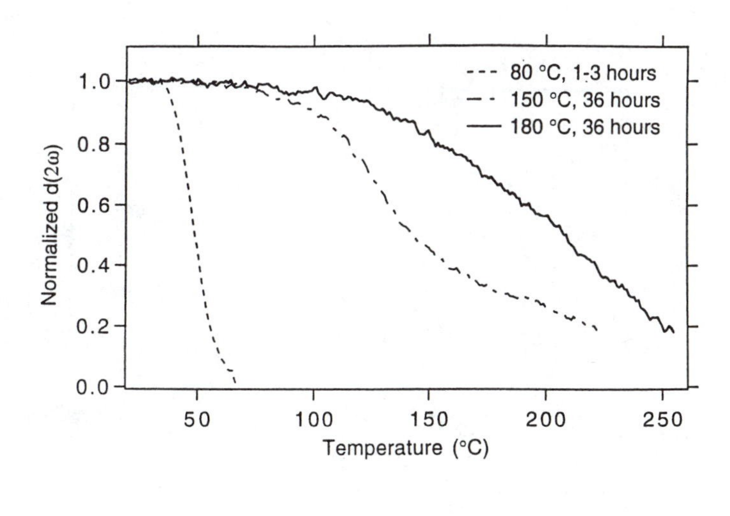

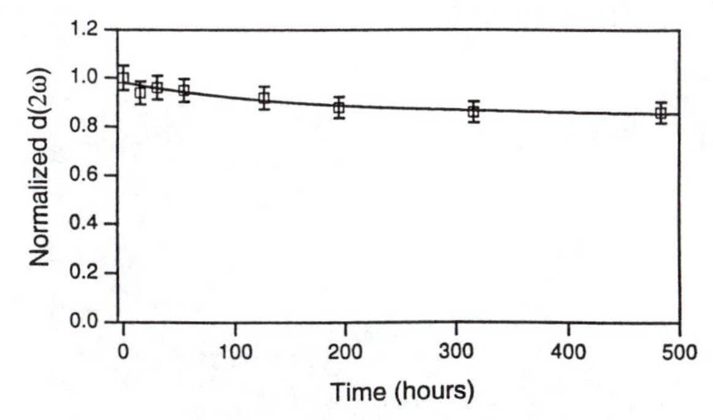

Figure 4. Upper. Dynamic assay of the thermal stability of second harmonic generation for the sol-gel materials of scheme 5 as a function of curing conditions. Lower. The long term thermal stability of the 180°C cured material is shown at 100°C.

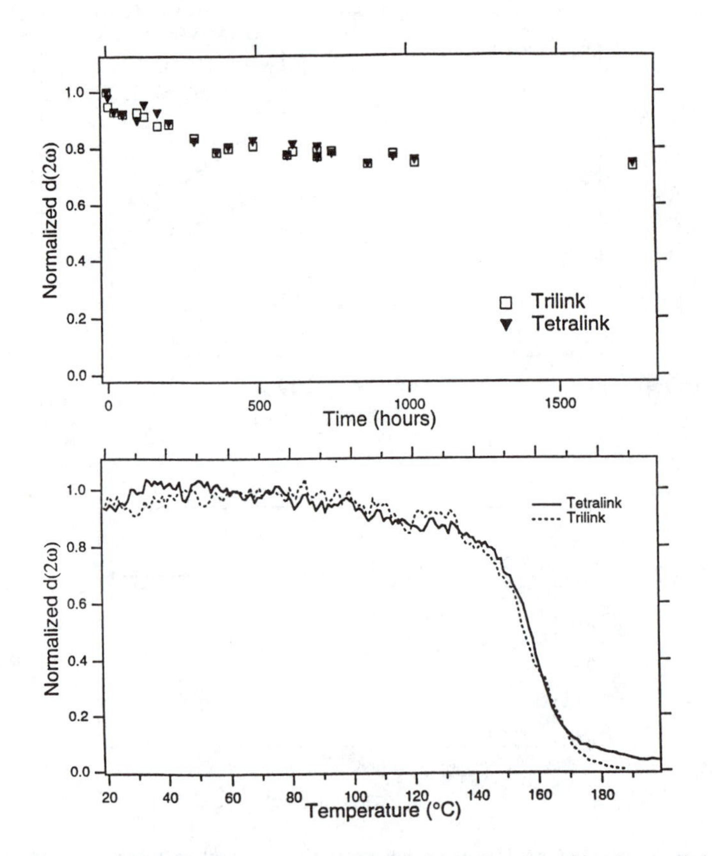

Figure 5. NLO thermal stability data is given for the tri and tetralink thermosetting materials.

Scheme 5.

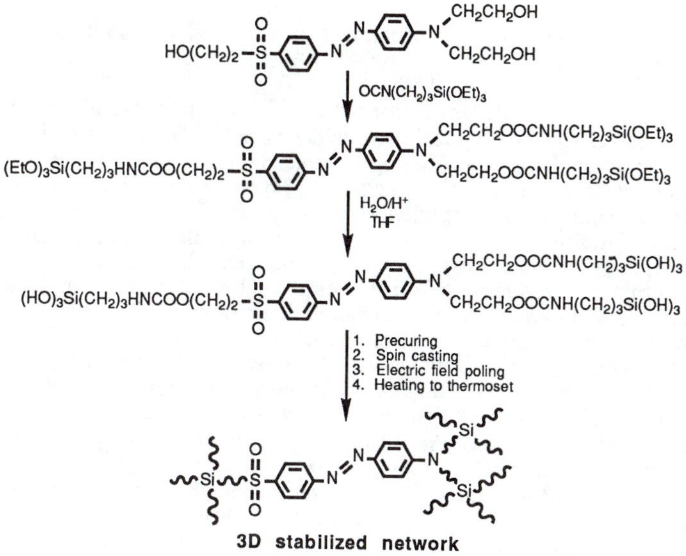

As with other schemes, recent efforts involving the thermosetting materials approach has focused upon incorporating improved $\mu\beta$ chromophores (see **scheme 6**). Unfortunately, adequate characterizations of optical nonlinearity and in particular the thermal stability of optical nonlinearity is not yet available to assess the success of this latest adaptation.

Scheme 6.

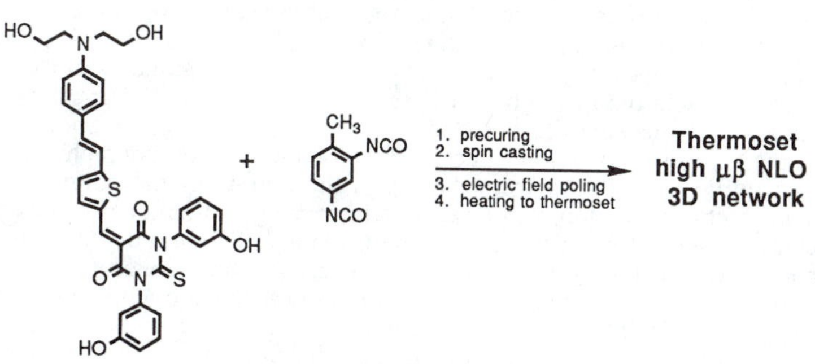

Fortunately, each of the methods demonstrated in this communication, together with the interpenetrating network approach of Tripathy and coworkers (*8*) appear capable of producing materials which satisfy the fundamental requirements of prototype device development. Let us now turn our attention to the processing of materials necessary for the fabrication and integration of polymeric buried channel waveguide modulators.

Processing of Hardened NLO Materials into Buried Channel Electro-Optic Modulator Waveguides.

The requirements for single mode propagation in fiber optic transmission lines and in buried channel nonlinear optical waveguides have been reviewed elsewhere (*13*). In particular, large dimension fiber optic cables require graded indices to achieve single mode guiding and significant cladding layers to minimize radiation losses. On the other hand, small dimension electro-optic modulator waveguides seek to avoid large cladding layers and corresponding large electrode separations since such separations mandate large drive (V_{π}) voltages. Obviously, problems exist both in terms of producing efficient coupling between small dimension EO modulator waveguide and large dimension fiber optic cable. One scheme for achieving systems integration of a polymeric modulator is shown in figure 6. Let us now review the procedures which have been used for producing buried channel EO modulator waveguides and let us consider how these protocols might be applied and adapted to address the issue of system integration. An abbreviated review will be given here; the reader is referred elsewhere (*13*) for a more extensive discussion, including relevant literature.

The three most commonly employed procedures for fabricating buried channel electro-optic modulator waveguides have been photochemical processing, reactive ion etching, and spatially selective poling (*13*). When we refer to photochemical processing, we typically mean inducing conformational changes of a photoactive species which alter the index of refraction of the material to produce a channel of high index (light guiding) material surrounding by a region of lower index material. The most commonly exploited photochemical process has been trans-cis isomerization (and accompanying chromophore reorientation) that can be induced in azobenzene and stilbene chromophores. As an aside, we note that photochemical processing has also been accomplished by inducing a photochemical polymerization reaction and then chemically removing the unpolymerized material to produce a rib waveguide structure of polymerized material.

With photochemical processing to produce channel waveguide structures by effecting index of refraction changes, a crucial issue is the spatial control of such changes by controlling light exposure. Although index changes can also be influenced by employing cladding layers containing reduced chromophore concentrations relative to the bulk EO material, the use of cladding layers adds complications which can be avoided by a two-step, two-color photochemical processing such as is shown in figure 7. In two-color photolithography, a channel of EO material is produced from the top to the bottom of the polymer film by protecting a channel region with a mask and exposing surrounding regions to light sufficiently removed from the chromophore (isomerization-inducing) resonance that good light penetrating is achieved. In the second step, the protecting mask is removed and the EO channel is exposed to light near the chromophore resonance; such light has a shallow penetration depth. The result is realization of a buried channel EO waveguide without application of cladding layers. In like manner, the tapered structures of figure 6 can be produced by exploitation of a knowledge of the kinetics of the photochemical changes which can be effected in a given material.

A concern with waveguides prepared by photochemical techniques is the long term stability of these structures with exposure to high light intensities and elevated temperatures. The stability of waveguides will likely vary with both

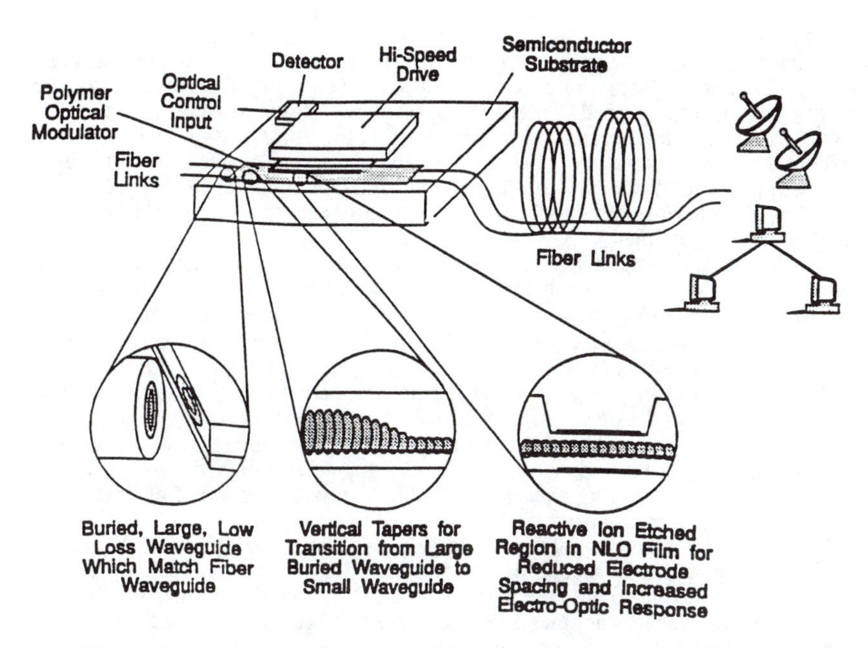

Figure 6. A scheme for integrating electro-optic modulators (configured as a directional coupler) into a switching network with fiber optic input/outputs is shown.

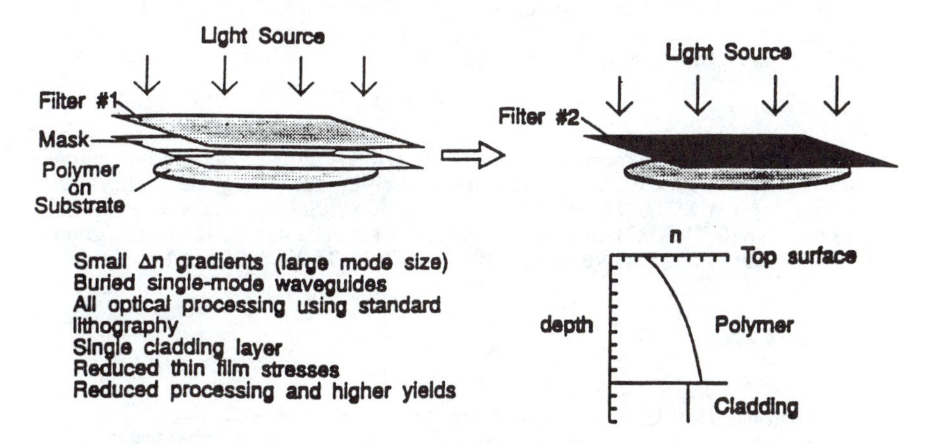

Figure 7. Two-color photochemical processing of a buried channel EO waveguide is shown.

photoactive chromophore and with the nature of the surrounding lattice (e.g., crosslink density). We have already observed dramatic improvements in the stability of photo-induced changes with intermolecular crosslinking *(2)*. The exact stability of a given waveguide must be determined by a study of that particular waveguide and to date there have been very few long term evaluations of waveguide structures used for electro-optic modulation. TACAN is currently conducting such an investigation for the electro-optic waveguide structures produced in our laboratory. Preliminary data are very promising in terms of suggesting that long term stability can be realized from photochemical processing; however, definitive statements concerning stability must await completion of this longitudinal evaluation.

Reactive ion etching (RIE) and electron cyclotron resonance (ECR) which have been widely used to fabricate electronic and optical circuits in inorganic materials have recently been used to fabricate EO modulator waveguides *(7,13)*. As we will discuss our recent efforts utilizing such techniques elsewhere in this volume, we will not review such processing here other than to note that such processing has the advantage of avoiding photochemically active species and are particularly useful in effecting tasks such as coupling of fibers and polymer waveguides exploiting V-groove techniques *(13)*. With reactive ion etching techniques controlled index variations can be accomplished by appropriate chemical design of cladding materials including the incorporation of claddings with reduced chromophore concentrations.

Spatially-selective poling has largely been pursued for the realization of quasi-phase matching in second harmonic generation. However, such a procedure can also be employed to produce a channel of EO material. With early efforts, the main limitation was the fact that optimized optical nonlinearity (which required the largest possible poling fields) prevented the realization of graded index variations. However, employing cladding material with reduced chromophore concentrations permits circumvention of the problem.

While all of the above techniques offer potential for the production of buried channel waveguides and the matching of waveguides to fiber optic transmission lines, the most commonly employed method will likely be RIE with other techniques used to achieve fine tuning of index variations necessary for optimizing mode coupling in transition regions.

Acknowledgments

The authors gratefully acknowledge support from the National Science Foundation (grant DMR-9107806), ARPA (National Center for Integrated Photonic Technology grant MDA 972-94-1-001), the Joint Services Electronics Program (contract F49620-94-0022), and the Air Force Office of Scientific Research (grant nos. F49620-94-1-0323, F49620-94-1-0201, F49620-94-1-0312, and F49620-91-0270).

Literature Cited

1. Zyss, J. *Nonlinear Opt.* **1991**, *1*, 3.
2. Dalton, L. R.; Sapochak, L. S.; Chen, M.; Yu, L. P. In *Molecular Electronics and Molecular Electronic Devices*; Sienicki, K., Ed.; CRC Press: Boca Raton, FL, 1992; pp. 125-208.
3. Dalton, L. R.; Xu, C.; Wu, B.; Harper, A. W. In *Frontiers of Polymers and Advanced Materials*; Prasad, P. N., Ed.; Plenum Press: New York, NY, 1994; pp. 175-185.

4. Teng. C. C. *Appl. Phys. Lett.* **1992**, *58*, 1538.
5. Girton, D.; Kwiatowski, S.; Lipscomb, G. F.; Lytel, R. *Appl. Phys. Lett.* **1991**, *58*, 1730.
6. Smith, B. A.; Jurich, M.; Moerner, W. E.; Volksen, W.; Best, M. E.; Fleming, W.; Swalen, J. D.; Bjorklund, G. C. *Proc. SPIE* **1993**, *2035*, 499.
7. Steier, W. H.; Shi, Y.; Ranon, P. M.; Xu, C.; Wu, B.; Dalton, L. R. Wang, W.; Chen D.; Fetterman, H. *Proc. SPIE* **1993**, *2025*, 535.
8. Marturunkakul, S.; Chen, J. I.; Li, L.; Jiang, X. L.; Jeng, R. J.; Sengupta, S. K.; Kumar, J.; Tripathy, S. K. *Polym. Prepr.* **1994**, *35*, 134.
9. Xu, C.; Wu, B.; Todorowa, O.; Dalton, L. R.; Shi, Y.; Ranon, P. M.; Steier, W. H. *Macromolecules* **1992**, *26*, 5303.
10. Becker, M. W.; Sapochak, L. S.; Dalton, L. R.; Shi, Y.; Steier, W. H., Jen, A. K.-Y. *Chem. Mater.* **1994**, *6*, 104.
11. Jen, A. K.-Y.; Drost, K. J.; Rao, V. P.; Cai, Y.-M.; Liu, Y. J.; Mininni, R. M.; Kenney, J. T.; Binkley, E. S.; Dalton, L. R.; Marder, S. R. *Polym. Prepr.* **1994**, *35*, 130.
12. Burland, D. M.; Miller, R. D.; Walsh, C. A. *Chem. Rev.* **1994**, *94*, 31.
13. Dalton, L. R.; Harper, A. W.; Ghosn, R.; Steier, W. H.; Ziari, M.; Fetterman, H.; Shi, Y.; Mustacich, R. V.; Jen, A. K.-Y.; Shea, K. J. *Chem. Mater.* submitted.

RECEIVED February 14, 1995

Chapter 13

Development of Functionalized Polyimides for Second-Order Nonlinear Optics

Dong Yu, Zhonghua Peng, Ali Gharavi, and Luping Yu[1]

Department of Chemistry, University of Chicago, 5735 South Ellis Avenue, Chicago, IL 60637

Several functionalized polyimides (both aromatic and aliphatic) with pendent nonlinear optical chromophores have been synthesized. The detailed characterization indicated very promising features for these polymers. The polyamic acids were soluble in several aprotic polar solvents, permitting thin film processing. The polyamic acid can be imidized at ca 200°C without damaging the NLO chromophore. Long term stability was observed at 170°C. The most interesting point is that the reaction approaches are versatile and allow for further developments of better materials.

The development towards practical applications of second order nonlinear optical polymers is gaining momentum.[1-3] However, in spite of great deal of research effort focused on these oriented polymer systems, hurdles towards that goal still remain, i.e. higher thermal stability and larger optical nonlinearity. It is known that in order to be useful in optical device applications, second-order NLO polymers should be chemically and physically stable. They should also maintain a significant bulk nonlinearity at possible device processing temperatures of 200 °C-250 °C and at elevated working temperatures that can reach 80 °C or higher. These practical considerations have led investigators to search for new polymeric systems which exhibit better temporal and thermal NLO stabilities.

Polyimides have been extensively studied as high performance materials for applications in integrated electronic circuits and aerospace devices due to their exceptional thermal, mechanical, optical and dielectric properties.[4,5] Recently, the unique properties of polyimides have aroused much attention from the optical community for their photonic applications. [6-8] The high glass transition temperature of these materials have been utilized to prepare second order nonlinear optical (NLO) composite materials which exhibit exceptionally high temperature stability in the dipole orientation.[6] While these systems enjoyed a certain degree of success, several problems also occurred. For example, NLO chromophore bleaching was observed in the polyimide composite systems and the glass transition temperature decreased because of the plasticizing effect of the small chromophore at prolonged high

[1]Corresponding Author

0097–6156/95/0601–0172$12.00/0

working temperatures, and so on. These problems severely hindered the further development of these materials for practical applications.

Recently, we synthesized a series of functionalized, nonlinear optical polyimides exhibiting large and exceptionally stable second harmonic generation (SHG) coefficients.[9-11] The rationale to synthesize these polymers was that the motion of the free volume in the polymer matrix can be frozen due to a high glass transition temperature. Also, the covalent attachments of the nonlinear optical chromophore can eliminate several problems associated with the composite systems, such as chromophore sublimation, limited effective chromophore number density due to the incompatibility between the polymer host and the chromophore guest etc.. This paper summarizes our synthetic efforts and reports detailed results on physical characterizations.

Experimental Section

THF was purified by distillation over sodium chips and benzophenone. DMF was purified by distillation over phosphorous pentoxide. 1,2,4,5-Benzenetetracarboxylic dianhydride (PMDA) and 3, 3', 4, 4'-benzophenone tetracarboxylic dianhydride (BTDA) were purified by recrystallization from acetic anhydride and then dried under a vacuum at 150 °C overnight before use. All of the other chemicals were purchased from Aldrich Chemical Company and used as received unless otherwise stated. The synthetic procedure for the intermediates used to produce the monomers can be found in references 9-11.

Compound 3: Diethyl azodicarboxylate (DEAD, 1.20 g, 6.9 mmol) in THF (5 ml) was added dropwise to the mixture of 4-(N-ethyl-N-β-hydroxyethyl)-4'-(methylsulfonyl)-stilbene (Compound 1, 1.83 g, 5.3 mmol), 2, 5-dinitrophenol (Compound 2, 0.78 g, 5.3mmol) and triphenyl phosphine (1.81 g, 0.69 mmol) in 35 ml THF at room temperature. After the reaction mixture was stirred for 24 hr, water (50 ml) was added. The resulting solid was collected by filtration and recrystallized from acetone/methanol to yield dark black crystals (2.30 g, 85 %, m.p. 176-177 °C). ^{1}H NMR ($CDCl_3$, ppm): δ 1.22 (t, J=6.9 Hz, $-CH_2\underline{CH_3}$, 3 H), 3.04 (s, $-SO_2CH_3$, 3 H), 3.52 (q, J=7.0 Hz, $-\underline{CH_2}CH_3$, 2 H), 3.86 (t, J=5.2 Hz, $-OCH_2\underline{CH_2}N-$, 2 H), 4.36 (t, J=5.2 Hz, $-O\underline{CH_2}CH_2N-$, 2 H), 6.68 (d, J=8.6 Hz, ArH, 2 H), 6.87 (d, J=16.2 Hz, =CH, 1 H), 7.13 (d, J=16.2 Hz, =CH, 1 H), 7.39 (d, J=8.5 Hz, ArH, 2 H), 7.56 (d, J=8.1 Hz, ArH, 2 H), 7.82 (d, J=8.3 Hz, ArH, 2 H), 7.87 (m, ArH, 3 H). Anal. Calcd. for $C_{25}H_{25}N_3O_7S$: C, 58.71; H, 4.89; N, 8.22; Found: C, 58.63; H, 4.96; N, 8.18.

Monomer A: Stanneous chloride (3.71 g, 19.6 mmol) was added to compound 3 (1.0 g, 2.0 mmol) in 50 ml hydrochloric acid (36.5 % - 38 %). The mixture was stirred at room temperature for 24 hr and then warmed to 40 to 50 °C for 2 hr until the solution became completely transparent. Deoxygenated sodium hydroxide solution (40 %) was then added to the mixture at an ice bath until the pH value of the solution reached 12 ~ 13. Yellow solid was collected by filtration under nitrogen and then recrystalized from $MeOH/H_2O$ to yield bright yellow crystals (0.5 g, 57 %, m.p. 166 - 168 °C). ^{1}H NMR ($CDCl_3$, ppm): δ 1.22 (t, J=6.9 Hz, $-CH_2\underline{CH_3}$, 3 H), 3.03 (s, $-SO_2CH_3$, 3 H), 3.30 (brs, $Ar-NH_2$, 4 H), 3.48 (q, J=7.0 Hz, $-\underline{CH_2}CH_3$, 2 H), 3.77 (t, J=5.8 Hz, $-OCH_2\underline{CH_2}N-$, 2 H), 4.09 (t, J=5.8 Hz, $-O\underline{CH_2}CH_2N-$, 2 H), 6.16 (d, J=8.2 Hz, ArH, 1 H), 6.19 (s, ArH, 1 H), 6.51 (d, J=8.1 Hz, ArH, 1 H), 6.72 (d, J=8.6 Hz, ArH, 2 H), 6.86 (d, J=16.2 Hz, =CH, 1 H), 7.12 (d, J=16.2 Hz, =CH, 1 H), 7.37 (d, J=8.3 Hz, ArH, 2 H), 7.56 (d, J=8.1 Hz, ArH, 2 H), 7.81 (d, J=8.0 Hz, ArH, 2 H). Anal. Calcd for $C_{25}H_{29}N_3O_3S$: C, 66.52; H, 6.43; N, 9.31. Found: C, 66.67; H, 6.49; N, 9.16.

Monomer B: To a two-neck flask containing compound 4a (0.800 g, 1.36 mmol) and THF (20ml) was added the hydrazine monohydrate (0.389g, 6.80 mmol). The mixture was refluxed for 5 hours under nitrogen and then cooled down to room temperature. The precipitate was filtered out and the filtrate was dried under a vacuum to yield a red solid. Ethanol (20ml) was then added to this solid and the resulting mixture was stirred for 15 minutes at ca. 50°C. The insoluble part was removed by filtration. Deoxygenated water was then added dropwise to the filtrate and a dark red product crystallized out. The crude product was recrystallized from MeOH/water to yield monomer B (dark-red plate crystals, 0.2g, 46.2%, mp, 160.2 °C (DSC result)). ^{1}H NMR (DMSO-d_6, ppm): δ, 1.50 (d, J = 4.07 Hz, $-NH_2$, 4 H), 2.67 (t, J = 6.90 Hz, $-C\underline{H_2}NH_2$,4 H), 3.31 (t, J=6.98 Hz, $-CH_2N-$, 4 H), 6.71 (d, J = 8.37 Hz, ArH, 2 H), 7.03 (d, J = 16.31 Hz, =CH, 1 H), 7.36 (d,J = 16.32 Hz, =CH, 1 H), 7.41 (d, J = 8.33 z, ArH, 2 H), 7.70 (d, J = 8.36 Hz, ArH, 2 H), 8.12 (d, J = 8.40 Hz, ArH, 2 H). Anal. Calcd for $C_{18}H_{22}N_4O_2$: C, 66.23; H, 6.80; N, 17.17. Found: C, 66.04; H, 6.75; N, 17.06.

Monomer C: The hydrazine monohydrate (3.0 g, 60 mmol) in THF (10 ml) was added slowly to a suspension of compound 4b (3.05 g, 5.0 mmol) in 40 ml of THF under stirring. The mixture was refluxed for 3 hr under nitrogen and then cooled down to room temperature. The solid was removed by filtration and the filtrate was concentrated under a vacuum to ca. 5 ml. Ethanol (10 ml) was added to the concentrated filtrate and the solution was placed in a refrigerator overnight. The yellow crystals were collected and recrystallized twice from ethanol, yielding monomer C (0.77g, 42.8 %, m.p. 105-106 °C). ^{1}H NMR (DMSO-d_6): δ, 2.66 (t, J = 6.55 Hz, $-NH_2$, 4 H), 3.18 (s, $-SO_2CH_3$, 3 H), 3.31 (t, J = 6.64 Hz, $-NCH_2-$, 4 H), 6.70, 7.38, 7.70, 7.79 (d, ArH, 2 H), 7.00, 7.27 (d, =CH, 1 H). Anal. calcd. for $C_{19}H_{25}N_3O_2S$: C, 63.49; H, 7.01; N, 11.69. Found: C, 63.54; H, 6.97; N, 11.49.

Polymerization. The polymerization procedure is examplified as follows: to a 25-ml two-necked, round-bottom flask was added monomer A (0.2148 g, 0.476 mmol) and the anhydrous NMP (6 ml) at -2 to -5°C under nitrogen. The 3, 3', 4, 4'-benzophenone tetracarboxylic dianhydride (BTDA, 0.1533 g, 0.476 mmol) was added at once. The dianhydride was completely dissolved in ca. 20 min. The mixture was stirred at -2 - -5°C for 4 hr after the addition. The polyamic acid was precipitated into methanol and collected by filtration (almost quantitative yield). To further purify the polymer, it was ground in acetone to remove the residual solvent and then washed with chloroform in a soxhlet extractor for 2 days. The ^{1}H NMR of polymer A2 in DMSO is shown in Figure 1.

Characterization: The ^{1}H NMR spectra were collected on a varian 500 MHz FT NMR spectrometer. The FTIR spectra were recorded on a Nicolet 20 SXB FTIR spectrometer. A Perkin-Elmer Lambda 6 UV/VIS spectrophotometer was used to record the UV/VIS spectra. The thermal analyses were performed by using the DSC-10 and TGA-50 systems from TA instruments under a nitrogen atmosphere. The melting points were obtained with open capillary tubes using a Mel-Temp apparatus. Elemental analyses were performed by Atlantic Microlab, Inc.

The NMP solution of the polyamic acid was filtered through 0.2-μm syringe filters and then spin cast onto indium-tin oxide (ITO) coated glass slides or normal glass slides. The films were dried in an oven at 40 °C for 48 hr under nitrogen, then vacuum dried at 60 °C overnight. The films were poled/ cured at a corona-discharge setup with a tip-to-plane distance of 1.0 cm at elevated temperatures from 60 to 200 °C within two hours. The second order NLO properties of the poled polymeric films

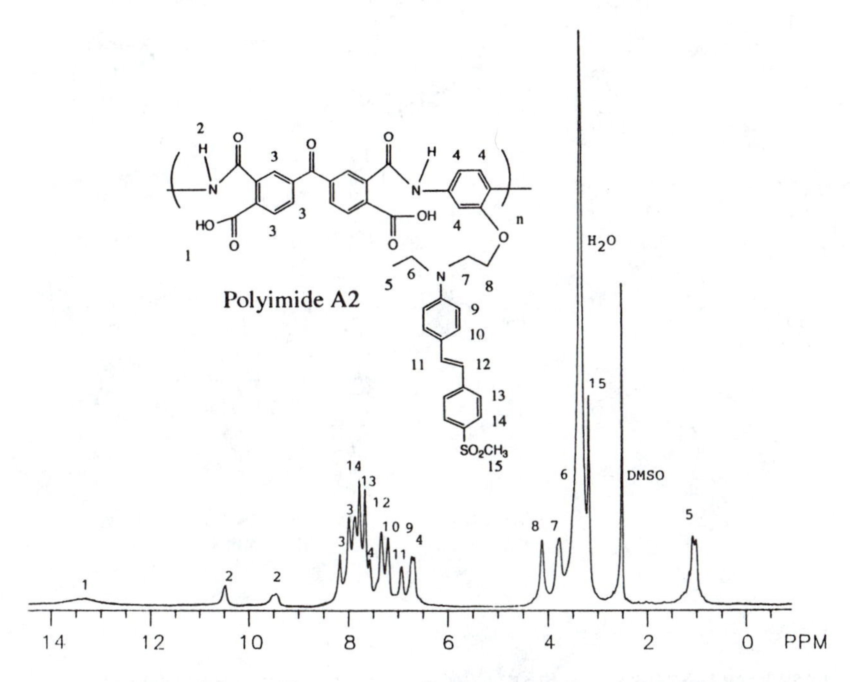

Figure 1. ^1H NMR spectrum of polyamic acid A2 in DMSO-d_6.

were characterized by second harmonic generation (SHG) measurements. A mode-locked Nd:YAG laser (Continuum-PY61C-10, 25 ps, 10 Hz repetition rate) was used as a fundamental light source (1,064 μm). The second harmonic signal was detected and amplified by a photomultiplier tube (PMT) and then averaged in a boxcar integrator. The temporal stability of the SHG signal of the polyimide was monitored at the following temperatures: room temperature, 90 °C, 150 °C and 170 °C respectively. The temperature reading error was within ± 5 °C. Detailed descriptions of the measurements are given in reference 12.

Scheme 1. Syntheses of diamino monomers.

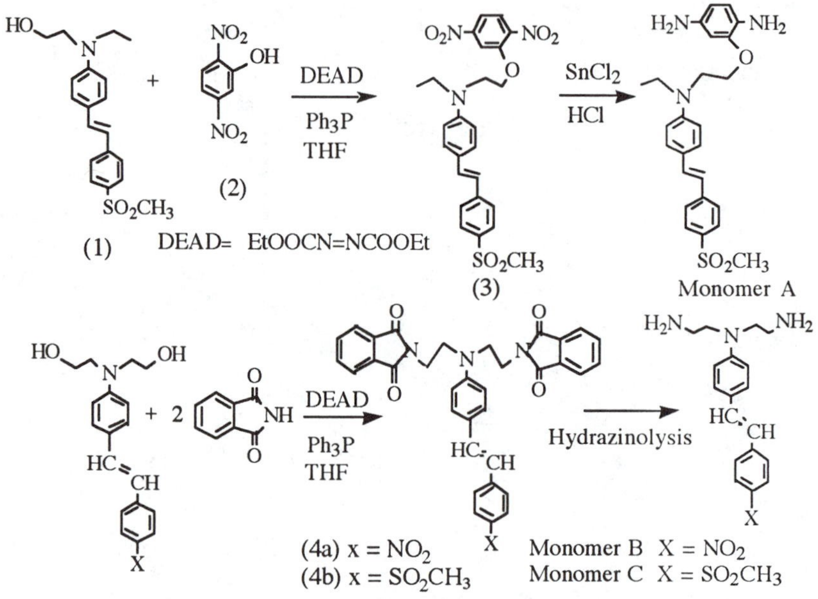

(1) + (2) → DEAD, Ph_3P, THF → (3) → $SnCl_2$, HCl → Monomer A

DEAD= EtOOCN=NCOOEt

+ 2 phthalimide → DEAD, Ph_3P, THF → (4a), (4b) → Hydrazinolysis → Monomer B, Monomer C

(4a) x = NO_2
(4b) x = SO_2CH_3

Monomer B X = NO_2
Monomer C X = SO_2CH_3

Results and Discussion

Our strategy in synthesizing the functionalized polyimides was to functionalize the diamine monomers. The key step in these syntheses was to utilize the Mitsunobu reaction to attach an NLO chromophore to 2,5-dinitrophenol and then to reduce the nitro groups (Monomer A) or to convert dihydroxy groups into diamino groups (Monomers B and C). The synthetic approaches for these monomers are shown in scheme 1. The monomers obtained were easy to purify. Both spectroscopic and microanalysis results supported the structures as proposed.

Polymerization can be easily effected in aprotic polar solvents (Such as DMF, NMP and DMAc, etc.) with different dianhydrides. The polyamic acids obtained usually possessed relatively high molecular weights (with intrinsic viscosities in the range of ca 0.5-1.0dl/g) and were soluble in aprotic polar solvents such as NMP, DMSO, DMAc and DMF. The polyamic acid solution can be directly used to prepare films or precipitated into acetone to obtain solid materials. The films can be further cured at high temperature (>200°C) to form polyimides. Scheme 2 shows the structures of different polyimides synthesized.

Spectroscopic studies, including ^1H NMR, FTIR and UV/vis, were in full agreement with the expected structures of the polymers. All of the polyamic acids exhibited an endothermic process at around 200 °C as shown by the DSC studies. This process was clearly due to imidization as correlated with the FTIR, UV/vis spectroscopic and TGA studies. The polyimides obtained exhibited a high glass transition temperature (all larger than 200 °C) which resulted in the high stability of the dipole orientation at elevated temperatures.

Scheme 2. Structures of second order NLO polyimides.

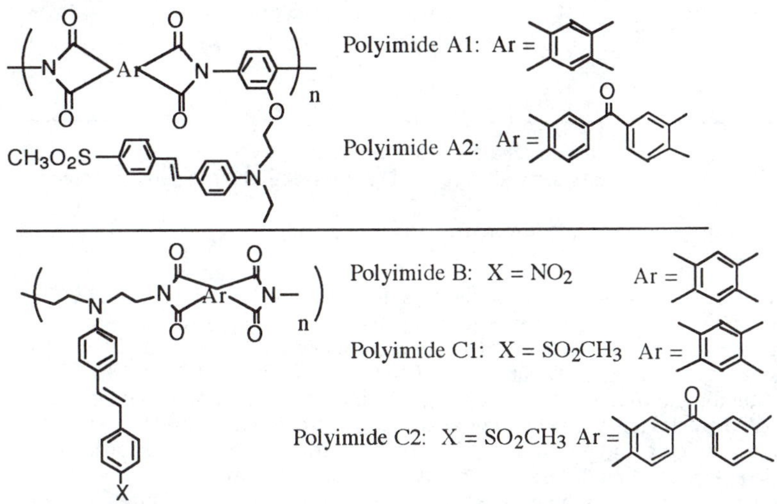

CH_3O_2S

Polyimide A1: Ar =

Polyimide A2: Ar =

Polyimide B: X = NO_2 Ar =

Polyimide C1: X = SO_2CH_3 Ar =

Polyimide C2: X = SO_2CH_3 Ar =

The UV/vis spectra of the polyamic acids showed a typical absorption maxim due to the dialkylamino-sulfonyl (or -nitro) stilbene chromophore. After thermal curing, the absorbance did not change significantly except for a slight blue shift caused by the matrix change due to the imidization (See Figure 2). However, after the polyamic acid was corona-poled as it was imidized, a large decrease in absorbance was observed due to the birefringence induced by the dipole alignment. Table I summarizes some of the physical data of these polymer systems.

Second harmonic generation measurements were performed at a wavelength of 1064 nm. A wide range of d_{33} values were obtained by the careful design of the NLO chromophore (See Table I). Listed in Table I are the dispersionless d_{33} values derived from the two level model. Due to the insufficient poling ($\Phi = 0.12$), the d33 value of polymer C1 is significantly lower than polymers A1, A2 and C2. Latest study after we submitted the paper showed comparable d33 value (43 pm/V) after improve the poling. The d33 value for polymer B1 is strongly resonance enhanced at 532 nm.

To evaluate the high temperature stability of our polymers, we studied the temporal stability of the second harmonic generation (SHG) signals at various temperatures. It was found that the SHG signals of all of these polyimides exhibited no decay at room temperature, 90 °C or at 120°C for more than 1200 hours. For polyimide A2, long term stability at 150 °C was observed (see Figure 3). Decay was

Table I. Physical data for different polyimides.[a]

Poly-imides	Tg (°C)	λ_{max} (nm)	Φ^b	n (532 nm)	d_{33} (532nm, pm/V)	d_{33} (∞, pm/V)[c]
A1	240	384	0.20	1.781	51	18
A2	230	383	0.18	1.815	46	16
B1	230	440	0.30	1.833	115	27
C1	250	372	0.12	1.769	29	10
C2	209	372	0.21	1.781	41	14

a. See Scheme 2 for structures of the polyimides.
b. Order parameters derived from the UV/vis spectra before and after electric poling.
c. The values estimated from the two level model.

noticeable at the initial stage when the samples (Polymer A1, B1 C1 and C2) were kept at 150°C and the signal was then stabilized at ca. 85% of its initial value. Several very promising features of these polyimides can be extracted from these results: first, its nonlinearity can be stabilized at 90 °C and 120 °C and 150°C for a significant long term (more than 1000 hr), which is the possible working temperature anticipated by device engineers. Second, the films can withstand high processing temperatures for a short term (e.g. 220 °C). Third, the reaction approaches are versatile in synthesizing different materials. We can obviously vary several structural parameters to obtain new materials. For example, the NLO chromophore can be changed from the stilbene to other structures which will lead to an improved NLO activity and higher thermal stability. Different dianhydride monomers can be utilized to synthesize polyimides so that the glass transition temperature can be finely tuned. Cross-linking units can be introduced into the polymers to further enhance its stability at high temperatures. Finally, due to the good processibility of polyamic acids, it is possible to fabricate several device elements.

Conclusions

Functionalized polyimides with pendent nonlinear optical chromophores can be synthesized by the careful design of suitable NLO chromophores. The polyamic acids obtained are soluble in several aprotic polar solvents, permitting thin film processing. The polyimides exhibited a high glass transition temperatures (>200 °C) after the polyamic acid was imidized at ca 200°C. The high glass transition temperatures of the polyimides resulted in a high stability of the optical nonlinearity at elevated temperatures. Long term stability was observed at 150°C.

Acknowledgments

This work was supported by the Office of Naval Research grants N00014-93-1-0092. The support from the Arnold and Mabel Beckman Foundation is gratefully acknowledged.

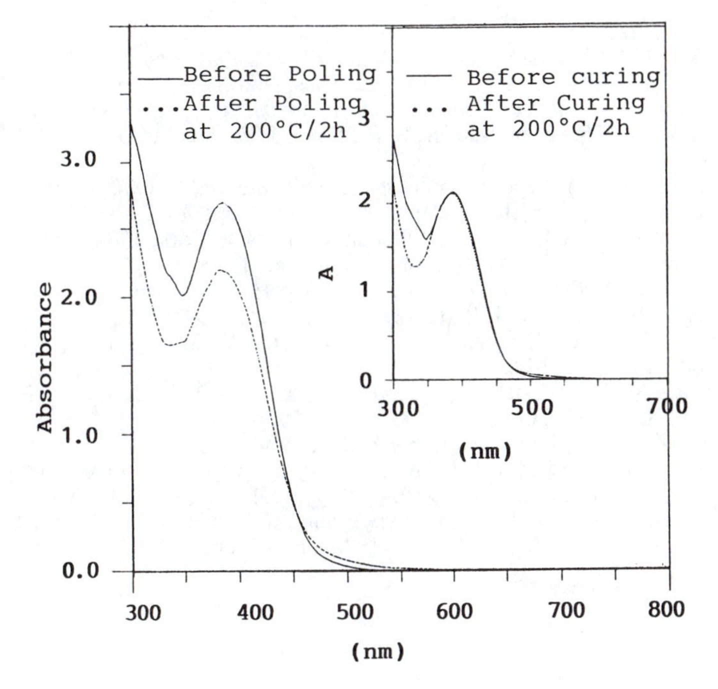

Figure 2. UV/vis spectra of the polyamic acid and polyimide under different conditions.

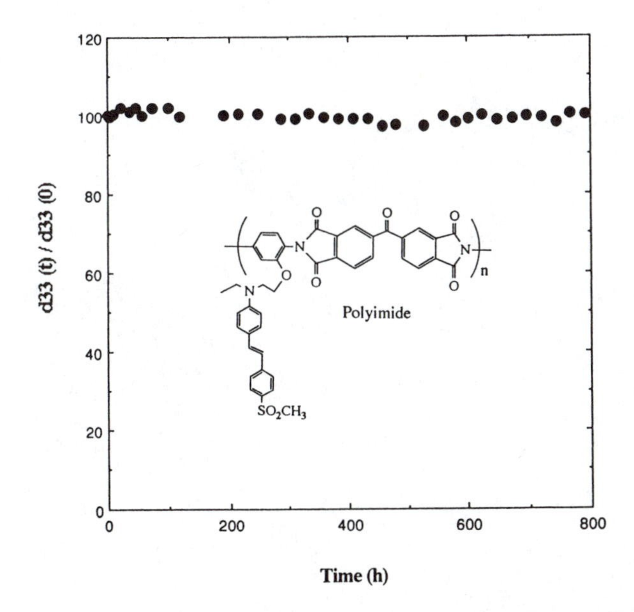

Figure 3. Temporal stability of the SHG signals at 150 °C for polyimide A2.

Literature Cited

1. Prasad, P. N., Williams, D. J., *Introduction to Nonlinear Optical Effects in Molecules and Polymers,* J. Wiley and Sons, New York, 1991.
2. *Materials for Nonlinear Optics: Chemical Perspectives,* ACS Symposium Series No. 455, ed. by Marder, S. R., Sohn, J. E., Stucky, G. D., Washington, 1991.
3. *Nonlinear Optical Properties of Organic Molecules and Crystals,* ed. by Chemla, D. S., Zyss, J., Academic Press, New York, 1987.
4. *Polyimides*, Mittal, K. L., Ed.; Plenum Press, New York, **1984**, Vols. *1, 2.*
5. Scroog, C. E., *J. Polym. Sci., Macromol. Rev.* **1976**, *11*, 161.
6. Wu, J. W., Valley, J. F., Ermer, S., Binkley, E. S., Kenney, J. T., Lipscomb, G. F., Lytel, R. J., *Appl. Phys. Lett.*, **1991**,*58*, 225.
7. Lin, J. T., Hubbard, M. A., Marks, T. J., Lin, W., Wong, G. K., *Chem. Mater.* **1992**, *4,* 1148.
8. Becker, M. W., Sapochak, L. S., Ghosen, R., Xu, C. Z., Dalton, L. R., Shi, Y. Q., Steier, W. H., *Chem. of Mater.*, **1994**, *6*, 104.
9. Peng, Z. H., Yu, L. P., *Macromolecules* , **1994**, *27*, 2638.
10 Yang, S. Y., Peng, Z. H., Yu, L. P., Macromolecules, **1994,** *27*, 5858.
11. a). Yu, D., and Yu, L. P., Macromolecules, **1994,** *27*, 6718. b). Yu, D., Gharavi, A., and Yu, L. P., *Macromolecules.* to Appear in Jan. 1995.
12. Yu, L. P., Chan, W. K., Bao, Z. N., Macromolecules, **1992**, *25*, 5609

RECEIVED January 30, 1995

Chapter 14

Advances in Main-Chain Syndioregic Nonlinear Optical Polymers

J. D. Stenger-Smith[1], R. A. Henry[1], A. P. Chafin[1], L. H. Merwin[1], R. A. Nissan[1], R. Y. Yee[1], M. P. Nadler[1], G. A. Lindsay[1], L. M. Hayden[2], S. Brower[2], N. Mokal[2], D. Kokron[2], W. N. Herman[3], and P. Ashley[4]

[1]U.S. Navy, Code 474220D, NAWCWPNS, China Lake, CA 93555–6001
[2]Department of Physics, University of Maryland, Baltimore, MD 21228
[3]Naval Air Warfare Center, Aircraft Division, Warminster, PA 18974–0591
[4]U.S. Army Missile Command, Redstone Arsenal, AL 35898–5248

Solution and solid state nuclear magnetic resonance were used to study a mainchain syndioregic polycyanocinnamamide. Photooxidation at elevated temperatures has been identified as the cause of previously unexplained low d_{33} measurements of this polymer. Corona poling in the dark was shown to give much better results. The electrooptical coefficient (r_{33})of this cyanocinnamamide polymer was measured in a Mach-Zehnder intensity modulator. Several new mainchain syndioregic polymers were synthesized and characterized. Preliminary d_{33} measurements were performed on one promising polymer and were found to be close to the value calculated using a simple semi-empirical procedure.

Incorporation of molecules with high nonlinear optical (NLO) activity into polymers (NLOP's) has attracted considerable interest for photonics applications due to their fast NLO response time, low cost, ease of fabrication, and large nonlinear second-order optical susceptibility, $\chi^{(2)}$. One focus of current research is the enhancement of the thermal and temporal stability of $\chi^{(2)}$. The stability of the resulting order in the thermodynamic nonequilibrium glassy state varies from polymer to polymer. Chemical attachment of the chromophore to the polymer as a sidechain,[1]as part of the mainchain,[2] or in a crosslinked structure[3] has been found to retard the relaxation of orientational order.

Mainchain NLOP's can be divided into three categories: isoregic (head-to-tail), syndioregic (head-to-head and tail-to-tail) and aregic (random). Syndioregic polymers appear to give the highest d_{33} coefficients (for a given chromophore density) of all three catagories (see Chapter 1 herein). Promising results on a syndioregic mainchain NLOP system[4] encouraged us to scale up and make waveguide measurements on this polymer (Polymer **1**) and to target new chromophores with higher glass transition temperatures and hyperpolarizabilities and improved photothermal stabilities.

Polymer **1** has a syndioregic mainchain cinnamamide repeat unit, and, depending on molecular weight, configuration of the cyclohexane ring, and substituents on the amine donor, has a glass transition temperature of 173° to 212°C. The *trans*-cyclohexane ethyl amine version of polymer **1**, used in this study, is given in Figure 1. Carbon atoms are labeled for reference.

0097–6156/95/0601–0181$12.00/0

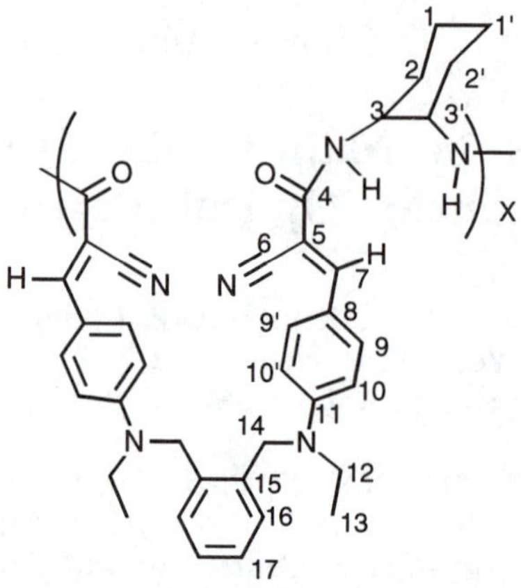

Figure 1: Polymer **1** Structure. *Cis* and *trans* refer to the isomer of cyclohexane ring; ethyl and methyl refer to alkyl substituent on amine nitrogen (carbons 12 and 13 when appropriate).

The dependence of glass transition temperature upon molecular weight is given in Figure 2. The *trans*-methyl version of polymer **1** shows the highest glass transition temperature, however it was difficult to obtain higher molecular weights of the *trans*-methyl polymer **1**. The *cis*-ethyl version of polymer **1** had lower glass transition temperatures, poorer solubility and stability than *trans*-ethyl polymer **1**.

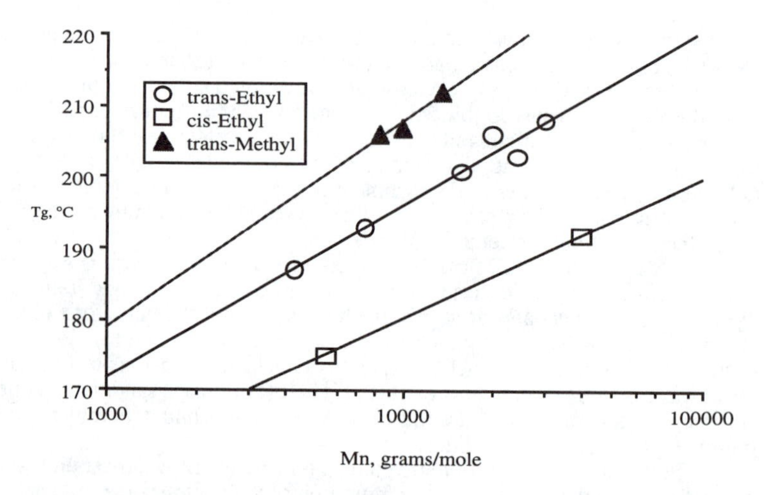

Figure 2: Glass transition temperature vs molecular weight for Polymer **1**. Measured by DSC in nitrogen at 10°C/min.

The first part of this paper describes the use of Nuclear Magnetic Resonance (NMR) to verify the structure and conformation of the substituents on cyclohexane ring in polymer **1**. Solid state NMR was used to make predictions about the temperature dependence of bulk motions of the chromophore, which should provide insight into the stability of NLO activity in a poled polymer film.

Due to exposure to UV and visible light during poling at 200°C, preliminary measurements on Polymer **1** indicated unexpectedly low nonlinear optical properties[5]. Subsequently, it has been found that poling Polymer **1** in the dark gives superior results which are in line with expectations based on the calculated molecular hyperpolarizability. The second part of this paper describes the measurement of the second-harmonic coefficients (d_{33} and d_{13}), the linear electro-optic coefficient (r_{33}) in poled films, and the stability of d_{33} with temperature.

The third part of this paper gives preliminary results on the incorporation of several new chromophores into the mainchain syndioregic topology.

NMR Characterization of Cyanocinnamamide Polymer 1.

Characterization of Polymer 1: Configuration of the Cyclohexane Ring

NMR spectroscopic techniques were used to identify the *cis* and *trans* isomers of polymer **1**, and of a model chromophore pair. Identification of *cis* or *trans* 1,2-cyclohexanes is based upon the magnitude of the C-H coupling constant, $^3J_{HC}$, which is expected to be 1-3 Hz for an axial proton and 3-8 Hz for an equatorial proton, and on the dynamic exchange of the cyclohexane ring which exists for the *cis* isomer but not for the *trans*. 1-D ^{1}H and ^{13}C as well as Heteronuclear Multiple Quantum Coherence, (HMQC)[6] and Heteronuclear Multiple Bond Coherence, (HMBC)[7] measurements were made on *cis* and *trans* diamide monomers as well as on model compounds and on polymers.

Initially, differentiation of the mixed *cis* and *trans* isomers was the objective and this was possible based on chemical shift and coupling constant considerations.[8] Axial protons tend to be upfield of their equatorial neighbors, owing to shielding effects from C-C bonds, and equatorial protons tend to display larger long range correlations to carbons two and three bonds away in accordance with the Karplus relation. The *trans* materials have two axial protons in the 1,2 position and the *cis* materials have one axial and one equatorial proton. The *cis* materials should thus be undergoing chair-chair interconversions, while the *trans* materials should be relatively rigid since a chair-chair interconversion would lead to two sterically disfavored axial groups. At later stages in the project, when separate samples of pure *cis* and *trans* 1,2-diaminocyclohexane became available, the differentiation between *cis* and *trans* isomers became elementary, although it was still worthwhile to verify that isomerization had not taken place.

A demonstration of the *cis* configuration in the model compound is observed in the long range proton-carbon correlation experiment, or HMBC. Figure 3 displays the HMBC spectrum for the *cis* model compound. The methine proton peak located at slightly greater than 4 ppm is assigned to the H-3 position of the cyclohexyl ring (see Figure 1). If we draw a vertical line down from this position we note that there are 2 spots or correlations with carbons. These two correlations represent a long range coupling between H-3 and the C-1 and C-2 positions on the cyclohexyl ring. We should only observe these correlations if the two and three bond coupling constants are greater than 2-3 Hz and this will only be the true when the H-3 proton is equatorial. The same correlations are present in the HMBC spectrum for the *cis* polymer. In the HMBC spectra for the *trans* model compound and *trans* polymer do not show these two correlations.

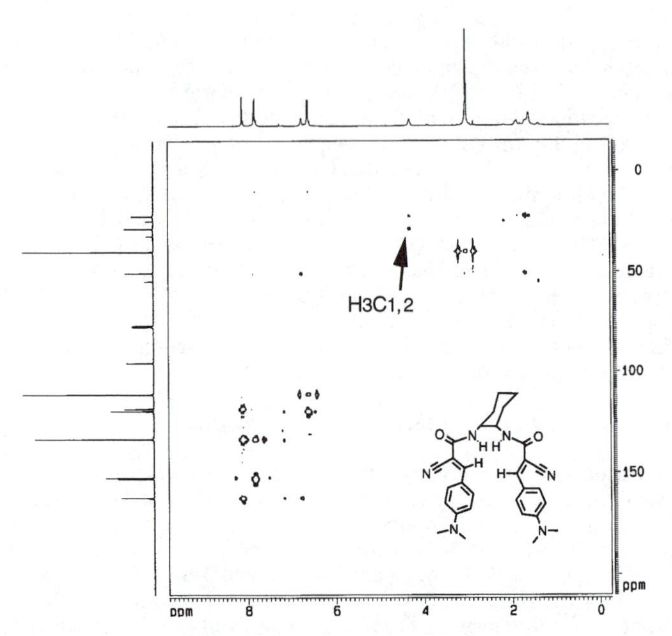

Figure 3: Heteronuclear Multiple Bond Correlation Spectrum for the *cis* model compound in $CDCl_3$.

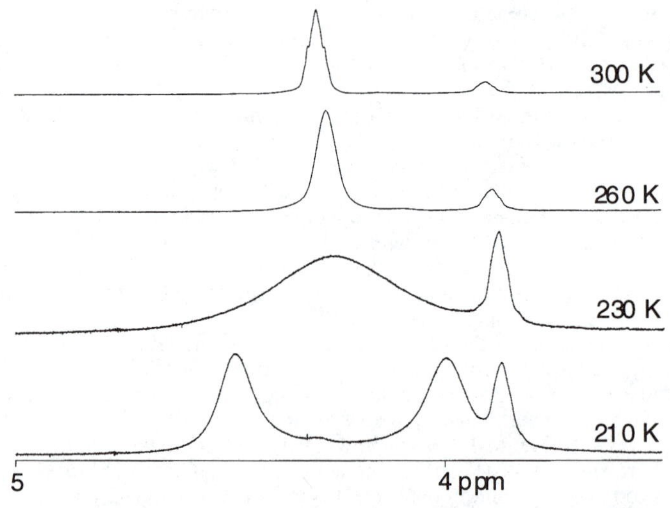

Figure 4: Variable Temperature ^{1}H NMR of cyclohexyl methine region for the *cis* model compound in $CDCl_3$.

In Figure 4 a variable temperature ^{1}H NMR study of the cyclohexyl methine region (H-3) for a *cis* model compound with 5% *trans* impurity is presented. At ambient temperatures, the methine protons for *cis* and *trans* isomers appear as broad resonances at 4.31 and 3.91 ppm respectively. The *cis* resonance broadens significantly as the temperature decreases and at 210 K is split into an axial proton at 4.01 and an equatorial proton at 4.58 ppm, providing evidence that the *cis* isomer rapidly exchanges between two conformations at room temperature. Attempts to perform this experiment on polymers were unsuccessful as polymers immediately precipitated from even dilute solutions when the samples were cooled.

In addition to the solution-state NMR work reported elsewhere[9], the *cis* and *trans* isomers of polymer **1** were also evaluated by ^{13}C solid-state NMR techniques. The use of solid-state NMR spectroscopy allowed us to begin to investigate the decay of the NLO properties over time from a molecular and/or mechanistic viewpoint. Relaxation of nuclear magnetic order for a given site in a molecule takes place via small local magnetic field variations. These local field variations, in turn, generally result from molecular motion, exchange processes or other physical processes which give rise to local field fluctuations. Without going into significant detail, it will be noted that different NMR experiments can be used to select the frequency of motion of interest. The NMR spin-lattice relaxation time (T_1), spin-lattice relaxation in the rotating frame ($T_{1\rho}$), and resonance line widths (T_2) can provide relative indication of molecular motion on the megahertz, kilohertz, and hertz timescales respectively[10]. Comparison of these data for various groups within a particular compound or between isomers of unpoled polymers can provide data on the sources of local motion which cause relaxation of NLO properties.

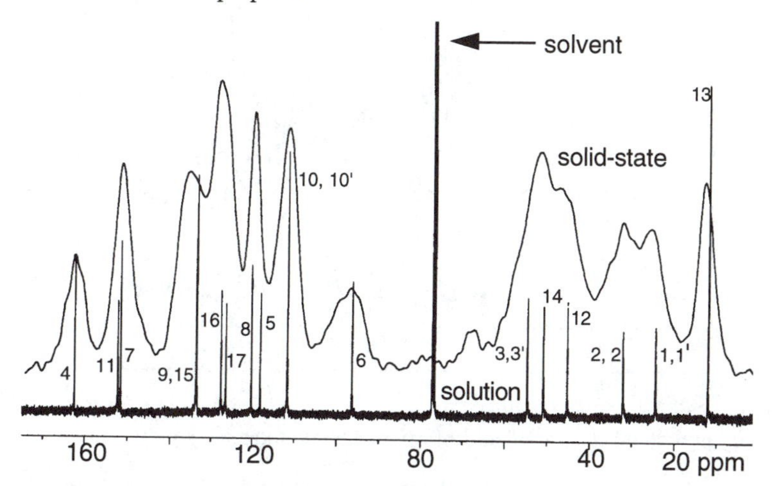

Figure 5: ^{13}C solid- and solution-state NMR spectra of *trans*-polymer **1**. The solid-state spectrum was obtained using cross-polarization and magic angle spinning at a frequency of 4.2 kHz. The solution spectrum was obtained in $CDCl_3$.

Solid-state NMR spectra were run on a Bruker MSL-200 NMR spectrometer operating at a field of 4.7 Tesla with a corresponding ^{13}C frequency of 50.3 MHz. All spectra were obtained using magic-angle spinning at speeds of ca. 3 to 4 kHz. 90° pulse widths of 5 µs were used throughout. For the cross-polarization (CP) experiments, contact times of 1 ms and recycle delays from 5 to 10 seconds were required. ^{13}C shift values are reported with respect to TMS and were measured by replacing the sample with the secondary reference, adamantane (high-frequency line = 38.5 ppm with respect to TMS)[11].

Figure 5 shows a comparison of the solution- and solid-state ^{13}C NMR spectra for the *trans* isomer of polymer **1**. This figure illustrates the relative resolution of the two techniques. The cross-polarization magic angle spinning (CP/MAS) spectrum is relatively well resolved for an amorphous polymer. The lower resolution of the solid spectrum as compared to the solution spectrum is a result of the fact that each individual molecular site has a range of local environments due to the amorphous nature of the glassy polymer. This results in each site giving rise to a range of chemical shift values and is apparent in the relative broadening of the resonance lines. In an attempt to observe changes in local motion with temperature, CP/MAS spectra of the *cis* and *trans* isomers of polymer 1 were run at temperatures up to 110°C. (The upper temperature was limited by instrumental considerations.) Increased molecular motion in the solid caused by elevated temperatures should be detected by NMR relaxation methods. In fact, over the temperature range investigated, there was no significant change in T_1, $T_{1\rho}$ or line width for the *trans* polymer. This is illustrated in Figure 6, which shows small changes in line position resulting from the temperature dependence of the chemical shift, but no measurable changes in line width which would indicate increased mobility of a particular group or increased overall mobility with increased temperature. Similar results were obtained in variable temperature studies of the *cis* isomer. (However, differences in NMR relaxation do exist between the *trans* and *cis* isomer, as will be discussed below).

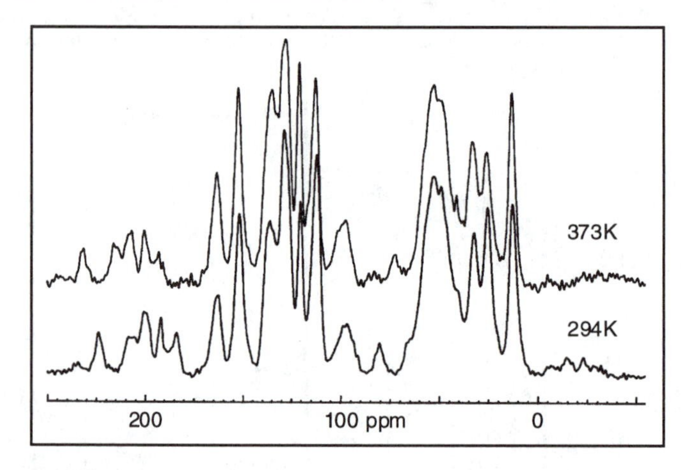

Figure 6: The ^{13}C CP/MAS NMR spectrum of the *trans*-polymer **1**. Resonances above 175 ppm and between 75 and 110 ppm are spinning sidebands.

Clearly the instrument dictated upper temperature limit of 110°C (~60° below the T_g of the *cis* isomer and ~100° below the T_g of the *trans* isomer) did not allow us to reach a temperature where we could observe increased molecular motion. However, we can definitively state that up to 110°C there was no observable increase in molecular motion compared with room temperature data.

Comparison of room temperature T_1 relaxation data for the *trans* and *cis* isomers of polymer **1**, however, provides molecular-level evidence for the higher T_g and stability of the *trans* isomer. Figure 7 shows a comparison of the T_1 values between isomers for several sites in the molecule (see Figure 1 for numbering scheme). Positions 5 and 8 (which overlap), and 3 and 14 (which overlap) show marked increase in T_1 value which would indicate lower molecular motion and increased rigidity in this region (which includes the entire chromophore) for the *trans* isomer polymer. This, in turn, correlates with the comparatively higher Tg of the *trans* isomer. The remaining positions have T_1 values which fall within experimental error in the relaxation measurement, suggesting that MHz motion in these groups is of the same order in the two systems.

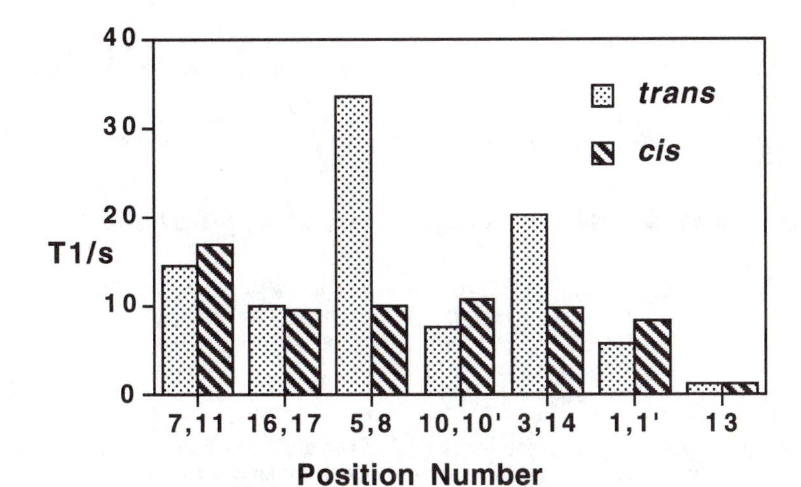

Figure 7: Comparison of ^{13}C solid-state NMR spin-lattice relaxation data for selected sites in the *trans* (sample 1460-27; MW 18,000) and *cis* (1460-28, MW 14,000) isomers of polymer **1**. For numbering scheme, refer to Figure 1.

Film Casting, Corona Poling, Optical Measurements and Stability.

Thin films of polymer **1** were bladed onto microscope slides from 8-10% solutions of polymer in solvent, air-dried under a Petri dish, then baked in vacuum above the glass transition temperature for 18 hours to remove residual solvent. Both pyridine and a mixture of 5% chlorobenzene in pyridine were used as solvents. The addition of chlorobenzene tended to inhibit cracking. The films used in the SHG decay studies were baked in a vacuum oven with a glass door, while all the films for d_{33} measurements were baked in a dark oven. These films were corona poled above the glass transition temperature for 10 - 30 minutes under nitrogen, then slowly cooled to room temperature with the poling field on. The poling schedule for a film poled in the dark in a Plexiglas box with an overpressure of nitrogen gas is shown in Figure 8 (as will be discussed later, the polymer photobleaches in air at high temperature).

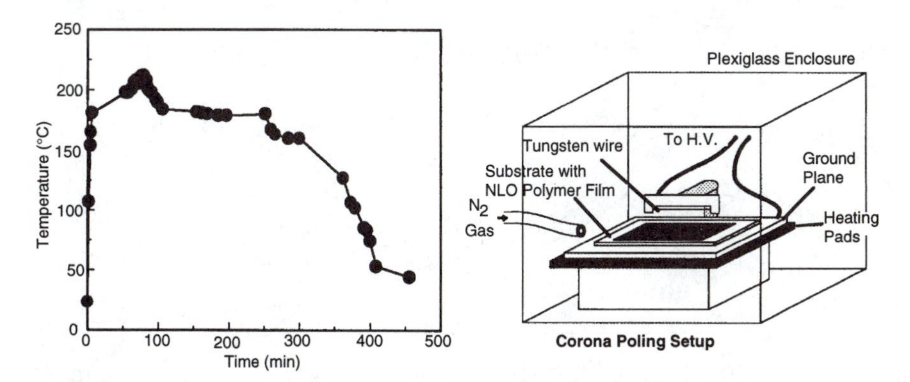

Figure 8. Corona poling schedule and poling setup for an accordion polymer film poled in an overpressure of nitrogen gas with the room lights off.

The glass transition temperatures and d_{33}-coefficients for several samples of this polymer with either a *cis* or *trans* configuration of the 1,2-cyclohexyl groups are given in Table I.

Table I: Glass transition temperatures and d_{33} coefficients for several samples of Polymer **1**.

Sample Number	Configuration	Tg (°C)	d_{33} (pm/V)	Room Lights (During Poling)
1400-43	*Cis*	192	nil	on
1325-55	*Cis*	173	10	on
1400-18	*Trans*	206	3	on
1400-58	*Trans*	200	10	on
1400-95	*Trans*	203	30	off

Quite evident is the variability in d_{33}. When considered together with variable results from SHG decay studies, this led us to search for an environmental factor as a reason for the inconsistent results. Light exposure (i.e., room light) at high temperatures in air during poling and during SHG stability studies is the cause for the variable results. The d_{33} for the last entry in Table I, a film made from synthesis 1400-95, was obtained by both baking and poling in the dark.

Dispersion in the refractive indices of the NLO polymer films was determined by the prism-coupled waveguiding technique.[12] The experimental set-up was identical to that used previously.[5b] A gadolinium-gallium-garnet prism was used in conjunction with polarized light of wavelength 543 nm and 633 nm from He-Ne lasers, 836 nm from a solid state laser, and 1064 nm from a Nd:YAG laser. The refractive indices at these four wavelengths were fit to a single-oscillator Sellmeier equation of the form shown in Figure 9 where λ_{max} = 400 nm was obtained from the UV-VIS spectra. Unpoled films that were bladed from the pyridine/chlorobenzene solutions showed very little birefringence ($n_{TM} - n_{TE}$ = -0.002), unlike the case of a 1 µm thick film of a somewhat lower molecular weight polymer (1400-16), spun-cast from chloroform, that exhibited a large birefringence of ~ -0.04 in the unpoled state.[13] In the latter case, the larger birefringence may have been due to a faster drying rate in the thinner film that did not allow chain reorientation.

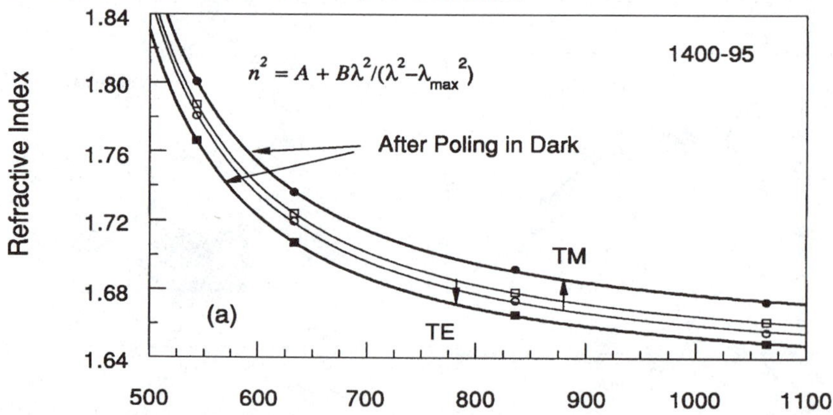

Figure 9: Refractive index dispersion curves before and after poling for a film poled in the dark under nitrogen purge.

The nonlinear optical d-coefficients were measured using the experimental setup described in reference 5b. The sample was rotated on a computer-controlled Oriel rotation stage at the beam waist of a focused (100 mm lens) fundamental beam (1064 nm) produced by a Q-switched Nd:YAG laser. The pulse width was 150 ns and the repetition rate 1 kHz. A half-wave plate was used to control the polarization of the fundamental beam to produce s- and p-polarized light incident on the sample. The second-harmonic signal was detected by a Hamamatsu R928 photomultiplier tube in conjunction with a Stanford Research SR250 boxcar averager, which was operated using active baseline subtraction. Data was collected by computer as a function of the angle of incidence of the fundamental optical beam. The reference quartz crystal was X-cut and 0.381 mm thick. To account for laser power fluctuations, the fundamental was monitored along with the second harmonic in order to normalize the second harmonic signal.

The second harmonic data shown in Figure 10 was obtained 17 hours after poling a film that was poled in the dark in an overpressure of nitrogen gas. To aid in eliminating charge effects due to poling, the poling field was removed when the film cooled to ~45 °C and an aluminum plate, shorted to the ground plane, was placed on the film surface. As a check against charge injection effects, second harmonic data taken four days after poling showed no measurable difference from that taken 17 hours after poling. A 0.4 OD neutral density filter was used in the collection of the p-p data. The d-coefficients were then obtained from a computer fit to an expression[14] for the second-harmonic power, using the measured refractive index and thickness data. Assuming Kleinman symmetry, since the UV-VIS spectra indicated negligible absorption at 532 nm, we obtained $d_{33} = 30 \pm 5$ pm/V with $d_{31}/d_{33} = 0.35$. No correction was made for absorption.

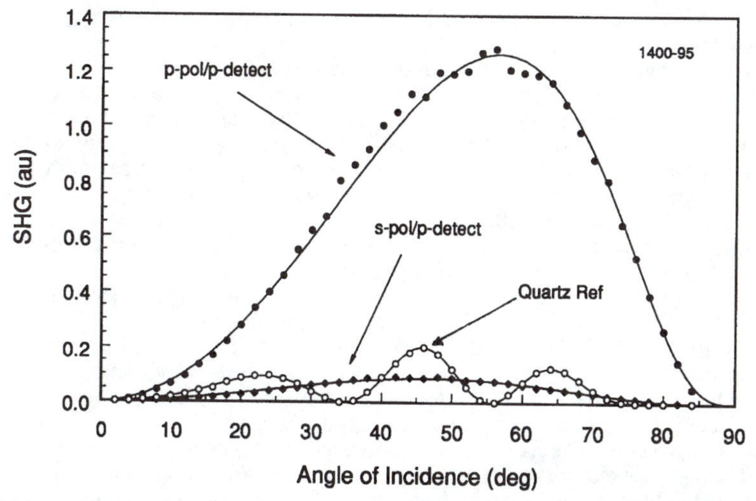

Figure 10: SHG data for a 1.8 µm-thick film poled in the dark in an over-pressure of nitrogen gas together with p-polarized data from a 0.381 mm-thick X-cut quartz.

Molecular orbital calculations using MOPAC 6.0 with the PM3 Hamiltonian on model compounds of the *trans*-isomer model compounds indicate that the diequatorial conformation is approximately 7 kcal/mole lower in energy than the diaxial conformation.[5] The zero-frequency hyperpolarizability, β_0, for the diequatorial and diaxial conformations was calculated to be 24×10^{-30} cm^5/esu and

12×10^{-30} cm^5/esu, respectively.[15] To obtain a rough estimate of the agreement of our SHG measurements with the molecular hyperpolarizabilities, we used a two-level model[16] (with λ_{max} = 400 nm) to estimate β at 1064 nm, and a rigid oriented gas model[17] together with $\langle cos^3\theta \rangle$ obtained from refractive index measurements. Using the method of Page et al.,[18] we obtain from the refractive index data of Figure 9 an order parameter, Φ = 0.027, which implies $\langle cos^3\theta \rangle$ = 0.35. The calculated d_{33} values of 32 pm/V, using the calculated β value and density for a single chromophore, and 26 pm/V, using the calculated β value and density for a pair of *trans* diequatorial chromophores, give good estimates of the measured d_{33} of a poled film of polymer **1**.

Table II: Calculated d_{33} coefficients based upon calculated hyperpolarizabilities and chromophore densities.

N (units/cm³)	β_0 (x 10^{-30} cm^5 /esu)[15]	β(1064 nm) (x 10^{-30} cm^5 /esu)[17]	d_{33} (pm/V)
2.4×10^{21} (Single NLO-phore)	15	40	32
1.2×10^{21} (NLO-phore Pair)	24	64	26

To evaluate orientational stability, poled films of polymer **1** were placed on a hot stage in a setup designed to measure the second-harmonic decay during isothermal aging. With the temperature maintained at 125 °C, p-polarized light at 1047 nm from a diode-pumped Nd:YLF laser was incident on the sample at an angle of incidence of 52 degrees. The pulse width was 12 ns and repetition rate was 1 kHz. The fundamental beam in this case was not focused.

The second harmonic signal was again normalized to the fundamental to compensate for any laser power fluctuations. The results of this experiment for a time period of 120 hours in air are given in Figure 11; all data were normalized to $d_{eff}(0)$. The top curve in Figure 11 (solid circles) shows the decay of d_{eff} as a function of time taken with the room lights off. The bottom curve (open circles) shows decay in the dark for about 70 hours, at which time the room lights were turned on and a fluorescent desk lamp was used to illuminate the film from a distance of about 1 foot for the duration of the experiment. This clearly shows that the light exposure at high temperature introduces an additional decay mechanism.

To characterize this decay mechanism, a series of UV-VIS spectra with the temperature of the film maintained at 125°C were taken, as shown in Figure 12. After 147 hr in the dark at 125 °C, the sample film was exposed to light from a fluorescent desk lamp for the duration of the test. Figure 12 clearly shows the absorbance peak decreasing in height and blue-shifting on prolonged exposure to the light at high temperature. This also indicates that photobleaching techniques can be used to define waveguides of polymer **1**. The fact that the absorbance at λ_{max} decreases 20% from 1 hour to 147 hours concurrently with a 20% decrease in d(t)/d(0) in the dark (top curve Figure 11) indicates that the decay mechanism is a chemical degradation (perhaps oxidation), and not loss of polar alignment.

Further characterization of the photochemical change at elevated temperature was obtained from IR spectra. After 24 hrs at 175 °C in a nitrogen blanket in the dark, there was no perceptible change in the IR spectra. However, when exposed to light in air at 150°C, the IR spectra exhibited appreciable change of the methylenes (carbonyl formation) attached to the anilino nitrogen after less than one hour. This possible oxidation would drastically reduce the electron donating power of the anilino nitrogen and greatly blue-shift the charge transfer band, thus effecting a d_{eff} decay at the molecular level. There was no evidence for a 2+2 cycloaddition of the cinnamoyl groups in the IR spectra.

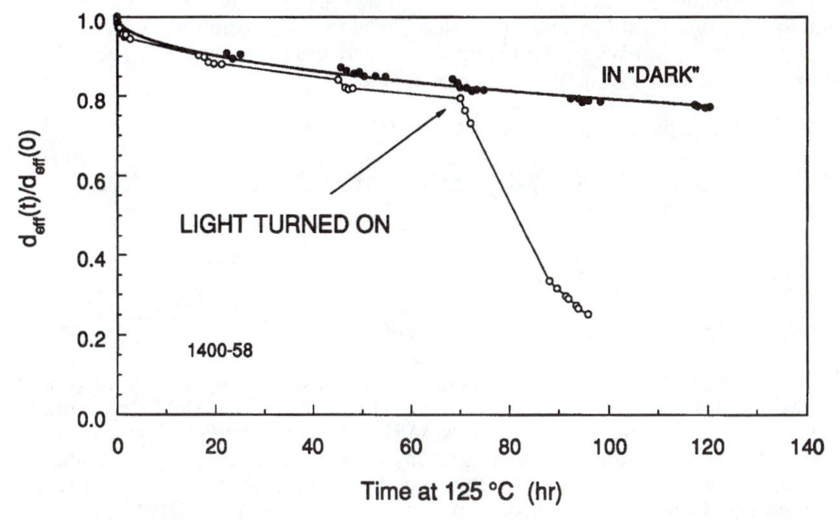

Figure 11. Effect of fluorescent light exposure on the decay of d_{eff} at 125 °C.

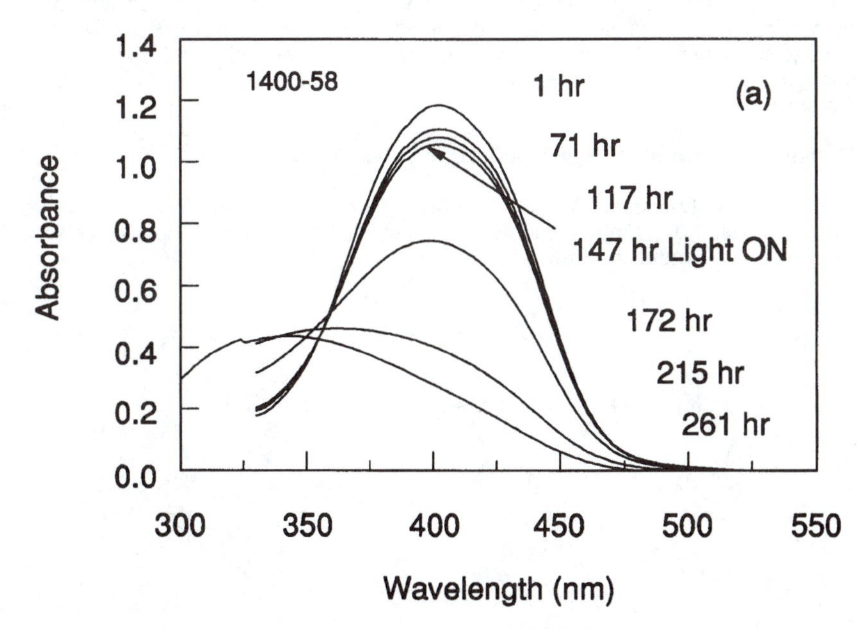

Figure 12. Effect of light exposure on absorbance with the sample film maintained at a constant elevated temperature of 125°C.

Waveguide construction and r_{33} measurement

Refractive index and propagation loss measurements of polymer **1** were made by preparing a thin film 1.4 μm thick on a glass substrate. Prism coupling was used to excite the guided modes at 0.859 and 1.3 μm wavelengths. Measurements were made for an unbleached film and a film bleached at room temperature for 4 hrs with a UV light source (17 mW/cm^2 @360 nm in air). The results are shown in Table III. The Δn was 0.052 at 0.859 μm and 0.003 at 1.3 μm.

Table III: Refractive Index and loss data for Polymer **1**.

Bleach time(hours)	n^a @ 0.859 μm	loss(dB/cm)	n^a @ 1.32 μm	loss(dB/cm)
0	1.666 ±0.005	3.3	1.608 ±0.002	<1
4	1.614 ±0.005	3.3	1.605 ±0.002	<1
Δn	0.052		0.003	

(a) Calculated values based on measurements of n_{eff}.

A Mach-Zehnder intensity modulator was constructed with polymer **1** as the core layer (2.3 μm thick) of a three layer stack. The cladding material (3.5 μm thick) used was UV curable polymer (Norland NOA81). A gold ground plane formed the lower electrode along with a patterned gold layer for the upper electrode. Channel waveguides were formed by photobleaching with a 5 μm wide mask. Poling was accomplished using the device electrodes with an average field of 107 V/μm at 200°C. The device was overcoated with a polymer layer and cut/polished for end fire coupling. The measured EO coefficient was found to be 8.5 pm/V and has remained unchanged at room temperture for 209 days, which was our most recent meaurement in this ongoing experiment. Using equation 38 in reference 17, d_{33} = 30 pm/V from SHG measurements, the index values from Figure 9 (extrapolating to 1.32 μm), and correcting for dispersion and local field effects, one obtains a calculated r_{33} value of 6.5 pm/V.

Design and Synthesis of New Syndioregic Polymers

The structures of polymers **2-8** are given below. For the calculations of hyperpolarizability and dipole moment the amine donor bridging group was replaced by two dimethylamino groups.

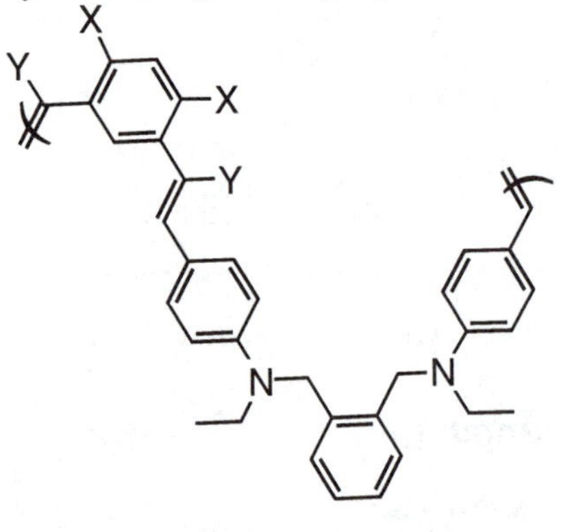

Polymer	X	Y
2	H	CN
3	NO_2	CN
4	H	SO_2CF_3
5	NO_2	SO_2CF_3
6	NO_2	H

Figure 13: Structure of Polymers **2-6**

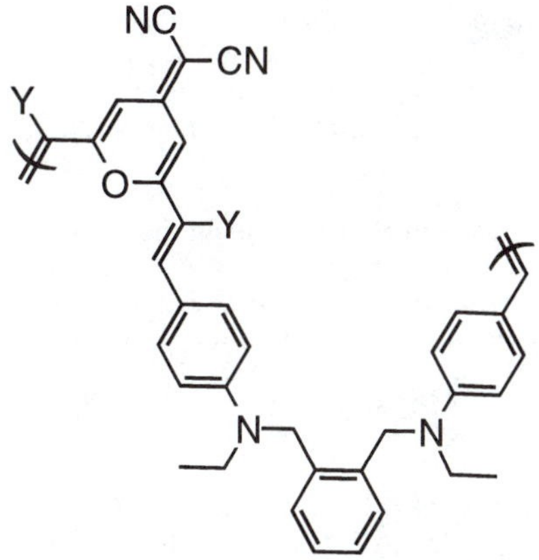

Polymer	Y
7	H
8	CN

Figure 14: Structure of Polymers **7** and **8**

Several considerations must be taken into account when designing chromophore pairs to be incorporated into the syndioregic NLO polymer architecture. From an electrooptical standpoint, one needs to design pairs with the highest hyperpolarizability per volume. From a synthetic standpoint, the chromophore pairs must be amenable to incorporation into a syndioregic polymer backbone. This requires the synthesis of a pair of monomers with two functional groups that are amenable to polymer synthesis. For example, polymer **1** required a bis aldehyde functional electron donating comonomer and a bis activated hydrogen functional electron withdrawing comonomer. Other types of complementary comonomers are usable so long as they give high molecular weight polymer with no side reactions.

MOPAC V.6[19]calculations were used to determine which chromophore pairs might give high second order nonlinearity.

Table IV: Calculated[a] dipole moments and hyperpolarizabilities for energy minimized pairs.

Model for Polymer	μ(Debye)	β($\times 10^{-30}$cm^{5}esu)	β/V[b]($\times 10^{-30}$cm^2/esu)
1	6.3	24.2	0.050
2	6 (7.0)	18.3 (31.6)	0.052(0.089)
3	9.3 (14.6)	20.5 (84.3)	0.052(0.215)
4	9.7 (11.0)	5.8 (48.5)	0.013(0.108)
5	13.9 (15.2)	38.8 (101.2)	0.077(0.202)
6	8.5	46.3	0.128
7	10.8	45.7	0.126
8	11.8(14.9)	35.7(69.6)	0.091(0.177)

(a) calculated using the PM3 Hamiltonian per pair of chromophores[19]. (b) Volume was calculated using additivity data[20]. Numbers in parentheses were calculated by constraining the chromophore to be planar.

The model compounds for polymers **2**, **3**, **4**, **5** and **8** were non-planar in their energy minimized conformation. The difference between the energy minimized conformation and the planar conformation was +3 kcal/mole for polymer **2**, +29 kcal/mole for polymer **3**, +17 kcal/mole for polymer **4**, +51 kcal/mole for polymer **5**, and +8 kcal/mole for polymer **8**. Due to these significant differences in heat of formation, it is unlikely that the chromophore pairs of polymer **3**, **4** and **5** will be planar. The low probability of planarity, coupled with synthetic challenges for polymers **3**, **4** and **5**, make further efforts on these particular polymers unattractive.

Synthesis of New Syndioregic Polymers

Dicyanomethylene-2,6-dimethyl-γ-pyrone was synthesized using the method of Woods[21], the yield was 56%. 1,3-Benzene diacetonitrile was obtained from Aldrich. 1,2-bis(N-ethyl,-N-(4-formyl phenyl)amino)xylene, the bis-aldehyde, was synthesized in 63% yield as reported previously by our group[6]. The anil of the bis aldehyde was made by refluxing the bis aldehyde with 2 equivalents of 4-chloroaniline in benzene (yield 71%). See Table V for a summary of polymer properties.

Preparation of *m*-di(α-cyano-4'-aminostyryl)benzene polymer, Polymer 2. 0.3610 g (2.31 mmol), of 1,3-benzene diacetonitrile, 0.9216 g (2.30 mmol) of bis aldehyde[4] and 0.5 g of dimethylaminopyridine (4.1 mmol) and 2 ml of piperdine were added to 25 ml of pyridine. After 4 days at reflux, no aldehyde endgroup was detected by ^{1}H NMR (200 MHz). The resulting material was dissolved in chloroform and precipitated into methanol. This polymer was soluble in chloroform and pyridine and has a λ_{max} of 414 nm. The molecular weight was determined using ^{1}H NMR (400 MHz) by the ratio of aldehyde protons to benzyl amine protons and found to be 91,000. The yield was 0.89 g (74%). The polymer was dried at 200°C in the TGA before analysis. Polymer **2** appears thermally stable up to 365°C.

Preparation of *m*-di(trifluoromethanesulfonylaminostyryl)benzene polymer, Polymer 4. 1,3 bis(trifluoromethanesulfonylmethyl) benzene was made by refluxing a mixture of 5.00 g α,α'-dibromo-*m*-xylene (19 mmol), 6.90 g potassium trifluoromethylsulfinate (40 mmol) and 0.05 g potassium iodide in 150 mL acetonitrile overnight. The mixture was cooled and concentrated in vacuum. The residue was taken up in 150 mL ether. This was washed with water then dried with magnesium sulfate and concentrated in vacuum to give 6.94 g product as a brown solid (99%). This was recrystallized from 50 mL absolute ethanol to give 5.76 g of tiny white needles (82%). mp.134-135°C.

0.2424 g (0.65 mmol)of α,α'-di(trifluoromethanesulfonyl)-*m*-xylene, 0.2621 g (0.65 mmol) of bis-aldehyde[4] and 5 drops of piperdine were placed in 12 ml of dry pyridine. After 2 days at reflux, the sample was precipitated into 300 ml of a 30/70 mixture of chloroform/methanol. The yield was approximately 33%. This polymer was slightly soluble in pyridine and insoluble in chloroform and methanol.

Preparation of *m*-di(aminostyryl)-4,6-dinitrobenzene polymer, Polymer 6. Condensation of dinitro xylene with acetals has been reported[22]. 0.1088 g (0.55 mmol)of 2,4-dinitro-*m*-xylene, 0.2220 g (0.55 mmol) of bis-aldehyde[4] and 5 drops of piperdine were placed in 11 ml of dry pyridine. After 2 days at reflux, a sample was taken for NMR analysis. The results showed mostly dimers. After 7 days at reflux, all the pyridine evaporated. A fraction of the polymer was redissolved in pyridine and precipitated into methanol. The insoluble fraction was washed with methanol and dried. The overall yield was 75%. The decomposition temperature of this polymer is very low (173°C), due possibly to inherent instability of the polymer, low molecular weight, or both. In any case, efforts on synthesis of this polymer were discontinued.

Preparation of 1,3-di(aminostyryl)dicyanomethylene-γ-pyrone polymer, Polymer 7. The synthesis of Polymer **7** was attempted using a variety of conditions.

Most of the attempts gave dimers and trimers. What follows is the most successful set of conditions to date: 1.2922 g (2.08 mmol) of the chloroanil of the bis aldehyde[4], 0.359 g (2.08 mmol) of 4-dicyanomethylene-2,6-dimethyl-γ-pyrone and 0.07 g of 1,4-diazobicyclo[2.2.2]octane (DABCO) were placed in 10 ml of 1,2-dimethoxybenzene (veratrole). After 4 days at 160°C, ^{1}H NMR indicated an average degree of polymerization of about 6. After 7 days at 160°C the solution was precipitated into methanol. This polymer was slightly soluble in veratrole and m-cresol, insoluble in pyridine, methanol, and chloroform. The yield was 70%.

Towards the synthesis of Polymer 8: Preparation of 4-dicyanomethylene-2,6-bis(bromomethyl)-γ-pyrone. 4 g (23 mmol)of 4-dicyanomethylene-2,6-dimethyl-γ-pyrone , 8.27 g (46 mmol) of N-bromo-succinimide (NBS), 0.05 g azo-bis(isobutylnitrile) (AIBN) and 150 mL of CCl_4 were placed in a 250 ml flask with condenser and nitrogen purge. The contents were refluxed for 4 weeks and the degree of conversion was monitored by ^{1}H NMR. Additional NBS and AIBN were added (a total of 8.5 g of NBS) when the degree of conversion was unchanged after a week. The reaction was stopped after the degree of conversion of the methyl groups was approximately 80%. The NMR of the crude product show a mixture of products (dimethyl; bromomethyl methyl; bis(bromo-methyl) and possible ring bromination products). the crude product was cooled, extracted 3 times with 100 ml water, once with 100 ml saturated sodium bicarbonate, dried with magnesium sulfate, filtered, then rotovapped and dried. The product was then eluted through a silica gel chromatography column using 30/70 ethyl acetate/hexane. Seven 50 ml fractions were collected. Fractions 1-3 were found to be similar by NMR and therefore combined and fractions 4- 7 were also similar and were combined also (separate from fractions 1-3). The combinations of fractions 4-7 were rotovapped and recrystallized from ethyl acetate/hexane. Yielding 2.0 g (26%) of pure product (^{1}H NMR $CDCl_3$: 6.6 ppm (2H); 4.5 ppm (4H); M.P. 138-142°C. Recrystallization of fractions 1-3 gave a crystalline material that was not the desired product.

Several attempts were made towards the synthesis of 4-dicyanomethylene-2,6-bis(cyanomethyl)-γ-pyrone. Reaction with sodium cyanide in DMF, DMSO, methanol, dichloromethane with 18-crown 6 ether, were unsuccessful as were reaction with copper cyanide in DMF, and reaction with trimethylsilyl cyanide in dichloromethane.

Analysis of Polymers

NLOP's with four different electron accepting groups were prepared, Table V summarizes the analysis results for each of the polymers.

Table V: Yields, glass transition temperatures, decomposition temperatures[a] and molecular weights[b].

Polymer	Yield (%)	T_g(°C)	T_d(°C)	M_n
1	80	~205	330	~25,000
2	74	165	365	~90,000
4	33	-	200	-
6	75	-	<173[c]	-
7	70	173	310	

(a)2% weight loss in nitrogen at 10°C/min heating rate by TGA. (b) by ^{1}H NMR end group analysis. (c) soluble fraction.

As seen from TablesIVand V, Polymer **2** is the most promising from a thermal stability and processability standpoint, whereas polymers **6** and **7** are more likely to give films with higher d_{33} coefficients. However, the thermal stability of polymer **6** is too low to warrant further study. Synthesis of the dinitro derivative of polymers **2** was abandoned due to synthetic difficulties. Polymer **5** (the dinitro derivative of polymer **4**) was not pursued because polymer **4** decomposes at 200°C. It is likely that polymer **5** will be even less stable.

The d_{33} coefficient of a lower molecular weight (13,000) version of Polymer **2**[23] was measured using the methods described in the previous section. After poling at 170°C with a similar time-temperature protocol, the d_{33} was found to be 27 pm/V, showing that it is close to the value predicted from the calculated β.

Summary

Solid state NMR indicates no significant increases in bulk chromophore movement in Polymer **1** at temperatures up to 110°C. Comparison of NMR T_1 relaxation data between the *cis* and *trans* isomer show the central region of the chromophore to be significantly more rigid in the case of the *trans* isomer.

A photooxidation reaction at elevated temperatures has been identified as the cause of unexplained variability in our previous NLO stability studies on a novel folded mainchain polymer containing α-cyanocinnamoyl chromophores. Processing in the dark (baking and poling) resulted in a d_{33} of 30 pm/V which is three times the best previous measurements obtained from poling in ambient light. An achieved alignment factor $\langle \cos^3\theta \rangle$ of 0.35, together with retention of 80% of NLO activity after 120 hrs at 125°C in the dark, indicate that the syndioregic architecture can be as well ordered as side chain polymers by electric field poling and is a promising stable structure for NLO polymers. The electrooptical coefficient (r_{33})of this polymer was found to be 8.5 pm/V, as measured in a Mach-Zehnder intensity modulator.

Several new mainchain syndioregic polymers were synthesized, demonstrating the possibility of incorporating new chromophores into the syndioregic mainchain polymer architecture. With the appropriate chemistry, it will be possible to make polymers with higher d_{33} coefficients, glass transition temperatures and thermal stabilities.

Acknowledgements

Financial support of ARPA and the NAWCWPNS Independent Research is gratefully appreciated. The authors also gratefully acknowledge the assistance of Ms. D. Paull for the thermal analysis data.

References

(1) (a)Singer, K.D.; Kuzyk, M.G.; Holland, W.R.; Sohn, J.E.; Lalama, S.L.;. Comizolli, R.B; Katz, H.E.; and Schilling, M.L.; *Appl. Phys. Lett.* **1988**, 53, 1800; (b) Lindsay, G.A.; Henry, R.A.; Hoover, J.M.; Knoeson, A..; and Mortazavi, M.A.; *Macromol.* **1992**, 25, 4888; (c) Herman, W.N.; Rosen, W.A.; Sperling, L.H. ; Murphy, C. J.; and Jain, H.; *Proceedings of SPIE*, **1992**, 1560, 206.

(2) (a) Lindsay, G.A.; Stenger-Smith, J. D.; Henry, R.A.; Hoover, J.M.; Nissan, R.A.; and Wynne, K. J.; *Macromol.* **1992**, 25, 6075; (b) Ranon, P.M.; Shi, Y.; Steier, W.H.; Xu, C.; Wu, B.; and Dalton, L.R.; *Appl. Phys. Lett.* **1993**, 62, 2605; (c) Lindsay, G. A.; Stenger-Smith, J. D.; Henry, R. A.; Nissan, R.A.; Chafin, A.P.; Herman, W.N.; and Hayden, L.M.; *ACS Polymer Preprints* **1993**, 34, 796.

(3) (a) Eich, M.; Reck, B.; Yoon, D.Y.; Wilson, C.G.; and Bjorklund, G.J.; *J. Appl. Phys.* **1989**,66, 3241; (b) Mandal, B.K.; Chen, Y.M.; Lee, J.Y.; Kumar, J.; and Tripathy, S.; *Appl. Phys. Lett.* **1991**, 58, 2459.
(4) Stenger-Smith, J. D.; Henry, R. A.; Hoover, J. M.; Lindsay, G. A.; Nadler, M. P.; and Nissan, R. A.; *J. Poly. Sci. Part A: Chem. Ed.*, **1993**, 31, 2899.
(5) (a) Lindsay, G. A.; Stenger-Smith, J. D.; Henry, R. A.; Nissan, R.A.; Merwin, L.H.; Chafin, A.P.; Yee, R.Y.; Herman, W.N.; and Hayden, L.M.; *Organic Thin Films for Photonic Applications*, ACS/OSA Technical Digest Series, **1993**, 17, 14; (b) Herman, W.N; Hayden, L.M.;Brower, S.;Lindsay, G. A.; Stenger-Smith, J. D.; and Henry, R. A.; *Organic Thin Films for Photonic Applications*, ACS/OSA Technical Digest Series, **1993**, 17, 18.
(6) Bax, A., Subramanian, G., *J. Magn. Reson.*, **1986**, 67, 575.
(7) Bax, A., Summers, M. F., *J. Am. Chem. Soc.*, **1986**, 108 2093.
(8) "Nuclear Magnetic Resonance Spectroscopy, Volume II", Pal Sohar, CRC Press Inc., Boca Raton, FL **1983**.
(9) Nissan, R. A., Stenger-Smith, J. D., and Merwin, L. H., in preparation.
(10) "*Spectroscopy of Polymers*, Chapter 10", Koenig, J. L., NMR Relaxation Spectroscopy of Polymers, American Chemical Society, American Chemical Society, Washington, D. C. **1992**
(11) Hayashi, S.; and Hayamizu, K.; *Bull. Chem. Soc. Japan*, **1989**, 62, 2429.
(12) Ulrich, R.; and Torge, R.; *Applied Optics* **1973**, 12, 2901.
(13) Wang, C.H.; and Guan, H.W.; *Proceedings of SPIE* **1992**, 1775.
(14) Hayden, L. M.; Sauter, G. F.; Ore, F. R.; P.L. Pasillas, Hoover, J. M.; Lindsay, G. A.; and Henry, R. A.; *J.Appl. Phys.* **1990**, 68, 456. See also Chapter XX?? in these proceeding.
(15) Lindsay, G. A.; Stenger-Smith, J. D.; Henry, R. A.; Nissan, R.A.; Chafin, A.P.; Herman, W.N.; and Hayden, L.M.; *ACS Polymer Preprints* **1993**, 34, 796.
(16) Oudar, J.L.; and Chemla, D.S.; *J.Chem. Phys.* **1977**, 66, 2664.
(17) Singer, K.D.; Kuzyk, M.G.; and Sohn,.J.E.; *J. Opt. Soc. Am.* **1987**, B4, 968.
(18) Page, R.H.; Jurich, M.C.; Reck, B.; Sen, A.; Twieg, R.J.; Swalen, J.D.; Bjorklund, G.J.; and Wilson, C.G.; *J. Opt. Soc. Am.* **1990**, B7, 1239.
(19) (a) Stewart, J.P.; "MOPAC, A Semi-Empirical Molecular Orbital Program", QCPE, **1983**, 455; (b) Stewart, J.J.P.; *J. of Computational Chemistry*, **1989**, 10(2), 221-264.
(20) D.A. Cichra, J.R. Holden and C. Dickinson, Naval Surface Weapons Center, Dahlgren, Virginia 22448, NSWC TR 79-273, Feb. 19, 1980 .
(21) Woods, L. L.; J. Am. Chem. Soc., **1958**, 81, 1440.
(22) Berlin, A., Bradamante, R., Ferraccioli, R., Pagani, G. A., and Sannicolo', R., *J. Chem. Soc, Chem. Commun.,* **1987**, 1176.
(23) Stenger-Smith, J. D.; Henry, R. A.; Chafin, A.P.; Lindsay, G. A.; *ACS Polymer Preprints* **1994**, 35(2), 140.

RECEIVED March 2, 1995

Chapter 15

Second-Order Nonlinear Optical Interpenetrating Polymer Networks

S. Marturunkakul[1], J. I. Chen[1], L. Li[1], X. L. Jiang[2], R. J. Jeng[2,3], S. K. Sengupta[1], J. Kumar[2], and S. K. Tripathy[2]

[1]Molecular Technologies Inc., Westford, MA 01886
[2]Center for Advanced Materials, Departments of Chemistry and Physics, University of Massachusetts, Lowell, MA 01854

An interpenetrating polymer network (IPN) incorporating nonlinear optical (NLO) chromophores as a second-order NLO material is reported. This IPN system consists of an NLO active epoxy-based polymer network incorporating a thermally crosslinkable NLO chromophore, and an NLO active phenoxy-silicon polymer network. Characterization of the thermal and optical properties of the IPN samples are reported. The optical nonlinearity of the poled/cured IPN samples is stable after being heated at 100 °C for 168 h. The nonlinear optical coefficient, d_{33}, of a poled/cured IPN sample is measured to be 32 pm/V at 1.064 μm. The electro-optic coefficients, r_{33}, of this sample are determined to be 18 pm/V at 633 nm and 6.5 pm/V at 1.3 μm. Waveguide optical loss measurements have been carried out at 0.83 and 1.3 μm.

Second-order nonlinear optical (NLO) materials are currently of interest to a large number of research groups as they can be used for a number of photonic applications, for example, doubled-frequency laser sources, electro-optic modulation, optical signal processing, optical interconnects, etc. A practical second-order NLO polymer should possess large second-order nonlinearity, excellent temporal stability at elevated temperatures, and low optical loss (*1*). A number of NLO polymers have been developed to exhibit large second-order NLO coefficients comparable to those of the inorganic NLO materials which are currently in use in devices (*2,3*). However, the major drawback of NLO polymers is the decay of their electric field induced second-order optical nonlinearities. This decay is a result of the relaxation of the NLO chromophores from the induced noncentrosymmetric alignment to a random configuration. Numerous efforts have been made to minimize this decay through different approaches (*4*).

Recently, we have reported on an approach to prepare stable second-order NLO polymers using a full interpenetrating polymer network (IPN) structure (*5,6*).

[3]Current address: National Chung-Hsing University, Department of Chemical Engineering, Taichung, Taiwan

0097–6156/95/0601–0198$12.00/0

An IPN is a structure in which two or more networks are physically combined (*7,8*). The IPN is known to be able to remarkably suppress the creep and flow phenomena in polymers. The motion of each polymer in the IPN is reduced by the entanglements between different networks. These properties of the IPN have been shown to be able to restrict the mobility of the aligned NLO chromophores, and greatly enhance the stability of the NLO properties (*5,6*). The attractive feature of this approach is that it combines high *Tg* polymers and highly crosslinked networks, elements known to enhance the stability, into a single system. The temporal stability of the IPN is synergistically enhanced as it is better than that of the parent polymers. The second-order NLO properties of an IPN at 110 °C is found to be extremely stable for over 1000 h.

In this paper, investigation of an IPN system with a high degree of crosslinking density and NLO chromophore density is reported. This IPN system consists of two NLO-active networks formed by a phenoxysilicon polymer and a thermally crosslinkable epoxy prepolymer incorporating a thermally crosslinkable NLO chromophore. The synthesis and characterization of the thermal and optical properties of the IPN are described. Waveguide properties of this IPN system are also reported.

Experimental

The epoxy prepolymer (namely BPDOMA, see Figure 1a) is based on the diglycidyl ether of bisphenol A and 4-(4'-nitrophenylazo)aniline functionalized with thermally crosslinkable methacryloyl groups. The crosslinkable NLO chromophore, namely DRMA (Figure 1b, Disperse Red 19 (DR19) functionalized with methacryloyl groups), is synthesized by coupling the corresponding acid chloride with the hydroxyl groups of DR19. The mixture of BPDOMA and DRMA forms a network through a radical addition reaction. The phenoxysilicon polymer, which is prepared through a sol-gel reaction (*5*), is based on an alkoxysilane dye (ASD) of (3-glycidoxypropyl) trimethoxysilane and 4(4'-nitrophenylazo)aniline, and the multifunctional phenoxyl compound 1,1,1-tris(4-hydroxyphenyl)ethane (THPE). The NLO properties and characterizations of the phenoxysilicon polymers have been reported earlier in ref 9. Figure 2 shows the structures of ASD and THPE. The sol-gel process which involves sequential hydrolysis and condensation reactions is shown in Scheme 1.

hydrolysis

$$\equiv SiOR + HOH \rightleftharpoons \equiv SiOH + ROH$$

water condensation

$$\equiv SiOH + HOPh- \rightleftharpoons \equiv SiOPh- + HOH$$

alcohol condensation

$$\equiv SiOR + HOPh- \rightleftharpoons \equiv SiOPh- + ROH$$

Scheme 1. The sol-gel reaction between ASD and THPE.

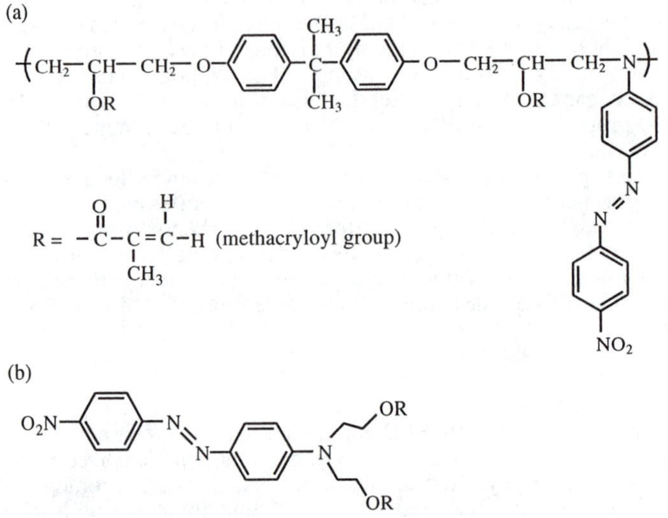

(a)

$$-\!\!\left[CH_2-CH(OR)-CH_2-O-C_6H_4-C(CH_3)_2-C_6H_4-O-CH_2-CH(OR)-CH_2-N(C_6H_4-N{=}N-C_6H_4-NO_2)\right]\!\!-$$

$$R = -C(=O)-C(CH_3)=C(H)-H \text{ (methacryloyl group)}$$

(b)

$$O_2N-C_6H_4-N{=}N-C_6H_4-N(CH_2CH_2OR)_2$$

Figure 1. Chemical structures of (a) BPDOMA and (b) DRMA.

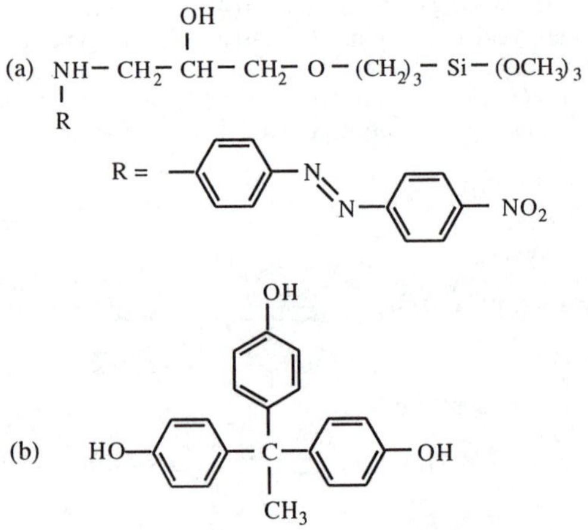

(a)

$$NH(R)-CH_2-CH(OH)-CH_2-O-(CH_2)_3-Si-(OCH_3)_3$$

$$R = -C_6H_4-N{=}N-C_6H_4-NO_2$$

(b)

$$(HO-C_6H_4)_3C-CH_3$$

Figure 2. Chemical structures of (a) ASD and (b) THPE.

Thin films were prepared as described previously (*5*). The procedure is summarized in Scheme 2. IPN samples with various compositions were prepared. The network of each polymer was formed simultaneously when the thin films were heated for 1 h at 200 °C on a hot stage. The typical thickness of the cured films was approximately 1 μm. The reactions of both networks were monitored by FTIR photometer (1760X, Perkin-Elmer). A differential scanning calorimeter (DSC 2910, T.A. Instrument Co.) was used to study the crosslinking reaction and determine thermal properties of the samples.

The corona poling technique (*9,10*) was employed to align the NLO chromophores. The corona field was applied as the temperature was raised to 80 °C. The temperature was then increased to 200 °C with the corona field on. The corona current was maintained at 2 μA with a potential of 4 kV while the poling temperature was kept at 200 °C for 60 min. The formation of the networks and the molecular alignment for the poled order proceeded simultaneously during this period. The sample was then cooled down slowly to room temperature before the corona field was removed.

The second-order NLO properties of the poled IPN samples were measured by second harmonic generation from 1.064 μm laser radiation (*9,10*). Electro-optic coefficients, r_{33}, of the IPN samples were measured at 633 nm and 1.3 μm, using the reflection method (*11*). The relaxation behavior of the second-order NLO properties was studied by monitoring the decay of the second harmonic (SH) intensity as a function of time at 100 °C after poling and curing. The waveguide optical loss of slab uncladded polymer films were determined at 0.83, 1.064 and 1.3 μm. The measurement scheme has been described in detail elsewhere (*12*).

Result and Discussion

The design rationale to incorporate the thermally crosslinkable methacryloyl group on the epoxy polymer, BPDOMA, is that a high degree of functionalization is possible (close to 100%). In order to further increase the chromophore- and the crosslinking densities of the IPN system, a thermally crosslinkable NLO chromophore, DRMA, is added. An onset point of an exotherm at 175 °C was observed for the methacryloyl groups from the BPDOMA and the functionalized chromophore by differential calorimetric analysis. As the samples are heated at 200 °C for 1 h during the curing process, inter- and intramolecular crosslinking reactions between the polymer and the functionalized dye take place. The cured materials are no longer soluble in THF which is a good solvent for both the dye and the polymer due to the extensive degree of the reaction. The SH intensity as a function of temperature was monitored for both poled/cured BPDOMA and BPDOMA/DRMA (2:1 by wt) samples. The onset temperature of the decay of the SH intensity for the BPDOMA/DRMA is higher than that of the BPDOMA. This demonstrated that the addition of the thermally crosslinkable dye, DRMA, increased the crosslinking density of the material leading to enhanced stability.

Figure 3 shows the IR spectra of an IPN sample before and after curing at 200 °C. The THPE/ASD network was formed through a sol-gel process. After the sample was heated at 200 °C for 1 h, the broad O-H stretching band at 3383 cm^{-1} and the bands at 1456 and 1376 cm^{-1} for the O-CH_3 bending decreased significantly due to the reaction of THPE with ASD. In addition, a new band at 948 cm^{-1} for the Si-OPh stretching in the cured sample confirms the formation of phenoxysilicon bonds (*9*). The epoxy-based polymer network was formed through a free radical polymerization of the methacryloyl groups. The C=C stretching at 1677 cm^{-1}

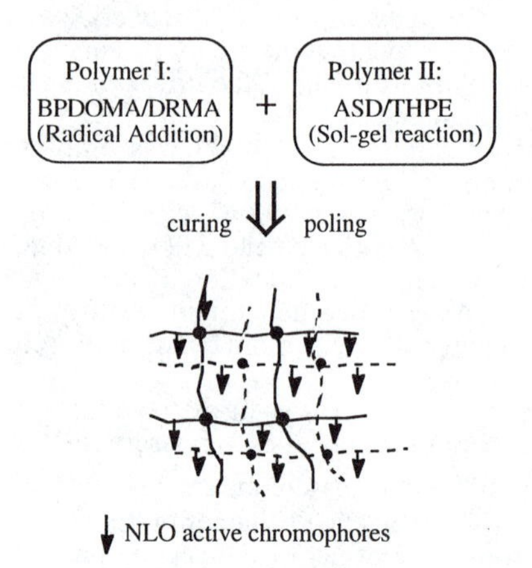

Scheme 2. Schematic diagram for the formation of an IPN.

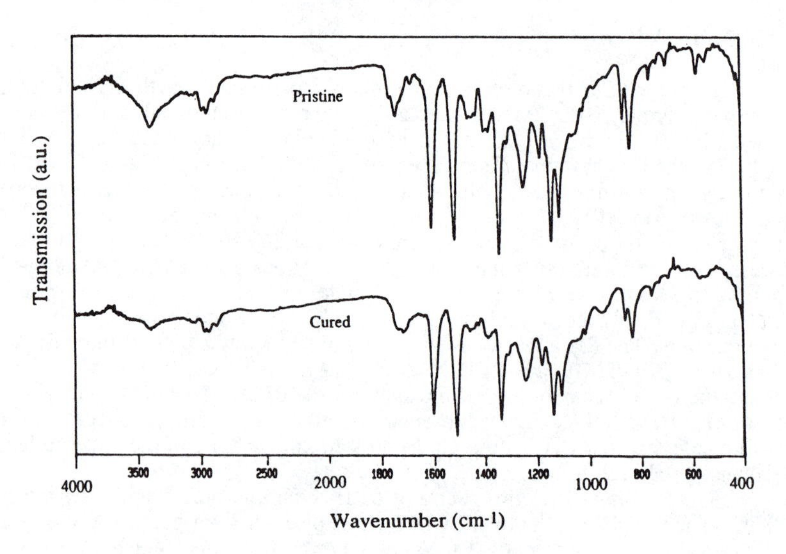

Figure 3. Infrared spectra of the IPN sample before (top) and after (bottom) curing.

disappeared in the cured sample. The strong carbonyl stretching band at 1735 cm^{-1} due to conjugation with the C=C group reduced after curing. The results from the IR spectra indicate that these two polymers each form their own networks through different reaction mechanisms upon heating.

The poled and cured IPN samples showed excellent optical quality. The homogeneity of this full IPN system is also suggested by the single T_g at 141 °C observed from the DSC thermogram with a 10 °C/min scanning rate. The T_g of the cured phenoxysilicon polymer (*9*) and BPDOMA were determined to be 110 and 145 °C respectively. The single T_g of the IPN sample implies that phase separation did not occur in this molecular composite. Optical microscopy reveals a clear transparent featureless film.

The temporal stability of the second-order optical nonlinearity for the poled/cured IPN, phenoxysilicon polymer, and BPDOMA samples are shown in Figure 4. The SH intensity of the sample is stable after a small initial decay as the sample is subjected to thermal treatment at 100 °C for up to 168 h, whereas the BPDOMA and the phenoxy-silicon samples show faster initial reduction of the SH intensity at the same heat treatment condition. The synergistically enhanced stability of the poled order in this IPN system is on account of the permanent entanglement within the interpenetrating network structure. The second-order NLO coefficient, d_{33}, of the poled/cured IPN samples are summarized in Table I. The IPN sample with 1:1 weight ratio of each network component exhibits a d_{33} of 32 pm/V at 1.064 μm. The r_{33} values of this sample were determined to be 18 pm/V at 633 nm and 6.5 pm/V at 1.3 μm.

Table I. Optical properties of the poled and cured IPN samples processed at 200 °C for 1 h.

	Phenoxysilicon (ASD/THPE)	Polymer-11/ DRMA (2:1 w/w)	d_{33} at 1.064 μm
Weight	1	1	32.0
ratio	2	1	26.0

BPDOMA exhibits relatively low optical loss even when the processing conditions have not been optimized. Measurements on unclad slab waveguides gave values of 4.7 dB/cm at 830 nm and 2 dB/cm at 1.3 μm. Loss measurement on the IPN samples was performed at 1.3 μm. The waveguide optical loss was determined to be 4 dB/cm.

Conclusion

A new second-order NLO IPN system has been developed and the samples exhibit large and stable optical nonlinearities. The nonlinearity of this IPN system is affected by the compositions. Waveguide optical loss measurement was possible on this system. Further improvements of the linear and nonlinear optical properties of this class of materials is currently underway from the perspective of practical device applications.

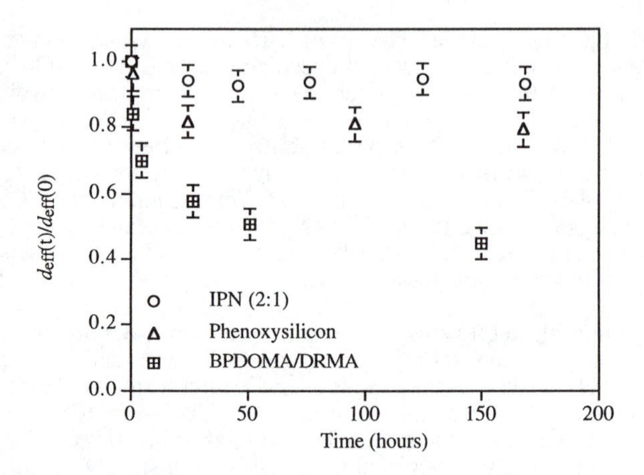

Figure 4. Temporal Stability of the IPN sample (Phenoxysilicon/(BPDOMA/DRMA), 2:1), phenoxysilicon and BPDOMA/DRMA (2:1) samples as subjected to thermal treatment at 100 °C.

Acknowledgments

Funding from NSF and ONR at UMass Lowell, and AFOSR (Contract # F49620-93-C-0039) at MTI is gratefully acknowledged.

References

1. Ermer, S.; Valley, J. F.; Lytel, R.; Lipscomb, G. F.; Van Eck, T. E.; Girton, D. G. *Appl. Phys. Lett.* **1992**, *61*, 2272.
2. Prasad, P. N.; Williams, D. J. in *Introduction to Nonlinear Optical Effects in Molecules and Polymers;* John Wiley & Sons: New York, 1991; pp 59-260.
3. Eaton, D. F. *Science* **1991**, *253*, 281.
4. Burland, D. M.; Miller, R. D.; Walsh, C. A. *Chem. Rev.* **1994**, *94*, 31.
5. Marturunkakul, S.; Chen, J. I.; Li, L.; Jeng, R. J.; Kumar, J.; Tripathy, S. K. *Chem. Mater.* **1993**, *5*, 592.
6. Chen, J. I.; Marturunkakul, S.; Li, L; Jeng, R. J.; Kumar, J.; Tripathy, S. K. *Macromolecule* **1993**, *26*, 7379.
7. Manson, J. A.; Sperling, L. H. *Polymer Blends and Composites*; Plenum Press: New York, 1976; Chapter 8.
8. Klempner, D.; Berkowski, L. In *Encyclopedia of Polymer Science and Engineering*, 2nd ed.; Mark, H. F., Bikales, N. M., Overberger, C. G., Menges, G., Kroschwitz, J. I., Eds.; Wiley, New York, 1986; Vol. 8, pp 279-341.
9. Jeng, R. J.; Chen, Y. M.; Chen, J. I.; Kumar, J.; Tripathy, S. K. *Macromolecules* **1993**, *26*, 2530.
10. Singer, K. D.; Sohn, J. E.; Lalama, S. J. *Appl. Phys. Lett.* **1986**, *49*, 248.
11. Teng, C. C.; Man, H. T. *Appl. Phys. Lett.* **1990**, *56*, 1734.
12. Chen, Y. M.; Jeng, R. J.; Li, L.; Zhu, X.; Kumar, J.; Tripathy, S. K. *Mol. Cryst. Liq. Crsyt. Sci. Technol. - Sec. B: Nonlinear Optics* **1993**, *4*, 71.

RECEIVED January 30, 1995

Chapter 16

Aromatic Heterocyclic Rings as Active Components in the Design of Second-Order Nonlinear Optical Chromophores

Bruce A. Reinhardt[1], Ram Kannan[2], Jay C. Bhatt[2], Jarek Zieba[3], and Paras N. Prasad[3]

[1]Polymer Branch, Materials Directorate, U.S. Air Force Wright Laboratory, WL/MLBP, 2941 P St. Ste 1 Wright-Patterson Air Force Base, OH 45433–7750
[2]Systran Corporation, 4126 Linden Avenue, Dayton, OH 45432
[3]The Photonics Research Laboratory, State University of New York, Buffalo, NY 14214

A series of second-order NLO model chromophores has been synthesized which contain only aromatic heterocyclic rings as the donors and acceptors. Experimentally determined values by EFISHG show that reasonably large values of the first hyperpolarizability β can be obtained with a much larger transparency window than normally demonstrated by convential chromophores with comparable β values. The dipole moments of these molecules can be controlled independently from β. These types of chromophore molecular structures can be incorporated into acetylene-terminated thermoset monomers which when incorporated into thin films of high Tg thermoplastics can be cured to produce composite films with stable second-order activity at 100°C.

The search for practical second-order nonlinear optical polymeric materials for use in frequency conversion, and integrated optics applications has been at the forefront of organic polymer research in recent years. The advantages of using organic polymers rather than organic or inorganic single crystalline materials include evironmental and optical stability, ease of fabrication and potential low cost. The figures of merit which a polymer system must meet to be device useable for second-order NLO applications have been summarized (1). Although some promising systems have been reported, all are based on polymer systems which incorporate NLO active elements containing standard electron donating and withdrawing groups such as dialkylamino and nitro or variations of other aliphatic polarizing functionality. Initially these types of polymers, many of which were based on acrylate type backbones, seemed to have the thermal stability necessary to meet the proposed device requirements. It is only recently that it has been realized that for an organic polymer to be truly device useable and ultimately commercially competitive especially for "on chip" applications it must be able to maintain a resonable second-order response (> 30 pm/V) during routine microelectronic circuit fabrication procedures (*1,2*). Such fabrication procedures include exposure to the high temperatures of solder baths which can be as high as 320°C for 20

0097–6156/95/0601–0205$12.00/0

minutes. Long term thermal stability requirements are also necessary with military applications demanding that a material maintain 95% of it's original EO coefficient after 10 years at 125° C. In conjunction with the necessary increased thermal stability the optical quality of the polymer film must also be preserved with losses below 1 dB/cm. Processing ease is also required for a polymer material to compete favorably in the commercial sector. The required processibility includes spin coatability and the necessary physical properties to either be self-assembled or electrically field poled into the noncentric bulk materials forms necessary for second-order activity. With the need for higher performance, the earlier high optical quality acrylate based polymers are no longer viable candidates for use in potential commercial "on chip" devices unless changes in the standard commercial processing procedures are instituted. Since changes in the current processing techniques would be extremely costly, the more logical approach involves the design of polymer second-order materials which have much greater thermal stability.

New Approaches to Improved Thermal Stability

Over the past two years there have been some promising approaches to synthesize new EO polymer systems which have increased thermal stability. Most of these are based on guest-host or chromophore side chain polymer systems involving thermoplastic polymer matricies with higher glass transition temperatures (Tg). Chromophore design involved the synthesis of new structures which have standard electron donor and acceptors connected by heterocyclic π electronic bridging systems with improved thermal stability (*3,4*). Additional studies have centered on the modification of standard donor groups to improve their thermal stability (5).

It remains to be seen how successful any of the approaches will eventually be but certain key problems currently exist with all of them which must be solved if any of the alternate approaches to improved thermal stability will ultimately lead to practical EO devices.

It must always be remembered that the underlying factor for any system to be commercially successful involves cost/performance considerations. Whatever the system turns out to be it must be of reasonable cost to produce, perform better than the state of the art materials, have minimal environmental impact, and answer the requirements of disposal and/or recyclability.

Thermal stability of a second-order NLO material can be subdivided into 2 types: (I) The **intrinsic thermal stability** (ITS) of the NLO activity moeity which is dependent on the temperature above which thermal degradation of the active portion of the molecules reduces the NLO activity. (II) The **alignment thermal stability** (ATS) of the bulk polymer system which is dependent on the glass transition temperature of the aligned NLO active material. At temperatures approaching Tg thermal relaxation of nonthermodynamically stable states will be greatly accelerated. Both types of thermal stability must be considered when designing new molecules for use in practical devices.

Our approach to the design of new molecules which will address both types of thermal stability issues centers around the synthesis of new NLO second-order active chromophores which have a higher degree of aromaticity and which can be incorporated into high temperature thermoset resins which are highly compatible with both NLO active or inactive high molecular weight thermoplastics.

Theory and Molecular Design.

It is generally recognized that organic ring compounds that possess (4n +2) π electrons (where n = 0, 1, 2, 3, etc.) have increased chemical and thermal stability over organic ring structures which do not meet this electronic criteria. Aromatic rings have long been the integral structural units in the design of high temperature organic polymers (*6*). It appears logical that the design of new second-order NLO chromophores should include structural units which are highly aromatic. A second-order NLO chromophore no matter what it's thermal stability requires an asymmetric charge distribution. If the structural building blocks are to be limited only to aromatic rings, charge asymmetry can only be accomplished by using **aromatic heterocyclic rings**. Here the presence of heteroatoms and the size of the rings can produce structural units which are either electron rich or electron poor when compared to benzene. Benzene contains six π electrons distributed equally over six atoms. The π electron densities at each of the atoms can be calculated using simple molecular orbital theory and in the case of the symmetrical benzene is found to have a value of 1.0. In five-membered heterocyclic rings containing one hetero atom, two of the 6π electrons which make up the aromatic sextet reside in a p orbital provided by the heteroatom. There are 6 π electrons somewhat unsymmetrically distributed over only 5 atoms and thus the HMO calculated π electron densities are greater than 1.0. Heterocyclic rings of this type are termed **electron excessive** (*7*) and exhibit higher reactivity to electrophilic aromatic substitution when compared to benzene. Six-membered rings containing one or more nitrogen atoms have their π electron density unsymmetrically distributed due to the higher electronegativity of the nitrogen atom. At carbon atoms which are ortho or para to the more electronegative nitrogen the π electron density is calculated by HMO theory to be less than 1.0. These rings are considered to be **electron deficient** (*7*) and behave as highly deactivated benzene rings toward electrophillic aromatic substitution. Figure 1 illustrates the relative electron excessive (electron donating) or electron defficient nature of various aromatic heterocyclic rings. Calculated π electron densities are also included for comparison.

Previous to our earlier reports (*8-10*) aromatic heterocyclic rings have been investigated as electron rich and electron poor centers in conductive polymers (*11*), and as donor (*9*), bridging (*12*) and acceptor (*13*) groups in second-order NLO molecules. In this report we describe our continuing studies which systematically attempt to evaluate the affect of electron excessive and electron deficient heterocyclic aromatic rings on the molecular first hyperpolarizability in a given series of structurally similar molecules.

Model Chromophores. As an initial attempt to ascertain the feasibility of using aromatic heterocyclic rings as donors and acceptors five general types of molecules were synthesized. The compounds synthesized within these 5 general types contained various combinations of electron rich heterocyclic (ERH) and electron deficient heterocyclic (EDH) rings. The ERH and the EDH rings were connected either directly together or with the use of a simple carbon carbon double bond. Although a simple double bond would not provide the most thermally stable bridge, it did provide ease of synthesis of a wide variety of compounds for proof of concept. Upon concept validation a polarizable aromatic bridge would be substituted for the double bond for improved thermal stability. The five general types of compounds synthesized are depicted graphically in Figure 2.

Synthesis. The syntheses of the model chromophores were carried out via the acetic anhydride or base catalyzed condensation of thiophene or bithiophene carboxaldehyde with the appropriate heterocyclic methyl compound. Yields of

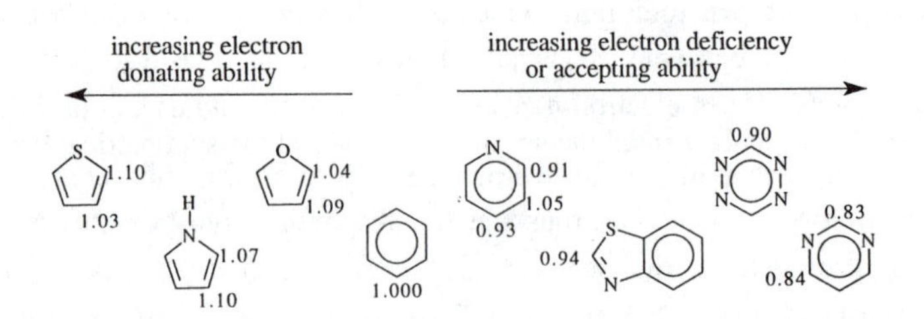

Figure 1. Relative electron donating and accepting ability for aromatic heterocyclic rings.

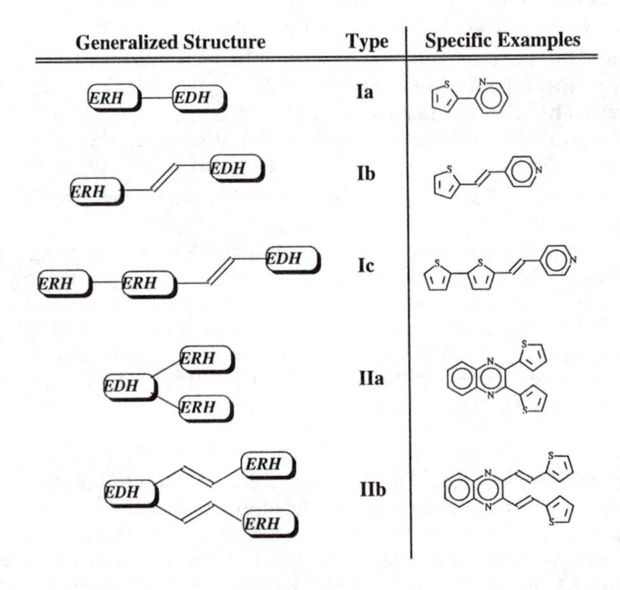

Figure 2. Generalized structures for model chromophores

purified products ranged from 30% to quantitative. The 3 synthetic reaction pathways and the structures of the compounds synthesized are depicted in Schemes A-C and Figures 3-5.

Model Chromophore Nonlinear Optical Characterization. The model chromophores were characterized using electric field induced second harmonic generation (EFISH) at 1.063 μm. The effective γ determined by EFISH measurements is the sum of the scaler orientationally averaged electronic part of the second hyperpolarizability and $\mu\beta/5kT$, the contribution to the total γ from aligned dipoles (**Equation 1**). In many cases the electronic part of γ was

$$\gamma_{eff} = \gamma_{elec} + \frac{\mu\beta}{5kT} \qquad (1)$$

considered to be negligible when compared to the contribution from the aligned dipoles. When the size of γ_{elec} was in question it was determined experimentally by degenerate four wave mixing (DFWM). The following approximations were made: For a *p*-nitroaniline (PNA) standard; $\gamma(-2\omega;\ \omega,\omega) = \gamma(-\omega;\ \omega,\omega,-\omega) = 4.8E\text{-}35$ esu. For Compound <u>2</u> ; $\gamma(-2\omega;\ \omega,\omega) = \gamma(-\omega;\ \omega,\omega,-\omega) = 5.0E\text{-}35$ esu, and for Compound <u>12</u>, $\gamma(-2\omega;\ \omega,\omega) = \gamma(-\omega;\ \omega,\omega,-\omega) = 3.16E\text{-}34$ esu. For all other compounds $\gamma(-2\omega;\ \omega,\omega) = 0$. The dipole moments μ were measured independently in xylene solution for all compounds investigated.

Model Chromophore Results and Discussion. Earlier studies (*10*) indicated that when single EDH rings and ERH rings are connected directly together without a bridge of polarizable π electrons(Type Ia and IIa chromophores) the values of the first hyperpolarizability β are relatively small. When a highly polarizable bridging carbon-carbon double bond is added (Type Ib, Ic, and IIb chromophores) the values of β increases substantially depending upon the ring system and the position of the molecular dipole with respect to the direction of conjugation. From the data in Figures 3, 4 and 5 some insights into the structure/NLO property relationships for these types of chromophores can be obtained. In the case of compound <u>1</u> the value calculated for β from EFISH measurements is much lower than that obtained for compound <u>2</u> due to the change in magnitude and direction of the net dipole of the molecule brought about by the change in point of attachment to the bridging double bond. Chromophore <u>2</u> has a β value which is slightly larger than that measured for *p*-nitroaniline (PNA) measured under identical conditions. Compound <u>2</u> crystallizes in noncentric almost colorless needles and has a much larger optical transparency window when compared to PNA. If one adds a second nitrogen atom ortho to the position of attachment to the double bond to form the six-membered pyrimidine ring (compound <u>6</u>) the β increases only slightly. The addition of a fused ring on the side of the pyridine ring in compound <u>2</u> produces the quinoline ring system (compound <u>4</u>). This again increases β only slightly but the addition of a second nitrogen atom and a change in linkage position to extend the conjugation into the fused benzene ring to form the quinoxaline ring system compound <u>11</u> increases β by more than 2 fold when compared to compound <u>2</u>. The experimental results for Type Ic chromophores in which thiophene is replaced by bithiophene compounds

Scheme A

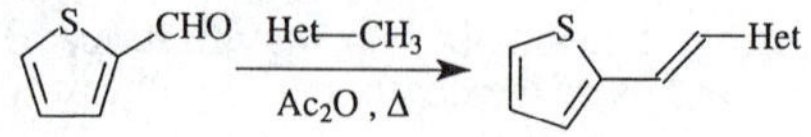

Compound	Het	mp °C	μ (D)	β x 10^{-30} esu	μβ x 10^{-47} esu	λ_{max} nm	β/MW esu / g MW
1	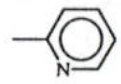	104-104.5	1.5	<3	<0.4	333	<0.2
2	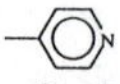	151-152	2.80	33	9.24	328	0.18
3	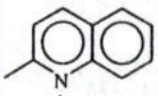	99-100.5	1.6	12	1.91	363	0.05
4	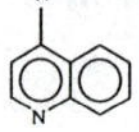	91-92	2.62	38	9.92	351	0.16

Figure 3. NLO Data for compounds synthesized by Scheme A

Scheme B

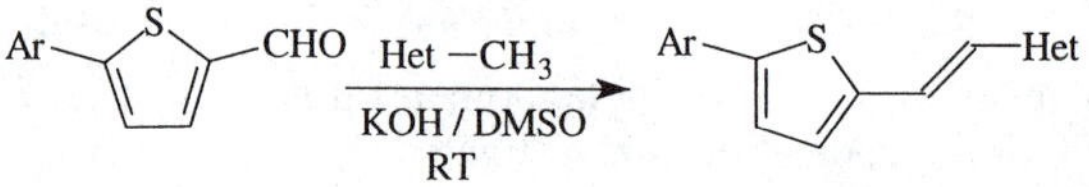

Compound	Ar	Het	mp°C	μ (D)	β x 10^{-30} esu	μβ x 10^{-47} esu	λ_{max} nm	β/MW esu/ g MW
5	H	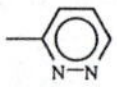	144-145	3.7	11	3.95	329	0.06
6	H	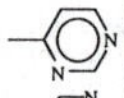	82-83	2.65	35	9.32	343	0.19
7	H	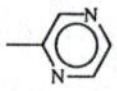	92-93	1.10	32	3.48	351	0.17
8	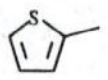	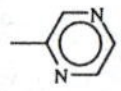	127-128	1.19	121	14	395	0.45

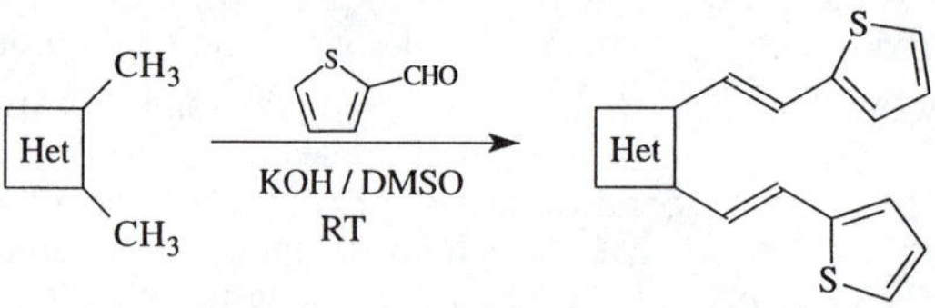

Compound	Het	mp°C	μ (D)	β x 10^{-30} esu	μβ x 10^{-47} esu	λ_{max} nm	β/MW esu/ g MW
9	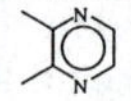	115-116	1.09	68	7.4	365	0.23
10	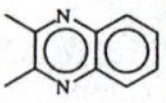	210-212	0.6	167	10	410 323	0.48

Figure 4. NLO Data for compounds synthesized by Scheme B

8 and **12** large increases are observed over the values obtained for the compounds containing only thiophene compounds **7** and **11**. It appears that this increase is due to both the increase in conjugation length and an amplification of the donating ability of the thiophene ring when a second thiophene is present. This is not surprising since it has been substantiated in reaction rate studies that bithiophene is greatly activated toward electrophilic substitution when compared to thiophene (7).

Values of β much higher than expected were obtained when two thienyl-substituted double bonds were joined to either a single or a fused EDH ring (Type IIb compounds **9** and **10**). This is surprising because these ring systems are not expected to be highly planer. Compounds **9** and **10** have β values which are approximately 2 times as large as their singularly substituted analogs.

Thermoset Polymers for Electrooptic Thin Film Applications.

Noncrosslinked thermoplastic EO polymer systems of all types (guest-host, side chain, main chain etc.) all suffer from the same deficiency. The name "thermoplastic" means that these systems can be formed and reformed using heat and/or pressure. A thermoplastic system will always tend to relax to the most thermodynamically stable state thus causing an electrical field poled thermoplastic material to ultimately relax to its unpoled configuration. Research on crosslinkable thermoplastics has had only moderate success at improving the second-order nonlinear optical stability of poled thermoplastic systems. However, at the present time no system has been described which contains structures capable of maintaining the needed EO activity at the 320°C processing temperature.

Thermoset resin systems offer an alternative to thermoplastics which as of this time has received far less attention. One possible reason for lack of interest in these systems is the realitive scarcity of thermoset systems known to have the necessary optical properties needed for waveguide applications. Extensive data on waveguide loss exists only for one class of thermoset resins, optical grade epoxies (*14*).

A class of thermoset resins originally designed to be high temperature moisture resistant alternatives to epoxies for structural applications are acetylene terminated or AT resins (*15*). These materials are known to undergo thermal and catalytic polymerization to form a highly crosslinked polyene matrix which is essentially insoluble in organic solvents (*16*). In such a tightly formed matrix a large number of NLO active sites could be incorporated in a relatively small volume. Our approach is to explore the possibility of using a monomer containing 2 acetylenic groups of different reactivity which when poled to produce an ordering of monomers where the triple bonds have reasonable registry could undergo multiple thermal reactions to form a ladder type crosslinked polyene (Figure 6) which would have greatly improved ITS and ATS. It should be noted that the thermal polymerization of acetylene containing monomers in the presence of an electric field has not as yet, to the best of our knowledge, been investigated. Previous studies (16-18) have been focused on the ratios and structures of the products obtained in the thermal polymerization of undiluted monomers in the absence of any external electric field. Such investigations have considered the formation of substituted phenyl rings via trimerization of acetylene groups as a competing reaction with polyene formation. These studies have been inconclusive in establishing any general relationship between amount (if any) of cyclic trimer formed, molecular size and electronic configuration. The formation of any cyclic trimer at the expense of polyene formation would decrease the ultimate second-order response of the subject composite system upon thermal cure. It is postulated that if a high degree of alignment of the acetylenic monomers is attained in the

Scheme C

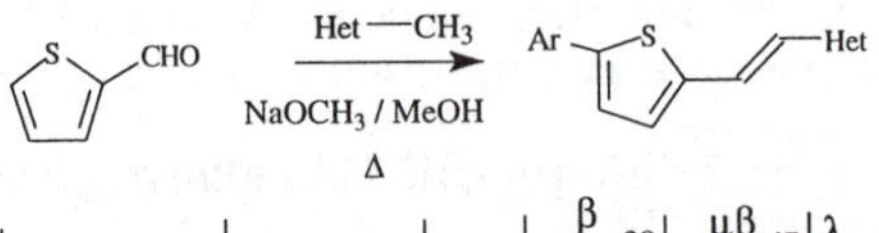

Compound	Ar	Het	mp °C	μ (D)	β x 10^{-30} esu	μβ x 10^{-47} esu	λ_{max} nm	β/MW esu / g MW
11	H	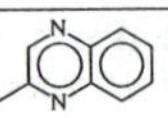 2-quinoxalinyl	103-104	0.92	78	7.17	382	0.33
12	2-thienyl	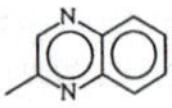 2-quinoxalinyl	134.2-135.2	1.09	224	24	420	0.70

Figure 5. NLO Data for compounds synthesized by Scheme C

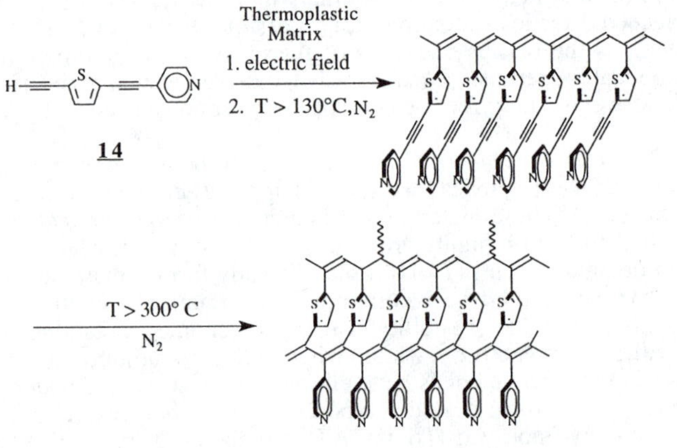

Figure 6. Idealized structures from thermal cure of aligned compound **14**

presence of the electric field, any trimerization will be held to a minimum, thus maximizing polyene formation and second-order NLO activity.

Acetylene Containing Model Compound and Monomer. Two different thermally reactive acetylene-containing monomers were prepared. The first containing only an internal acetylenic group (compound 13) and the second containing an internal and a terminal acetylenic group(compound 14).

Synthesis. Compound 13 was made in 55% overall yield by the synthetic pathway depicted in Scheme D, Figure 7. The second thermoset monomer, compound 14, was synthesized in 35% overall yield using the series of reactions illustrated in Scheme E, Figure 8.

Thermal Analysis. A differential scanning calorimetry (DSC) thermogram of 13 under 750 psi of nitrogen at a heating rate of 10°C/min (Figure 9) showed a sharp endothermic transition maximizing at 73.29 °C characteristic of a crystalline melting point and a somewhat broader exothermic transition with an onset at approximately 270° maximizing at 335° C. The transition is thought to be characteristic of the thermal polymerization of the internal triple bond (*19*). The DSC analysis was carried out under high pressure due to rapid sublimation of compound 13 at 100°C.

The DSC thermogram of compound 14 showed an initial endothermic transition maximizing at 132.54°C which is characteristic of a crystalline melting point followed immediately by a large exothermic peak maximizing at 140° C characteristic of the polymerization of the primary acetylenic group.

NLO Characterization of Thermoset monomer. The experimentally determined values for compounds 13 and 14 (Figure 10) show that compound 13 has a value for β slightly lower than that obtained for it's double bonded analog 2. This is expected since it has been observed that the second-order NLO properties for molecules with triple bonds are lower than identical structures containing double bond bridging groups. The somewhat reduced values for β indicated the decreased availability of the π electrons in a triple bond when compared to it's double bonded analog (*20*). The β value calculated from EFISH data for compound 14 is almost identical to that calculated for compound 13.

Thin films from Blends of Thermoset Monomer 14 with 6F-PBO. Thin films with excellent optical quality were prepared by casting from solutions containing mixtures of the thermoset monomer 14 and high molecular weight hexafluoroisopropylidene-containing polybenzoxazole (6F-PBO) thermoplastic (*21*) (Figure 11). The film casting solutions were made by dissolving 0.026g of 14 and 0.174g of 6F PBO in 10ml of chloroform (2% weight solids/volume concentration), filtering the resulting solution and casting in a 5 cm casting dish. The concentration of thermoset in the cast film was 13% by weight. The films had excellent optical quality and showed no signs of phase separation. FT infrared analaysis (FTIR) of the films showed a C≡C-H absorption at 3306 cm^{-1} characteristic of the terminal triple bond and an absorption at 2206 cm^{-1} characteristic of the C≡C bond stretching vibrations from the terminal and the internal triple bonds.

Scheme D

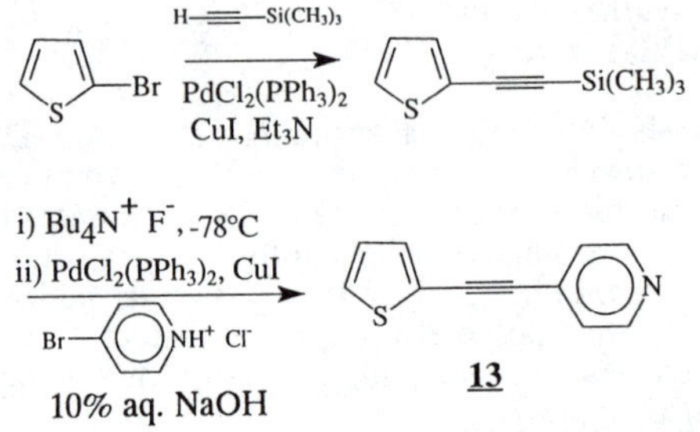

H—≡—$Si(CH_3)_3$

—Br $PdCl_2(PPh_3)_2$ CuI, Et_3N —≡—$Si(CH_3)_3$

i) $Bu_4N^+ F^-$, -78°C
ii) $PdCl_2(PPh_3)_2$, CuI
Br— $NH^+ Cl^-$
10% aq. NaOH

13

Figure 7. Scheme for the synthesis of compound 13

Scheme E

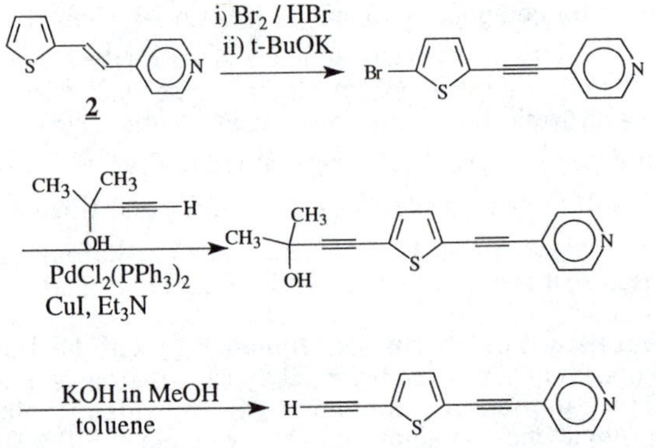

2

i) Br_2 / HBr
ii) t-BuOK

Br

CH_3 CH_3 —≡—H OH

$PdCl_2(PPh_3)_2$
CuI, Et_3N

CH_3 CH_3 OH

KOH in MeOH
toluene

H—≡—

14

Figure 8. Scheme for the synthesis of compound 14

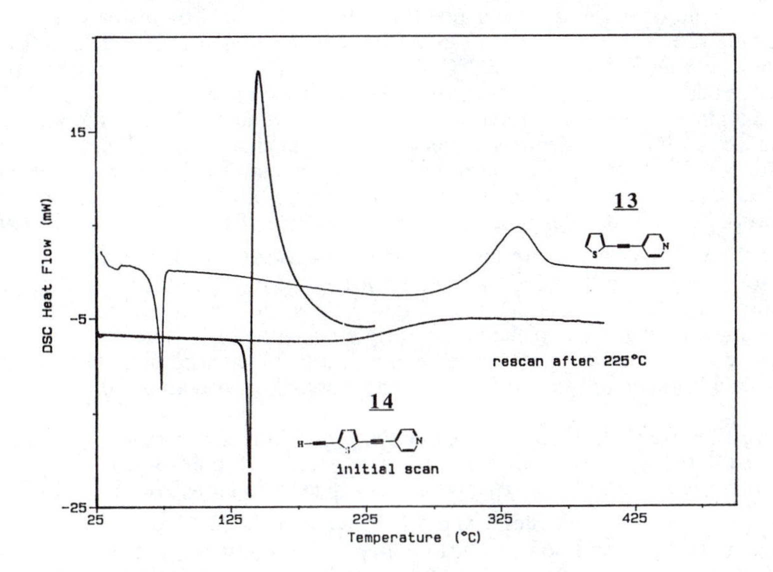

Figure 9. DSC scans for compounds **13** and **14**

Compound	mp °C	μ (D)	β x 10^{-30} esu	μβ x 10^{-47} esu	λ_{max} nm	β/MW esu / g MW
	71.5-72.5	2.65	20	5.2	308	0.11
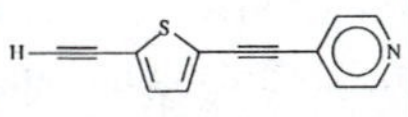	138 [a]	3.29	22.4	7.38	375	0.11

[a] Melts with reaction

Figure 10. NLO results for compounds **13** and **14**

Thermal Analysis of Films of Thermoset - 6F-PBO Blends. The DSC analysis of an uncured blend film shows a broad exotherm centering around 175°C. A rescan of the same film sample after heating to 250°C shows a baseline shift at approximately 235°C which is characteristic of a glass transition. Thermogravimetric analysis (TGA) in a nitrogen atmosphere of an uncured film sample shows an initial weight loss of 9% starting at approximately 80°C due to trapped chloroform in the polymer film. This is followed by major weight loss begining at 508°C. TGA analysis on a film sample cured at 250°C under nitrogen shows no weight loss until 508°C showing that the residual solvent has been removed during the thermal curing process. Thermal mechanical analysis (TMA) of the same film using a force of 0.01N showed that the 250°C thermal cure produces a film that undergoes a major dimensional change (stretching) beginning at 237°C indicative of a glass transition. The film cured at 250°C was somewhat darker in color than the uncured film characteristic of the formation of the polyene network but still maintained its good optical quality. FTIR analysis of the cured film showed that the absorption at 3306 cm^{-1} indicative of the terminal C≡C-H stretch was now absent due to reaction of the terminal acetylene groups. The absorption at 2206 cm-1 exhibited a slight decrease in intensity of approximately 10% showing that most of the internal triple bonds had not reacted at the 250°C temperature. This was substantiated by the fact that the entire film after cure was still soluble in chloroform indicating that no crosslinking had occurred.

Insitu Electric Field Poling Experiments. Thin films of monomer **14** (0.03g) in 6F-PBO (0.15g) were spin cast from mixtures of chloroform (7.0g) and chlorobenzene (4.0g) at 500 rpm on an ITO sputtered glass slide. Film thickness measured using an Alpha Step 100 profilometer were 0.6 to 0.7 μm . The film was dried at 60-65°C for 3 hours prior to poling. The film was poled using a corona discharge produced by imposing +3.5 kV (positive corona polarity) of DC current to a 25 μm thick tungsten wire. The distance between the film surface and the corona wire was 8 mm. A flow of argon at the rate of 10 ml/min was directed onto the film surface during poling. The SHG signal was monitored continuously during the poling sequence. The poling (Figure 12) was carried out by first turning on the field and maintaining the sample at room temperature for 45 minutes. The temperature was then raised to 150-160°C (± 5°C) and held there for 3 hours. The temperature was then increased again to 270°C for 30 minutes. The film was then allowed to cool to room temperature and the field turned off (Figure 13). The SHG signal was monitored for the next 12 hours. Upon reaching a stable SHG signal the sample was removed from the in situ poling stage and it's SHG angular dependence measured. The obtained SHG signal values were used to calculate the $\chi^{(2)}$ value by using the Y-cuted quartz plate as a standard. The increase in SHG intensity as a function of time at temperature is shown in Figure 12. Upon cooling the SHG signal decreases rapidly until room temperature is reached. When the field is removed the signal decays slowly until a stable value was reached at about 12 hours (Figure 13). This SHG intensity remained constant with a stable $\chi^{(2)}$ value of 2 x 10^{-8} esu being retained at 100°C for 22 hours (Figure 14). The magnitude of the SHG signal was found to decay rapidly upon raising the temperature above 125°C.

It is postulated that the rapid decay of the SHG signal was due primarily to the resin being cured at a temperature below what is required for the development of the tightly crosslinked ladder type network depicted in Figure 6. DSC analysis of

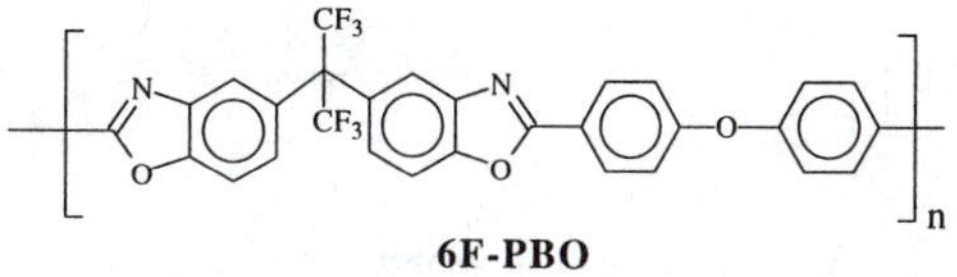

6F-PBO

Figure 11. Structure of 6F-PBO thermoplastic

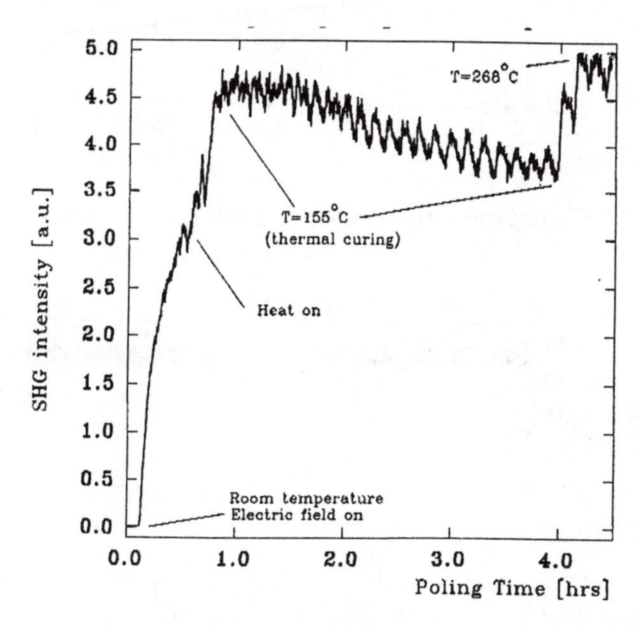

Figure 12. Heating curve from in situ poling experiment: Compound **14** in 6F-PBO

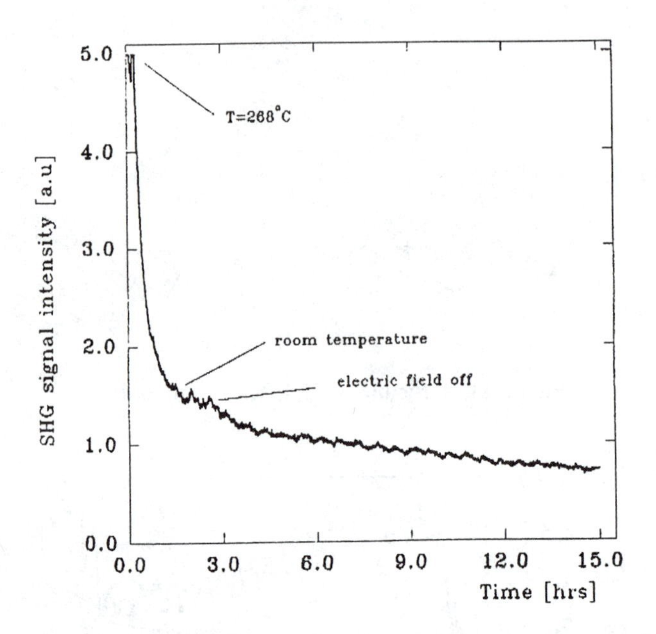

Figure 13. Cooling curve from in-situ poling experiment: Compound **14** in 6F-PBO

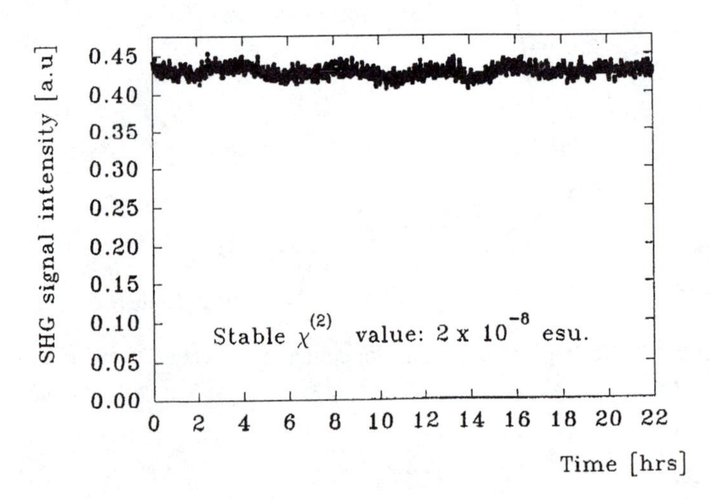

Figure 14. SHG intensity as a function of time for poled compound **14** in 6F-PBO at 100^0C

monomer 14 (Figure 9) indicates that a temperature greater than 320°C would be required for maximum cure to be accomplished.

Conclusions

The NLO characterization of model compounds containing aromatic heterocyclic donors and acceptors indicate that these types of molecules can produce reasonable nonlinearity while maintaining larger optical transparency windows for a given value of β when compared to chromophores containing standard donor and acceptor groups. The values of μ and β can be controlled independently depending on the type of heterocyclic rings involved and the position of attachment in relationship to the π electron bridge. The versatility of the chemical structures offers many ways of incorporating these types of chromophores into polymer systems. The initial system chosen for investigation was based on the use of an acetylene terminated thermoset monomer in a host matrix of 6F-PBO thermoplastic polymer. Data obtained from in situ poling experiments indicate that these systems have a potential for producing bulk materials with reasonable second-order activity and alignment thermal stability. Further studies are necessary to optimize chromophore composition, poling, and thermal cure conditions in order to realize the full potential of these systems as electrooptic materials.

Synthetic Experimental

Generalized Synthesis Procedures for Model Chromophores 1-12

Generalized Procedure - Scheme A. A mixture of aldehyde (0.05 mol), the appropriate methyl-substituted heterocycle (0.055 mol), acetic anhydride (6 ml) and glacial acetic acid (3 ml) were heated at reflux under nitrogen for 24 to 48 hr. The reaction mixture was then cooled and poured into distilled water. The water mixture was then made basic with sodium hydroxide and extracted with toluene, the organic layer washed with water, dried over anhydrous magnesium sulfate, and concentrated under reduced pressure. The crude product is purified either by recrystalization, or column chromatography on silica gel followed by recrystallization.

Generalized Procedure - Scheme B. A mixture of the appropriate aldehyde (0.025 mol.), the appropriate methyl-substituted heterocycle (0.025mol.), powdered potassium hydroxide (0.1 mol.), and dimethyl sulfoxide (15 ml) was stirred at RT for 18 hr and poured into distilled water. The separated solids were filtered, dried, and purified by column chromatography on silica gel followed by recrystallization.

Generalized Procedure - Scheme C. A solution of sodium methoxide was prepared by reacting sodium metal (0.80g) with methanol (25 ml). To this solution was added the appropriate aldehyde (0.025mol) and the appropriate methyl-substituted heterocycle (0.0125 mol) and the resulting mixture heated at reflux for 2 hr. The reaction mixture was then cooled to room temperature, the separated solids filtered and purified by column chromatography on silica gel followed by recrystalization.

Analytical Data for Compounds 1-12. Compound 1: This compound was prepared in 30% yield using Scheme A, mp 104-104.5°C (hexane) Lit (22) mp 78-79°C. Elem Anal. Calcd. for $C_{11}H_{19}NS$: C 70.57, H 4.84, 7.49%. Found: C 70.71,

H 4.86, N 7.35%. EIMS 187 M^+. **Compound 2:** This compound was prepared in 66% yield using Scheme A, mp 151-151.5°C (ethyl acetate-hexane). Lit (*22*) mp 149-151°C. **Compound 3**: This compound was prepared in 61% yield using Scheme A, mp 99-100°C (hexane-ethyl acetate). Lit (*23*) mp 89°C. Elem Anal. Calcd. for $C_{15}H_{11}NS$: C 75.94, H 4.64, N 5.90, S 13.49. Found: C 75.84, H 4.60, N 5.87, S 13.78. EIMS 237 M^+. **Compound 4**: The compound was prepared in 50% yield using Scheme A, mp 91-92°C (hexane-ethyl acetate). Elem Anal. Calcd. for $C_{15}H_{11}NS$: C 75.94, H 4.64, N 5.90, S 13.49. Found: C 75.85, H 4.67, N 5.94, S 13.35. EIMS 237 M^+. **Compound 5**: This compound was prepared in 48% yield using Scheme B, mp 144-145°C (ethyl acetate). Lit (*24*) mp 139-141°C. **Compound 6**: This compound was prepared in 56% yield using Scheme B, mp 82-83°C (ethyl acetate-hexane). Elem. Anal. Calcd. for $C_{10}H_8N_2S$: C 63.82, H 4.28, N 14.89, S 17.00. Found: C 63.76, H 4.24, N 14.88, S 16.79. EIMS 188 M^+. **Compound 7**: The compound was prepared in 67% yield using Scheme B, mp 92-93° C (hexane). Elem. Anal. Calcd. for $C_{10}H_8N_2S$: C 63.82, H 4.28, N 14.89, S 17.00. Found C 63.87, H 4.26, N 14.79, S 16.93. EIMS 188 M^+. **Compound 8**: The compound was prepared in 52% yield using Scheme B, mp 127-128°C (hexane-ethyl acetate). Elem Anal. Calcd. for $C_{14}H_{10}N_2S_2$: C 62.22, H 3.73, N 10.37, S 23.68. Found: C 62.31, H 3.68, N 10.39, S 24.02%. EIMS 270 M^+. **Compound 9**: The compound was prepared in 22% yield using Scheme B, mp 115-116.3°C (hexane-ethyl acetate). Elem Anal. Calcd. for $C_{16}H_{12}N_2S_2$: C 64.86, H 4.08, N 9.46, S 21.60. Found: C 64.61, H 3.95, N 9.51, S 21.79%. EIMS 296 M^+. **Compound 10**: The compound was prepared in 52% yield using Scheme B, mp 210-212°C (ethyl acetate). for $C_{20}H_{14}N_2S_2$: C 69.36, H 4.07, N 8.09, S 18.50. Found: C 69.15, H 4.05, N 8.04, S 18.71%. EIMS 346 M^+. **Compound 11**: The compound was prepared in quantitative yield using Scheme C, mp 103-104°C (hexane-ethyl acetate). Lit (*23*) mp 98-99°C. Elem Anal. Calcd. for $C_{14}H_{10}N_2S$: C 70.57, H 4.23, N 11.76, S 13.43. Found: C 70.55, H 4.07, N 11.68, S 13.42%. EIMS 238 M^+. **Compound 12:** The compound was prepared in 61% yield using Scheme C, mp 134.2-136.1°C (hexane-ethyl acetate). Elem Anal. Calcd. for $C_{18}H_{12}N_2S_2$: C 67.50, H 3.78, N 8.75, S 19.98. Found: C 67.60, H 3.59, N 8.74, S 19.65%. EIMS 320 M^+

Preparation of 1-(2-thienyl)-2-(4-pyridyl)ethyne 13. Scheme D. Copper iodide [0.11g, 0.56 mmol], $Pd(PPh_3)_2Cl_2$ [0.79g, 1.12mmol] and trimethylsilylacetylene [5.00g, 50.90 mmol] were added to a stirred degassed solution of 2-bromothiophene [9.13g, 56.00 mmol] in diethylamine [60 ml]. The solution was stirred at room temperature for 18 hr, diluted with hexane [500 ml], stirred for an additional 1/2 hr, filtered and the solvent removed under reduced pressure. Purified 2-(2-trimethylsilyl)ethynyl-thiophene was isolated by distillation under reduced pressure [6.84g, 75%], bp 58-60°C at 1.6 mm Hg. Tetrabutylammonium fluoride [80 ml of 1M solution in THF] was added dropwise to a stirred solution of 2-(2-trimethylsilyl)ethyne [10.00g, 55.45 mmol] in degassed THF at -78°C. The solution was allowed to warm to 0°C and CuI [0.53g, 2.77 mmol], $Pd(PPh_3)_2Cl_2$ [1.30g, 1.85 mmol] and 4-bromopyridium hydrochloride [10.80g, 55.45 mmol] were added. Sodium hydroxide [45 ml of a 5.5N aqeous solution] was added dropwise to the above solution and the mixture stirred ar room temperature for 24 hr. The resulting dark solution was filtered, the residue washed with methylene chloride [200 ml] and the filtrate concentrated. Water [500 ml] was added to the concentrate, stirred for 1/2 hr and then extracted with methylene chloride. The organic extract was dried over anhydrous magnesium sulfate, filtered and the filtrate concentrated under vacuum to obtain a dark oil [28.29 g]. Purified 1-(2-

thienyl)-2-(4-pyridyl)ethyne [7.65g, 76%] was isolated by chromotography on silica gel using 1:1 hexane/methylene chloride as the eluting solvent. mp 72-73°C. Elem Anal. Calcd. for $C_{11}H_7NS$: C 71.32, H 3.81, N 7.56. Found: C 70.98, H 3.92, N 7.03%. EIMS 185 M^+

Preparation of 2-ethynyl-5-(4-pyridyl)ethynylthiophene 14. Scheme E. Bromine was added dropwise to a solution of compound **2** [2.00g,10.7 mmol] in HBr [20ml] kept cold in a salt-ice bath. After the addition of bromine was complete, the solution was then warmed to 35-50°C for 3 hr. The reaction mixture was then cooled, 2N sodium hydroxide added, and stirred for 1 hr. The resulting suspension was filtered and dried at 50°C at 20 mm Hg for 16 hr. The dried powder was added to a mixture of potassium t-butoxide [2.40g, 21.40 mmol] in t-butanol [50 ml] and the resulting reaction mixture heated at reflux for 2.5 hr. The reaction mixture was cooled to room temperature and most of the t-butanol removed by heating under reduced pressure. Water [100 ml] was then added to the concentrated residue, stirred for 1 hr and filtered to isolate the crude product. The crude material was purified by column chromatography on silica gel using 4:1 THF/hexane as the eluent to give 1.56g, 58% of 2-bromo-5-[2-(4-pyridyl)ethynyl]thiophene. mp 128-130.5°C. Elemental Analysis Calculated for $C_{11}H_6BrNS$: C 50.02, H 2.29, Br 30.25, N 5.30, S 12.14. Found: C 50.01, H 2.51, Br 30.84, N 5.40, S 11.75%. EIMS 263,265 (M+), 184 (M-Br). A mixture of 2-bromo-5-[2-(4-pyridyl)ethynyl]thiophene [4.00g, 15.15 mmol], 3-methylbutyn-3-ol [1.59g, 18.94 mmol], $PdCl_2(PPh_3)_2$ [0.21g,0.30 mmol], and CuI [0.14g, 0.76 mmol] in degassed triethylamine [50 ml] was stirred at room temperature for 40 hr. and then heated at reflux for 5 hr. The reaction mixture was cooled, filtered and the filtrate concentrated under reduced pressure to afford a brown oil. The oil was purified by column chromatography on silica gel using 7:3 heptane/THF as the eluent followed by recrystallization from heptane to yield [3.00g, 74%] of 2[3-methyl-3hydroxybutynyl]-5-[2-(4pyridyl)ethynyl]thiophene mp 129.5-130.1°C. FTIR (KBr) , (cm^{-1}) 3556-3016 (OH), 2977, 2926 (aliph. C-H), 2201 (C=C). EIMS 267 (M+), 252 ($M-CH_3$), 210 ($252-CH_2CO$). A 8.50g [31.84 mmol] sample of 2[3-methyl-3hydroxybutynyl]-5-[2-(4pyridyl)ethynyl]thiophene, KOH [1.78g, 31.84 mmol] in methanol [25 ml] and toluene [200 ml] were heated at 90°C during which time the methanol and the acetone formed during the reaction were distilled off (total heating time 1.5-2.0 hr). The toluene solvent was then removed under reduced pressure and the reddish residue purified by column chromatography on silica gel using 7:3 heptane/THF to yield 4.65g (70%) of purified 2-ethynyl-5-(4-pyridyl)ethynyl thiophene **14**, mp 138°C with reaction, Elem Anal. Calcd. for $C_{13}H_7NS$: C 74.61, H 3.37, N 6.70, Found: C 75.54, H 3.19, N 6.72%. EIMS 209 (M^+),

Acknowledgments

The authors would like to thank Dr. Alan T. Yeates of the Polymer Branch WL/MLBP for the calculation of the free p electron densities of the heterocyclic rings and Ms. Marlene Houtz of the University of Dayton Research Institute for the thermal analysis data.

Literature Cited

1. Burland, D. M.; Miller, R. D.; Tweig, R. J.; Volkson, W.; Walsh, C. A. *SPIE Proc.* **1993,** *1852* , 186.

2. Wu, J. W.; Valley, J. F.; Ermer, S.; Binkley, E. S.; Kenney, J. T.; Lipscomb, G. F. *Appl. Phys. Lett.* **1991,** *58* , 225.
3. Wu, W. H.; Yamada, R. F.; Shi, R. F.; Cai, Y. M.; Garito, A. F. *Amer. Chem Soc. Polym. Preprints* **1994**, *35 (2)*, 103.
4. Girton, D. G.; Ermer, S.; Valley, J.F.; Van Eck, T. E. ; Lovejoy, S. M.; Leung, D. S. ; Marley, J. A. *Amer. Chem. Soc. Polym. Preprints* **1994**, *35 (2)*, 219.
5. Moylan, C. R.; Tweig, R. J.; Lee, V. Y.; Swanson, S. A.; Betterton, K. M.; Miller, R. D. *J. Amer. Chem .Soc.* **1993**, *115*, 12599.
6. Mark, H. F.; *Macromolecules* **1977**, *10*, 881.
7. Kauffmann, T. *Angew.Chem.Int. Ed. Engl.* **1979**, *18*, 1.
8. Reinhardt, B. A. *SPIE Proceedings* **1993**, *1853*, 50.
9. Karna, S. P.; Zhang, Y.; Samoc, M.; Prasad, P. N.; Reinhardt, Bruce A., Dillard, A. G. *J. Chem Phys.* **1993**, *99*, 9984.
10. Reinhardt, B. A.; Kannan, R., Bhatt, J. C., *SPIE Proceedings* **1994**, *2229*, 24.
11. Zhou, Z.; Maruyama, T.; Kanbara, T.; Ikeda, T.; Ichimura, K.; Yamamoto, T.; Tokuda, K. *J. Chem. Soc. Chem. Commun.* **1991**, 1210.
12. Dirk, C. W.; Katz, H. E.; Schilling, M. L.; King, L. A. *Chem. Mater*. **1990**, *2*, 700.
13. Gao, J. P.; Darling, G. D. *J. Amer. Chem. Soc.* **1992**, *114*, 3997.
14. Hartman, D. H. *App Opt*. **1989,** *28*, 40.
15. Hergenrother, P. M. *J. Macromol.Sci.-Rev. Macromol. Chem.* **1980**, *C13 (1)* , 1.
16. Pickard, J. M.; Jones, E. G.; Goldfarb, I. J. *Macromolecules* **1979**, *12* , 895.
17. Sefeik, M.D.; Stejskal, E. O.; McKay, R. A.; Schaefer, J. *Macromolecules* **1979**, *12*, 423.
18. Pickard, J. M.; Chattoraji, S. C.; Loughran, G. A.; Ryan, M. T. *Marcomolecules* **1980**, *13*, 1289.
19. Unroe, M. R.; Reinhardt, B. A. *J. Polym. Sci: Part A Polym. Chem.* **1990**, *28*, 2207.
20. Prasad, P. N.; Williams D. J. *Nonlinear Optical Effects in Molecules and Polymers:* John Wiley and Sons: New York, NY, 1991, Chapter 7, p 39.
21. Denny, L. R.; Evers, R. C.; Reinhardt, B. A.; Unroe, M. R.; Dotrong, M.; Houtz, M. D. *SAMPE Preprints, 22nd Inter SAMPE Tech Conf.* **1990**, *22*, 186.
22. Pampalone, T. R. *Org Prep Proceed. Int.* **1969**, *1*, 209.
23. Reid, W.; Hinsching, S. *Liebigs Ann. Chem.* **1956**, *600*, 47.
24. Kauffmann, T.; Mitschker, A.; Streitberger, H. J. *Angew. Chem Intl. Ed.* **1972**, *11*, 847.

RECEIVED January 30, 1995

Chapter 17

Nonlinear Optical Properties of Poly(1,4-phenylenevinylene) and Its Derivatives

Jung-Il Jin[1] and Hong-Ku Shim[2]

[1]Department of Chemistry, Korea University, Seoul 136-701, Korea
[2]Department of Chemistry, Korea Advanced Institute of Science and Technology, Taejon 305–701, Korea

Poly(1,4-phenylenevinylene) (PPV) is one of the representative poly-conjugated polymers. Since PPV contains phenylene units, it can be easily altered structurally to modify its electronic structure. Moreover, PPV and its derivatives are easily prepared via either water-soluble or organic-soluble precursor polymers. One can pole the precursor polymer films obtained from those carrying the electron-donating as well as electron-attracting groups on phenylene rings. And then they are subjected to thermolysis to obtain the final polyconjugated polymers in poled states that are noncentrosymmetrical. These polymers display extremely stable second-order optical nonlinearities ($\chi^{(2)}$) in the range of 10^{-8} esu. When the phenylene rings carry two alkoxy electron-donating groups, the substituted PPVs show the third-order nonlinear optical properties up to value of $\chi^{(3)}$ of 10^{-9} esu.

Nonlinear optics (NLO) is concerned with the interaction of electromagnetic field, generally in the optical frequency range, with a medium, which results in the alteration of the phase, frequency, or other propagation characteristics of the incident light. Developments in the field of NLO hold promise for such important applications as in optical information processing, telecommunications and integrated optics. The early experimental and theoretical investigations were primarily concerned with inorganic materials because of this field was evolved from solid-state physics of inorganic crystals. But interests in organic and polymeric systems have been growing rapidly recent years, since organic and polyene-like structures with delocalized π-electron systems lead to extremely large second- and third-order nonlinear coefficients. Especially, π-conjugated polymers are currently attracting much interest as materials for a wide variety of applications because of their large nonlinear responses, in many

0097–6156/95/0601–0223$12.00/0

cases, much larger than their inorganic counterparts (*1*). The large optical nonlinearity coupled with the fast response time for conjugated polymers implies that they can be of great importance in optical signal processing (*2,3*). In addition, polyconjugated polymers are easy to modify structurally in order to optimize materials properties such as processability, mechanical and thermal stability, and laser damage threshold (*4,5*).

The starting point for nonlinear optics is the relationship between the polarization (P) induced in a molecule and the electric field components (E) of incident electromagnetic waves.

$$P = \alpha \cdot E + \beta \cdot EE + \gamma \cdot EEE + \cdots$$

where the vector quantities P and E are related by the tensor quantities α, β and γ, which are often referred to as the polarizability, first hyperpolarizability and second hyperpolarizability, respectively. Similarly the polarization induced in macroscopic or bulk media can be expressed as

$$P = \chi^{(1)} \cdot E + \chi^{(2)} \cdot EE + \chi^{(3)} \cdot EEE + \cdots$$

where the coefficients $\chi^{(2)}$ and $\chi^{(3)}$ are referred to as the second- and third-order nonlinear optical susceptibilities, respectively. As is well known, the contribution of $\chi^{(2)}$ to nonlinear polarization is zero in centrosymmetric media. Therefore, organic polymers consisting of structural units having high dipole moments are poled to align the dipoles and, thus to attain a noncentrosymmetric structure.

Conjugated polymers showing extensive π-electron delocalization are one of the most important classes of third-order nonlinear optical materials because of the large π-electron contribution to the optical nonlinearity. The nonlinear optical response time for these materials has been shown to be in the subpicosecond region. Especially, interests in poly(1,4-phenylenevinylene) (PPV) and its derivatives have been stimulated in recent years by the recognition that these polymers may possess high second-order (*6-8*) and third-order (*9-12*) optical nonlinearities, good processability and high film quality, etc. These polyconjugated polymers can be readily prepared in high molecular weight either through the water-soluble (*13-16*) or organic-soluble (*11,17,18*) precursor route. Conventional solvent casting or spin coating techniques are applicable for fabricating thin films from these precursor polymers.

Up to now, many second-order NLO polymers have been studied, but their temporal stability after poling has been one of the major problems that requires improvement for practical device applications. Poling is usually conducted near the glass transition temperature (T_g) of the polymers. But maintenance of the poled state is often difficult even at room temperature because of the relaxation of the polymer chains as well as side chain chromophore units, which destroys the orientation of poled dipoles. One of the attractive methods to prevent the relaxation of poled chains is to crosslink them during or after poling to form three-dimensional networks (*19,20*). This approach can produce material with very high $\chi^{(2)}$ values that retain their nonlinear optical characteristics for a long period of time. An unavoidable drawback to this method, however, lies in the fact that the crosslinking reaction is always very difficult to control, and achievement of uniform crosslinks is even more difficult. Therefore, crosslinking of polymer chains is considered inappropriate to obtain materials of high optical grade with uniform quality. Formation of interchain hydrogen bonds also has been proven to improve temporal stability of orientation of poled chromophores. (*21*)

Attaching electron donor and acceptor groups or NLO chromophores to PPV as either side chains or directly to the polymer backbone may result in highly stable materials after poling, because PPV has a highly rigid backbone structure. In this article, we would like to review the relationship between the structure of PPV derivatives and their nonlinear optical properties. We have recently studied second harmonic generation (SHG) and electro-optic (EO) properties of PPV derivatives, thus, experimental details on the measurements of SHG and EO activities of PPV derivatives have been included.

Second-Order Optical Nonlinearity of PPV Derivatives

Synthesis and Poling

PPV and its derivatives can be prepared by several different synthetic methods. Among them, the method originally developed by Wessling and Zimmermann (*22*) is very versatile and is presently most widely used;

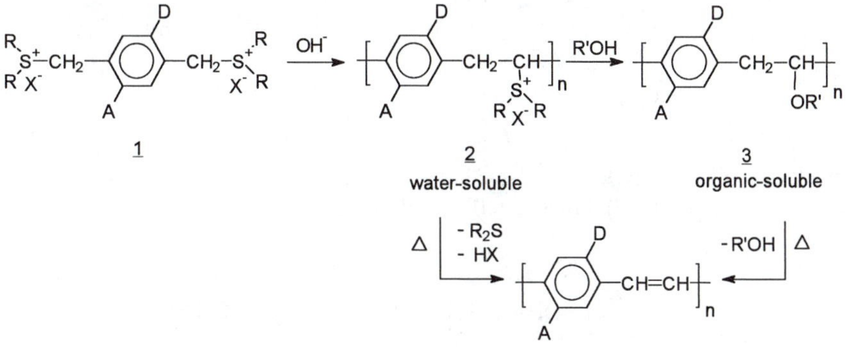

1

2 water-soluble

3 organic-soluble

4

The bis-sulfonium salt monomer 1 is polymerized to the water-soluble precursor polymer 2, which then is subjected to thermolysis to the final polymer 4. The precursor polymer 2 can also be transformed into organic-soluble precursor 3, which is then converted to 4. The substituents D and A refer to the donor and acceptor groups, respectively.

Since PPV and its derivatives consist of π-systems conjugated along the main chain, they, once prepared, are resistant to dipole orientation by poling. This difficulty could be circumvented by subjecting the precursor polymer, either 2 or 3, to thermolysis in the poling electric field. The following compositions were prepared by us in poled film form in order to study their second-order optical nonlinearity by measuring either their second-harmonic generation (SHG) or electro-optic (EO) coefficient. Synthetic details of the polymers can be found elsewhere (*6,7,23*).

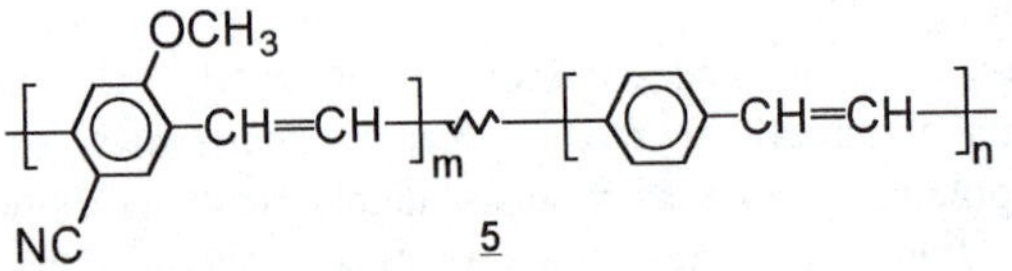

5

Poly[(2-cyano-5-methoxy-1,4-phenylenevinylene)-co-(1,4-phenylenevinylene)] (*6,7*) ; Poly(CMPV-co-PV)

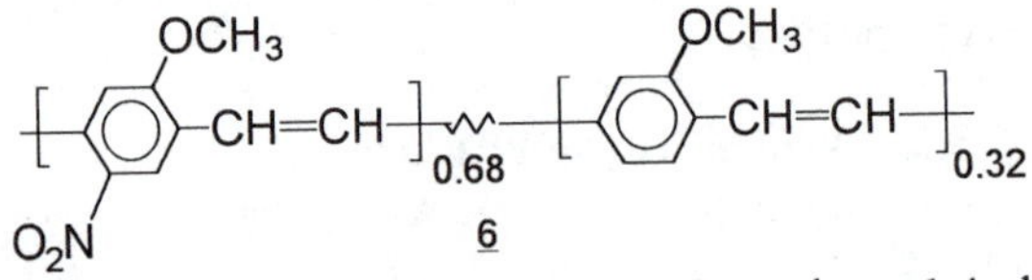

6

Poly[(2-methoxy-5-nitro-1,4-phenylenevinylene)-co-(2-methoxy-1,4-phenylene-vinylene)] (*8*); Poly(MNPV-co-MPV)

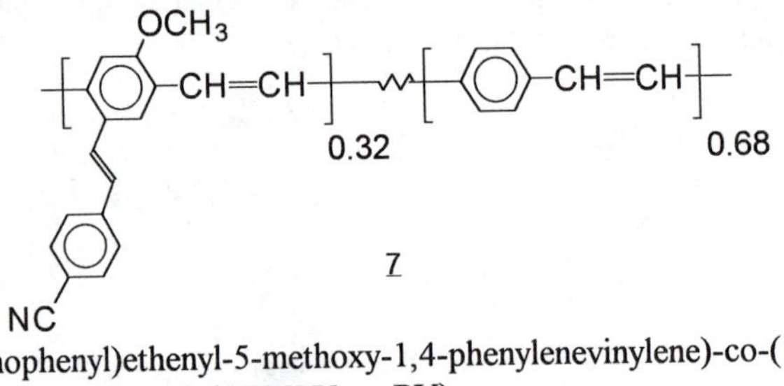

7

Poly[(2-(4-cyanophenyl)ethenyl-5-methoxy-1,4-phenylenevinylene)-co-(1,4-phenylenevinylene)] (*23*); Poly(CEMPV-co-PV)

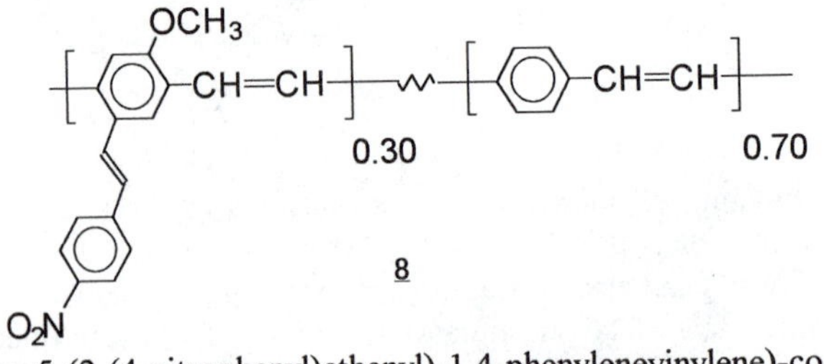

8

Poly[(2-methoxy-5-(2-(4-nitrophenyl)ethenyl)-1,4-phenylenevinylene)-co-(1,4-phenylene-vinylene)]; Poly(MNEPV-co-PV) (Jin, J.-I.; Lee, Y.-H.; Nam, B.-K.; Lee, M. *Chem. Mat.*, in press.).

As the final elimination reaction of poled precursor polymers proceeds in an electric field, the polymer chains will become stiffer due to the generation of vinylene(-CH=CH-) units along the chain. Therefore, one should try to avoid of a premature elimination in order to attain a high degree of dipole orientation by poling.

Second-Harmonic Generation (SHG)

The second-harmonic generation by a thin film (1.48 μm) of copolymer 6 was measured in transmission with a Nd:YAG laser beam of fundamental wavelength of 1.064 μm at an angle of 90° to the film. As described above, the corresponding organic-soluble precursor polymer was cast into a thin film and subjected to thermolysis while an electric field of 10^5 V/cm was constantly applied in a contact poling mode where the electrodes were arranged side-by-side. The laser beam was polarized parallel to the poling direction. The determination of the relative second-order nonlinear optical susceptibility (d_{33}) was performed by the analysis of the Maker fringes (*24,25*), and d_{33} was estimated using the following equation.

$$d_{33} = d_q\left(\frac{I_{M,s}(0)}{I_{M,q}(0)}\right)^{\frac{1}{2}}\left(\frac{I_{c,q}}{I_{c,s}}\right)\left(\frac{\eta_s}{\eta_q}\right)^{\frac{1}{2}}$$

where the subscripts s and q stand for the sample and quartz, respectively. $I_M(0)$ and I_C represent the relative envelope intensity at zero degree and coherence length for normal incidence, respectively. The η values were estimated from the following equation;

$$\eta = \frac{(n_\omega + 1)(n_{2\omega} + 1)(n_\omega + n_{2\omega})}{n_{2\omega} \cdot P^2(0)R(0)}$$

where the n's are the refractive indices at the fundamental (1.064 μm) and harmonic (532 nm) wavelengths. *P*(0) and *R*(0) are the projection factor and multiple reflection factor at zero degrees, respectively. These are equal to unity (*24,26*).
The d_{33} value measured for polymer 6 is 0.5 x 10^{-8} esu ($\chi^{(2)}$=1.0 x 10^{-8} esu). The value is close to the d_{22} value (0.74 x 10^{-8} esu) of $LiNbO_3$ and one order of magnitude smaller when compared with d_{33} value (9.8 x 10^{-8} esu) of $LiNbO_3$. The SHG signal remained constant over a period of more than one month for a sample kept at room temperature and also more than three months for the one having a thermal history up to 100 °C as represented in Figure 1. Figure 1a shows the variance of the d_{33} values as a function of time. The empty circles (O) represent data points for room temperature studies while the filled ones (●) those that have been obtained for a sample which has been thermally cycled as shown in Figure 1b. The results certainly demonstrate that there is no decay in the SHG properties of the samples. Since thermal elimination of the poled precursor polymer generates vinylene linkages along the main chain, the

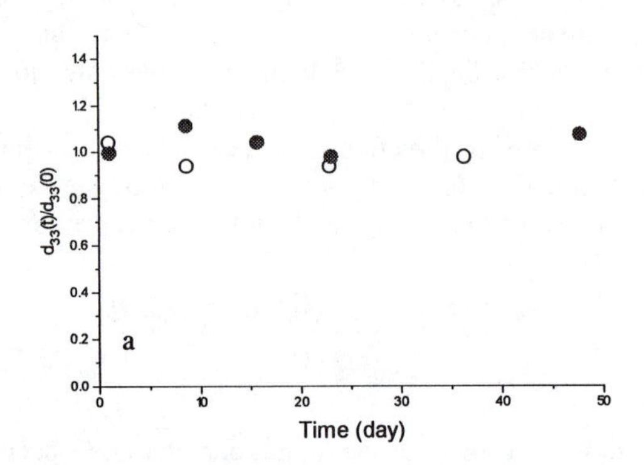

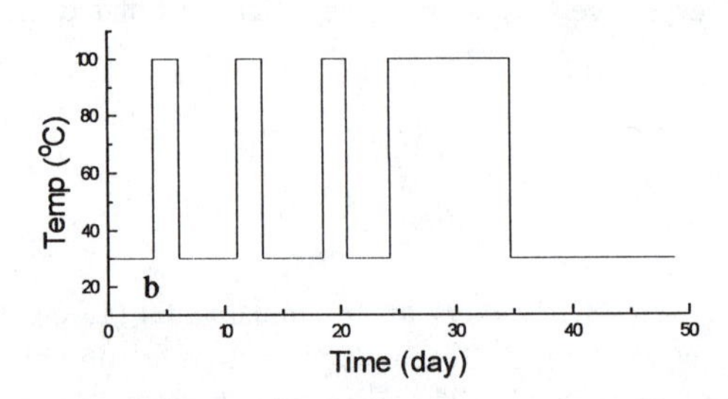

Figure 1. a) SHG intensity of poly(MNPV-co-MPV) at room temperature (o) and after heat treatment (●) and b) thermal cycle for the sample heat treated.

final polymer is expected to be extremely rigid due to polyconjugation. This rigidity of the polymer chains will impart a very high rotational energy barrier to the oriented dipoles maintaining the alignment.

We also have recently reported the synthesis and nonlinear optical properties of the polymer 8, that has an electron-withdrawing nitro group attached to its β-styryl side-groups (Jin, J.-I.; Lee, Y.-H.; Nam, B.-K.; Lee, M. *Chem. Mat.*, in press.). The electron-donating methoxy group is located on the same phenylene ring carrying the substituted styryl group. A thin film (0.133 μm) of poly[(2-methoxy-5-(2-(4-nitrophenyl)ethenyl)-1,4-phenylenevinylene)-co-(1,4-phenylenevinylene)], 8, poly-(MNEPV-co-PV), was poled by carrying out thermolysis of its organic-soluble precursor polymer in an electric field (10^5 V/cm). The precursor polymer (3) caried the benzyloxy group (R'O = $C_6H_5CH_2O$) along the chain. The value of d_{33} for poly(MNEPV-co-PV) was estimated using the method described above. The d_{33} value thus obtained is $1x10^{-8}$ esu. This value is twice the d_{33} value of the previous polymer 6. There are at least three major reasons for the less than expected improvement upon changing the structure from polymer 6 to 8: 1) the overall content of the 2-methoxy-5-(p-nitrostyryl)phenylene unit in 8 is lower than that of the 2-methoxy-5-nitrophenylene unit in 6 (30 vs. 68 %), 2) polymer 6 also contains electron-donating methoxy-substituted phenylenevinylene comonomers, whereas polymer 8 contains only unsubstituted phenylenevinylene comonomer units, and 3) the dipole reorientation during poling may be more difficult for polymer 8 than for polymer 6, because 8 requires the rotation of much bulkier structural units. Nevertheless, the d_{33} value measured for polymer 8 is the largest ever reported for a PPV derivative. The SHG signal of polymer 8 also remained constant for more than one month at room temperature and was unchanged after repeated thermal cycling at 100 °C as shown in Figure 2b (Figure 2)⊠ Furthermore, polymers 6 and 8 did not show any glass transition up to their decomposition temperatures as determined by DSC analysis. We believe, therefore, that the poled polymers would show a stability in their SHG properties even at higher temperatures.

Electro-Optic Properties

We have recently reported the synthesis and NLO properties of poly[{2-(4-cyanophenyl)ethenyl-5-methoxy)-1,4-phenylenevinylene}-co-(1,4phenylenevinylene)], 7, poly(CEMPV-co-PV) (*23*). In this case, the electro-optic properties of corona poled polymer were investigated by using reflection measurement technique (*27*). For electro-optic coefficient measurements, an ITO coated glass plate was used as substrate and corona discharge poling was employed to orient the NLO chromophores. We adopted a two stage poling process based on the TGA thermogram. By keeping the film at 120 °C for 2 h under corona field, the NLO chromophores are expected to be orient with slow elimination. The spacial dipole alignment generated was stabilized by the thermal elimination at 210 °C. A He-Ne laser beam (632.8 nm) was employed as the fundamental incident beam. Further experimental details can be found in one of our recent papers (*6,23*).

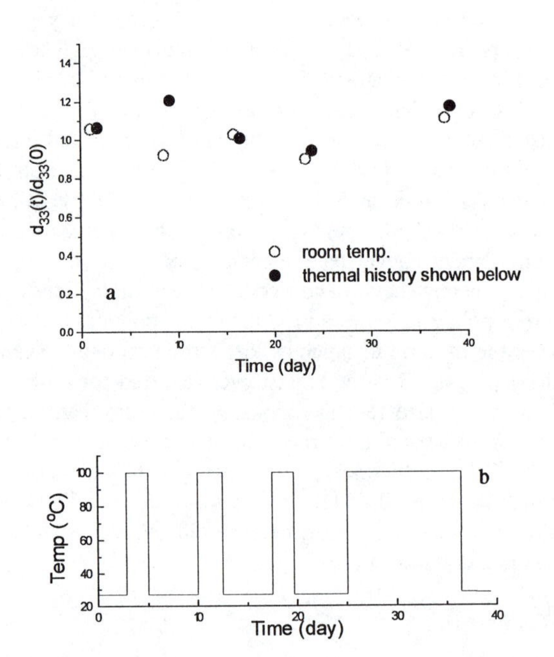

Figure 2. a) SHG intensity of poly(MNEPV-co-PV) at room temperature (○) and after heat treatment (●) and b) thermal cycle for the sample heat treated.

Elecro-optic coefficient was calculated by the following equation ;

$$r_{33} = \frac{3\lambda I_m}{4\pi V_m I_c n^2} \frac{(n^2 - \sin^2\theta)^{\frac{3}{2}}}{(n^2 - 2\sin^2\theta)} \frac{1}{\sin^2\theta}$$

where λ is the optical wavelength, I_m the amplitude of the modulation, I_c the half intensity point, V_m the modulating voltage and n the refractive index.

Poly(CEMPV-co-PV), 7, showed the electo-optic coefficient of 1.2 pm/V ($\chi^{(2)} = 3.0 \times 10^{-9}$ esu). Although the structure of polymer 7 resembles that of polymer 8, the $\chi^{(2)}$ value of polymer 7 is one order of magnitude smaller than that of polymer 8, which means that the electron-withdrawing power of nitro group is stronger than that of cyano group. It is, of course, very possible that their degrees of poling or the degrees of dipole alignment of the poled films are quite different. Polymer 7 also showed good temporal and thermal stability of the electro-optic responses, and no significant decay of the electro-optic responses was observed for over three months at room temperature. Most significant is the fact that this polymer shows no significant detectable decay at room temperature and no tendency of relaxation even after being left one week at 100 °C (Figure 3 and 4).

The measured electro-optic coefficients of 5 poled at 1 MV/cm are plotted in Figure 5 and 6 versus CMPV content in the polymer (*6,7*). The EO coefficient increases almost linearly up to 35 wt% of CMPV and remains almost same at the higher concentrations with the value of r_{33}=1.2 pm/V. The EO response was stable even at 100 °C.

This approach to obtain stable, poled polyconjugated polymer films via the polymer precursor route discussed in this article will certainly provide us with a wide variety of materials having improved second-order NLO properties when the chemical structure of the polymer and the poling conditions are properly optimized. Electro-optic properties of PPV derivatives have mainly been studied recently only by our group. Thus, the discussion on SHG and EO activities of PPV derivatives is rather limited.

Third-Order Optical Nonlinearity of PPV Derivatives

Conjugated polymers, especially PPV derivatives, reveal large third-order optical nonlinearities because of the large π-electron contribution. These conjugated polymers, however, are difficult to process because they are rarely fusible and are insoluble in most solvents.

Usually $\chi^{(3)}$ of π-conjugated polymers is enhanced through lengthening of the delocalization length. It has been reported that $\chi^{(3)}$ is proportional to the 6th power of the delocalization length (*29*). Substitution on the phenylene rings of PPV with different types of electron-donating groups has a significant influence on the electronic band structure, and thus, is expected to affect the electrical and nonlinear optical properties of the resulting polymers.

Synthetic Strategy

Various PPV derivatives have been prepared in thin films via the water-soluble precursor method and the general synthetic schemes are shown below. The synthetic details can be found elsewhere (*11,13-18*).

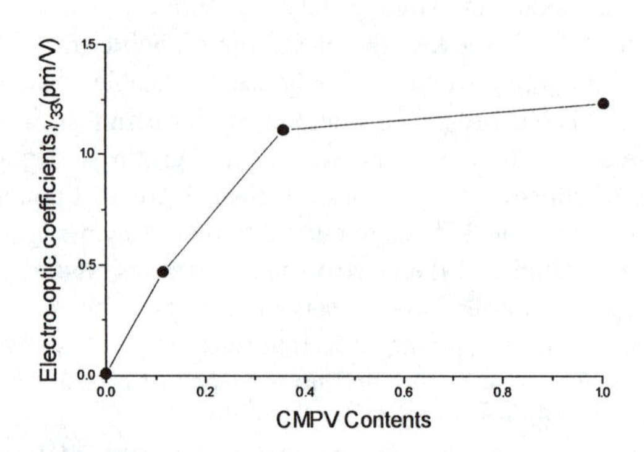

Figure 3. EO coefficient, γ_{33}, of poly(CEMPV-co-PV) measured as a function of time at room temperature.

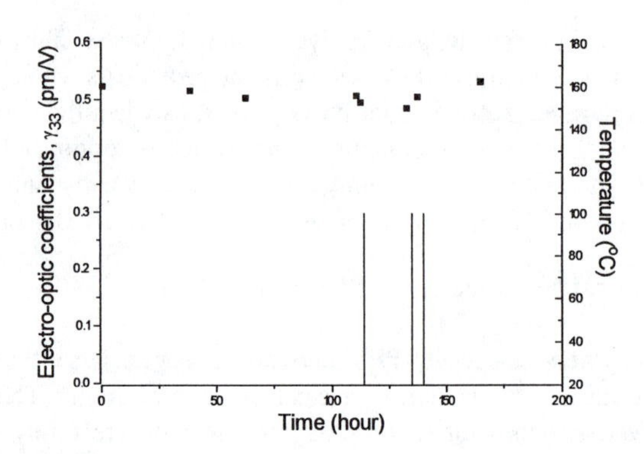

Figure 4. EO coefficient, γ_{33}, of poly(CEMPV-co-PV) measured as a function of time at 100 °C.

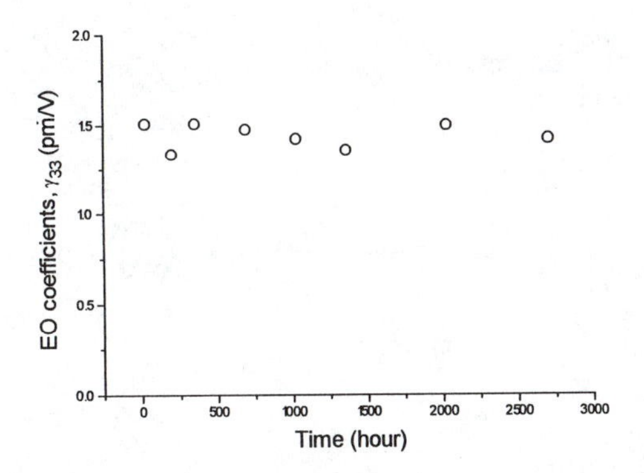

Figure 5. EO coefficients of poled poly(CMPV-co-PV) at room temperature versus the CMPV content of the polymer.

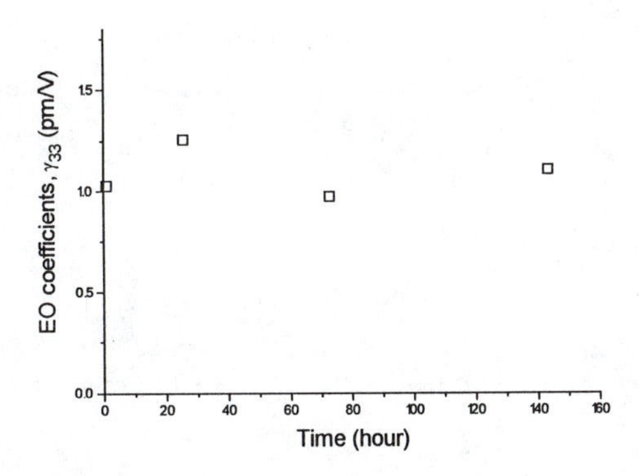

Figure 6. Decay of the EO coefficient γ_{33} of poly(CMPV-co-PV) poled initially at 130 °C for 1hr followed by at 200 °C for 1hr under vacuum.

Scheme 1 (Chloromethylation method)

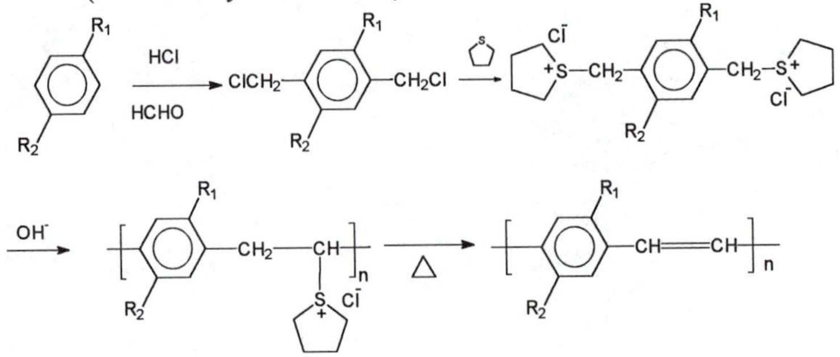

HCl
HCHO
$ClCH_2$ CH_2Cl
Cl^- $^+S-CH_2$ CH_2-S^+ Cl^-
R_1 R_2
OH^-
CH_2 CH n S + Cl^-
Δ
CH=CH n

Scheme 2 (NBS method)

CH_3 CH_3 NBS $BrCH_2$ CH_2Br S
$^+S-CH_2$ CH_2-S^+ Br^- Br^-
R_1 R_2
OH^-
CH_2 CH n S + Br^-
Δ
CH=CH n

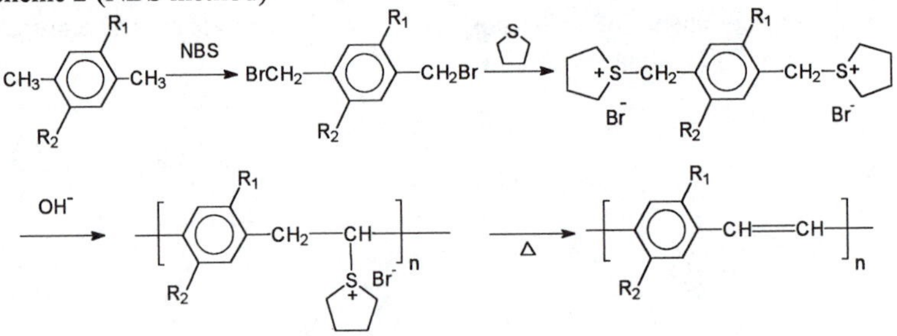

Scheme 3 (Organic soluble precursor route)

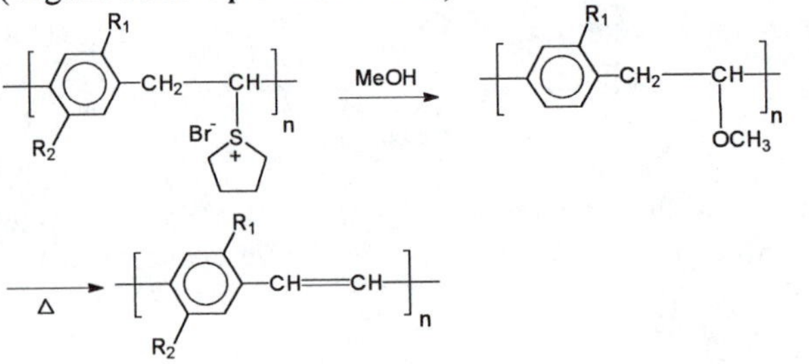

CH_2 CH n Br^- S +
MeOH
CH_2 CH n OCH_3
Δ
CH=CH n
R_1 R_2

Table I shows some of the substituted PPVs whose optical third harmonic properties have been studied.

Table I. Substituted PPVs

Polymers	Structures	Ref.
PMPV	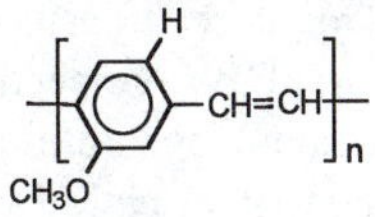 H; CH_3O; –CH=CH–; n	*11*
PBPV	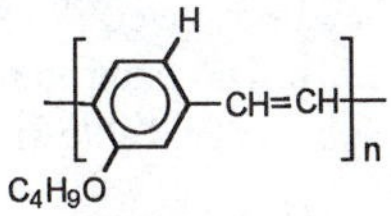 H; C_4H_9O; –CH=CH–; n	*30*
PDPV	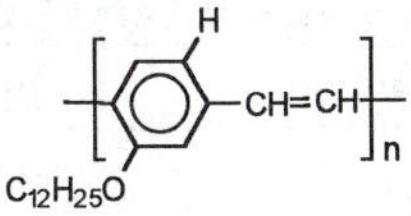 H; $C_{12}H_{25}O$; –CH=CH–; n	*31*
PNMPV	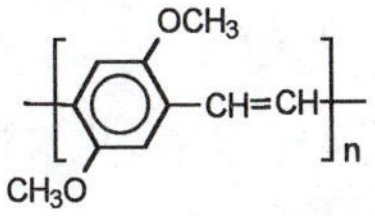 OCH_3; CH_3O; –CH=CH–; n	*32*
PIMPV	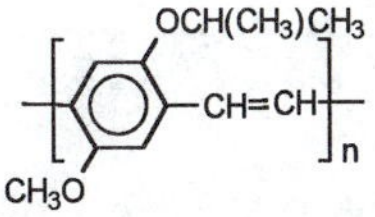 $OCH(CH_3)CH_3$; CH_3O; –CH=CH–; n	*a*

a. Shim, H.-K.; Yoon, C.-B.; Lee, J.-I.; Hwang, D.-H. *Polymer Bulletin*, in press.

Third-Harmonic Generation (THG)

In order to determine the third-order nonlinear optical susceptibilities of the polymers, the third harmonic intensities of polymers and fused quartz were measured as a function of changing angle of incidence. As the polymers were coated on fused quartz, it was necessary to calibrate the interference effect between film and substrate.

To obtain I_{max} and I_{min}, maximum and minimum points of zero degree, Maker fringe patterns werecurve-fitted. The third-order nonlinear optical susceptibility, $\chi^{(3)}$, of the thin polymer film sample (much thinner than the coherence length) was calculated using the following equation (*33-35*).

$$\chi^{(3)} = \frac{2}{\pi}\chi_s^{(3)}\frac{(I_{3\omega})^{1/2}}{l}\frac{l_{c,s}}{(I_{3\omega,s})^{1/2}}$$

where l is the thickness of the polymer sample, and $\chi^{(3)}$ and $l_{c,s}$ are the third-order nonlinear susceptibility and the coherence length of the fused quartz, respectively. I_{3w} and $I_{3w,s}$ are the third harmonic intensities of the film and the reference quartz, respectively. Measured $\chi^{(3)}$ values and film thicknesses are shown in Table II. These data show that as the alkoxy side chain length increases, the $\chi^{(3)}$ value decreases.

The UV maximum absorption for π–π* transitions occurs at 450 nm for PMPV, 440 nm for PBPV and 400 nm for PDPV, respectively. As the chain length of side alkoxy group increases, the maximum absorption gradually shifts to the shorter wavelength region. This result is consistent with the trend observed for the $\chi^{(3)}$ values of monoalkoxy-substituted PPVs. It is well known that $\chi^{(3)}$ value is inversely proportional to band gap energy (*36*). The band gap energy increases with increasing length of the alkoxy substituent as shown in Table II.

Table II. Band Gab Energies and Third Order Nonlinear Coefficient

	PMPV	PBPV	PDPV
I_{3w}	1.80	0.92	4.58
Thickness (mm)	0.017	0.025	0.132
Band gap energy (eV)	2.25	2.31	2.33
$\chi^{(3)}$ (10^{-12} esu)	12	5.8	2.5

*$\chi^{(3)}_s = 2.76 \times 10^{-14}$ esu, $I_{3\omega,s} = 4.57$, $l_{c,s} = 18.41$ μm

Recently, Kaino et al. (*37,38*) have examined the optical nonlinearity of a PPV thin film. The $\chi^{(3)}$ of PPV film reported by these authors is 7.8×10^{-12} esu at 1.85 μm. This $\chi^{(3)}$ value is almost the same as the value obtained by us for PBPV film. They have also reported the frequency dependence of the $\chi^{(3)}$ (3ω; ω, ω, ω) values of poly(2,5-dimethoxy-1,4-phenylenevinylene) (PDMPV) (*32*). The largest $\chi^{(3)}$ value of PDMPV thin film was 1.6×10^{-10} esu at the harmonic wavelength of 532 nm and the UV maximum absorption for the π–π* transition of PDMPV film appeared at 480 nm. Its $\chi^{(3)}$ value was a little lower, 5.4×10^{-11} esu, when the fundamental wavelength was 1.85 μm (*12*).

In case of PIMPV, the absorption maximum position for π–π* transition and the $\chi^{(3)}$ value were 460 nm and 2.5×10^{-12} esu, respectively. This $\chi^{(3)}$ value is one order of magnitude smaller than that of PDMPV. This result is rationalized by the two possible alternatives. One is the band gap energy difference of the polymers. The absorption edge decreases from 610 nm of PDMPV to 590 nm of PIMPV, which indicates an increase in the band gap energy of PIMPV. Another additional substituent effect possible is the steric effect exerted by the bulkier isopropoxy group in the later, which may affect the coplanarity between the phenylene and vinylene units. This will result in less efficient delocalization of π-electrons along the chain. Since, electronically, the substituent effect should be very similar, it seems more reasonable that the delocalization length for PIMPV is decreased by conformation twisting. This, in turn, will increase the band gap. The increased band gap should result in the decrease $\chi^{(3)}$ value as was observed.

Conclusion

In this article, we have tried to review the nonlinear optical properties of a number of PPV derivatives synthesized mainly by us. For the first time, we succeeded in the preparation of PPV copolymers in a poled state and found that the compositions carrying a strong electron-donating and electron-withdrawing groups show large second-order optical nonlinearities (~10^{-8} esu) and exhibit no relaxation for a prolonged period of time for the samples kept even at 100 °C. Optimization of the thermolysis and poling conditions, such as finding the best heating rate and use of a stronger poling field, is expected to enhance the dipole alignment leading to a significantly higher second-order optical nonlinearity. The remarkable stability of the poled dipoles must be due to the rigid nature of the polyconjugated main chain of the present polymers.

We have also examined the third order optical nonlinearities of various PPV derivatives. The introduction of electron-donating substituents to PPV is to be effective in decreasing the band gap energy and, thus, increasing $\chi^{(3)}$.

The approaches to obtain stable and processible polyconjugated polymer films through the water-soluble or organic-soluble precursor route will certainly provide us with a wide variety of materials having improved second- or third-order optical nonlinearities when the chemical structure of the polymer as well as processing conditions are properly optimized.

Acknowledgements

We are grateful to the Korea Science and Engineering Foundation for the support of this work.

Literature Cited

1. Prasad, P.N.; Williams, D.J., *Introduction to Nonlinear Optical Effects in Molecules and Polymers*, John Wiley & Sons, Inc.: 1991.
2. *Nonlinear Optical Properties of Organic and Polymeric Materials*, Williams, D.J., Ed.; American Chemical Society: Washington, D.C., 1983.
3. Prasad, P.N. *Nonlinear Optical and Electroactive Polymers*, Ulrich, D.R.; Eds., Plenum Press: New York, 1988.
4. Heeger, A.J.; Orenstein J. *Nonlinear Optical Properties of Polymers*, Ulrich, D.R., Ed.; Mat.Res.Soc.: Pittsburgh, PA., 1988.
5. Heeger, A.J.; Moses, D.; Sinclair, M., *Synth. Met.*, **1986**, *15*, 85.
6. Kim, J.-J.; Hwang, D.-H.; Kang, S.-W.; Shim, H.-K., *Mater. Res. Soc. Symp. Proc.*, **1992**, *277*, 229.
7. Kim, J.-J.; Kang, S.-W.; Hwang, D.-H.; Shim, H.-K., *Synth. Met.*, **1993**, *55-57*, 4204.
8. Jin, J.-I.; Lee, Y.-H. Mol. Cryst. *Liq. Cryst.*, **1994**, *247*, 67.
9. Prasad, P.N. Frontiers of Polymer Research , Nigam J.K., Ed.; Plenum Press: New York, 1991.

10. Wung, C.J.; Pang, Y.; Prasad, P.N.; Karasz, F.E. *Polymer*, **1991**, *32*, 605.
11. Shim, H.-K.; Hwang, D.-H.; Lee, K.-S. *Makromol. Chem.*, **1993**, *194*, 1115.
12. Kaino, T.; Kobayashi, H.; Kubodera, H.; Kurihara, K.; Saito, T., Tsutsui, T.; Kokito, S. *Appl. Phys. Lett.*, **1989**, *54*, 1619.
13. Murase, I.; Ohnishi T.; Noguchi, T.; Hiroka, M. *Synth. Met.*, **1987**, *17*, 639.
14. Han, C.-C.; Lenz, R.-W.; Karasz, F.-E. *Polym. Commun.*, **1987**, *28*, 261.
15. Jin, J.-I.; Park, C.-K.; Shim, H.-K.; Park, Y.-W. *J.Chem. Soc. Chem. Commun.*, **1989**, 1205.
16. Gagnon, D.K.; Capistran, J.D.; Karasz, F.E.; Lenz, R.W.; Antoun, S. *Polymer*, **1987**, *28*, 567.
17. Murase, I.; Ohnishi, T., Noguchi, T. ;Hirooka, M. *Polym. Commun.*, **1987**, *28*, 229.
18. Tokito, S.; Momii, T.; Murata, H., Tsutsui, T. ;Saito, S., *Polymer*, **1990**, *31*, 1137.
19. Eich, M.; Reck, B.; Yoon, Do Y.; Wilson, C.G.; Bjorklund, G.C. *J. Appl. Phys.*, **1989**, *66*, 3241.
20. Jungbauer, D.; Reck, B.; Twieg, R.; Yoon, Do Y.; Wilson C.G.; Swalen, J.D. *Appl. Phys. Lett.*, **1990**, *56*, 2610.
21. Chen, M; Yu, L.; Dalton, L.R.; Shi, Y.; Steier, W.H. *Macromolecules*, **1991**, *24*, 5421.
22. Wessling, R.A.; Zimmermann R. *U.S.Pat.*, 3401152 (1968); 3706677 (1972).
23. Hwang, D.-H.; Lee, J.-I., Shim, H.-K., Hwang, W.-Y., Kim, J.-J.; Jin, J.-I., *Macromolecules*, **1994**, *27(21)*, 6000.
24. Jerphagnon, J.; Kurtz, S.K. *Phys. Rev. B.*, **1970**, *1(4)*, 1739.
25. Maker, P.D.; Terhune, R.W.; Nisenoff, M.; Savage, C.M. *Phys. Rev. Lett.*, **1962**, *8*, 21.
26. Jerphagnon, J.; Kurtz, S.K. *Phys. Rev.*, **1964**, *134*, A1416.
27. Teng, C.C.; Man, H.T. *Appl. Phys. Lett.*, **1990**, *56*, 1934.
28. Man, H.T.; Yoon, H.-N. *Adv. Mater.*, **1992**, *4*, 159.
29. Bradley, D.D.C.; Mori, Y. *Polym. Prepr. Jpn*, **1988**, *37*, 3151.
30. Lee, J.-I.; Hwang, D.-H.;Shim, H.-K.;Lee, M.; Yu, S.-K.; Lee, G.-J., *Mol. Cryst. Liq. Cryst.*, **1994**, *247*, 121.
31. Kang, I.-N.; Lee, G.-J.; Kim, D.-H.; Shim, H.-K. *Polymer Bulletin*, **1994**, *33*, 89.
32. Murata, H.; Takada, N.; Tsutsui, T.; Saito, S.; Kurihara, T.; Kaino, T. *Synth.Met.*, **1992**, *45-50*, 131.
33. Kubodera, K. *Nonlinear Optics,* **1991**, *1*, 71.
34. Kajzar, F.; Messier, J.; Rosilio, C. *J. Appl. Phys.*, **1988**, *60*, 3040.
35. Tomaru, S.; Kubodera, K.; Zembutsu, S.; Takeda, K.; Hasegawa, M. *Electronics Letters*, **1987**, *23*, 595.
36. Santeret, C.; Herman, J.P.; Fray, R.; Pradere, F.; Duching, J.; Baughman, R.H.; Chance, R.R. *Phys. Rev. Lett.*, **1976**, *36*, 956.
37. Kaino, T, Kubodera, K., Tomaru, S., Kurihara, T., Saito, S., Tsutsui, T. ;Tokito, S., *Electron. Lett.*, **1987**, *23*, 1095.
38. Kaino, T.; Kubodera, K.; Kobayashi, H.; Kurihara, T.; Saito, S.; Tsutsui, T.; Tokito, S.; Murata, H. *Appl. Phys. Lett.*, **1988**, *53*, 2002.

RECEIVED April 19, 1995

Chromophore Alignment

Chapter 18

All-Optical Poling of Polymers for Phase-Matched Frequency Doubling

J. M. Nunzi, C. Fiorini, F. Charra, F. Kajzar, and P. Raimond

Groupe Composants Organiques, Laboratoires d'Electronique et de Technologies de l'Information, (Commissariat a l'Energie Atomique–Technologies Avancées), Centre d'Etudes Saclay, F91191 Gif sur Yvette, France

The resonant nonlinear optical process which permits the all-optical preparation of noncentrosymmetric materials for frequency doubling is described in terms of a polar orientational photoinduced effect occuring into the centrosymmetric distribution of molecules. A permanent optical poling of acrylic copolymers is efficiently achieved by seeding-type preparation with a dual frequency laser. Nonlinear chromophores undergo a net light-induced rotation resulting from reversible izomerization cycles. Optimisation of the preparation conditions in thin films yields a frequency doubling second-order susceptibility as high as using the corona-poling technique. The polarization lies in the film plane. This opens a new direction towards the molecular engineering of transparent phase-matched oriented micro-structures for efficient frequency conversion.

The multi-functionality of organic materials which is offered by organic synthesis diversity as well as by physico-chemical tailoring yields an infinity of possibilities for the optimisation of material properties from the molecule to the device. However, the practical realisation of devices for frequency conversion, with an aim at building a compact blue laser source, is still a major technical problem. The main routes explored for the realization of compact blue-light sources are electro-luminescence, fluorescence up-conversion and frequency doubling (SHG), but up-to now, only frequency doubling has proved useful for an efficient blue light generation. In this respect, a large collection of non-centrosymmetric molecules has been engineered and tested [*1*], but there still remains major difficulties with frequency doubling.

1. A polar orientation of molecules is essential in order to achieve noncentrosymmetric assemblies of molecules with a non-zero $\chi^{(2)}$.
2. The second harmonic wavelets generated at different places along the propagation direction of the fundamental beam must all interfere constructively.

0097–6156/95/0601–0240$12.00/0

Polar ordering is not a universal tendency of molecules. Molecules indeed generally arrange into centrosymmetric structures which minimise the electrostatic energy of dipoles. Polar ordering is naturally achieved by means of the short range interactions between molecules which take place in some specific supra-molecular assemblies such as crystals, Langmuir-Blodgett films, liquid crystals and ferroelectrics. Those effects have already been reviewed with the aim at optimising the related nonlinear optical properties. This led for instance to ideas such as using so called *octupolar* molecules which, by symmetry, have zero-permanent dipole moment [2].

For the sake of processability, compatibility with other technologies and ease of tailoring, it is interesting to use polymers for nonlinear optics [3]. However, polar orientation is not spontaneous in dispersed media such as liquid and polymer solutions. It is usually achieved by using the so called electric field poling method. It can be performed with molecules having a strong permanent dipole moment. Molecules align their dipole moment parallel to an applied static-electric field E_{DC} in order to minimise electrostatic energy. Starting from an isotropic solution, this leads to a net molecular orientation such as the one pictured in Figure 1 in which noncentrosymmetry manifests simply as the physical distinction between upward and downward orientations.

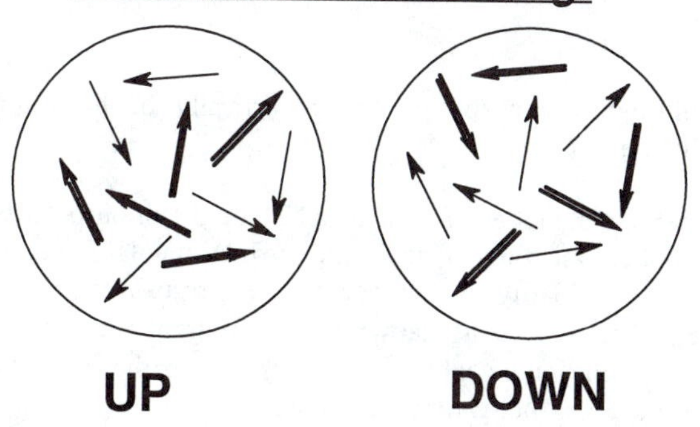

Figure 1: Polar orientation resulting from an oriented selection among an isotropic set of arrows. Thin and bold arrows represent molecules with different second order hyperpolarizabilities β.

An interesting question is related to the possibility of using light-matter interactions in order to orient molecules. For instance, light beams can be used to draw wave guides using their bleaching effect, as well as to promote local heating in order to pole locally using the electric field poling technique [4]. However, no polar orientation is achieved by monochromatic optical fields E_{ω}. They change sign periodically at rates as high as 10^{15} Hz. Either polarized, they only induce optical anisotropy in molecules which have an anisotropic polarizability (Fig. 2). This leads to photo-induced dichroïsm or birefringence. No polar orientation is achieved because the net effect is quadratic in field. Indeed, the field itself has no polarity. The excitation process is called polarized hole-burning [5]. Another question now is: how can we perform an orientation hole-burning such as in Figure 1 ?

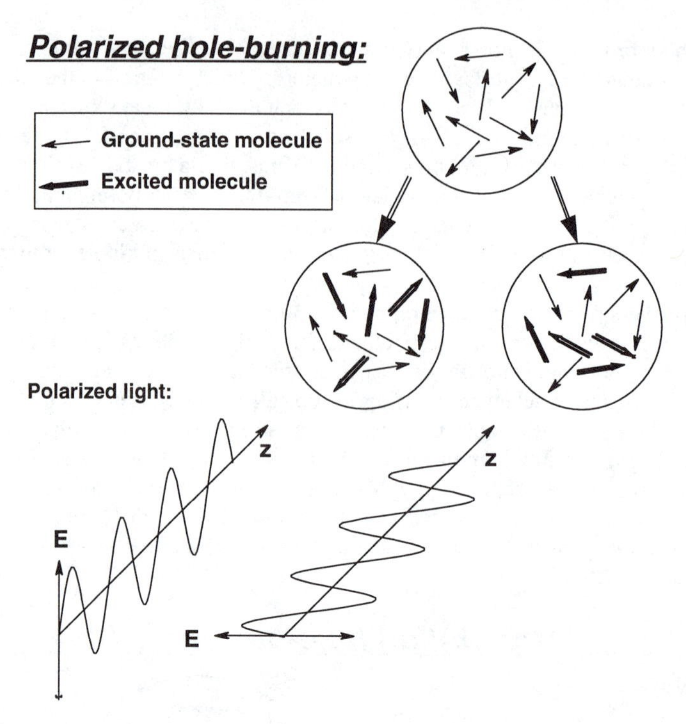

Figure 2: Photo-induced anisotropy using vertically or horizontally-polarized monochromatic light-beams.

The discovery that combinations of optical fields could exhibit polarity opens up new possibilities for optical handling of molecules [*6*]. A standard situation in which an optical field exhibits polarity is the coherent superposition of a field E_{ω} at a fundamental frequency ω with its harmonic $E_{2\omega}$ at frequency 2ω. It is pictured in Figure 3. Polarity of the cubic interferences appears in the non-zero average cube $\langle E^3 \rangle_t$ where the subscript *t* holds for time average.

The sign of the polar interference $\langle E^3 \rangle$ is determined by the relative phase between ω and 2ω beams. The fact that in condensed materials fundamental and harmonic frequency propagation-speeds b different make the optical polarity alternate between positive and negative extrema. In a dispersive material with index $n(\omega)$, the periodicity of this alternance is $\lambda/2(n(\omega)-n(2\omega))$ were λ is fundamental wavelength. This is precisely the periodicity which is in principle necessary in order to achieve a quasi phase-matching for frequency doubling [*7*].
The observation of a second-harmonic generation in optical fibres prepared by an intense light at 1064 nm [*8*] or with a simultaneous *seeding* light at double frequency (532 nm) [*9*] revealed the possibility of inducing a noncentrosymmetry using light in a centrosymmetric material. It is the first reported application of a mechanism involving cubic interferences with $\langle E^3 \rangle \neq 0$. In order to build the molecular engineering rules which permit a control of this effect, it is necessary to identify efficient processes by which condensed materials record $\langle E^3 \rangle$.

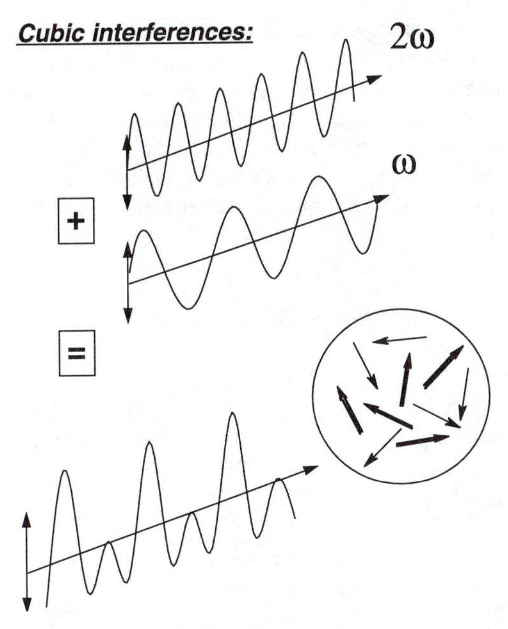

Figure 3: Polar cubic interference between fields at fundamental and harmonic frequencies.

$\langle E^3 \rangle$ effects find a simple description in the case of a solution of molecules excited under resonant conditions. The origin of the nonlinearity can then be pictured in terms of a hole burning into the distribution of molecular orientations. Let us consider a two-level molecule. Under one-photon excitation at 2ω, the projection of its wave function on the excited state is proportional to $(\vec{\mu}_{01}.\vec{E}_{2\omega})$ where μ_{01} is the transition dipole-moment between ground and excited states. Under two-photon excitation at ω, the projection of its wave function on the excited state is proportional to $(\vec{\mu}_{01}.\vec{E}_{\omega})(\delta\vec{\mu}.\vec{E}_{\omega})/\hbar\omega$ where $\delta\mu$ is the dipole-moment difference between ground and excited states. A molecule which is irradiated at resonance with fields E_{ω} and $E_{2\omega}$, respectively at fundamental and harmonic frequencies, experiences a transition probability to the excited state which contains the one and two-photon interference contribution: $(\vec{\mu}_{01}.\vec{E}_{2\omega})(\vec{\mu}^*_{01}.\vec{E}^*_{\omega})(\delta\vec{\mu}.\vec{E}^*_{\omega})/\hbar\omega$ [*10*]. In this expression, brackets represent scalar products between vectors. This interference term which is proportional to E^3 depends on the polar orientation of the molecule via $\delta\vec{\mu}$. The process involves the microscopic polarizability tensor β of the molecule. The average value of this cubic interference over molecule orientations is zero in isotropic mixtures. That means that the cubic interference is not associated with an index grating. However, absorption is preferentially directed towards molecules oriented either upward or downward, as in Figure 1, depending on the phase of the product $E_{2\omega}E^{*2}_{\omega}$. There results a net noncentrosymmetry. The overall process of second harmonic generation which is pictured in Figure 4 can be viewed as a six-wave mixing processes [*11*]. The process consists in the record and readout of a $\chi^{(2)}$ grating caused by the $<E^3>$ interference [*12*]. The grating is a quasi phase-matched one and unlike with usual quasi phase-matching techniques, there is no simultaneous induction of an index grating [*13*]. Hence, using dual-frequency optical excitation, both polar orientation and phase matching conditions can be achieved in polymers. Six-wave mixing effects are usually assumed to be small but it can be proved that they are the dominant contributions in degenearte wave-mixing experiments performed with polymers excited under two-photon absorption conditions [*14-16*].

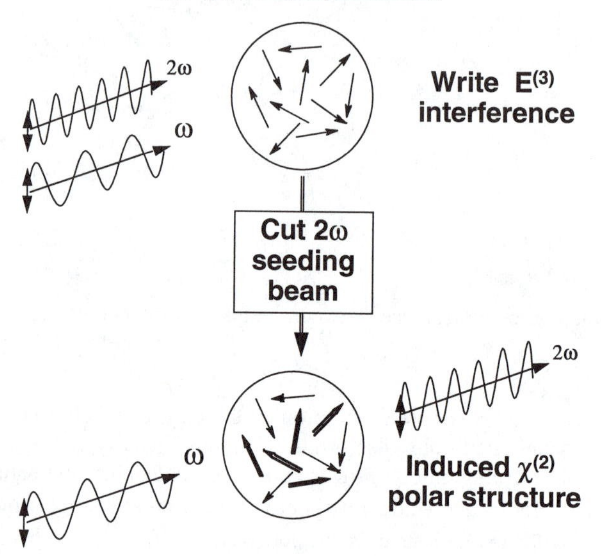

Figure 4: Preparation process for all-optical poling of materials using orientational hole-burning.

The first experimental observation of an optical selection of molecular orientations is the transient induction of non-centrosymmetry recorded using phase-matched frequency doubling in a solution of the polar molecule 4-diethylamino- 4'-nitrostilbene in tetrahydrofuran [*11*]. The same procedure has been applied to the orientation of non-polar molecules such as the triphenyl-methane dye ethyl-violet which exhibits octupolar symmetry [*17*]. It has also been applied to ionic molecules and used to study their dipole moments [*18*]. Unlike electric field poling in which molecules undergo an electric field assisted rotation, the optical selection of orientations corresponds to a bleaching of molecules [*19*]. A cause of relaxation of the induced non-centrosymmetry is thus recovery of the bleaching. However, in liquid solutions, loss of orientation has been identified as the fastest relaxation process. Loss of orientation is related to viscosity; it completes within a nanosecond, leading to isotropy. Such a fast relaxation effect is removed in glassy polymers [*10*].

The experimental set-up for the optical preparation of plastic materials is pictured in Figure 5. It consists in a pulsed picosecond Nd:YAG laser delivering both fundamental (1064 nm) and harmonic (532 nm) wavelengths at 10 Hz repetition-rate. Energies are 500 μJ and 2.5 μJ, respectively at 1064 nm and 532 nm. Beam diameter is 2 mm at the sample location.

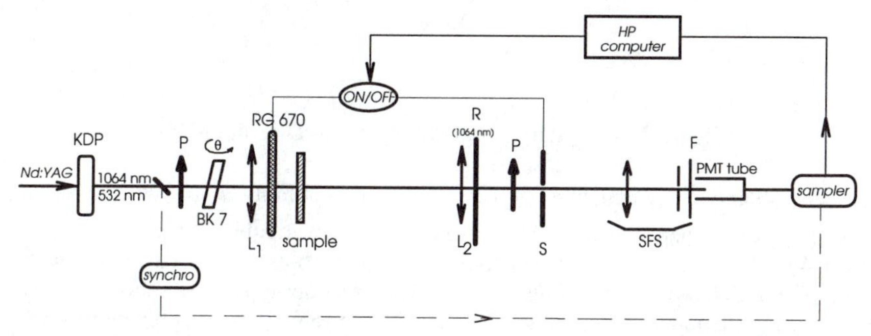

Figure 5: Experimental set-up. A fast photodiode synchronizes the sampler. SHG intensity is measured with the photo-multiplier tube (PMT). P: polarizers; F: interferential filter at 532 nm; SFS : spatial filtering system; R: dielectric mirror for fundamental rejection; S: shutter synchronised with insertion of the green blocking RG670-Schott filter. BK7 glass plate is fixed on a rotating stage for phase adjustment between writing beams.

Poling is performed using dual excitation with fundamental and harmonic beams. During preparation time in Figure 6, the shutter in front of the photo-multiplier-tube is closed. Each minute, a test of the second-harmonic efficiency is performed. A red-cutting filter (Schott RG-670) is then inserted in front of the sample in figure 6 and the shutter is opened. The experiment is fully automated.

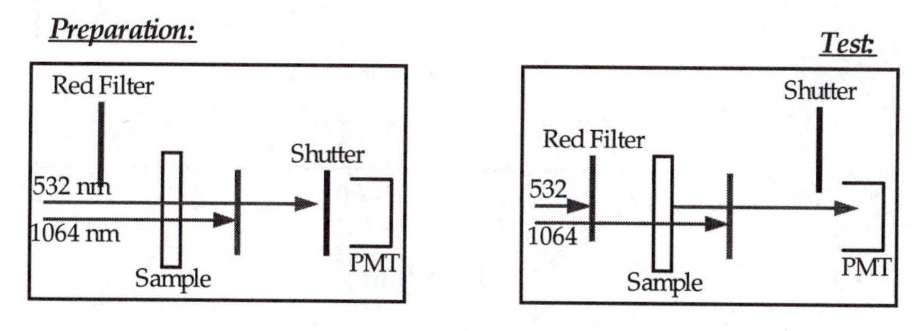

Figure 6: Beam arrangement for seeding-type preparation and test experiments.

Polymer samples where spin-coated films of a random methacrylic copolymer with nonlinear 4-nitro-4'-amino-diazobenzene chromophores grafted as pendant side groups. Molar concentration was n = 35% (Fig. 7).

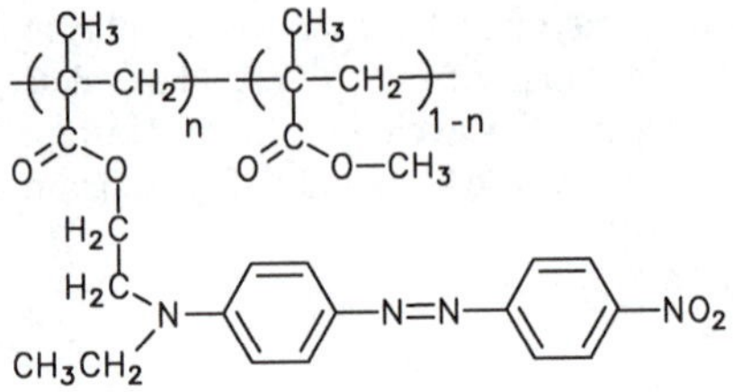

Figure 7: Side-chain 35/65-copolymer of DR1 in PMMA.

The growth and decay dynamics of the efficient susceptibility $d_{33} = \frac{1}{2}\chi^{(2)}_{yyy}$ induced in this polymer is displayed in Figure 8. The frequency doubling coefficient reaches $d_{33} = 45$ pm/V after 2 hours of preparation. It is as high as what is currently obtained using corona electric-field poling [*20,21*]. After seeding-type preparation, the decay dynamics of the induced d_{33} is the same as with the same polymer prepared using the corona-poling technique. This corresponds to an electrooptic coefficient r_{33} close to 10 pm/V at 1.3 μm-wavelength. Molecules are oriented in the plane of the film. Order parameter $<\cos^3(\theta)>$ is close to 0·2.

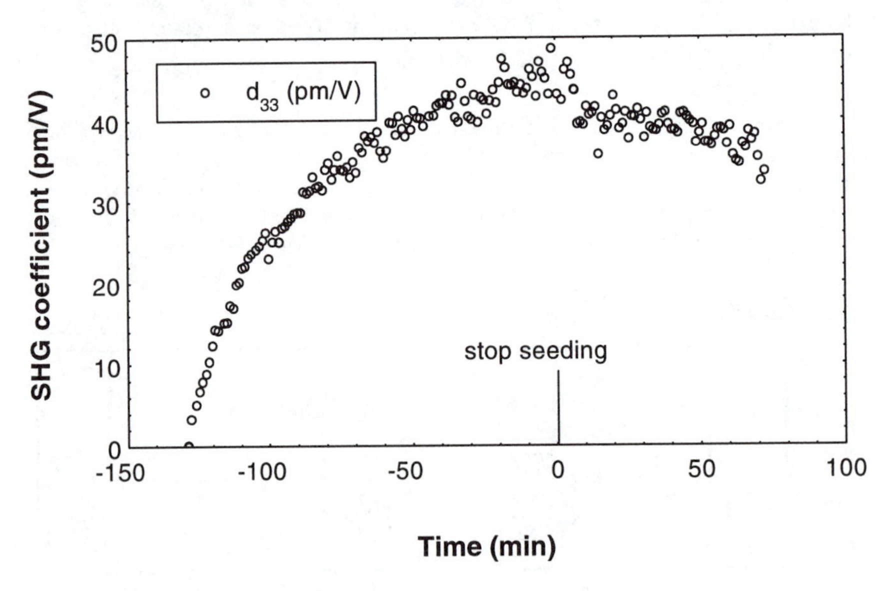

Figure 8: Growth and decay dynamics of the efficient susceptibility d_{33}.

The permanent orientation achieved in Figure 8 may be attributed to a net light-induced rotation effect [22]. It is related to *trans-cis* izomerization (Fig. 9) in the same way as the one which promotes the light-assisted electric-field poling effect [*23*].

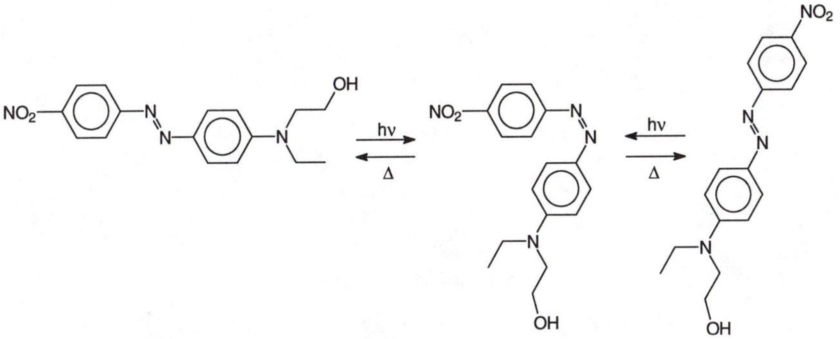

Figure 9: Illustration of the photo-induced molecular-axis rotation by *trans/cis* izomerization. Spontaneous reverse *cis* to *trans* reactions are thermally activated.

We have studied the spatial profile of the induced second-harmonic efficiency. It is pictured in Figure 10. Unlike what is currently observed in seeded bulk glasses [*24-26*], the spatial profile along the polymer-film plane is uniform. This confirms that the optical poling of polymers results in a local effect. Our preparation technique thus permits patterning of micro structures using the laser spot.

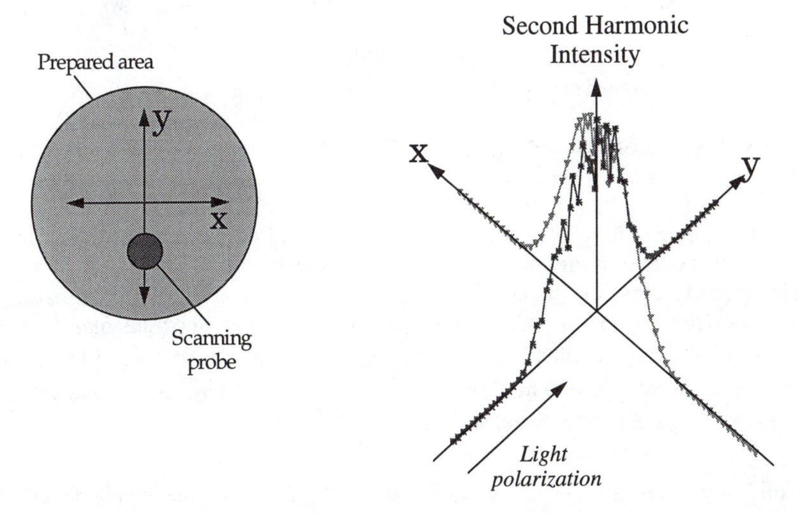

Figure 10: Spatial profile of seeded second-harmonic efficiency along film plane.

In this respect, an interesting question concerns the effect of accumulated space-charge on the nonlinear optical coefficient. Space-charge fields are indeed known to dominate SHG in seeded optical glass-fibres [*27,28*]. Considering that we get the same nonlinearity as using corona-poling, we can expect internal fields as large as 1 MV/cm, the usual poling field, in the optically poled region. We thus prepared a polymer film using various angles of incidence (Fig. 11). It experimentally results that the SHG signal does not depend on the angle of incidence, proving that space-charge field effects are not the dominant effects with this polymer.

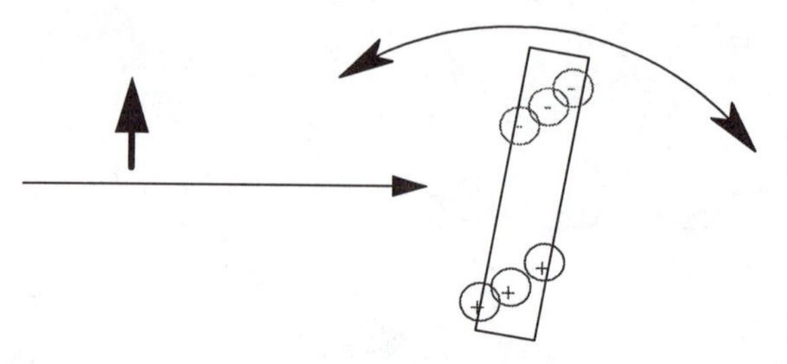

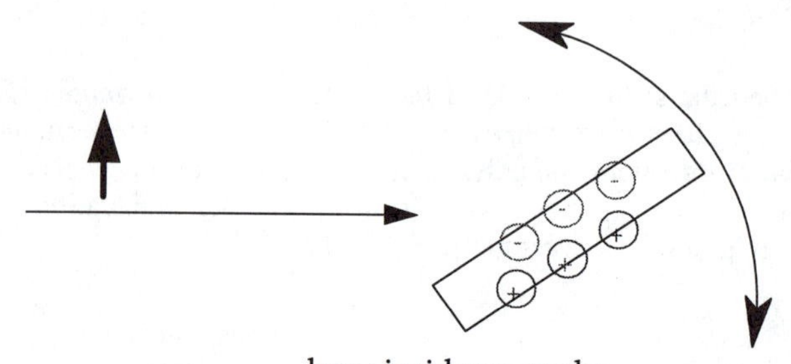

large incidence angle:
Charge separation ~ active-layer thickness (~µm)

Figure 11: Probe of space-charge field effects using different angles of incidence during preparation.

The polarization profile of the induced second-harmonic coefficient in the film-plane is reported in Figure 12. For this measurement, the film was prepared using parallel polarizations for fundamental and harmonic beams. The second-harmonic signal was detected through an analyser parallel to the reading fundamental beam. Figure 12 represents the square-root of the signal as a function of the film angle θ around the laser-beam axis. The fitting function is a cos(θ). That means that we get a good dipolar alignment of molecules with $\chi^{(2)}_{yyy} = 3\chi^{(2)}_{yxx}$ [*19*].

The optically induced $\chi^{(2)}$ is proportional to $E_{2\omega}E^{*2}_{\omega}$. As the fundamental and harmonic beams have a relative phase $\Delta\Phi$, $\chi^{(2)}$ is proportional to $\cos(\Delta\Phi + \Delta k.z)$ where $\Delta k = 2\omega(n(\omega)\text{-}n(2\omega))/c$. The relative phase $\Delta\Phi$ can be controlled by varying the tilt angle θ of a dispersive BK7 glass which is inserted in front of the sample in Figure 5. The optically-induced second harmonic signal in thin films thus depends periodically on this phase, with a 180° period, as reported in figure 13 [29].

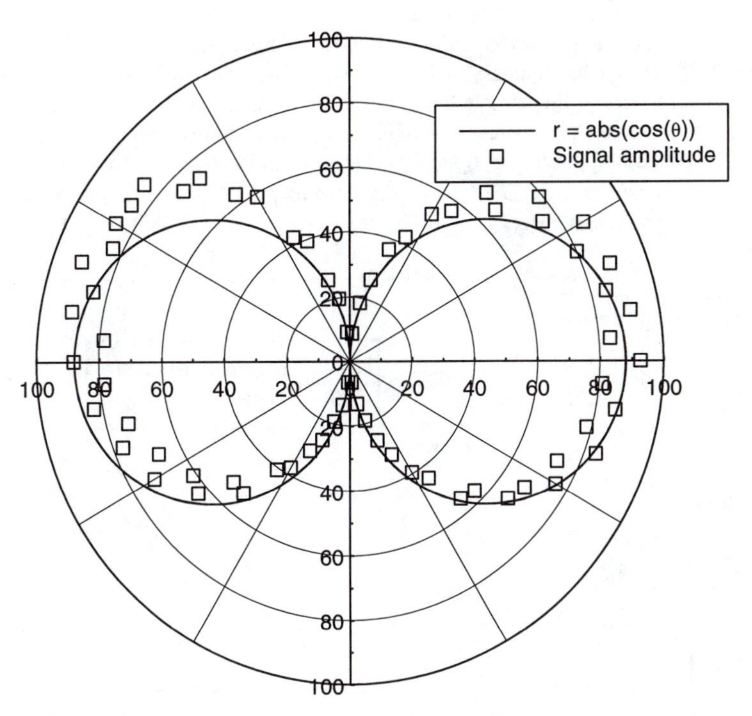

Figure 12: Polarization dependence of the seeded SHG coefficient. Angle θ is measured relative to the polarization axis around the beam-propagation direction.

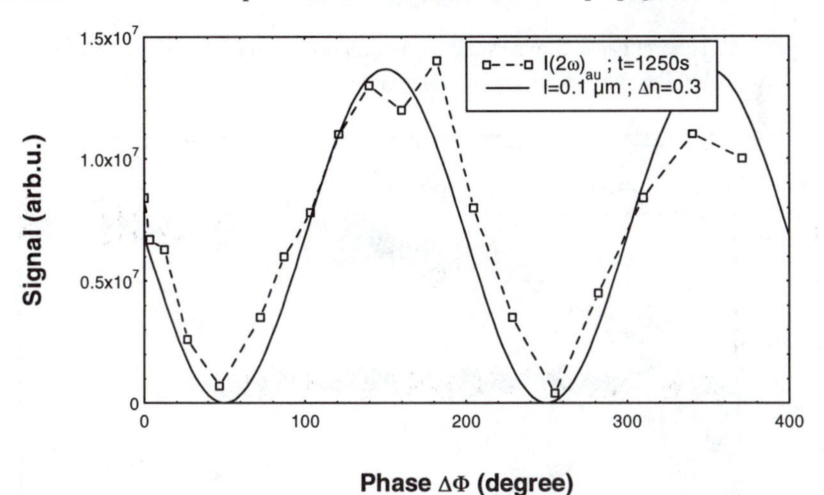

Figure 13: Experimental dependence of the SHG intensity, induced after a 20-minutes preparation-time, with the relative phase $\Delta\Phi$ of the ω and 2ω writing beams. Solid line corresponds to a theoretical dependence with $\Delta n = 0.3$-index mismatch. Sample was 0.1 µm thick with 0.3-optical density at 532 nm.

While the sample-thickness is increased, oscillations with the phase $\Delta\Phi$ are damped [*30*]. It is what appears in Figure 14 in which the contrast of modulation is plotted with respect to the sample thickness. It represents the transition from the SHG in a thin film to the phase-matched SHG in a thick sample. Phase-matched SHG-growth is illustrated with the optical-poling of centimetre-size copolymer rods in Figure 15. In these samples, DR1 loading is close to $3 \cdot 10^{-3}$ % per monomer in order to maintain a low optical density.

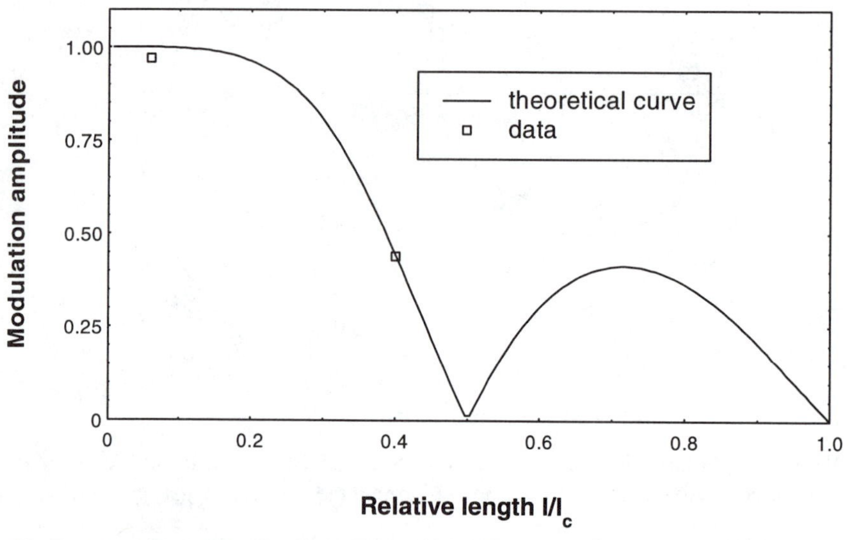

Figure 14: Theoretical and experimental modulation-amplitude of the SHG signal with respect to the sample thickness l relative to the coherence length $l_c = \lambda/2\Delta n$.

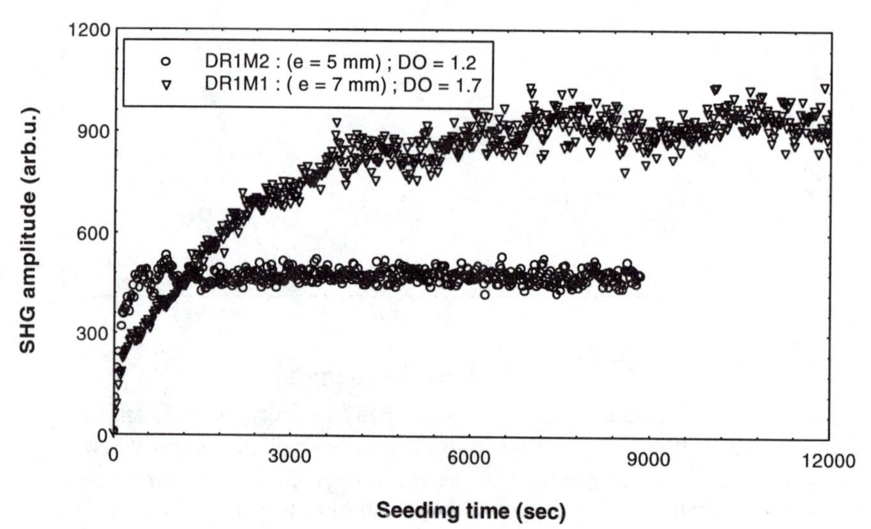

Figure 15: Phase-matched SHG in thick PMMA rods with low DR1 doping-level.

An optimised poling efficiency also requests a good interference between one and two-photon absorption effects, in order to maximise the polar E^3-term. It requests equalisation of one and two-photon absorptions; that is $\frac{1}{2}\varepsilon_0 ncI_{2\omega} = (\delta\mu \times I_\omega/\hbar\omega)^2$. Obviously, the $\chi^{(2)}$-initial growth-rate is proportional to E^3 which is also $I_\omega \times I_{2\omega}{}^{1/2}$. $\chi^{(2)}$ saturates as $I_\omega I_{2\omega}{}^{1/2}/\left(\frac{1}{2}\varepsilon_0 ncI_{2\omega} + (I_\omega \delta\mu/\hbar\omega)^2\right)$, when the poling becomes as efficient as the loss of polar orientation resulting from axial excitation processes [*23*]. Such behaviour is observed in Figure 16 in which the $I_{2\omega}/I_\omega$-ratio was varied while keeping the phase-$\Delta\Phi$ constant. It is clear that optimisation of the optical-preparation conditions can lead to orders of magnitude gain on the poling efficiency.

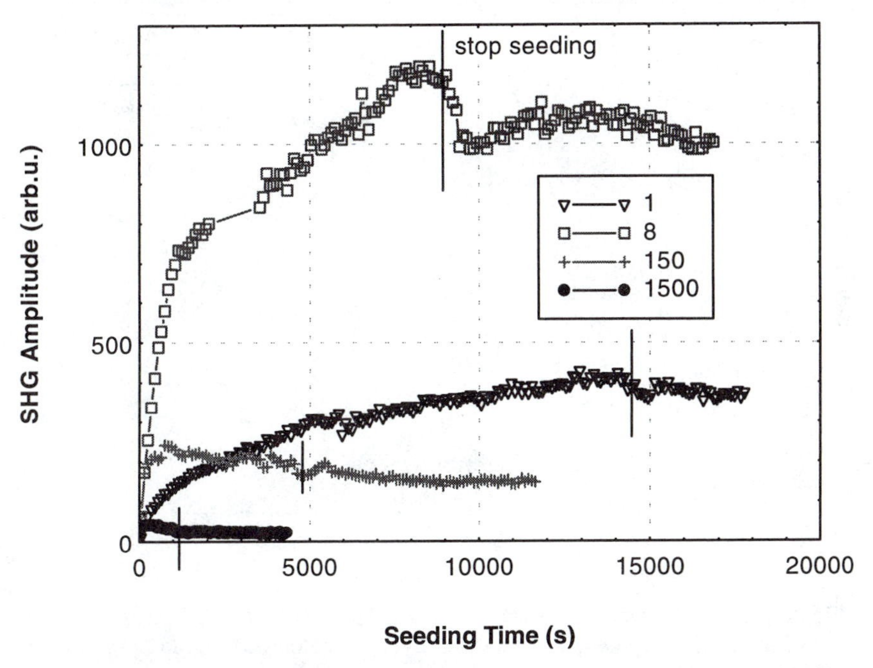

Figure 16: SHG-amplitude growth at different seed intensity-ratios ($I_{2\omega}/I_\omega \times 2000$).

In an orientational hole-burning regime, the all-optical poling technique suffers from the lack of transparency of the doped polymer. A possibility in order to improve on this nonlinearity-transparency trade-off is to use photochromic molecules such as the spiropyran in Figure 17. Such molecule which can be oriented in its coloured merocyanine form which exhibits a large β is transparent in its bleached form [*31*]. The doped PMMA-polymer sample was irradiated with a UV-lamp during the optical preparation. Signal growth and decay are illustrated in Figure 18. Without any optimisation, we get a maximum nonlinearity $\chi^{(2)} \approx 0.6$ pm/V. After preparation, the SHG signal drops as fast as the decoloration process which is illustrated in the inset, down to a quasi-constant ≈ 0.3 pm/V-level. At this stage, the polymer is transparent.

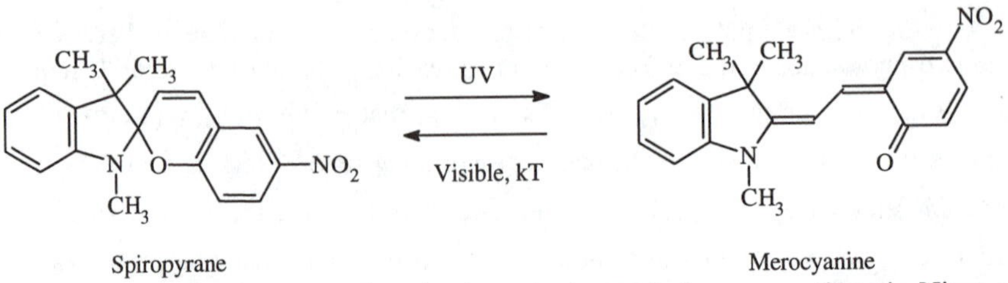

UV

Visible, kT

Spiropyrane Merocyanine

Figure 17: Reversible photochromic ring-opening and closure reactions in Nitro-BIPS spiropyran.

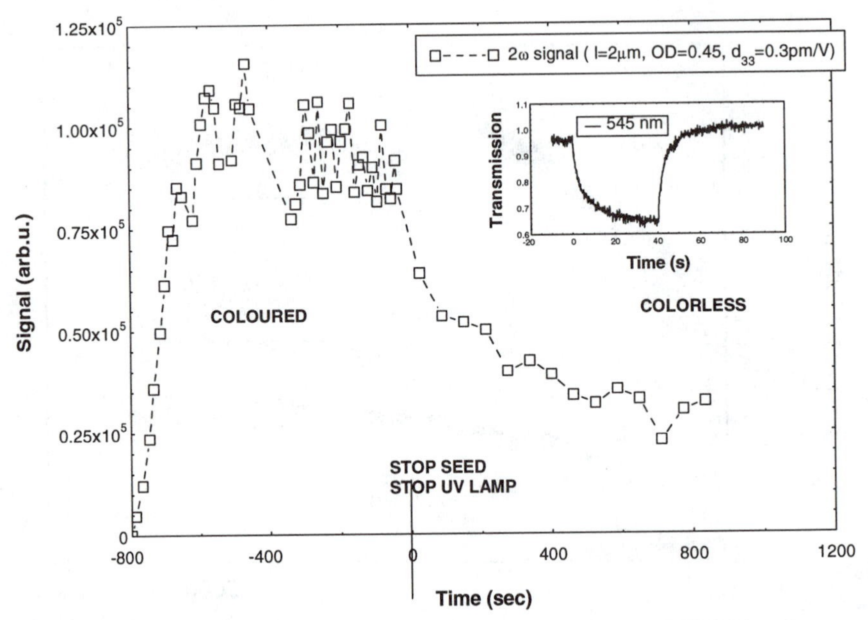

Figure 18: SHG signal growth and decay in a spiropyran doped PMMA-polymer with 2 μm-thickness and 0.45-optical density in the coloured form. Inset represents UV-assisted coloration and bleaching dynamics at 545 nm.

The possibility of inducing a long-lived optical orientation in organic materials opens a door to the microscopic control of organised structures suitable for frequency conversion. The ratio of oriented molecules in seeded plastics is larger than 0.2. The technique compares well in efficiency with the electric-field poling method and as optically-induced orientation and excitation have the same symmetries, polar orientation can in principle be improved with this technique (Table I). The magnitude of the induced $\chi^{(2)}$ is as large as achieved using other poling techniques (Table II), orders of magnitude larger than in seeded optical fiber materials. In addition to the possibility of patterning micro structures with the laser beam, a major breakthrough with this new technique is the natural possibility of inducing a quasi-phase-matched orientation over large distances. Experiments are now in progress in order to improve in this direction, with an aim at orienting molecules in a plastic wave guide. With the use of organic materials, we have the possibilities offered by molecular engineering to tailor their optical properties in order to optimise materials for frequency conversion devices. In addition, the rich tensor properties of E^3 permits the preparation of materials with new symmetries such as the octupoles [*32*].

Table I. Comparison of the physical mechanisms involved into electric-field poling (EFP) and orientational hole-burning (OHB) techniques.

Poling Technique	**EFP**	**OHB**
Nonlinear process	$\chi^{(3)}$	$\chi^{(5)}$
Efficiency	$\mu_0.\beta$	$\delta\mu.\delta\beta$
Orientation effect	E_{DC}	E^3
Excitation effect	kT, E_ω^2	E^3
Relaxation effect	kT	kT

Table II. Nonlinear optical characteristics of some selected electrically poled polymers: *a*) calculated; *b*) see ref.[*20*]; *c*) polymer used in this study; *i*) initial value; *f*) final value after relaxation.

Polymer	**d_{33} (pm/V)**	**r_{33} (pm/V)**	**λ (μm)**	**ref.**
HCC-1232	75	38	1.3	*33*
HCC-1238	18.2	20	1.3	*33*
DANS		28	1.34	*34*
DCV-MMA	46^f	15	1.58	*35*
PPNA	19^f		1.06	*36*
3RDCVXY	80	40	1.5	*37*
(PS)ONPP	18		1.06	*38*
PLBP-73	13		0.63	*39*
PVCN-DR	15-30		1.06	*40*
Bis A-NPDA	14		1.06	*41*
Bis A-DO3	25		1.32	*42*
Bis A-DIAM^b	42^i	26^a	1.32	
Bis A-DIAM^b	28^f	17^a	1.32	
DRGPMMA^{bc}	74^i	45^a	1.32	
DRGPMMA^{bc}	29^f	18^a	1.32	
Polyimide	73	18	1.31	*43*

References

1. Chemla, D.S.; Zyss, J.; *Nonlinear Optical Properties of Molecules and Crystals*; Academic press: Orlando, 1987.
2. Zyss, J.; *Nonlinear Optics,* **1991**, *1*,3.
3. Hornak, L.A.; *Polymers for Lightwave and Integrated Optics;* Marcel Dekker: New York, 1992.
4. Yilmaz, S.; Bauer, S.; Gerhard-Multhaupt, R.; *Appl. Phys. Lett.,* **1994**, *64,* 2770.
5. Moerner, W.E.; *Persistent Spectral Hole-Burning: Science and Applications*; Springer: Berlin, 1988.
6. Baranova, N.B.; Zel'dovitch, B.Ya; *JETP Lett.*, **1987**, *45*, 716.
7. Armstrong, J.A.; Bloembergen, N.; Ducuing, J.; Pershan, P.S.; *Phys. Rev.*, **1970**, *127*, 1918.
8. Osterberg, U.; Margulis, W.; *Opt. Lett.*, **1986,** *11*, 516.

9. Stolen, R.H.; Tom, H.W.K.; *Opt. Lett.*, **1987,** *12*, 585.
10. Charra, F.; Kajzar, F.; Nunzi, J.M.; Raimond, P.; Idiart, E.; *Opt.Lett.*, **1993**, *18*, 941.
11. Charra, F.; Devaux, F.; Nunzi, J.M.; Raimond, P.; *Phys. Rev. Lett.*, **1992,** *68*, 2440.
12. Yeh, P.; *Introduction to Photorefractive Nonlinear Optics;* Wiley: New York, 1993, 377.
13. Khanarian, G.; Norwood, R.A; Hass, D.; Feuer, B.; Karim, D.; *Appl. Phys. Lett*, **1990**, *57*, 977.
14. Nunzi, J.M.; Grec, D.; *J. Appl. Phys.*, **1987**, *62*, 2198.
15. Nunzi,J.M.; Charra, F.; *Nonlinear Optics,* **1991**, *1*, 19.
16. Charra, F.; Nunzi, J.M.; *J. Opt. Soc Am. B*, **1991,** *8*, 570.
17. Nunzi, J.M.; Charra, F.; Fiorini, C.; Zyss, J.; *Chem. Phys. Lett.,* **1994**, *219*, 349.
18. Markovitsi, D.; et al.; *Chem. Phys.,* **1994**, *69*, 219.
19. C.Fiorini, F.Charra, J.M.Nunzi, *J. Opt. Soc Am. B*, **1995**, *12* (1), in press.
20. Chollet, P.A.; Gadret, G.; Kajzar, F.; Raimond, P.; Zagorska, M.; *SPIE Procs.*, **1992,** *1775,* 121.
21. Broussoux, D.; et al.; *Rev. Tech. Thomson-CSF*, **1989,** *20-21*, 151.
22. Jones, C.; Day, S.; *Nature*, **1991,** *351,* 15.
23. Sekkat, Z.; Dumont, M.; *Appl. Phys. B*, **1992,** *54*, 486.
24. Dianov, E.M.; et al; *Sov. Lightwave Com.*, **1992,** *2,* 83.
25. Dominic, V.; Feinberg, J.; *Phys. Rev. Lett.*, **1993,** *71*, 3446.
26. Driscoll, T.J.; Lawandy, N.; *J. Opt. Soc. Am. B*, **1994,** *11*, 355.
27. Ouellette, F.; *Photoinduced Self-organization Effects in Optical Fibers, SPIE Procs.*; vol.*1516,* 1991.
28. Ouellette, F.; *Photosensitivity and Self-Organization in Optical Fibers and Waveguides; SPIE Procs.*, vol.*2044,* 1993.
29. Fiorini, C.;Charra, F.; Nunzi, J.M.; Raimond, P.; *Nonlinear Optics*, **1994,** *9*, in press.
30. Fiorini, C.;Charra, F.; Nunzi, J.M.; Raimond, P.; *EQEC'94, Amsterdam*, **1994,** paper QThF5.
31. Nakatani, K.; Atassi, Y.; Delair, J.A.; Guglielmetti, R.; *Nonlinear Optics*, **1994,** *8*, 33.
32. Fiorini, C.;Charra, F.; Nunzi, J.M.; Zyss, J.; *CLEO Europe'94*, Amsterdam, **1994,** paper CMK3.
33. Haas, D.; et al.; *SPIE Procs.*, **1990,** *1147*, 222.
34. Mohlmann, G.R.; et al.; *SPIE Procs.*, **1990,** *1147*, 245.
35. Singer, K.D.; et al.; *Appl. Phys. Lett.*, **1988,** *53*, 1800.
36. Eich, M.;et al.; *J. Appl. Phys.*, **1989,** *66*, 2559.
37. Shuto, Y.; et al.; *SPIE Procs.*, **1991,** *1560,* 184.
38. Ye, C.; et al.; *Nonlinear Optical Effects in Organic Polymers*, J.Messier et al. eds., NATO ASI Series E, vol. *162*, Kluwer: Dordrecht, 1989, p.173.
39. Broussoux, D.; et al.; *Proc. Int. Conf. on Fine Chemicals for the Electronic Industry*, Univ. of York, 1990, April 18-20.
40. Mandal, B.K.; et al.; *Makromol. Rapid Commun.*, **1989,** *12*, 63.
41. Reck, B.; et al.; *SPIE Procs.*, **1990,** *1147*, 74.
42. Gadret, G.; et al.; *SPIE Procs.*, **1991,** *1560*, 226.
43. Zysset, B.;et al.; *SPIE Procs.*, **1993,** *2025*, 70.

RECEIVED February 14, 1995

Chapter 19

Photo-Induced Poling of Polar Azo Dyes in Polymeric Films

Toward Second-Order Applications

Z. Sekkat[1], E. F. Aust[1], and W. Knoll[1,2]

[1]Max-Planck-Institut für Polymerforschung, Ackermannweg 10, 55021 Mainz, Germany
[2]Frontier Research Program, Institute of Physical and Chemical Research (RIKEN), Wako, Saitama, 351-01 Japan

The application of a DC field across a polymer film containing polar azo dye chromophores, at a temperature far below that of its glass transition, leads to an appreciable polar order when the azo dyes undergo cis⇔trans isomerization. This paper reports on room temperature photo-electro poling or photo-induced poling (PIP) of Disperse Red 1 (DR1) molecules in different polymeric environments. We compare the effect of PIP when DR1 molecules are chemically linked to flexible or to rigid main chains polymer, and we discuss the ability of the polymer main chains to move with the photoinduced movement of the azo chromophores they are covalently linked to. The evolution of the polar orientation during PIP is probed by the electro-optic Pockels effect using an Attenuated Total Reflection (ATR) setup. The experimental findings are discussed within the framework of our phenomenological theory of the PIP process. This is based on the enhanced mobility of the azo chromophres during the isomerization process, and allow a real physical insight into the phenomena which appear on PIP.

Much interest has been focused on organic materials for nonlinear optics since Davydov et al. (*1*) established the correlation between enhanced nonlinear activity and charge transfer character in conjugated molecules. This interest was driven by needs in the field of optical signal processing and telecommunications (*2*). For second order processes, e.g. second harmonic generation and electro-optical modulation, films of poled molecular polymeric materials take advantage of the high optical quality of polymer glasses and allow the use of organic molecules presenting high molecular optical response (hyperpolarizability) (*3*). However, centrosymmetry or misorientation of the nonlinear molecules may reduce or cancel the second order coefficient of the bulk material.

Orienting polar molecules with a large microscopic second order polarizability within the polymer host, is therefore a necessary step for creating second order susceptibility. This is usually performed by DC electric field poling, in which the temperature of the polymer is elevated above its glass transition temperature so that the nonlinear molecular units are free to rotate in the presence of an external DC electric field (*4*). On cooling through the phase transition to room temperature, the orientational order of the molecules is preserved after removing the orienting field. Poling at room

0097–6156/95/0601–0255$12.00/0

temperature using CO_2 as a swelling agent was reported by Barry and Soane (*5*). A new method of poling polymeric films containing photoisomerizable polar azo dyes at room temperature, was reported recently (*6-10*). By applying a DC electric field at room temperature during the photoisomerization process, the centrosymmetry is broken and an important and stable electro-optic Pockels and second-harmonic signals have been obtained using a polymer film of poly(methylmethacrylate) (PMMA) functionalized by Disperse-Red 1 (DR1) chromophores (*6,8,9*). The photoisomerization reaction allows the polar azo dye molecules a certain mobility, thus making polar orientation in the direction of the DC poling electric field possible. In contrast to flexible main chain polymers, PIP of rigid main chain polymers would be of a particular interest since it may provide valuable information about the capability of the polymer main chains to hinder the movement of the azo dye chromophores induced by the photoisomerization reaction.

In the next section we recal briefly some aspects of the isomerization reaction in polymeric environment. The third section reports on PIP and TP of a new rigid main chain polyester and compares the experimental findings with PIP of a PMMA-DR1 copolymer. The evolution of the induced polar order during PIP is probed with the electro-optic Pockels effect obtained by using an ATR setup. The fourth section presents the general equations of our phenomenological theory of the PIP process, and describes the transient properties of the photo-induced polar order. The conclusion and some remarks on the study are given in the last section.

Cis⇔Trans isomerization

The isomerization reaction of azobenzene derivatives is a light or heat induced interconversion of their cis and trans geometric isomers (Figure 1a). The thermal isomerization proceeds generally in the "cis⇒trans" direction because the trans form is generally more stable than the cis form (50 $KJ.mol^{-1}$ in the case of azobenzenes (*11*)), light induces both "trans⇒cis" and "cis⇒trans" isomerizations. When the isomeric forms absorb a photon they are raised to electronically exited states from which a nonradiative decay brings them back to the ground state either in the trans form or in the cis form. Once the molecule is in the cis form it recorvers the trans form either by the back thermal reaction or the inverse photoisomerization cycle. Figure 1b shows a simplified model of the molecular states.

The photoisomerization of the push-pull azobenzene derivative DR1 occurs efficiently in PMMA thin films at room temperature (*12*), the quantum yields of the direct and reverse photoisomerizations have been measured and the absorption spectra for the thermally unstable cis form has been deduced (*12*). As classified by Rau (*11*), the DR1 molecule is a pseudo-stilbene type molecule which means that the high energy π-π^* transition is overlapping the low energy n-π^* transition and leads to a large structureless band in the trans isomer with an absorption maximum strongly dependent on the polarity of the host material which may be either a polymer or a solvent. Frequently the geometrical change of the azo molecules during the isomerization process leads to a loss of their initial orientation, then anisotropy can be induced in a system containing anisometrically shaped photosensitive molecules which are known to align perpendicular to the polarization of the irradiating light. The resulting anisotropy depends on the viscosity of the surrounding medium and can be negligible in low viscosity solutions where the reorientation time of the molecules is very short and allows the thermal agitation to restore instantaneously the isotropy destroyed by the light. In azo dye doped polymeric films where the mobility of the guest molecules is appreciable, photoisomerization leads to reversible polarization holography (*13*). In azo dye functionalized polymeric films where the mobility of the azo chromophores is very much reduced, the photoisomerization creates a permanent alignment which may lead to writing erasing optical memories (*14,15*) or to permanent second order nonlinear-optical effects at room temperature in the presence of a poling DC electric field (*6-10*).

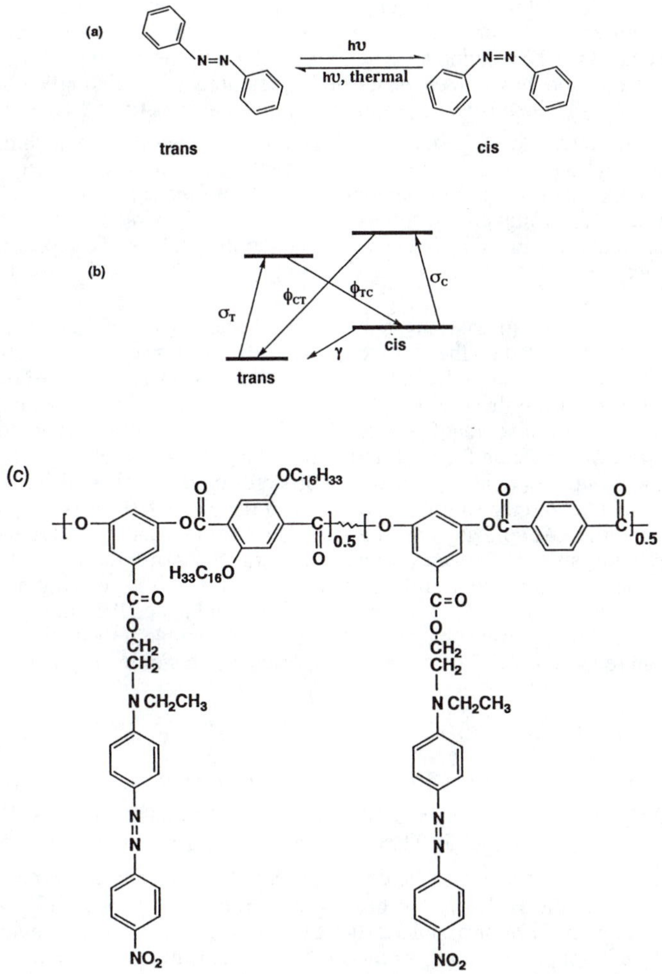

Figure 1. **(a)** Trans⇔cis isomerization of azobenzenes. **(b)** Simplified model of the molecular states. Only two excited states have been represented, but each of them may represent a set of actual levels (*11*) : we only assume that the lifetime of all these levels is very short. σ_t and σ_c are the cross sections for absorption of one photon by a molecule in the trans or the cis state, respectively. γ is the thermal relaxation rate. Φ_{ct} and Φ_{tc} are the quantum yields of photoisomerization. **(c)** structure of the polymer.

Photo-Electro Poling Experimental

The structure of the DR1-polyester copolymer which we have used for this srudy is shown in Figure 1c. The intrinsic viscosity of this polymer obtained from 1,1,2,2-tetrachloroethane (TCE) solution at 25 °C is 0.47 dl/g. Details of the polymer synthesis and characterization have been described elsewhere (*16*). The glass transition temperature (T_g) obtained from differential scanning calorimetry (DSC) at the heating rate of 10 °C/min was 42 °C. This polymer is a random copolymer which means that the terephthaloyl and the 2,5-dihexadecylterephthaloyl comonomers are randomly distributed in the main chain. This random distribution of the comonomers can produce large amount of free volume in the polymer matrix (*17-19*). Moreover the inclusion of the kink resorcinol unit in the main chain may hinder the packing of the polymer chains in solid state (*20*). Both of these reasons may cause the low T_g of the present polymer.

For the electro-optic measurements by the ATR method, a 2 to 3 nm thick layer of chromium was evaporated on the cleaned glass substrate followed by 30 to 50 nm thick layer of gold as a lower electrode. Polymer films (1-2.5μm) were spin-cast on the gold layer. These films were dried at 65 °C for 30 to 50 hours using a vacuum oven to remove the last traces of solvent from the films. Finally, an upper gold layer of 100 nm thickness was evaporated on the polymeric film. The gold layers act as electrodes. The glass slide was put in optical contact with a prism through which a HeNe (632.8 nm) laser was reflected onto the sample. The power of the HeNe laser was reduced to few microwatts in order to avoid any absorption from the sample and our ATR and electro-optic modulation spectra showed perfect shapes (at 632.8 nm the absorption of the DR1 molecules is almost negligible as shown by the UV-VIS aborption spectra in Figure 2, but the waveguiding resonance condition enhances the energy inside the polymer film and may cause some absorption from the sample if the incident intensity is not enough reduced (*21*)). The waveguiding resonance condition is given by:

$$k_{zm}d + \Psi_1 + \Psi_2 = m\pi \qquad (1)$$

where m is the order of the mode, k_{zm} is the normal component of the wavevector of mode m, d is the film thickness, and Ψ_1 and Ψ_2 are the phase shifts at reflection on the inner faces of the waveguide. They are given by Fresnel formulae, and depend on light polarization. The prism and the detector were placed on a Θ-2Θ goniometer and the reflectivity of the HeNe probe beam with Transverse Electric (TE or s) and Transverse Magnetic (TM or p) polarizations was studied for incidence angles ranging from 20° to 82° (see Figure 3a). Some of these fulfill the waveguiding resonance condition (equation 1) and dips occur in the reflectivity corresponding to TM (p) and TE (s) guided modes in the polymer film (Figure 3b). The experimental reflectivity curves are fitted by a computer programm which calculates theoretical reflectivities of layered media and gives, with a very good accuracy, the thickness and the indices of refraction n_x, n_y and n_z in the three principal directions of the polymer film. x and y correspond to the principal directions in the plane of the film and z is the normal to the sample.

When an external perturbation is applied to the polymer film, the ATR guided modes shift their angular positions and the reflectivity is modulated (Figure 3c). These angular shifts are very small in the case of electro-optic experiments: they correspond to refractive index variations of the order of 10^{-5}. One has then to modulate the measuring electric field at a low frequency Ω ($E = E_1\cos\Omega t$) and to detect the modulated signal with lock-in amplifiers. The lock-in signals detected at the modulation

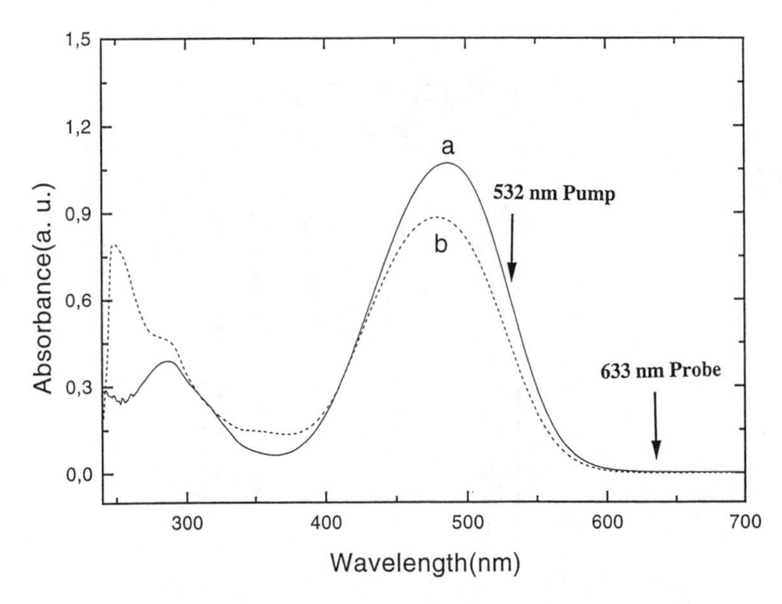

Figure 2. UV-Visible spectra of **(a)** DR1 compound and **(b)** polymer in TCE solution.

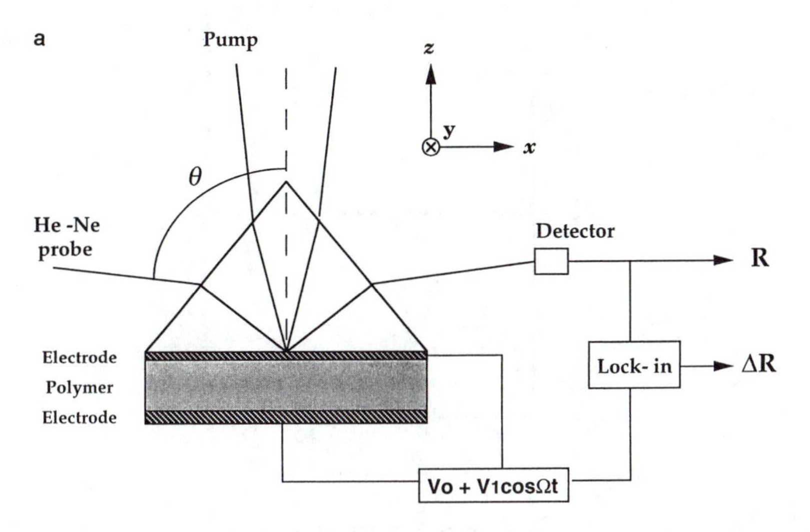

Figure 3. **(a)** The ATR experimental setup and **(b)** an example of the record of the reflectivity and **(c)** electro-optic modulation

Continued on next page

b

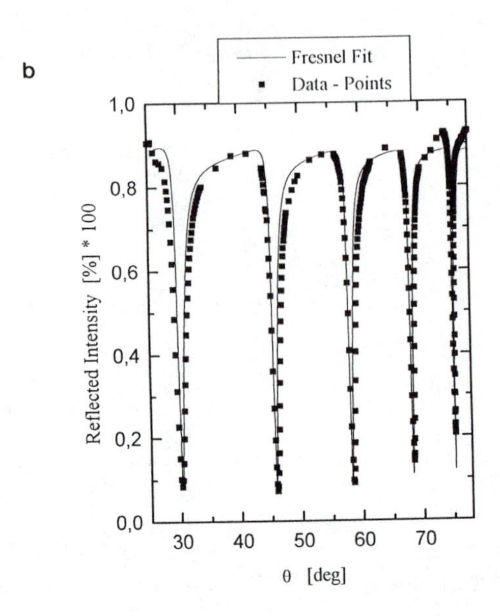

c

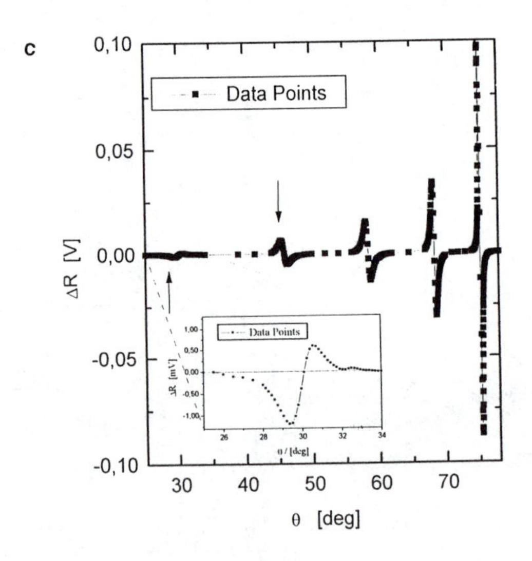

Figure 3. *Continued*

frequency and its second harmonic give respectively the linear (or Pockels) and the quadratic (or Kerr) electro-optic effects. The amplitude of the modulation of the thickness and the refractive indices is evaluated by a computer fit, and allows the determination of Pockels (r) and Kerr (s) coefficients. These are related to the electro-optic susceptibilities by the following formulae (equations 2):

$$2\chi_{iiz}{}^{(2)} = - r_{iz}\, n_i^4 = 2n_i\Delta n_i(\Omega)\, (d\,/\,V_1) \tag{2a}$$

$$3\chi_{iizz}{}^{(3)} = - s_{iz}\, n_i^4 = 4n_i\Delta n_i(2\Omega)\, (d\,/\,V_1)^2 \tag{2b}$$

The factors 2 and 3 in the left hand side of, respectively, equations 2a and 2b refer to the number of permutations of the electric fields in the definition of the macroscopic polarizability when Kleinmann symmetry holds true. Sometimes these factors are included in the definitions of the nonlinear susceptibilities (*2*). The factor 4 in the right hand side of equation 2b originates from $\cos^2\Omega t$ (see also equation 3) assuming the general definition for the electro-optic Kerr susceptibility (e.g. $3\chi^{(3)}E_1{}^2 = 2n\Delta n$). V_1 is the voltage corresponding to E_1, and $\Delta n_i(\Omega)$ and $\Delta n_i(2\Omega)$ are the index variations measured from the Ω and 2Ω lock-in modulated signals, respectively and i refers to x, y and z. The piezoelectric effect and the electrostriction are estimated from the thickness modulation (*22*), and their contribution to the index variations is almost negligible in our material.

For PIP experiments the sample is irradiated by the green light of a frequency doubled laser diode (532nm, at this wavelenght the absorption cross section allows a good penetration in the sample) through the prism and the gold layer as shown in Figure 3a. The power of the pump beam incident on the sample is estimated to be 70 mW/cm^2, taking into account the reflection on the gold sample. The variations of $\chi^{(2)}$ during PIP are followed by setting the goniometer at a fixed angle of incidence corresponding to the TM mode at high incidence angles, which provides a high sensitivity to the n_z refractive index variations, and recording the signal modulated at Ω from the lock-in amplifier. When the sample is irradiated the mode shifts its angular position and must be followed manually. This is not a very precise procedure but it allows valuable information about the dynamics of the PIP process to be obtained. The recording of the whole set of TM and TE modes at a stationary state of the pumping or of the relaxation process gives a precise measurement of the electro-optic susceptibilities and the higher the number of the modes the better the precision (*23*).

Figures 4 and 5 show the evolution of the electro-optic Pockels coefficient r_{33} of our polymeric film (thickness ~ 1.8 μm) during typical PIP cycles at room temperature. These curves were obtained as explained before by recording continuously the lock-in signal modulated at Ω near the largest incidence TM mode which gives the direct measurment of Δn_Z. When the measuring AC (E_1) and poling DC (E_0) electric fields are applied simultaneously to the polymer film, the Δn_z (z is equivalent to 3 which refers to the symmetry axis of our system) variation is given by (equation 3):

$$\begin{aligned} 2n_z\Delta n_z(\Omega) &= 2\chi_{333}{}^{(2)}\,(E_0 + E_1\cos\Omega t) + 3\chi_{3333}{}^{(3)}\,(E_0 + E_1\cos\Omega t)^2. \\ &= (2\chi_{333}{}^{(2)} + 6\chi_{3333}{}^{(3)}\,E_0)\,E_1\cos\Omega t \\ &\quad + (3/2)\chi_{3333}{}^{(3)}\,E_1{}^2\cos 2\Omega t + \text{unmodulated terms.} \end{aligned} \tag{3}$$

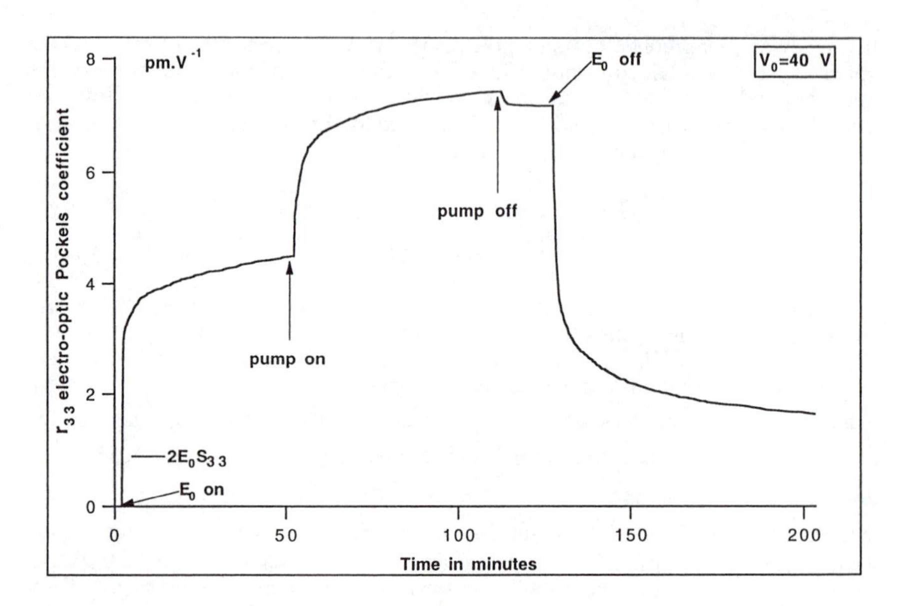

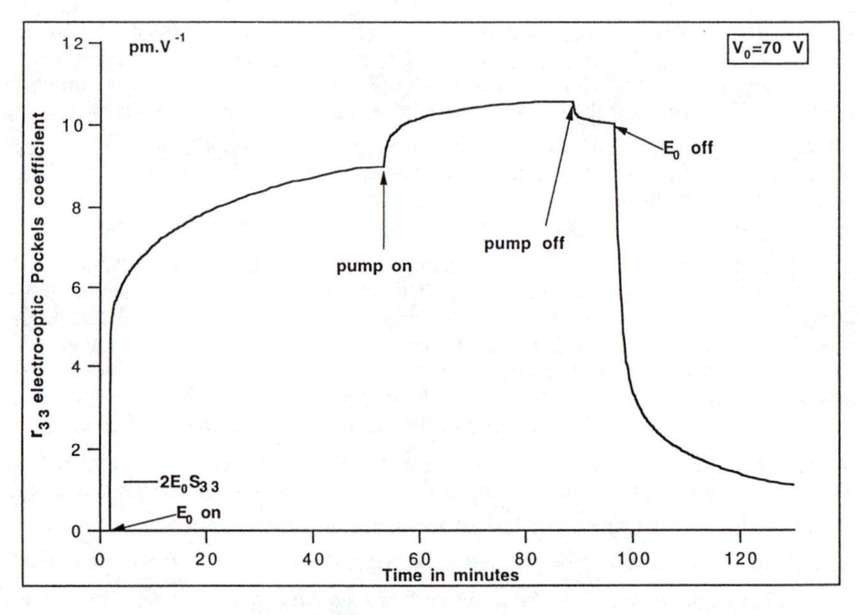

Figures 4-5. PIP of a DR1-polyester 1.8 μm thick film, with **(4)** 40V and **(5)** 70V as a poling voltage. The moments of turning this latter and the pumping light (70 mW/cm^2) on and off are indicated by arrows. The horizontal line gives the estimated value of the Kerr effect contribution ($2E_0S_{33}$) to the apparent Pockels effect. The signal above this line characterizes a pure polar orientation of the azo chromophores.

In this case, the apparent r_{33} electro-optic coefficient given by the signal modulated at Ω contains a contribution from the electro-optic Kerr effect given by $\chi^{(3)}$. This can be measured easily using the signal modulated at 2Ω. At the time t=0 where the sample is centrosymmetric ($\chi^{(2)} = 0$), the DC poling field E_0 is applied; then the Pockels effect appears rapidly and grows slowly in the long time range. The contribution of the Kerr effect $2E_0S_{33}$ ($3\chi_{3333}^{(3)} = 110.10^{-21}$ m^2 V^{-2}) to the apparent Pockels coefficient is indicated by the horizontal line in Figures 4 and 5, and shows the existence of a large and appreciable pure $\chi^{(2)}$ signal above this line resulting from a polar orientation of the nonlinear-optical azo chromophores in the direction of the poling DC field. This is the Electric Field Induced Pockels Effect (EFIPE) which is the analogue of the Electric Field Induced Second Harmonic (EFISH). This indicates that the mobility of the majority of the chromophores is significant even at room temperature ("free molecules"), since they can rotate under the action of the DC poling field alone without heating. The onset and the decay of the EFIPE is fast in the short time range and indicates that the mobility of the "free molecules" is significant at room temperature. The slow effect observed at long times originates from molecules situated in a more viscous environment ("non free" molecules). When the sample is illuminated with a pump beam, the electro-optic Pockels signal increases strongly. When the pump beam is switched off in the presence of the DC field, the $\chi^{(2)}$ signal decreases slightly and keeps an r_{33} value higher than that obtained with the DC field alone. When this is removed the electro-optic Pockels signal drops rapidly but keeps an appreciable r_{33} value corresponding to 1.65 pm.V^{-1}. Using the theoretical expression of the first odd order parameter A_1 at the steady-state of the EFIPE, one can compare the experimetal results with Maxwell-Boltzmann statistics (*24*).

The PIP cycle performed on our copolymer shows a quite similar behavior to that obtained with DR1 doped PMMA thin films (*7*) (see Figure 6) and shows that the photoisomerization process allows a certain mobility to the "non free" molecules since they can experience a polar orientation with the simultaneous application of the DC poling field. This can be seen as an enhancement of the mobility of the azo chromophores during the isomerization reaction and can be described theoretically as we will discuss later. Another important point in these experiments is that the photo-induced polar order obtained with 70V as a poling voltage is at least two times smaller than that obtained with 40V as a poling voltage. This indicates that a certain fraction of the azo chromophores which can not rotate under 40V can rotate under the action of the 70V poling voltage, and demonstrates that the isomerization process lowers the threshold for the viscous polar orientation in the direction of the poling field. This change of the apparent viscosity during the photoisomerization reaction can be introduced formally in a phenomenological theory as we will show in the next Section.

Next we discuss the thermal poling results and compare them with the PIP experiments. We poled a 2.4 μm thick film of the copolymer following the TP procedure at 60 ^{o}C with 79V/μm poling voltage. The r_{33} electro-optic Pockel coefficient obtained just after the TP was 15.5 pm.V^{-1} and is much higher than the r_{33} coefficient obtained just after PIP when the DC field is removed (1.65 pm.V^{-1}). The relaxation process after TP shows the behavior reported in Figure 7 where we can see that 30 minutes after removing the poling DC field, only 10% of the initial signal obtained just after removing the DC field (15.5 pm.V^{-1}) disappears. In the case of PIP, the majority of the signal drops rapidly when the DC field is removed. Keeping in mind that the copolymer used in this study has rigid main chains, the data suggests that the thermal poling process performed above T_g of the polymer affects the conformation of the main chains and thus can change their orientation, but the PIP affected only the orientation of the azo chromophores through the free volume of the polymer matrix

without an appreciable effect on the main chain motion. This may originate from the rigidity of the polymer main chains which hinder the photoinduced motion of the azo dye chromophores. This is in contrast to the results we obtained with a PMMA-DR1 copolymer where the PIP order was stable over several months (*8*) (see also Figure 6) because the PMMA main chains are flexible and do not hinder the photoinduced motion of the covalently linked azo chromophores. This may originate from a very much reduced free volume in the DR1-PMMA copolymer in which the DR1 chromophores can only produce a change in the conformation of the polymer main chains during the photoisomerization reaction in order to experience an alignment. The free volume distribution seems to be very important for producing a stable polar order either by TP or by PIP, as we can remark from the above behavior of the present DR1-polyester. These results suggest that further PIP and TP investigations in a high-Tg rigid main-chain polymer should be considered.

Theory of Photoisomerization Induced Polar Order

It has been shown in PIP experiments that on applying a DC electric field at room temperature during the photoisomerization process, centrosymmetry was broken and an important and stable $\chi^{(2)}$ signal (EOPE and SHG) was obtained using a DR1-PMMA copolymer (*6,8,15*) (see Figure 6). This implies that the polymer chains are flexible and do not hinder the movement of the covalently linked pendent DR1 group. This conformational change in the polymer chain is induced by the photoisomerization reaction and allows the polar azo dye molecules a certain mobility, which makes possible polar orientation in the direction of the DC poling electric field. This can be seen as a change in the local viscosity of the molecules during the cis⇔trans isomerization reaction, since the poling field alone does not induce any orientation at room temperature, consequently the viscosity seen by the nonlinear-optical molecules is very high and there is nearly no spontaneous angular relaxation at room temperature. The free volume distribution in polymeric films leads to different magnitude of mobility for chromophores situated at different polymer sites. In PMMA doped with DR1, photoisomerization has been shown to allow a certain mobility to the "non free" chromophores situated in an environment with an appreciable viscosity (*7*) (Figure 6), and in a DR1-polyester copolymer photoisomerization has been shown to lower the threshold for the viscous polar orientation in the direction of the poling field (*24*) (Figures 4 and 5).

In the basis of these observations we developpe a phenomenological theory of photo-induced poling. This assumes that the mobility of the molecule in the trans configuration is very low, and when it absorbs a photon and undergoes a trans⇔cis photoisomerization, the mobility of the molecule in a hypothetical excited state is high enough to allow a certain polar orientation in the direction of the DC poling field during the lifetime of this state. This could be assigned with the cis lifetime because the lifetimes of all the other excited states are very short in comparison (*11*). Once the mobility of the chromophores has been enhanced, polar orientation occurs naturally by rotational diffusion in the presence of a polar external torque exerted by the DC field. Polar orientation of the trans state is then possible via the cis⇒trans back thermal isomerization or optical pumping, provided that the memory of the polar orientation in the cis state is not completely lost during these back reactions. Subsequently, an efficient polar orientation may be built after successive cycles of trans⇔cis photoisomerization as shown by the experiments (*6-10*) (Figures 4-6). One can imagine that polar orientation may be achieved during the transitions, but the relative magnitude of the reorientation time and the time of the optical transition seems to rule this hypothesis. The theory we develop here reproduces exactly the steps of the photoisomerization induced poling and allows a real physical insight into the observed experimental facts.

The phenomenological theory of photo-induced poling combines the usual thermal poling theory and general equtions which describe the photoisomerization and photo-induced reorientation (*6,9*). When azo molecules are optically pumped by a polarized light beam, the probability of pumping is proportional to the square of the cosine of the angle θ between the transition dipole moment of the molecule and the pumping light direction of polarization (angular hole burning). Thus the dye molecules which have their transition dipole moment parallel to the direction of polarization of the pump beam present the highest probability for pumping and undergoing a trans⇒cis isomerization. This changes the orientation of the molecules (angular redistribution). When the molecules are in the cis state they posses sufficient mobility for alignment in the direction of the DC poling electric field. This creates a polar order and an anisotopy and can be described by the rotational Brownian motion accounting for the thermal agitation in the presence of a polar external torque exerted by a DC poling electric field.

To derive the time dependent expression of photo-induced poling, we consider the elementary contribution to the polar order made by the fraction of the molecules $dn(\Omega)$ whose representative moment of transition are present in an elementary solid angle $d\Omega$ around the direction $\Omega(\theta,\varphi)$ relative to the different axis of the system. This elementary variation results from the angular hole burning, angular redistribution and rotational diffusion in the presence of an external torque exerted by a poling electric field along the direction 3. We assume that the molecules are anisotropically shaped in both trans and cis configurations, then if the pump light beam is linearly polarized in the 3 direction, the elementary variation is given by :

$$\begin{aligned}
\frac{dn_t(\Omega)}{dt} =& -I\phi_{tc}\left[\sigma_\perp^t + \left(\sigma_{//}^t - \sigma_\perp^t\right)\cos^2\theta\right] n_t(\Omega) \\
&+ I\phi_{ct}\iint n_c(\Omega')\left[\sigma_\perp^c + \left(\sigma_{//}^c - \sigma_\perp^c\right)\cos^2\theta'\right] P^{ct}(\Omega' \to \Omega)\ d\Omega' \\
&+ \frac{1}{\tau_c}\iint Q(\Omega' \to \Omega) n_c(\Omega')\ d\Omega' \\
&+ D_t\mathbf{R}.\left\{\mathbf{R}n_t(\Omega,t) + n_t(\Omega,t)\mathbf{R}U_t/kT\right\} \\
\frac{dn_c(\Omega)}{dt} =& -I\phi_{ct}\left[\sigma_\perp^c + \left(\sigma_{//}^c - \sigma_\perp^c\right)\cos^2\theta\right] n_c(\Omega) - \frac{1}{\tau_c} n_c(\Omega) \\
&+ I\phi_{tc}\iint n_t(\Omega')\left[\sigma_\perp^t + \left(\sigma_{//}^t - \sigma_\perp^t\right)\cos^2\theta'\right] P^{tc}(\Omega' \to \Omega)\ d\Omega' \\
&+ D_c\mathbf{R}.\left\{\mathbf{R}n_c(\Omega,t) + n_c(\Omega,t)\mathbf{R}U_c/kT\right\}
\end{aligned} \tag{4}$$

$P^{tc}(\Omega' \to \Omega)$, $P^{ct}(\Omega' \to \Omega)$ and $Q(\Omega' \to \Omega)$ are the probabilities that molecule will rotate in the "redistribution" process of trans⇒cis and cis⇒trans optical transition and cis⇒trans thermal recovery respectively. The "angular hole burning" is represented by a probability proportional to $\cos^2\theta$. The last terms in the right hand side of eqs.4 describe the rotational diffusion due to Brownian motion under the action of an external torque exerted by a poling electric field. This is a Smoloshowski equation for the rotational diffusion characterized by a constant of diffusion D_c (D_t) for the cis (trans) configuration (*25*). This is given by the following Einstein relation.

$$D = \frac{kT}{\xi} \tag{5}$$

where ξ is the friction coefficient which depends on the viscosity of the polymer and the molecular shape (25). k is the boltzmann constant and T is the absolute temperature. $\boldsymbol{R}$ is the rotational operator.

$$\boldsymbol{R} \equiv \mathbf{v} \times \nabla \tag{6}$$

where ∇ is the nabla operator and $\mathbf{v}$ is the unit vector parallel to the long molecular axis which, in turn, is parallel to the permanent dipole moment of the molecule. As a first approximation, the interaction energy U is given by (26):

$$U_{t,c} = -\ \mu_{t,c} E \cos\theta \tag{7a}$$

where μ_t and μ_c are, respectively, the permanent dipole moments of the trans and cis forms.

If the pump light is unpolarized or circularly polarized in the {1, 2} laboratory plane, we change $\cos^2\theta$ to $(\sin^2\theta)/2$ in eqs.4 and we can introduce a parameter e characterizing the polarization state of the pumping light. Indeed the expression between the square brackets can be written in the form

$$\sigma^{t,c}\left[1 + e\imath^{t,c}\left(3\cos^2\theta - 1\right)\right],$$

where (7b)

$$\imath^{t,c} = \left(\sigma_{//}^{t,c} - \sigma_{\perp}^{t,c}\right) / \left(\sigma_{//}^{t,c} + 2\sigma_{\perp}^{t,c}\right) \quad \text{and} \quad \sigma^{t,c} = \left(\sigma_{//}^{t,c} + 2\sigma_{\perp}^{t,c}\right) / 3$$

represents the molecular anisotropy and the isotropic absorption cross section, respectively. $\sigma_{//}^{t,c}$ and $\sigma_{\perp}^{t,c}$ are the absorption cross sections in the parallel and perpendicular directions of the trans and cis long molecular axes respectively. e = 1 corresponds to linearly polarized light in direction 3 and e = - 1/2 corresponds to an unpolarized light or to a circularly polarized light in the {1, 2} laboratory plane. I is the intensity of the pumping light expressed in flux of photons per square of centimeter. $n_t(\Omega)d\Omega$ and $n_c(\Omega)d\Omega$ are the number of trans and cis molecules whose representative moment of transition along the long axis is present in the elementary solid angle $d\Omega$ around the direction $\Omega(\theta, \varphi)$ respectively. τ_c is the cis lifetime. The normalizations are :

$$\iint n_t(\Omega)\ d\Omega = N_t \qquad \iint n_c(\Omega)\ d\Omega = N_c$$

$$N_t + N_c = N$$

$$\iint P^{ct,tc}(\Omega' \to \Omega)\ d\Omega' = 1 \qquad \iint Q(\Omega' \to \Omega)\ d\Omega' = 1 \tag{8}$$

where N_t and N_c are the molecular densities at the trans and cis states respectively, and N is the total molecular density.

Considering that the molecular distribution is axially symmetric, the symmetry axis is defined by the direction of the DC field, it is obvious that the Legendre polynomials are the eigen functions for the system (27). The trans and cis populations can then be written in the following form :

$$n_t(\Omega)=\frac{1}{2\pi}\sum_{n=0}^{\infty}\frac{2n+1}{2}\;T_nP_n(\cos\theta);\; n_c(\Omega)=\frac{1}{2\pi}\sum_{k=0}^{\infty}\frac{2k+1}{2}\;C_kP_k(\cos\theta) \tag{9}$$

As the cis and trans populations change with the pump intensity, it is convenient to define non normalized expansion coefficients T_n and C_n for Legendre polynomial $P_n(\cos\theta)$. These are given by :

$$T_n=\int_0^{\pi} n_t(\theta)P_n(cos\theta)\;sin\theta\,d\theta \quad and \quad C_n=\int_0^{\pi} n_c(\theta)P_n(cos\theta)\;sin\theta\,d\theta \tag{10}$$

The redistribution processes $P^{tc}(\Omega'\rightarrow\Omega)$; $P^{ct}(\Omega'\rightarrow\Omega)$ and $Q(\Omega'\rightarrow\Omega)$ depend only on the rotation angle χ between the directions Ω and Ω', therfore we can expand them in terms of Legendre polynomials $P_n(\cos\chi)$:

$$P^{ct}(\chi)=\frac{1}{2\pi}\sum_{q=0}^{\infty}\frac{2q+1}{2}\;P_q^{ct}P_q(\cos\chi) \qquad P^{tc}(\chi)=\frac{1}{2\pi}\sum_{q=0}^{\infty}\frac{2q+1}{2}\;P_q^{tc}P_q(\cos\chi)$$

$$Q(\chi)=\frac{1}{2\pi}\sum_{m=0}^{\infty}\frac{2m+1}{2}\;Q_mP_m(\cos\chi), \tag{11}$$

with these definitions : $T_0=N_t$; $C_0=N_c$ and $P_0^{tc}=P_0^{ct}=Q_0=1$ (12).

The general formalism given in equations (4) can be simplified when Legendre formalism is used, and the variations of the cis and trans populations are given by the variations of their expansion parameters T_n and C_n respectively. These variations are given by the system of equations 13.

$$\begin{cases} \dfrac{dT_n}{dt}=-I_t\{T\}+I_cP_n^{ct}\{C\}+\gamma_0Q_nC_n \\ \qquad +n(n+1)D_t\left[-T_n+\dfrac{u_t}{2n+1}(T_{n-1}-T_{n+1})\right] \\ \dfrac{dC_n}{dt}=I_tP_n^{tc}\{T\}-I_c\{C\}-\gamma_0C_n \\ \qquad +n(n+1)D_c\left[-C_n+\dfrac{u_c}{2n+1}(C_{n-1}-C_{n+1})\right] \end{cases} \tag{13}$$

where: $\{T\}=\left\{3er^t\kappa_{n+}.T_{n+2}+\left[1+er^t(3\kappa_n-1)\right].T_n+3er^t\kappa_{n-}.T_{n-2}\right\}$

$$\{C\}=\left\{3er^c\kappa_{n+}.C_{n+2}+\left[1+er^c(3\kappa_n-1)\right].C_n+3er^c\kappa_{n-}.C_{n-2}\right\}$$

with $\kappa_{n+} = \frac{(n+1)(n+2)}{(2n+1)(2n+3)}$; $\kappa_n = \frac{2n^2+2n-1}{(2n-1)(2n+3)}$ and $\kappa_{n-} = \frac{n(n-1)}{(2n-1)(2n+1)}$;

$\gamma_0 = 1/\tau_c$ and $u_{t,c} = \mu_{t,c} E/kT$; $I_{t,c} = \overline{\sigma}^{t,c} I$.

We see that the temporal behaviour of the nth order is related to the temporal behaviour of the (n - 2)th, (n + 2)th, (n + 1)th and (n - 1)th orders. This demonstrates that the anisotropy given by the even orders and the polar orientation given by the odd orders are coupled. This anisotropy results from the pumping which aligns the molecules perpendicularly to the pump beam polarization, and the poling which aligns them in the direction of the poling DC electric field.

A general analytic solution of the system of eqs (13) can not be found, but one can give an approximate numerical solution by truncating the system above any arbitrary order n and diagonalizing the corresponding matrix by numerical methods. The solution for each order n would be the sum of an infinite number of exponentials with time constants which depend on the pump intensity, the viscosity seen by both the cis and trans configurations and the strength of the poling electric field. In practice (*22*) $r_{33}/r_{13} \approx 3.3$ which consequently gives $A_3 \approx 0.05 A_1$. For analytically solving the system, it is then a good approximation to neglect the expansion parameters above the order 3. We did this for solving the system when the pump beam is switched off at the steady-state of PIP, in the presence of the poling DC field. For the onset of the PIP, we analytically solved the system, by expanding the orientational distribution function up to the first order as a first approximation, and by assuming that the trans⇒cis direct isomerization losses total memory of the initial orientation ($P_n^{tc} = 0$).

The tensorial components of the second order susceptibility of the whole molecular distribution are given by :

$$\chi^{(2)}_{333}(t) = 3\chi^{(2)}_{113}(t) = \frac{3}{5}\left\{\beta^{t,*}_{zzz} T_1(t) + \beta^{c,*}_{zzz} C_1(t)\right\} \tag{14}$$

where $\beta^{c,*}_{zzz}$ and $\beta^{t,*}_{zzz}$ are the molecular second order hyperpolarizabilities of the cis and trans azo molecules along their z-principal axis respectively (* indicate that local field effects have been included in the expressions of the hyperpolarizabilities). When the trans form isomerize to the cis form, the molecular hyperpolarizability β decreases. For the DR1 molecule, a semi-empirical MNDO calculation of β by a finite field method (*2, 25*) gave $\beta^0 = 44.6 \times 10^{-30} esu$ for the trans form and $\beta^0 = 8.4 \times 10^{-30} esu$ for the cis form (*12*). The ground state dipole moments were found to be 10.1 Debye and 5.2 debye for the trans and cis forms respectively (*12*). Considering this, we can simulate the temporal behaviour of the polar order in the cis (C_1) and trans (T_1) distributions and second order susceptibility normalized by the trans molecular hyperpolarizability ($\chi^{(2)}/\beta^{t,*}_{zzz}$). This is shown by a dashed line on the figures.

Recently, we have investigated in detail the steady-state and transients of the photo-induced polar order and its related anisotropy in both trans and cis molecular distributions (*28*). The effect of all the physical parameters involved in the PIP phenomena has been considered. These include molecular anisotropy, pump intensity, pump polarization, strength of the poling field, the molecular mobility and retention of memory of the molecular orientation. Here we show that PIP occurs only if the mobility of the azo chromophores is enhanced during the isomerization process, and we discuss the effect of retainment of molecular orientation memory during the cis⇒trans thermal back reaction. In Figure 8, the DC field is turned on at a negative time, and at the time t=0 the pumping light is switched on. This because the polar order remains zero when the DC field alone is switched on if the trans molecular mobility is very much reduced ($\tau_D^t = 10^4$). The numerical values of the physical parameters appear on the figures in a box in the following order $\left(\tau_D^c, \tau_D^t, \tau_p, e\lambda^t, Q\right)$. The time constants are normalized by the cis lifetime and $\mu_t E/kT = 1$. Photoisomerization allows polar orientation to be built in the presence of a DC field if the mobility of the cis molecules is sufficient to align them, while the mobility of the trans molecules is very much reduced (Figure 8). Photo-induced polar order in the trans state is possible only if a certain memory of the cis orientation is preserved during the cis⇒trans back reaction. The higher the memory of the molecular alignment, the higher the trans polar order (Figures 8a and 8b). In Figure 9 T_1 and C_1 are normalized by their respective values just before removing the pumping light, $\mu_t E/kT = 1$, the trans mobility is very much reduced $\left(\tau_D^t = 10^4\right)$ and the pump light beam is switched off at the time t=0. Figure 9a shows that cis molecules experience polar orientation during the cis lifetime when the pump beam is turned off in the presence of the DC field. This, of course, is possible if the optimum cis-polar orientation is not achieved during the PIP process, or if the steady-state of PIP is not achieved. When this is reached, the cis-polar order relaxes naturally as shown in Figure 9b. Retaining even partial memory during the cis⇒trans thermal back reaction is once again emphasized to be a very important physical phenomenon in the PIP process. Indeed, Figure 9c shows that a complete loss of molecular orientation memory during the thermal back reaction prevents trans molecules from building polar orientation while the second order susceptibility changes slightly but not significantly. For partial retention of molecular orientation memory, both the trans polar order and the second order susceptibility grow nearly in the same way (Figures 9a and 9b). This is highly relevant for the experimental data and shows that for PMMA-DR1 copolymer a certain molecular orientation memory was retained during the cis⇒trans thermal back reaction, when the pump beam was turned off with the DC field applied (see Figure 6).

In a practical PIP experiment, the orientation of the nonlinear-optical molecules is more or less preserved when the pump beam is switched off with the DC field applied. When this field is removed, the orientation of the molecules relaxes to another equilibrium state which depends only on the mobility of the trans molecules through the rotational diffusion constant D_t. Each order papameter A_n relaxes with $\tau_n = 1/n(n+1)D_t$ as a time constant (*26, 29*). In practice, PIP only creates the first two order parameters, then the decay of birefringence and second order processes should be monoexponential as predicted by the theory with $1/6D_t$ and $1/2D_t$ as time constants respectively. The very slow relaxation of the PIP $\chi^{(2)}$ signal of a PMMA-DR1 copolymer shown in Figure 6 after the removal of the poling field, confirms that the mobility of the trans molecules is very much reduced at room temperature. The relaxation of the orientation of nonlinear-optical molecules in polymeric films have shown multiexponential relaxations because of the free volume distribution that have been represented by Williams-Watt stretched exponentials and the William-Landel-Ferry law (3). Subsequently, a more rigourous theory of the complex relaxation phenomena which appear in polymer films should combine the brownian relaxation with a free volume theory (30). Finally, let us mention that this general theory contains both the theory of the usual TP process, and the general equations for photoisomarization and photo-induced reorientation. Another potential application of this theory would be writing erazing optical memories based on photoisomerization of azo chromophores in polymeric films or other host materials (e.g. sol-gels). Perhaps one of the most important applications of this theory is photoisomerization induced dichroism or linear

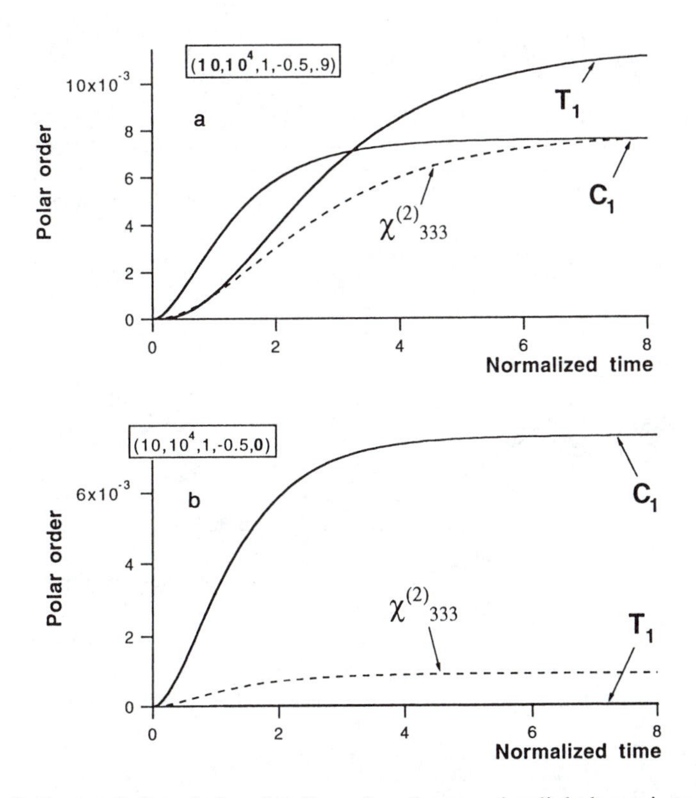

Figure 8. Onset of photo-induced poling when the pumping light beam is turned on ($\tau_p = 1$) with a trans mobility very much reduced ($\tau_D^t = 10^4$), and with ((**a**) $\tau_D^c = 10$, Q= 0.9) and without ((**b**) $\tau_D^c = 10$, Q= 0) retention of molecular orientation memory during the cis⇒trans back isomerization.

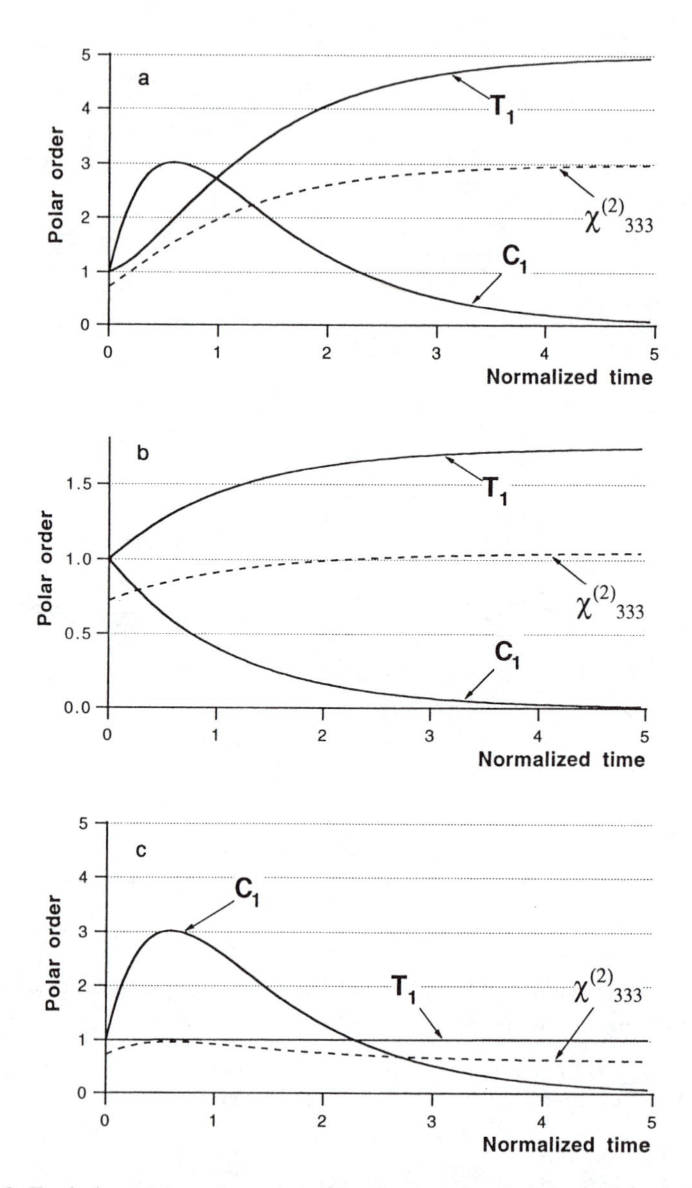

Figure 9. Evolution of the polar order when the pumping light beam is switched off with the DC field applied. **(a)** If the seady-state of PIP was not pursued to saturation and Q=0.8. **(b)** If the seady-state of PIP was pursued to saturation and Q=0.8. **(c)** If the seady-state of PIP was not pursued to saturation and Q=0.

Stark effect. In this case it would be possible to separate the trans from the cis contribution in the UV region as, in this region in the case of the push-pull azo molecules like DR1, only the cis state absorbs. Separation of photoinduced population changes from reorientation effects would then be possible.

Conclusion

In contrast to the usual thermal poling process (TP), PIP presents two major advantages for practical applications. First, it can be processed at room temperature and eliminates the heating which is necessary in TP, and second it allows the drawing of patterns of oriented molecules which could be useful for producing electro-optical integrated devices without complex electrode structures. The order created by PIP seems to be less efficient (by a factor 3) than that induced by TP (8, 9). While the stationary state and the transients of TP are well known (26, 31), a theory of PIP is necessary to allow a good understood of the physical processes which appear on PIP, and may serve to identify different physical phenomena which appear on PIP. These include the retention of molecular orientation memory during the cis⇒trans thermal back reaction when the pump beam is switched off with the DC field applied.

In summary, this paper contains both experimental and theoretical investigations of polar order induced by photoisomerization of polar azo dyes in polymeric films. In one hand, in contrast to flexible main chains polymer, we have investigated PIP and TP of a rigid main chain DR1-polyester copolymer. The data suggest that PIP of a high Tg rigid main chains polymer, could provide valuable information about the ability of the photoisomerization reaction to change the conformation of the stiff main chains through the photo-induced movement of the azo chromophores. In the other hand, we have studied theoritically the photoisomerization induced polar order. When compared with experiments, the model reproduces all the physical observations very well and allows a good physical insight in the phenomena which appear on PIP.

Acknowledgments

The DR1-polyester copolymer was synthetized by Dr. C.-S. Kang from the Prof. Wegner Group. We are particularly indebted to them as well as for Dr. D. Neher and Dr. M. Schulze for their very pleasant collaboration, and for fruitful discussions and valuable advice. Z.S. would like to acknowledge the Max-Planck society for providing a postdoctoral fellowship.

References

1. Davydov, B. D.; Derkacheva, L. D.; Dumina, V. V.; Zhabostinskii, M. E.; Zolin, V. F.; Koreneva, L. G.; Sanokhina, M. A. *Opt. Spectrosc.* **1971**, 30,274.
2. Zyss, J.; Chemla, D. S. In *Nonlinear Optical Properties of Organic Molecules and Crystals,* Chemla, D. S., Zyss, J., Eds., Academic Press: New York, 1987, Vol. 1, p 3.
3. Burland, D. M.; Miller, R. D.; Walsh, C. A. *Chem. Rev.* **1994**, 94, 31.
4. Mortazavi, M. A.; Knosen, A.; Kowel, S. T.; Higins, B. G.; Diens, A. *J. Opt. Soc. Am.* **1989**, B6, 733.
5. Barry, S.; Soan, D. *Appl. Phys. Lett.* **1991**, 58, 1134.
6. Sekkat, Z. Ph.D. Thesis, Paris-Sud university, Paris, **1992**.
7. Sekkat, Z.; Dumont, M. *App. Phys.* **1992**, B54, 486.
8. Sekkat, Z.; Dumont, M. *Mol. Cryst. Liq. Cryst. Sci. Tech. B: Nonlinear Opt .* **1992**, 2, 359.
9. Sekkat, Z.; Dumont, M. *Synth. Metal.* **1993**, 54, 373.
10. Blanchard, P. M.; Mitchell, G. R. *Appl. Phys. Lett.* **1993**, 63, 2038.

11. Rau, H. In *Photocemistry and Photophysics*, Rabeck, F. J., Ed.; CRC: Boca Raton, FL, 1990; Vol. **II**, Chapter 4, pp 119-141. This paper contains a large bibliography on photoisomerization.
12. Loucif-Saibi, R.; Nakatani, K.; Delaire, J. A.; Dumont, M.; Sekkat, Z. *Chem. Mater.* **1993**, 5, 229.
13. Todorov, T.; Nikolova, L.; Tomova, N. *Appl. Opt.* **1984**, 23, 4309.
14. Nathansohn, A.; Rochon, P.; Gosselin, J.; Xie, S. *Macromolecules.* **1992**, 25, 2268.
15. Dumont, M.; Sekkat, Z.; Loucif-Saibi, R.; Nakatani, K.; Delaire. J. A. *Mol. Cryst. Liq. Cryst. Sci. Tech. B: Nonlinear Opt.* **1993**, 73, 2705.
16. Kang, C.-S.; Winkelhahn, H.-J.; Schultze, M.; Neher, D.; Wegner, G. *Chem. Mater.* **1994**, 6, 2159 .
17. Jin, J.-I.; Chang, J.-H. *Macromolecules.* **1989**, 22, 4402.
18. Jin, J.-I.; Kang, C.-S.; Chang, J.-H. *J. Polym. Sci. Polym. Chem. Ed.* **1993,** 31, 229.
19. Jin, J.-I.; Kang, C.-S.; Lee, I.-H.; Yun, Y.-K. *Macromolecules.* **1994**, 27, 2664.
20. Jin, J.-I.; Lee, S.-H.; Park, H.-J. *Polym. Bull.* **1989**, 20, 19.
21. Sekkat, Z.; Morichere, D.; Dumont, M.; Loucif-Saibi, R.; Delaire, J. A. *J. Appl. Phys.* **1992**, 71, 1543.
22. Dumont, M.; Sekkat, Z. *SPIE Proc.* **1992**, 1774, 188.
23. Aust, E. F.; Knoll, W. *J. Appl. Phys.* **1993**, 73, 2705.
24. Sekkat, Z.; Kang, C.-S.; Aust, E. F.; Wegner, G.; Knoll, W. *Chem. Mater.* **1995**, 7, 142.
25. Doi, M.; Edwards, S. F. *The Theory of Polymer Dynamics*, Clarendon Press: Oxford, 1986.
26. Wu, J. W. *J. Opt. Soc. Am.* **1991**, B8, 142.
27. Kuzyk, M. G.; Singer, K. D.; Zahn, H. E.; King, L. A. *J. Opt. Soc. Am.* **1989**, B6, 742.
28. Sekkat, Z.; Knoll, W. *J. Opt. Soc. Am. B.*, in press.
29. Benoit, H. *Ann. Phys. (Paris)* **1951**, 6, 561.
30. Robertson, R. E. *J. Polym. Sci., Polym. Symp.* **1978**, 63, 173.
31. Sekkat, Z.; Knoll, W. *Ber. Bunsenges. Phys. Chem.* **1994**, 10, 1231.

RECEIVED February 14, 1995

Chapter 20

Determination of the Second Harmonic Coefficients of Birefringent Poled Polymers

Warren N. Herman[1] and L. Michael Hayden[2]

[1]**Naval Air Warfare Center, Aircraft Division, Warminster, PA 18974–0591**
[2]**Department of Physics, University of Maryland, Baltimore, MD 21228**

We present a new formulation of Maker fringes in parallel-surface films using self-consistent boundary conditions for reflections and allowing for any degree of refractive index dispersion. This treatment of the second harmonic reflections and dispersion, unlike a number of previous derivations, leads correctly to the expected form for the effective second harmonic *d*-coefficients. Complete expressions with physically meaningful factors are given for the generated second-harmonic power for birefringent poled polymer films including reflections for the case of no pump depletion. Comparison with the isotropic approximation and practical considerations in the use of these expressions for the fitting of experimental data are discussed.

Since 1970, the determination of the nonlinear optical (NLO) coefficients d_{ij} (i = 1-3, j = 1-6) associated with second harmonic generation (SHG) has been performed almost exclusively by the Maker fringe technique(*1*) as described in detail by Jerphagnon and Kurtz(*2*) (hereafter, JK). Over the years, this technique has been applied to inorganic and organic crystals and organic polymers. However, some of the original assumptions made in the derivations in JK, while applicable to the transparent crystals of the day, are not uniformly applicable to all material classes currently under investigation. Also, the theory presented in JK did not consider absorbing materials nor the general case of anisotropic media. Many of the second order NLO organic materials under investigation today such as liquid crystals, Langmuir-Blodgett films, poled polymers, and organic crystals are absorbing at the wavelength of the typical SHG experiment and exhibit a fair degree of anisotropy. Unfortunately, during the past 24 years, the results of JK have been applied, by many workers, to systems that were either anisotropic or absorbing, resulting in errors in the determination of d_{ij}.

In solving the boundary value problem for SHG, JK used approximate boundary conditions for the SH waves by ignoring the SH wave reflected from the second interface in the boundary conditions at the first interface, and assumed, in portions of their calculations, that the nonlinear material had the same index of refraction at the fundamental frequency ω as it had at the second harmonic frequency 2ω. As a result,

0097–6156/95/0601–0275$12.00/0

their expression for $p(\theta)$, the angular projection factor for the nonlinearity, is non-intuitive, and difficult to calculate in general. Furthermore, it is incorrect for materials where there is dispersion. These problems are not widely realized in the literature and continue to propagate, even recently (*3-5*). Some authors have realized the problem with $p(\theta)$ and have replaced it in the theory of JK, ad hoc, with the correct term (*6*) As a result, a combination of JK with an ad hoc correction for the projection factor is sometimes used to analyze SHG experiments.

Ignoring reflections, expressions for the sum-frequency fields generated by an isotropic transparent nonlinear parallel slab were given three decades ago by Bloembergen and Pershan (*7*). Subsequent to JK, Chemla and Kupecek (*8*) obtained Poynting vectors for nonlinear wedges in air for anisotropic nonabsorbing materials with 32 point symmetry and for absorbing isotropic materials, but, like JK, considered the boundary conditions for the input and output faces of the nonlinear material separately. More recently, Okamoto et al. (*3*) obtained expressions for Maker fringes in anisotropic parallel slabs with $C_{\infty v}$ space symmetry, but, again, they used the JK boundary conditions which lead to complicated expressions that are not only difficult to interpret physically, but are also incorrect if reflections are kept and no assumptions are made about the dispersion. Finally, we note that Kajzar and Messier (*9*) treated the reflected harmonic waves properly for the case of third harmonic generation and normal incidence, but for the case of SHG, the effect of ignoring the SH wave reflected from the output interface when treating the boundary conditions at the input interface has not been adequately discussed in the literature.

In this paper we provide a new derivation of the Maker fringes using complete boundary conditions for the SH waves in birefringent media and apply it to poled polymeric materials. The case of absorbing materials is covered elsewhere (*10*). Our theory, which includes the reflections of the SH wave in the nonlinear material, is also an extension of an earlier work which neglected those reflections (*11*). The new theory predicts the same results as JK for transparent, isotropic materials, but exhibits significant differences for materials with non-zero dispersion. These differences have implications for experimentalists investigating the order in poled polymers by measuring the ratio d_{31} / d_{33}. We also show that the effect of neglecting the birefringence in the analysis of Maker fringe data can result in large errors in the determination of d_{ij}.

Theory.

In this section we will present our general approach to the boundary value problem for SHG. Specifically, we derive the expression for the SH power generated in a birefringent nonlinear film assuming no depletion of the fundamental. We do not include the effects of multiple reflections of the fundamental here, although Neher (*12*) has included them for the case of third harmonic generation. They found that for the case where the nonlinear film is on the side of the substrate facing the incident beam (as in this paper), the effect of neglecting the multiple reflections of the fundamental is small. We will show that including reflections of the SH wave not only makes the final result for SHG less ambiguous than JK and simpler to use in practice, but also is essential to the correct derivation of d_{eff} and is needed in order to include the effects of any degree of dispersion. The major difference between the two approaches is that JK neglected to consider the amplitude at the input boundary of the SH wave reflected from the second interface and we do not.

Since the measurement of d_{ij} of a thin nonlinear film usually requires that the film be supported on a substrate, we consider a three layer system as shown in Figure 1. The linear region III is assumed to be much thicker than the nonlinear region II and ends at a substrate-air interface whose effect on the second harmonic beam can simply be taken into account by a transmission coefficient. We assume that an electromagnetic wave of frequency ω is incident upon the structure from the bottom at an angle θ (the *x-z* plane is the plane of incidence). For a frequency $m\omega$, the nonlinear layer of thickness L has an index $n_{\rm m}$ and the substrate has an index $n_{\rm ms}$. For free standing films or crystal plates, setting the substrate index n_{2s} equal one will suffice.

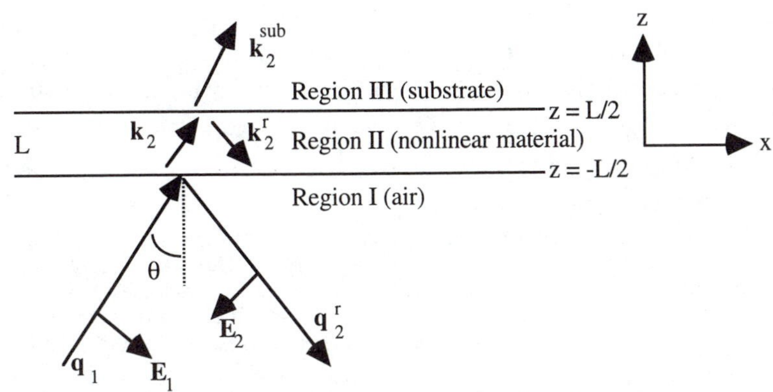

Figure 1. Three layer slab geometry for SHG due to a nonlinear layer in region II with the origin at the center of the nonlinear film.

In a uniaxial material, the dielectric permittivity is a second rank tensor and can be represented as,

$$\vec{\varepsilon}_i = \begin{pmatrix} n_{io}^{\ 2} & 0 & 0 \\ 0 & n_{io}^{\ 2} & 0 \\ 0 & 0 & n_{ie}^{\ 2} \end{pmatrix}, \quad i = 1,2 \tag{1}$$

where n_{io} and n_{ie} are the ordinary and extraordinary indices of refraction at the fundamental (i =1) and SH (i = 2) frequencies respectively. If we write the electric field of the incident fundamental wave as,

$$\mathbf{E}_{1\nu} = \hat{\mathbf{e}}_\nu E_1 \exp[i(\mathbf{q}_1 \cdot \mathbf{r} - \omega t)] \tag{2}$$

with $\hat{\mathbf{e}}_\nu = (0,1,0)$ or $(\cos\theta, 0, -\sin\theta)$ for $\nu = s$ or p respectively and $\mathbf{q}_1 = (\omega / c)(\sin\theta, 0, \cos\theta)$, then for a p -polarized input, the SH fields in the three regions shown in Figure 1 are given by,

Region I:

$$\begin{aligned} \mathbf{E}_2 &= R\hat{\mathbf{e}}_p^r \exp[i\mathbf{q}_2^r \cdot (\mathbf{r} + \tfrac{L}{2}\hat{\mathbf{z}})] \\ \mathbf{H}_2 &= -\frac{ic}{2\omega} \nabla \times \mathbf{E} = R\hat{\mathbf{e}}_s^r \exp[i\mathbf{q}_2^r \cdot (\mathbf{r} + \tfrac{L}{2}\hat{\mathbf{z}})] \end{aligned} \tag{3}$$

Region II:

$$\begin{aligned} \mathbf{E}_2 &= \mathbf{e}_b \exp[i2\mathbf{k}_1 \cdot \mathbf{r}] + A\hat{\mathbf{e}}_{2p} \exp[i\mathbf{k}_2 \cdot \mathbf{r}] + B\hat{\mathbf{e}}_{2p}^r \exp[i\mathbf{k}_2^r \cdot \mathbf{r}] \\ \mathbf{H}_2 &= \mathbf{h}_b \exp[i2\mathbf{k}_1 \cdot \mathbf{r}] + n_2(\theta_2)\cos\gamma_2 \left(A\hat{\mathbf{e}}_{2s} \exp[i\mathbf{k}_2 \cdot \mathbf{r}] + B\hat{\mathbf{e}}_{2s}^r \exp[i\mathbf{k}_{2r} \cdot \mathbf{r}] \right) \end{aligned} \tag{4}$$

Region III:
$$\begin{aligned}\mathbf{E}_2 &= T\hat{\mathbf{e}}_{2p}^{sub}\exp[i\mathbf{k}_2^{sub}\cdot(\mathbf{r}-\tfrac{L}{2}\hat{\mathbf{z}})] \\ \mathbf{H}_2 &= n_{2s}T\hat{\mathbf{e}}_{2s}^{sub}\exp[i\mathbf{k}_2^{sub}\cdot(\mathbf{r}-\tfrac{L}{2}\hat{\mathbf{z}})]\end{aligned}, \tag{5}$$

where $\hat{\mathbf{e}}_v^r = (0,1,0)$ or $(-\cos\theta, 0, -\sin\theta)$ is the polarization unit vector for the reflected SH field in the air. A and B are the complex amplitudes for the forward and backward SH free waves in the nonlinear media, respectively, while T and R are the corresponding SH amplitudes in the substrate and air, respectively. The polarization unit vectors for the forward and backward SH waves in the nonlinear media are

$$\begin{aligned}\hat{\mathbf{e}}_m &= \left(\cos(\theta_m-\gamma_m), 0, -\sin(\theta_m-\gamma_m)\right), \qquad m=1,2 \\ \hat{\mathbf{e}}_2^r &= \left(-\cos(\theta_2-\gamma_2), 0, -\sin(\theta_2-\gamma_2)\right).\end{aligned} \tag{6}$$

The walkoff angle γ_m is defined in Figure 2. For the substrate, $\hat{\mathbf{e}}_{2s}^{sub} = (0,1,0)$ and $\hat{\mathbf{e}}_{2p}^{sub} = (c_{2s}, 0, -s_{2s})$. The wave vectors are (outside the slab) $\mathbf{q}_2^r = (2\omega/c)(\sin\theta, 0, -\cos\theta)$ and (in the substrate) $\mathbf{k}_2^{sub} = (2\omega n_2^{sub}/c)(s_2^{sub}, 0, c_2^{sub})$, where $s_m = (1/n_m)\sin\theta$ and $c_m = \sqrt{1-s_m{}^2}$.

In matching the tangential components of the fields between regions I and II, JK neglects the third term on the right hand side of Eq. (4). They justify this approximation by assuming that the reflected free wave is small. By doing so, their solution for the transmitted SH power contains a small term that does not oscillate (which they neglect) and non-standard Fresnel transmission factors. Furthermore, their projection factor $p(\theta)$ is incorrect since they obtain it by assuming $n_1 \approx n_2$. Keeping those terms however is not difficult and results in an expression that includes the effects of reflection of the SH wave in the nonlinear media and the correct, general expression for the projection factor $d_{eff} = \hat{\mathbf{e}}_2 \cdot \overleftrightarrow{\mathbf{d}}:\hat{\mathbf{e}}_1\hat{\mathbf{e}}_1$.

Maxwell's Equations require that the SH field $\mathbf{E}_2$ satisfy

$$\nabla\times\nabla\times\mathbf{E}_2 - \left(\frac{2\omega}{c}\right)^2 \overleftrightarrow{\varepsilon}_2\cdot\mathbf{E}_2 = \left(\frac{2\omega}{c}\right)^2 4\pi\mathbf{P}^{NL}\exp[2i\mathbf{k}_1\cdot\mathbf{r}], \tag{7}$$

where use is made of the electric displacement vector $\mathbf{D}_2 = \overleftrightarrow{\varepsilon}_2\cdot\mathbf{E}_2 + 4\pi\mathbf{P}^{NL}\exp[2i\mathbf{k}_1\cdot\mathbf{r}]$, $\overleftrightarrow{\varepsilon}$ is the permittivity tensor given in Eq. (1), and $\mathbf{P}^{NL}$ is the nonlinear polarization. In the linear media (air, substrate), $\mathbf{P}^{NL}$ is zero and Eq. (7) is homogeneous.

In the nonlinear film (Region II of Figure 1), the propagation vectors are

$$\begin{aligned}\mathbf{k}_m &= \frac{m\omega n_m(\theta_m)}{c}(s_m, 0, c_m), \qquad m=1,2 \\ \mathbf{k}_2^r &= \frac{2\omega n_2(\theta_2)}{c}(s_2, 0, -c_2)\end{aligned} \tag{8}$$

with the angle-dependent refractive indices $n_m(\theta_m)$ given by

$$\begin{aligned}n_i(\theta_i) &= \left(\frac{\cos^2\theta_i}{n_{io}{}^2} + \frac{\sin^2\theta_i}{n_{ie}{}^2}\right)^{-\frac{1}{2}}, \quad i=1,2 \\ n_2(\theta_1) &= \left(\frac{\cos^2\theta_1}{n_{2o}{}^2} + \frac{\sin^2\theta_1}{n_{2e}{}^2}\right)^{-\frac{1}{2}}.\end{aligned} \tag{9}$$

While $\mathbf{D}_2$ is still perpendicular to $\mathbf{k}_2$, the electric field deviates from the perpendicular by the walk-off angle (*13*) γ, as shown in Figure 2.

The bound wave, $\mathbf{E}_{2b}$, is a particular solution to the inhomogeneous equation (7) in the nonlinear material (*14*) Inserting $\mathbf{E}_{2b} = \mathbf{e}_b\exp[i\mathbf{k}_b\cdot\mathbf{r}]$, with $\mathbf{k}_b = 2\mathbf{k}_1$, into Eq. (7) gives the following equations for the bound wave amplitude $\mathbf{e}_b$:

$$\begin{aligned}\left(-n_1^{\,2}(\theta_1)c_1^{\,2}+n_{2o}^{\,2}\right)e_{bx}+n_1^{\,2}(\theta_1)s_1c_1e_{bz}&=-4\pi P_x^{NL}\\ n_1^{\,2}(\theta_1)c_1s_1e_{bx}+\left(-n_1^{\,2}(\theta_1)s_1^{\,2}+n_{2e}^{\,2}\right)e_{bz}&=-4\pi P_z^{NL}.\end{aligned}\tag{10}$$

Upon solving for e_{bx} and e_{bz}, the solution can be conveniently written as,

$$\begin{aligned}\mathbf{e}_b&=\frac{4\pi n_2^{\,2}(\theta_1)}{n_1^{\,2}(\theta_1)-n_2^{\,2}(\theta_1)}\left[(\vec{\varepsilon}_2)^{-1}\cdot\mathbf{P}^{NL}-\frac{\mathbf{k}_b(\mathbf{k}_b\cdot\mathbf{P}^{NL})}{\left(\frac{2\omega}{c}n_{2o}n_{2e}\right)^2}\right]\\ \mathbf{h}_b&=n_1(\theta_1)\hat{\mathbf{k}}_1\times\mathbf{e}_b,\\ \mathbf{P}^{NL}&\equiv|E_1|^2\vec{\mathbf{d}}:\hat{\mathbf{e}}_1\hat{\mathbf{e}}_1.\end{aligned}\tag{11}$$

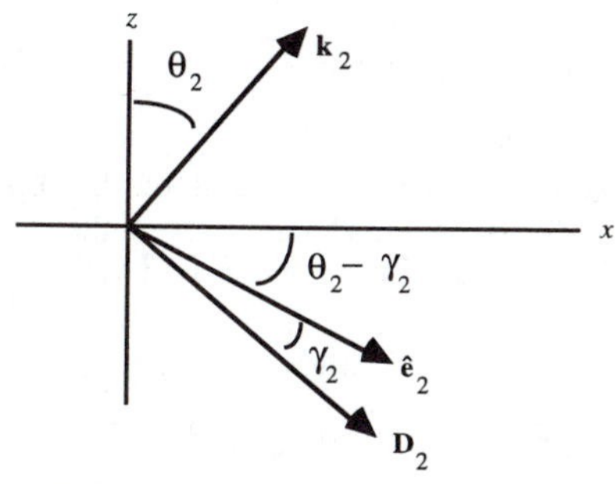

Figure 2. Definition of the walk-off angle γ in a birefringent medium.

Using $\mathbf{D}=\vec{\varepsilon}\cdot\mathbf{E}$ and $\mathbf{D}=\left[\frac{n(\theta)}{c}\right]^2\hat{\mathbf{k}}\times(\mathbf{E}\times\hat{\mathbf{k}})$ for the linear electric displacement field, we can derive the following useful relations,

$$\cos(\theta_m-\gamma_m)=\left(\frac{n_m(\theta_m)}{n_{mo}}\right)^2\cos\gamma_m\cos\theta_m,$$

$$\sin(\theta_m-\gamma_m)=\left(\frac{n_m(\theta_m)}{n_{me}}\right)^2\cos\gamma_m\sin\theta_m,\tag{12}$$

and

$$\tan\gamma_m=\tfrac{1}{2}n_m(\theta_m)^2\sin 2\theta_m\left(\frac{1}{n_{mo}^2}-\frac{1}{n_{me}^2}\right).\tag{13}$$

Furthermore, since $\hat{\mathbf{e}}_m\cdot\hat{\mathbf{e}}_m=1$ we have,

$$\cos\gamma_m=\frac{n_{mo}n_{me}}{n_m(\theta_m)\sqrt{n_{mo}^2+n_{me}^2-n_m(\theta_m)^2}}.\tag{14}$$

The angles θ_m in the birefringent film are obtained by using Eq. (9) in conjunction with Snell's Law, $n_m(\theta_m)\sin\theta_m=\sin\theta$, to get,

$$\sin\theta_m=\frac{n_{me}\sin\theta}{\sqrt{(n_{mo}n_{me})^2+(n_{me}^2-n_{mo}^2)\sin^2\theta}}.\tag{15}$$

For the $s\rightarrow p$ case, the fundamental beam is s-polarized, so that $\gamma_1=0$ and

$$\mathbf{k}_1 = \tfrac{\omega}{c} n_{1o}(s_1, 0, c_1)$$
$$\hat{\mathbf{e}}_1 = (0,1,0) \qquad , \tag{16}$$
$$\sin\theta_1 = \sin\theta / n_{1o}$$

while Eqs. (6), (8), and (12) - (15) remain valid for the SH wave ($m = 2$).

The boundary conditions at the I-II interface (z = $-L/2$) requiring the continuity of E_{2x} and H_{2y} result in the equations

$$-\cos(\theta)R = e_{bx}\exp(-i\phi_1) + \cos(\theta_2 - \gamma_2)[A\exp(-i\phi_2) - B\exp(i\phi_2)],$$
$$R = h_{by}\exp(-i\phi_1) + n_2(\theta_2)\cos\gamma_2[A\exp(-i\phi_2) + B\exp(i\phi_2)], \tag{17,18}$$

respectively. Continuity of these field components at the II-III interface (z = $L/2$) yields,

$$c_{2s}T = e_{bx}\exp(i\phi_1) + \cos(\theta_2 - \gamma_2)[A\exp(i\phi_2) - B\exp(-i\phi_2)],$$
$$n_{2s}T = h_{by}\exp(i\phi_1) + n_2(\theta_2)\cos\gamma_2[A\exp(i\phi_2) + B\exp(-i\phi_2)], \tag{19,20}$$

where $\phi_1 = k_1 c_1 L$ and $\phi_2 = k_2 c_2 L/2$. Note that in the JK approach, $B = 0$ in Eqs. (17 - 20). Solving Eqs. (17 - 20) for A and B and inserting the results into the equation for T, obtained by multiplying Eq. (19) by n_{2s} and Eq. (20) by c_{2s} and adding, we obtain the transmitted field

$$T = \tfrac{1}{\Delta}\{-u_a^+ v^+ \exp(\phi_1 - 2\phi_2) + u_a^- v^- \exp(\phi_1 + 2\phi_2)$$
$$+ 2n_2(\theta_2)\cos\gamma_2\cos(\theta_2 - \gamma_2)(h_{by}\cos\theta + e_{bx})\exp(-i\phi_1)\}, \tag{21}$$

where

$$\Delta = -u_a^+ u_s^+ \exp(-2i\phi_2) + u_a^- u_s^- \exp(2i\phi_2), \tag{22}$$

and

$$u_a^{\pm} = \cos(\theta_2 - \gamma_2) \pm n_2(\theta_2)\cos\gamma_2\cos\theta,$$
$$u_s^{\pm} = n_{2s}\cos(\theta_2 - \gamma_2) \pm n_2(\theta_2)c_{2s}\cos\gamma_2, \tag{23}$$
$$v^{\pm} = h_{by}\cos(\theta_2 - \gamma_2) \pm n_2(\theta_2)e_{bx}\cos\gamma_2.$$

When the identities

$$\exp i(\phi_1 \pm 2\phi_2) = 2i\sin(\phi_1 \pm \phi_2)\exp(\pm i\phi_2) + \exp(-i\phi_1), \tag{24}$$

are used in Eq. (21), the terms proportional to $\exp(-i\phi_1)$ cancel out (unlike when using JK-type boundary conditions). Furthermore, from Eqs. (1) and (18) we find

$$v^{\pm} = \frac{4\pi n_2(\theta_1)^2}{n_1(\theta_1)^2 - n_2(\theta_1)^2}\frac{[n_1(\theta_1)c_1 \pm n_2(\theta_2)c_2]}{n_{2o}^2} \times \begin{cases}\hat{\mathbf{e}}_2 \cdot \mathbf{P}^{NL} \\ \hat{\mathbf{e}}_2^r \cdot \mathbf{P}^{NL},\end{cases} \tag{25}$$

after considerable algebraic simplification requiring Eqs. (6), (9), and (12). Thus, use of Eqs. (24) and (25) in (21) gives

$$T = -\frac{8\pi i}{\Delta}\left[\frac{n_2(\theta_1)}{n_{2o}}\right]^2 \frac{1}{n_1(\theta_1)^2 - n_2(\theta_2)^2}$$
$$\times\{u_a^+[n_1(\theta_1)c_1 + n_2(\theta_2)c_2]\hat{\mathbf{e}}_2 \cdot \mathbf{P}^{NL}\sin(\phi_1 - \phi_2)\exp(-i\phi_2) \tag{26}$$
$$+ u_a^-[n_1(\theta_1)c_1 - n_2(\theta_2)c_2]\hat{\mathbf{e}}_2^r \cdot \mathbf{P}^{NL}\sin(\phi_1 + \phi_2)\exp(i\phi_2)\},$$

The p-polarized reflection coefficients at the air-film and film-substrate interfaces for the SH wave are

$$r_{af}^{(2p)} = -\frac{u_a^-}{u_a^+} \text{ and } r_{fs}^{(2p)} = \frac{u_s^-}{u_s^+}, \tag{27}$$

while the transmission coefficient for the SH wave at the film-substrate interface is

$$t_{fs}^{(2p)} = \frac{2n_2(\theta_2)\cos\gamma_2\cos(\theta_2 - \gamma_2)}{u_s^+}. \tag{28}$$

Then, the use of Eqs. (27), (28), and (22) in (26), together with

$$n_1(\theta_1)^2 - n_2(\theta_2)^2 = [n_1(\theta_1)c_1 + n_2(\theta_2)c_2][n_1(\theta_1)c_1 - n_2(\theta_2)c_2], \tag{29}$$

which holds because of the identity $c_1^2 = 1 - s_1^2$ and Snell's Law, gives

$$T = \frac{4\pi i t_{fs}^{(2p)}}{n_2(\theta_2)\cos\gamma_2\cos(\theta_2-\gamma_2)}\left(\frac{2\pi L}{\lambda}\right)\left[\frac{n_2(\theta_1)}{n_{2o}}\right]^2\left[\frac{n_1(\theta_1)^2 - n_2(\theta_2)^2}{n_1(\theta_1)^2 - n_2(\theta_1)^2}\right]$$

$$\times \frac{\left[\hat{\mathbf{e}}_2\cdot\mathbf{P}^{NL}\frac{\sin\Psi_{bi}}{\Psi_{bi}}\exp(-i\phi_2) - r_{af}^{(2p)}\hat{\mathbf{e}}_2^r\cdot\mathbf{P}^{NL}\frac{\sin\Phi_{bi}}{\Phi_{bi}}\exp(i\phi_2)\right]}{\left[\exp(-2i\phi_2) + r_{af}^{(2p)}r_{fs}^{(2p)}\exp(2i\phi_2)\right]}, \tag{30}$$

where

$$\Psi_{bi} = \phi_1 - \phi_2 = \frac{2\pi L}{\lambda}[n_1(\theta_1)c_1 - n_2(\theta_2)c_2],$$
$$\Phi_{bi} = \phi_1 + \phi_2 = \frac{2\pi L}{\lambda}[n_1(\theta_1)c_1 + n_2(\theta_2)c_2]. \tag{31}$$

The components of $\mathbf{P}^{NL}$ along the SH $\mathbf{E}$-fields are given by

$$\hat{\mathbf{e}}_2\cdot\mathbf{P}^{NL} = -\frac{8\pi}{c}d_{eff}I_1\left[t_{af}^{(1\gamma)}\right]^2$$
$$\hat{\mathbf{e}}_2^r\cdot\mathbf{P}^{NL} = -\frac{8\pi}{c}d_{eff}^r I_1\left[t_{af}^{(1\gamma)}\right]^2, \tag{32}$$

where I_1 is the intensity of the incident fundamental wave, $t_{af}^{(1\gamma)}$ is the transmission coefficient for the fundamental at the air-film interface,

$$t_{af}^{(1\gamma)} = \begin{cases} \dfrac{2\cos\theta}{\cos(\theta_1-\gamma_1) + n_1(\theta_1)\cos\gamma_1\cos\theta}, & \gamma = p \\ \dfrac{2\cos\theta}{\cos\theta_1 + n_{1o}\cos\theta}, & \gamma = s \end{cases} \tag{33}$$

and the effective d-coefficients for $C_{\infty v}$ space symmetry are,

$$d_{eff} = -\hat{\mathbf{e}}_2\cdot\overleftrightarrow{\mathbf{d}}:\hat{\mathbf{e}}_1\hat{\mathbf{e}}_1$$
$$= \begin{cases} d_{15}\cos(\theta_2-\gamma_2)\sin 2(\theta_1-\gamma_1) \\ \qquad + \sin(\theta_2-\gamma_2)[d_{31}\cos^2(\theta_1-\gamma_1) + d_{33}\sin^2(\theta_1-\gamma_1)], & p\to p \\ d_{31}\sin(\theta_2-\gamma_2), & s\to p \end{cases}$$

$$d_{eff}^r = -\hat{\mathbf{e}}_2^r\cdot\overleftrightarrow{\mathbf{d}}:\hat{\mathbf{e}}_1\hat{\mathbf{e}}_1 \tag{34}$$
$$= \begin{cases} -d_{15}\cos(\theta_2-\gamma_2)\sin 2(\theta_1-\gamma_1) \\ \qquad + \sin(\theta_2-\gamma_2)[d_{31}\cos^2(\theta_1-\gamma_1) + d_{33}\sin^2(\theta_1-\gamma_1)], & p\to p \\ d_{31}\sin(\theta_2-\gamma_2), & s\to p \end{cases}$$

The SH power transmitted through the substrate into air is then given by $\frac{c}{8\pi}\left[t_{sa}^{(2p)}\right]^2|T|^2 A$ where A is the cross-sectional area of the fundamental beam, and the transmission coefficient for the p-polarized SH wave from the substrate into air is

$$t_{sa}^{(2p)} = \frac{2n_{2s}c_{2s}}{n_{2s}\cos\theta + c_{2s}}. \tag{35}$$

Thus, we get

$$P_{2\omega}^{\gamma\to p} = \frac{128\pi^3}{cA}\frac{\left[t_{sa}^{(2p)}\right]^2\left[t_{fs}^{(2p)}\right]^2\left[t_{af}^{(1\gamma)}\right]^4 P_\omega^2}{n_2(\theta_2)^2\cos^2\gamma_2\cos^2(\theta_2-\gamma_2)}\left(\frac{2\pi L}{\lambda}\right)^2\left[\frac{n_2(\theta_1)}{n_{2o}}\right]^4\left[\frac{n_1(\theta_1)^2-n_2(\theta_2)^2}{n_1(\theta_1)^2-n_2(\theta_1)^2}\right]^2$$
$$\times d_{eff}^2\frac{\left\{\frac{\sin^2\Psi_{bi}}{\Psi_{bi}^2}+\left[r_{af}^{(2p)}\right]^2R^2\frac{\sin^2\Phi_{bi}}{\Phi_{bi}^2}-2r_{af}^{(2p)}R\frac{\sin\Psi_{bi}}{\Psi_{bi}}\frac{\sin\Phi_{bi}}{\Phi_{bi}}\cos 2\phi_2\right\}}{\left\{1+\left[r_{af}^{(2p)}r_{fs}^{(2p)}\right]^2+2r_{af}^{(2p)}r_{fs}^{(2p)}\cos 4\phi_2\right\}}, \tag{36}$$

where R is the ratio d_{eff}^r/d_{eff}. For the case of $s\to p$ generation, the bound waves $\mathbf{e}_b$ and $\mathbf{h}_b$ are given by Eqs. (11) with the substitution $n_1(\theta_1)\to n_{1o}$. This substitution in Eq. (36) gives the corresponding SH power. The isotropic result is obtained from Eq. (36) by letting $n_{1e}=n_{1o}\to n_1$ and $n_{2e}=n_{2o}\to n_2$.

A comparison of the Maker fringe pattern for X-cut quartz predicted by JK and the current method (HH) given by the isotropic version of Eq. (36) is shown in Figure 3. The high frequency oscillations on the HH curve are the result of the reflections of the SH wave in the nonlinear media. The magnitude of the oscillations is related to the index difference between the nonlinear media and regions I and III. Figure 3 supports the assertion of JK that the effect of SH reflections is small. In addition, in order to carefully map out the high frequency oscillations, an angular resolution far beyond that which is normally attempted is required.

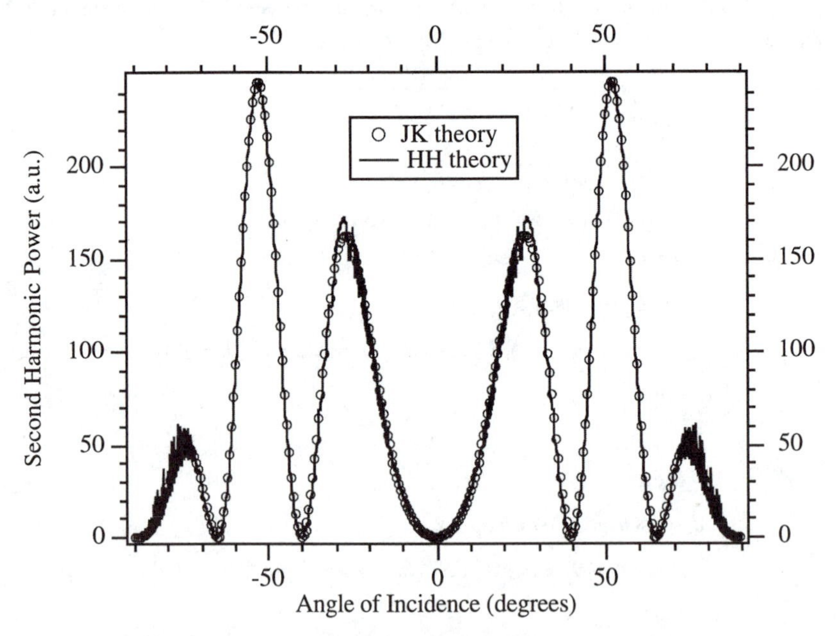

Figure 3. Maker fringe pattern predicted by JK and HH for a 300 μm thick piece of X-cut quartz. Rotation is about the z-axis. Reproduced with permission from Ref. 10.

In terms of the ability to measure d_{ij} accurately, we find that the difference between coefficients determined by considering reflections and neglecting them (<2%) is smaller than the typical experimental error in those measurements (≈ 10–20%). Because of the small difference predicted in the calculated d_{ij} we will henceforth neglect the reflections by setting $r_{af}^{(2p)} \approx 0$ in Eq. (36). This action is different from the method used by JK to neglect reflections. In JK, their approximation of no reflections in the boundary conditions results in the necessity to later assume no dispersion ($n_1 \approx n_2$) and incident angles $\theta_i \leq \pi/4$ in order to reduce the complexity of their result [see Eqs. (A36) and (A38) in JK]. The consequences of these assumptions can be significant for materials with even modest dispersion. In practice one usually measures d_{31} first and then uses that value in the calculation for d_{33}. This procedure is sometimes followed when trying to determine the order parameter in a poled polymer achieved via electric-field poling. By taking the ratios of the s-->p and p-->p transmitted SH powers from the theories of JK and HH, and assuming $d_{31} = d_{15}$ (Kleinman symmetry), we have derived the following relation,

$$\left(\frac{d_{31}}{d_{33}}\right)_{JK} = \left(\frac{2c_1(n_2c_2 - n_1c_1)}{n_1 s_1^{\ 2}} + \left(\frac{d_{33}}{d_{31}}\right)_{HH}\right)^{-1}. \qquad (37)$$

Figure 4 shows that as the dispersion increases, the ratio of d_{31}/d_{33} calculated with JK decreases. Since the standard rigid-gas model (*15*) describing the electric field induced orientation of dipoles predicts $d_{31}/d_{33} = 1/3$ for low poling fields and ratios smaller than 1/3 for high poling fields, measured SHG ratios of $d_{31}/d_{33} < 1/3$ are generally attributed to more order. In these cases, however, our results indicate that if JK is used to analyze SHG experiments in dispersive media, the implied high degree of order may be an artifact caused by the inability of JK to account properly for the effects due to dispersion.

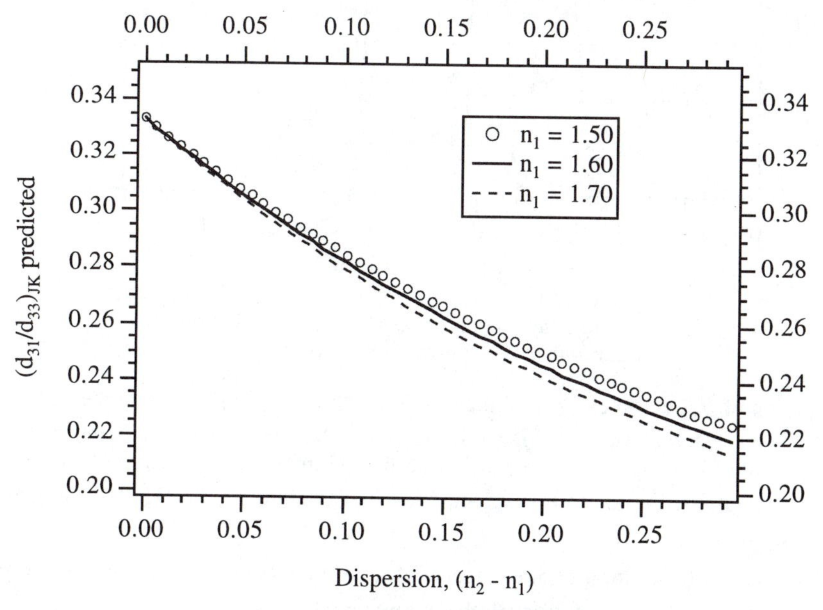

Figure 4. Effect of dispersion on the ratio d_{31}/d_{33} as calculated with JK. $(d_{31}/d_{33})_{HH}$ is assumed to be 1/3. Reproduced with permission from Ref. 10.

Results

We showed earlier that JK is a good approximation only for the case where $n_1 \approx n_2$. But, since we wish to include any degree of dispersion, when we examine the effects of birefringence in this section we will do so only in terms of the current theory (HH). We will concentrate on those effects from the standpoint of an experimentalist's ability to accurately determine d_{ij} using the Maker fringe technique. Further, because birefringence is nearly always present when studying organic films and because of the current interest in these materials, we will use representative values found in the current literature to illustrate the magnitude of the effects. The theory is not limited to application to these materials however.

We start our discussion on the effect of birefringence by examining the Maker fringe plots for each case for X-cut quartz. Figure 5 shows only a slight difference between the two cases.

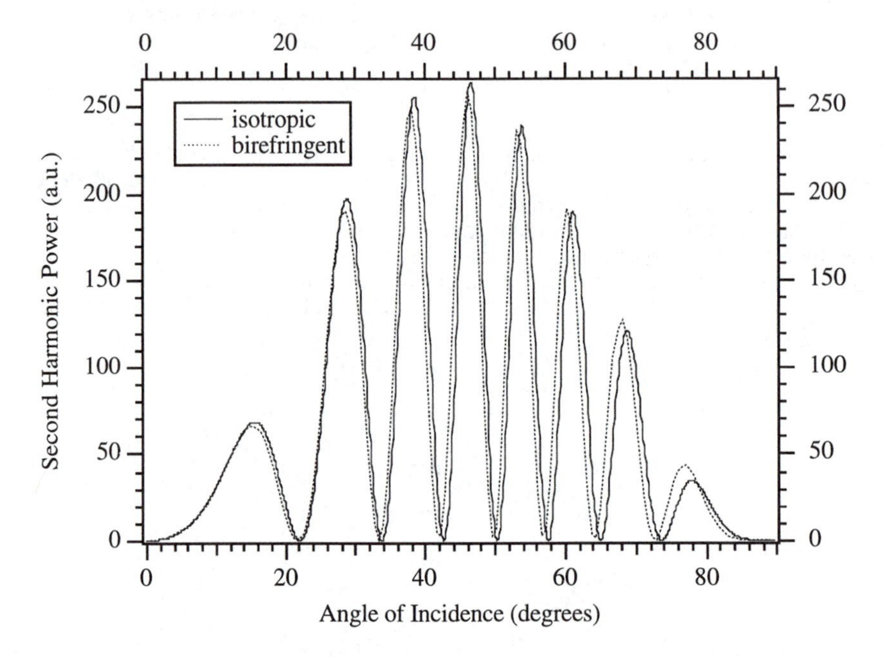

Figure 5. Maker fringe plots for 1000 μm thick X-cut quartz. Reproduced with permission from Ref. 10.

Calculation of d_{11} and the crystal thickness for each case yields a difference of 2% and 2 μm respectively for a 1000 μm thick crystal. The close agreement between the isotropic and birefringent treatments is due to the small birefringence of quartz. Quartz has a birefringence ($\Delta n_i \equiv n_{ie} - n_{io} = 3\sigma_i$) of 0.0088 and 0.0091 at 1.064 μm and 0.532 μm respectively.

The standard definition of birefringence Δn, can be related to the poling induced birefringence in polymers, σ_i, through the following relations,

$$\begin{aligned} n_{io} &= n_{iu} - \sigma_i \\ n_{ie} &= n_{iu} + 2\sigma_i \end{aligned} \tag{38}$$

where i =1,2 and n_{iu} is the unpoled index at the frequency $i\omega$. Equation (38) is generally valid for amorphous poled polymers. Recent data for poling induced birefringence in a polymer (BISA-ANT, $\sigma_i = 0.020 \rightarrow 0.025$) is given by Jungbauer et al. (*16*) In Figure 6, we have calculated, using Equation (36), the ratio of d_{33} (birefringent)/d_{33} (isotropic) as a function of film thickness and degree of birefringence for an incident angle of 50°, a fundamental wavelength of 1.064 μm, and assuming $d_{31} = d_{15}$ and $d_{33} = 3d_{31}$. It can be seen that large errors occur as the thickness approaches an integer multiple of twice the coherence length.

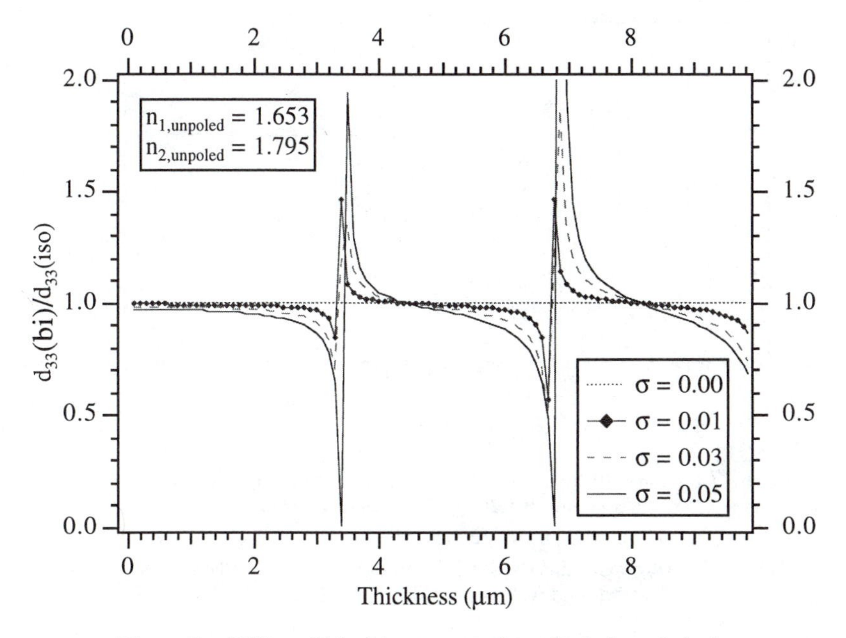

Figure 6. Effect of birefringence on the calculation of d_{33} in a poled polymer at of fixed angle of incidence = 50°. Extraordinary indices were used for the isotropic case. Reproduced with permission from Ref. 10.

The use of the ordinary indices gives similar qualitative results as in Figure 6 but the magnitude of the error is worse than use of the extraordinary indices for a $p \rightarrow p$ experiment. Figure 6 also shows us that experiments that attempt to determine d_{33}, while making comparisons at only a single angle and ignoring birefringence, are subject to very large errors depending on the thickness of the film, regardless of the degree of birefringence. We have also calculated d_{33} from isotropic fits of complete angular test data generated with the full birefringent theory. Figure 7 shows these results as a function of film thickness using the index data (σ =0.013) for a typical DR1-MMA side chain (*17*). We are making two different comparisons in this plot. First, we assume that the thickness is accurately known and let d_{33} vary alone (dotted line). This results in fairly significant differences from the true value of d_{33}. Next, we allow the film thickness to vary also (diamonds). The ratio of the fitted thickness to the true thickness (open box) is seen to be quite close to one for all thicknesses greater than about two coherence lengths. In this range, the fitted value of d_{33} is closer to the true value than for the first case when the thickness was fixed, but it is still appreciably different. For thicknesses less than two coherence lengths, there is a wide deviation from the true values for both the thickness and d_{33}. The use of the full birefringent theory is clearly called for if accurate determinations of d_{ij} are sought.

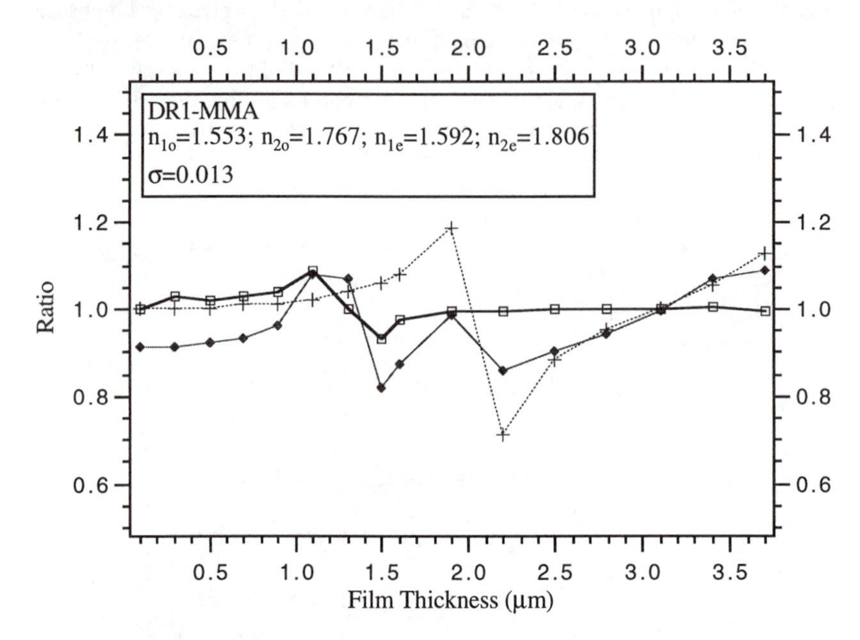

Figure 7. Ratio of the predicted d_{33} and film thickness for the isotropic and birefringent cases. The results are obtained from fits to Eq. (36) (with $r_{af}^{(2p)} \approx 0$) for d_{33} (crosses) and d_{33} and the thickness (diamonds, boxes). Reproduced with permission from Ref. 10.

Conclusions

In this paper, we have presented new expressions for analyzing data from Maker fringe SHG experiments. These expressions include the effects of reflections of the SH wave in the nonlinear material and are valid for any angle of incidence or degree of dispersion. We have shown that neglecting the reflections after solving for the SH power is preferred to doing so in the boundary conditions when setting up the problem, since the results are more generally applicable to systems with dispersion. Furthermore, we show that the use of JK as opposed to the present theory can result in an underestimation of the ratio d_{31}/d_{33} in experiments where dispersion is a factor.

We have also presented, for the first time, an accurate extension of the Maker fringe technique to birefringent materials. By applying that theory to representative materials studied today, we have shown that large discrepancies in the prediction of d_{ij} can result if birefringence is neglected.

Acknowledgments

WNH gratefully acknowledges the support of the Office of Naval Research. LMH would like to thank the U. S. Naval Academy for partial support for this work.

References

1. P. D. Maker, R. W. Terhune, M. Nisenhoff, and C. M. Savage, *Phys. Rev. Lett.* **8**, 21 (1962).
2. J. Jerphagnon and S. K. Kurtz, "Maker Fringes: A Detailed Comparison of Theory and Experiment for Isotropic and Uniaxial Crystals," *J. Appl. Phys.* **41**, 1667 (1970).
3. N. Okamoto, Y. Hirano, and O. Sugihara, "Precise estimation of nonlinear-optical coefficients for anisotropic nonlinear films with $C_{\infty v}$ symmetry," *J. Opt. Soc. Am. B* **9**, 2083 (1992).
4. M. Eich, B. Reck, D. Y. Yoon, C. G. Willson, and G. C. Bjorklund, "Novel second-order nonlinear optical polymers via chemical cross-linking-induced vitrification under electric field," *J. Appl. Phys.* **66**, 3241 (1989).
5. C. H. Wang and H. W. Guan, "Second harmonic generation and optical anisotropy of a spin cast polymer film," *J. Polymer Science B, Poly. Phys.* **31**, 1983 (1993).
6. M. G. Kuzyk, K. D. Singer, H. E. Zahn, and L. A. King, "Second-order nonlinear-optical tensor properties of poled films under stress," *J. Opt. Soc. Am. B* **6**, 742 (1989).
7. N. Bloembergen and P. S. Pershan, "Light waves at the boundary of nonlinear media," *Phys. Rev.* **128**, 606 (1962).
8. D. Chemla and P. Kupecek, "Analyse des expériences de génération de second harmonique," *Rev. Phys. Appl.* **6**, 31 (1971).
9. F. Kajzar and J. Messier, "Third harmonic generation in liquids", *Phys. Rev. A* **32**, 2352 (1985).
10. W. N. Herman and L. M. Hayden, "Maker fringes revisited: second harmonic generation from birefringent or absorbing materials," *J. Opt. Soc. Am. B*. **12**, 416 (1995).
11. L. M. Hayden, G. F. Sauter, F. R. Ore, P. L. Pasillas, J. M. Hoover, G. A. Lindsay, and R. A. Henry, "Second-order nonlinear optical measurements in guest-host and side-chain polymers," *J. Appl. Phys.* **68**, 456 (1990).
12. D. Neher, A. Wolf, C. Bubek, and G. Wegner, *Chem . Phys. Lett.* **163**, 116 (1989).
13. See, for example, M. Born and E. Wolf, *Principles of Optics* (Pergammon Press, New York, 1975), Section 14.2.
14. N. Bloembergen, *Nonlinear Optics*, (W. A. Benjamin, Inc., New York, 1965), Chapter 4.
15. D. J. Williams, "Nonlinear optical properties of guest-host polymer structures", in *Nonlinear Optical Properties of Organic Molecules and Crystals*, D. S. Chemla and J. Zyss, ed., (Academic Press, Inc., New York, NY, 1987), p. 405.
16. D. Jungbauer, I. Teraoka, D. Y. Yoon, B. Reck, J. D. Swalen, R. Tweig, and C. G. Willson, "Second-order nonlinear optical properties and relaxation characteristics of poled linear epoxy polymers with tolane chromophores," *J. Appl. Phys.* **69**, 8011 (1991).
17. A. Nahata, J. Shan, J. T. Yardley, and C. Wu, "Electro-optic determination of the nonlinear-optical properties of a covalently functionalized Disperse Red 1 copolymer," *J. Opt. Soc. Am. B* **10**, 1553 (1993).

RECEIVED March 2, 1995

Chapter 21

Stark Spectroscopy as a Tool for the Characterization of Poled Polymers for Nonlinear Optics

M. I. Barnik[1], L. M. Blinov[2], T. Weyrauch[3,4], S. P. Palto[2], A. A. Tevosov[2], and W. Haase[3]

[1]Organic Intermediates and Dyes Institute, 103787, B. Sadovaya 1-4 Moscow, Russia
[2]Institute of Crystallography, Russian Academy of Science, 117333, Leninsky pr. 59, Moscow, Russia
[3]Technische Hochschule Darmstadt, Institut für Physikalische Chemie, Peterenstrasse 20, 64287 Darmstadt, Germany

The Stark spectroscopy (electroabsorption) technique was developed for the case of a cylindrically symmetric polar molecular ensemble. The quadratic-in-field electroabsorpton was used to evaluate the difference of the dipole moments of the ground and excited state and subsequently the linear-in-field electroabsorption was applied to study the polar order parameter of the chromophores induced by electric fields in thermal poling processes as well as by photoassisted poling at room temperature. The polar order has been measured for several systems including guest-host polymer-dye solutions and side chain polymers with chromophore pending groups. With the polar order parameter measured in a corona poled side chain polymer using the linear Stark effect, the second order nonlinear optical susceptibility has been calculated, which agrees with the same value measured directly by the second harmonic generation (SHG) technique. It is shown, that the quadratic-in-field effect allows one to distinguish between soft and rigid polymeric matrices.

Due to their low cost, easy processibility and high performance characteristics polymers possessing nonlinear optical properties are of current scientific and technological interest (*1*). In certain cases, a glassy polymer may have a polar structure, e. g., when a polymeric ferroelectric liquid crystal is cooled down from the melt. Amorphous polymers may be prepared in the form of nonequilibrium but long living electrets when they are cooled down from the melt with a d.c. electric field applied (*2,3*). Such nonlinear optical materials may possess a rather high second order susceptibility $\chi^{(2)}$ and exhibit larger electro-optic coefficients than e. g. lithium niobate. Poled (polar) polymers containing nonlinear optical chromophores aligned by

[4]Corresponding author

0097–6156/95/0601–0288$12.00/0

an electric d.c. field are of great interest as media generating the optical second harmonic (*4-6*). The main advantage of such materials is that they can be easily prepared in the form of thin films of large area and good optical quality.

There are many problems relevant both to the physical mechanism and to the optimization of the poling process still to be solved. Among them of great importance is the problem of a quantitative description of the degree of the polar order and its relaxation with time for the chromophores responsible for the nonlinear properties of a material irrespective of the type of poling used. The measurement of the macroscopic polarization (the dipole moment of a unit volume) using the thermally stimulated depolarization current technique is strongly influenced by any charges trapped in the sample and is, in principle, a destructive technique since it requires heating of the sample. The intensity of the second harmonic generation (SHG) itself depends on many factors and needs a special calibration. With an electrooptic technique (e. g. the Pockels effect used in (*7,8*)) the field induced birefringence is measured, which is only indirectly related to the polar order parameter. In addition, a special geometry of the sample is necessary for electrooptic measurements (a Mach-Zehnder interferometer (*9*), reflecting prisms (*7*), etc.).

The Stark spectroscopy, i. e. the measurement of the field induced changes in the absorption spectrum (electroabsorption), allows the determination of the polar order parameters of dipolar chromophores. Its application to Langmuir-Blodgett films (*10,11*) has already been reported. Some preliminary results on Stark spectroscopy application to electrically poled polymers have also been published (*12*). The idea is to measure first the magnitude of the dipole moment difference $\Delta\mu$ for a relevant dye on an isotropic polymer solution from the quadratic-in-field Stark effect and then to measure polar order parameter for the same polymer poled in a d.c. electric field from the linear-in-field Stark effect characteristic of noncentrosymmetric (polar) media.

The paper is organized as follows: First a theory for the quadratic- and-linear-in-field electroabsorption is presented for a rigid and nonrigid molecular ensembles including polar ones. Then the simpliest case of two model isotropic solid solutions (guest-host systems) of the same dye in polycarbonate (PC, rigid ensemble) and poly(methylmethacrylate) (PMMA, nonrigid ensemble) are experimentally studied. After this, two other model systems representing polar molecular ensembles are studied: A dye solution in PMMA poled from electrodes and a side chain polymer poled by the corona discharge, both on cooling from the melt. The value of the polar order parameter for the corona poled sample is used for the calculation of $\chi^{(2)}$ to be compared with the same value measured in a SHG experiment. Finally, an example of the polar structure prepared by photoassisted poling is discussed. The latter was suggested originally in (*7,8*).

Quadratic- and Linear-In-Field Electroabsorption

The geometry of the Stark spectroscopy experiments is shown in Figure 1. A polymer film is sandwiched between two optically transparent electrodes (shaded areas), the direction of the poling and the measuring field $\boldsymbol{E}$ coincides with the wave vector of light (the z-axis). The light polarization vector $\boldsymbol{e}$ is always located in the x,y-plane. For rod-like molecules of dyes the direction of the dipole moment of the ground state of the chromophore μ_g usually coincides with its long molecular axis and the transition dipole moment μ_{ge} of the longest wavelength absorption band. In typical cases (e. g., for azo- or stilbene chromophores) the direction of the dipole moment of the excited

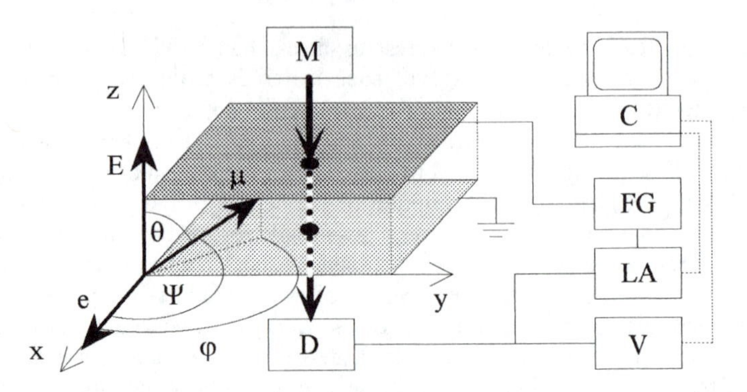

Figure 1. The geometry of a sandwich cell (***E***: d.c. electric field; ***e***: arbitrarily chosen electric vector of nonpolarized light; μ: chromophore dipole moments) and the sketch of the set-up for electroabsorption measurements (M: monochromator; D: photomultiplier; V: d.c. voltmeter; LA: lock-in amplifier; FG: function generator; C: computer).

state μ_e for the transition mentioned also coincides with the longitudinal axis. Thus all of the vectors μ_e, μ_g, μ_{ge} and $\Delta\mu=\mu_e-\mu_g$ are collinear (direction μ in the Figure).

Let us consider first an orientationally isotropic ensemble consisting of molecules with only two energy levels (i), corresponding to the ground and an electronic excited states. The two molecular states are characterized by their own dipole moments and polarizabilities. The corresponding energy levels of a molecule fixed rigidly in space are shifted in an electric field ***E*** by

$$\Delta W_i = (\mu_i \boldsymbol{E}) + (1/2)\alpha_i E^2. \tag{1}$$

Thus the wavenumber of the optical transition is changed by

$$c\,\Delta\nu_{ik} = h^{-1}\left[(\Delta\mu_{ik}\boldsymbol{E}) + (1/2)\Delta\alpha_{ik}E^2\right], \tag{2}$$

where c and ν are light velocity and wavenumber, h is Planck's constant.

Isotropic Molecular Ensemble. For a molecular ensemble with an isotropic distribution of dipoles over angles the optical absorption in the static electric field was calculated by Liptay (*13,14*). In the simplest case of a rigid ensemble with one absorption band and the electric field coinciding with the direction of light wave vector (the case of sandwich cells) the field induced change in absorption is

$$\Delta A = E^2\left[\frac{(\Delta\mu)^2}{10h^2c^2}\nu\frac{\partial^2(A/\nu)}{\partial\nu^2} + \frac{\Delta\alpha}{2hc}\nu\frac{\partial(A/\nu)}{\partial\nu}\right], \tag{3}$$

where A is the absorbance (optical density) of the sample. The field E is a local electric field at the chromophore. For a rigid molecular ensemble it may be taken in the Lorentz form $E = E_{ext}(\varepsilon+2)/3$, where ε is the dielectric constant of the substance at the frequency of the external field E_{ext} applied.

In the electroabsorption experiment a lock-in amplifier measures the effective value of the field induced change ΔT_{eff} of the light intensity transmitted by the sample. ΔT_{eff} is normalized by a computer to the transmission T itself, measured as the d.c. component of the photomultiplier output. Thus the field induced absorption is:

$$\Delta A_{eff} = -\Delta T_{eff} / (T \ln 10) \tag{4}$$

One usually measures the $\Delta A(\nu)$ spectrum at the second (2ω) or first harmonic (ω) of the applied a.c. field. If both, a d.c. and an a.c. field are applied simultaneously to a sample we have for the quadratic effect:

$$\Delta A \propto (E_0 + E_m \sin \omega t)^2 = E_0^2 + (E_m^2/2) + 2E_0 E_m \sin \omega t - (E_m^2/2)\cos 2\omega t \tag{5}$$

Thus, the measurement of the quadratic Stark effect may be carried out either at the second harmonic

$$\Delta A_{eff}(2\omega) = \frac{1}{\sqrt{2}} E_{eff}^2(\omega) \left[\frac{\Delta\alpha}{2hc} \nu \frac{\partial(A/\nu)}{\partial\nu} + \frac{(\Delta\mu)^2}{10h^2c^2} \nu \frac{\partial^2(A/\nu)}{\partial\nu^2} \right], \tag{6}$$

or at the first harmonic of the applied electric field

$$\Delta A_{eff}(\omega) = 2E_0 E_{eff}(\omega) \left[\frac{\Delta\alpha}{2hc} \nu \frac{\partial(A/\nu)}{\partial\nu} + \frac{(\Delta\mu)^2}{10h^2c^2} \nu \frac{\partial^2(A/\nu)}{\partial\nu^2} \right]. \tag{7}$$

In the latter case, in addition to the electric a.c. field a d.c. field E_0 has to be applied to the sample. For $E_0 = E_{eff}$, the ratio

$$\Delta A_{eff}(\omega) = 2\sqrt{2}\, \Delta A_{eff}(2\omega) \tag{8}$$

may be a criterion of a rigid molecular ensemble.

The absorbance $A(\nu)$ of a sample and its first and second derivatives are assumed to be known from independent measurements. Thus, as soon as the electroabsorption of an isotropic sample at a field E is measured, the parameters $\Delta\mu$ and $\Delta\alpha$ can be calculated and afterwards used in calculations of the polar order parameters for anisotropic ensembles.

In some cases an internal (built-in) electric field E_{int} may exist in a sample (in general due to an asymmetry of the film preparation, e.g., different electrode materials). The internal field linearizes the quadratic Stark effect (the quasi-linear effect) and may be found from equation 7 (where it should stand instead of E_0) as soon as all of the other parameters are known.

Polar Rigid Ensemble. For noncentrocymmetric (polar) molecular ensembles the linear Stark effect is not averaged out and the electroabsorption must be observed at the first harmonic of the applied a.c. voltage even in the absence of any (external or internal) d.c. field. This electroabsorption is due to the dipolar ($\Delta\mu$) contribution only,

$$\Delta A_{\Delta\mu} = \frac{\Delta\mu E}{hc} \nu \frac{\partial (A_m/\nu)}{\partial \nu} I_1 + \frac{\Delta\mu^2 E^2}{2(hc)^2} \nu \frac{\partial^2 (A_m/\nu)}{\partial \nu^2} I_2, \tag{9}$$

where

$$I_1 = \iint_{\varphi,\theta} \cos\theta \sin^2\theta \cos^2\varphi \; f(\varphi,\theta) \sin\theta \; d\varphi d\theta = \frac{1}{2}\left(\langle\cos\theta\rangle - \langle\cos^3\theta\rangle\right) \tag{10}$$

$$I_2 = \iint_{\varphi,\theta} \cos^2\theta \sin^2\theta \cos^2\varphi \; f(\varphi,\theta) \sin\theta \; d\varphi d\theta = \frac{1}{2}\left(\langle\cos^2\theta\rangle - \langle\cos^4\theta\rangle\right) \tag{11}$$

(see Figure 1 for the definition of the angles φ and θ). Here

$$A_m = \frac{2A}{1 - \langle\cos^2\theta\rangle} \tag{12}$$

relates the absorbance A in the geometry of Figure 1 to the absorbance A_m of a (theoretical) ensemble where all oscillators are parallel to the electric field vector of light. Note, that equations 10 and 11 are only valid for the geometry where the vectors $\Delta\mu$ and μ_{ge} are collinear and the light polarization vector is perpendicular to the electric field $\boldsymbol{E}$.

The linear-in-field electroabsorption observed at the field frequency ω is determined by the first term of equation 9

$$\Delta A_{eff,\omega} = E_{eff,\omega} \frac{\Delta\mu}{hc} \nu \frac{\partial (A/\nu)}{\partial \nu} S_1, \tag{13}$$

where

$$S_1 = \frac{\langle\cos\theta\rangle - \langle\cos^3\theta\rangle}{1 - \langle\cos^2\theta\rangle}. \tag{14}$$

The term S_1 may play the role of the polar order parameter and could be found from equation 13 if the $\Delta\mu$ value has already been found from the quadratic Stark effect. It should be noted that the same factor $\langle\cos\theta\rangle - \langle\cos^3\theta\rangle$ determines the second order nonlinear optical susceptibility $\chi^{(2)}_{zxx}$ and the intensity of the optical second harmonic generation (*12*). Thus the electroabsorption technique may be a good tool for a quantitative characterization of nonlinear-optical properties of materials.

For a weakly anisotropic ensemble with the Boltzmann distribution function $\exp(-\mu \boldsymbol{E}_p/kT) \cong 1$ we have

$$S_1 \cong \mu E_p / 5kT, \tag{15}$$

where μ is the static dipole moment interacting with the poling field E_p, k is the Boltzmann constant and T is the poling temperature.

The quadratic-in-field electroabsorption at the second harmonic 2ω of the applied field has also to include the contribution from the change in molecular polarizability $\Delta\alpha$. Let us assume the cylindrical shape of the chromophores, then $\Delta\alpha_L$ and $\Delta\alpha_T$ are changes of the longitudinal and transversal molecular polarizabilities upon excitation. For the cylindrical symmetry of the chromophore distribution we have

$$\Delta A_{\Delta\alpha} = \frac{1}{2hc}\left\langle (\mathbf{E},\Delta\alpha\mathbf{E})\cos^2\Psi \, \nu\frac{\partial(A_m/\nu)}{\partial\nu}\right\rangle = \nu\frac{\partial(A_m/\nu)}{\partial\nu}\times$$
$$\frac{1}{2hc}\iint_{\varphi,\theta}\left(\Delta\alpha_L E^2\cos^2\theta + \Delta\alpha_T E^2\sin^2\theta\right)\sin^2\theta\cos^2\varphi\, f(\varphi,\theta)\sin\theta\, d\varphi\, d\theta \tag{16}$$
$$= \frac{E^2}{2hc}\left[\Delta\alpha_L S_2 + \Delta\alpha_T(1-S_2)\right]\nu\frac{\partial(A/\nu)}{\partial\nu}$$

with angles Ψ, φ and θ as described in Figure 1. S_2 is given by

$$S_2 = \frac{\langle\cos^2\theta\rangle - \langle\cos^4\theta\rangle}{1-\langle\cos^2\theta\rangle}. \tag{17}$$

Finally, due to the contributions of $\Delta\mu$ and $\Delta\alpha$, the quadratic-in-field electroabsorption at the second harmonic of the applied field is:

$$\Delta A_{eff,2\omega} = \frac{E^2_{eff,\omega}}{2\sqrt{2}}\left[\frac{\Delta\alpha_L S_2 + \Delta\alpha_T(1-S_2)}{hc}\nu\frac{\partial(A/\nu)}{\partial\nu} + \frac{(\Delta\mu)^2 S_2}{h^2c^2}\nu\frac{\partial^2(A/\nu)}{\partial\nu^2}\right] \tag{18}$$

For an isotropic ensemble $<\cos\theta>=<\cos^3\theta>=0$, $<\cos^2\theta>=1/3$, $<\cos^4\theta>=1/5$, $S_1=0$, $S_2=1/5$ and the two equations 13 and 18 are reduced to one Liptay equation with $\Delta\alpha = \Delta\alpha_L = \Delta\alpha_T$ (equation 3).

Non Rigid Ensemble. Due to a finite rotational mobility of dipoles an electric d.c. field may induce a dipolar order even in the glassy state, which can be described by a dynamic polar order parameter S_1^{dyn}. It means that now a new contribution proportional to the first derivative to the absorption spectrum appears during the measurements of the quadratic Stark effect at the first harmonic. According to equation 7 it is indistinguishable from the $\Delta\alpha$ contribution and results in an enhanced apparent value $\Delta\alpha_{app}$ measured. From this apparent value the dynamic polar order parameter may be calculated

$$S_1^{dyn} = (\Delta\alpha_{app} - \Delta\alpha)E_0/\Delta\mu \tag{19}$$

as soon as the values of $\Delta\mu$ and $\Delta\alpha$ (found from equation 6) are known. In equation 19 E_0 is the same local field from the external source as that in equation 7. For a nonrigid ensemble care has also to be taken of the $\Delta\mu$ value, which may be found either from equation 6 or from equation 7, see below.

An other contribution to the electroabsorption of a nonrigid ensemble results from a slight modulation of the absorbance due to the a.c. field induced reorientation of the chromophores (the field induced dichroism). It is exactly the same effect which determines the so-called electrochromism in liquid solutions of dyes. Its spectral contribution is proportional not to absorbance derivatives but to absorbance $A(\nu)$ itself (*13,14*).

$$\Delta A_{eff} = kA(\nu)/\nu \tag{20}$$

Experimental Technique

Substances. Three different systems were studied experimentally (cf. Figure 2): (i) solid solutions of an azodye, p-diethylamino-p'-nitroazobenzene (DEANAB) in poly(methylmethacrylate) (PMMA) and polycarbonate (PC); (ii) the stilbene dye p-diethylamino-p′-nitrostilbene (DEANS) in PC and (iii) the side chain copolymer poly-[(ethylen)-co-(N-hexyl-N-(4-methacryloxy-butyl)-4-amino-4'-nitrostilbene)] (E/MA-ANS). For the latter the ratio of repetition units (ethylen: methacrylate) was about 9:1. This particular choice of substances allowed us to compare the behaviour of the guest-host systems with that of functional side-chain polymers on the one hand, and the systems with a stronger (azobenzene) and weaker (stilbene) tendency to the *trans-cis-trans* isomerisation on the other. In addition, the PMMA and PC matrices differ considerably in their rigidity.

Samples and Poling Conditions. For the DEANAB and DEANS guest-host system, solution of the dyes and PC were prepared in cyclopentanone, the concentration of dye with respect to the polymer being 2.4-2.5 wt%. The 0.4μm (DEANS) and 0.9μm

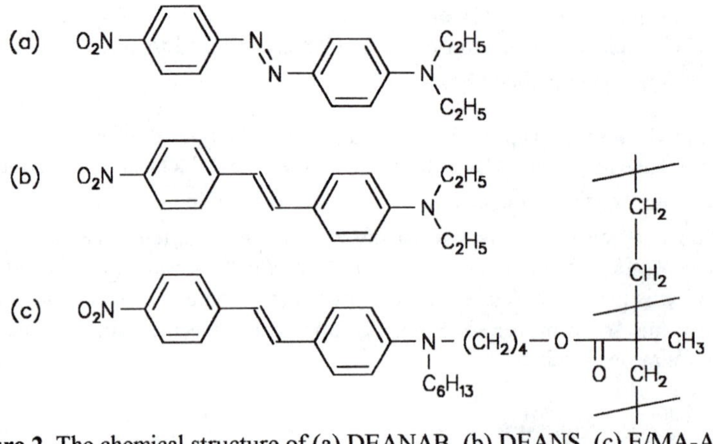

Figure 2. The chemical structure of (a) DEANAB, (b) DEANS, (c) E/MA-ANS.

(DEANAB) thick films were prepared on glass plates covered with transparent ITO electrodes using spin-coating of the polymer-dye solution. The E/MA-ANS polymer was also dissolved in cyclopentanone with a weight concentrations of 15% and a 0.27μm thick film was obtained by spin-coating. All of the films were dried under vacuum at elevated temperatures. Aluminium top electrodes were evaporated onto the film surfaces after drying. A part of each film, which were shown to be uniform in thickness, was left uncovered for the absorbance measurements. The area of electrode overlapping was about $5\times5mm^2$. In some cases (DEANAB in PMMA and PC) samples of thicknesses 10-20μm were prepared just by squeezing a polymer solution of the dye (0.2wt%) between two ITO covered glass plates at a temperature slightly above the glass point.

The thicknesses of the films were calculated from the cell capacities in the frequency range of 100Hz-10kHz (with dielectric constant assumed to be 2.5). In some cases the thicknesses might also be estimated from the interference fringes observed in absorption spectra and the results of the two techniques were consistent with each other. After measuring the capacities and the spectra of absorption and electroabsorption of the unpoled films and determining molecular parameters $\Delta\mu$ and $\Delta\alpha$ the samples were poled.

We used three types of poling: (i) on cooling from the melt with a d.c. field applied from the electrodes; (ii) on cooling from the melt in a corona discharge; and (iii) at room temperature under strong illumination (*7,8*). For poling from the melt a d.c. voltage of 150V was applied to a 15.5μm thick sample of DEANAB-PMMA. For corona poling a specially prepared (dried under vacuum at 110°C for 23h) film of E/MA-ANS was heated to 90°C. A voltage was applied to the corona needle (10kV) and to a grid between the needle and the sample (200V) and then the film was cooled down to room temperature within 35min. The photoassisted poling was carried out by illumination of the samples from the ITO electrode side for 15 min with filtered light of a 150W halogen lamp focused with a F=50mm lens (the power density at the sample was of $200mW/cm^2$ in the spectral range of 400nm-500nm). During the illumination an electric d.c. voltage was applied to the samples.

Measurements. Absorption spectra in the visible range (350-650nm) were measured with a Cary-17 spectrometer equipped with a computer. The Stark (electro-absorption) spectra were measured using home made spectrometers, based on the prism and grating monochromators Zeiss MM12 and the russian MDR-23, respectively. The sketch of the set-up is shown in Figure 1. Incandescent halogen lamps as light sources and photomultipliers as light detectors were used. The electric field of frequency 500Hz was applied to the samples and the signal of the quadratic and linear-in-field modulation of the optical transmission $\Delta T(\lambda)$ was detected either at the doubled or the fundamental frequency of the applied field by a lock-in amplifier also equipped with a computer. The light intensity level $T(\lambda)$, that is the optical transmission of the sample (with electrodes), was recorded simultaneously using the d.c. output of the same photodetector. Then a normalized spectrum of $\Delta A_{eff}(\nu)$ was calculated (equation 4) and fitted to the spectra of absorbance derivatives according to equations 6, 7, 13 and 18 in order to find the parameters of the excited state of the chromophores ($\Delta\mu$ and $\Delta\alpha$) and the polar order parameter S_1 induced by the electric field.

The measurements of the optical second harmonic generation were performed using a Nd/YAG laser at a fundamental of 1064nm. The Maker fringes technique (*15-17*) was used to estimate the second order susceptibility. $\chi^{(2)}_{zzz}$ values were evaluated from experiments with p-polarized incident light assuming $\chi^{(2)}_{zzz} = 3\chi^{(2)}_{zxx}$ (*18*).

Results on Electroabsorption

Isotropic Rigid and Nonrigid Guest-host Systems. Two model cells were prepared in order to compare the field behaviour of a rigid and a nonrigid isotropic ensemble: Solutions of the same dye (DEANAB) in PC (rigid matrix) and PMMA (nonrigid matrix). Figure 3 shows the absorption spectra of DEANAB in PC (curve 1) and the quadratic electroabsorption spectra of the sample measured at the second (curve 2) and first (curve 3) harmonic of the applied field. It is easily seen that the two spectra have the same shape and their amplitudes satisfy the relationship of equation 8. It means that our PC matrix is indeed rigid, which is consistent with the results on the SHG relaxation (*19*). The fitting of equations 6 and 7 using only two parameters to be varied gave us the following values for $\Delta\mu$ and $\Delta\alpha$: $\Delta\mu$=10.1D, $\Delta\alpha$=54Å^3. The value of $\Delta\mu$ is two times less than that measured for very diluted liquid solutions of the same dye (*14,20*), which may be accounted for by a certain compensation of dipole moments due to a partial aggregation of the dye molecules in 2.5% concentration polymer solutions. The other possible reason is a partial compensation of the $\Delta\mu$ value (that is a degree of the intramolecular charge transfer on excitation) due to a specific interaction of the dye with the PC matrix.

Figure 4 shows the quadratic electroabsorption spectra of the sample DEANAB-PMMA measured at the second (curve 2) and first harmonics (curve 3) of the applied field. The absorption spectrum (curve 1) of the PMMA solution of the dye (now concentration is only 0.2%) has the same shape as that of its PC solution. However,

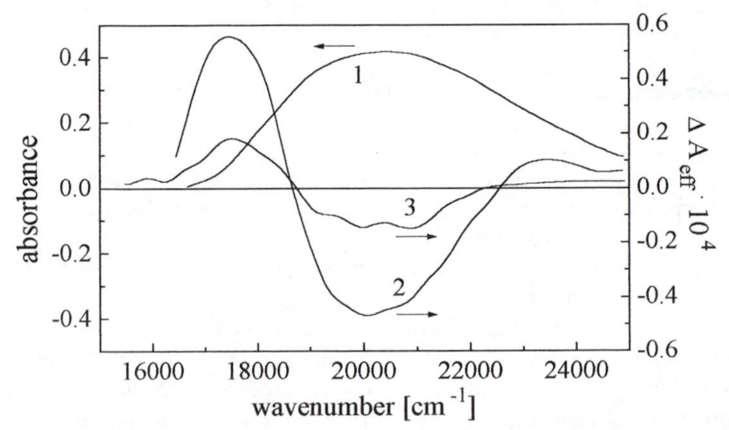

Figure 3. DEANAB in PC, isotropic 2.5% solution, film thickness 0.9µm: Absorbance spectrum (curve 1), quadratic electroabsorption at the second harmonic (U_{eff}=21.4V, curve 2) and first harmonic of the applied voltage (U_{eff}=10.2V, U_0=4.8V, curve 3).

now the two electroabsorption spectra have different shape (contribution of the term proportional to the first derivative of absorption is much higher for the spectrum 3) and their amplitudes are too far from each other to satisfy the relationship in equation 8. The softness of the PMMA matrix, especially not aged, has already been discussed (*21*). The fitting, in addition to equations 6 and 7, requires for the third parameter to be varied (equation 20), and gave us the following values: For measurements at the second harmonic $\Delta\mu$=6D, $\Delta\alpha$=43Å^3, k=–0.1; for the first harmonic with a d.c. voltage applied: $\Delta\mu$=15.8D, $\Delta\alpha_{app}$=672Å^3, k=–0.5. A smaller value of $\Delta\mu$=6D (d.c. field-off) in PMMA may be accounted for by a stronger association of DEANAB molecules in the latter matrix even at smaller dye concentration, or a stronger interaction of the dye with PMMA matrix, more polar than PC (note, that the $\Delta\alpha$ values in the two polymers are close to each other). In the d.c. field-on regime a considerable field induced dipole orientation occurs, the dynamic polar order parameter S_1^{dyn} calculated from equation 19 is equal to 0.02, which is very close to the value S_1=0.017 calculated from equation 15 for the Boltzmann distribution of dipoles μ_g=7D (*20*). Thus the dipoles rotate rather free in PMMA (at least in the limit of a small polar order). Not surprisingly that the value of $\Delta\mu$ in the d.c field-on regime increase from 6 to 15.8D due to a break of the dipole associates although this value is still smaller than that of individual DEANAB molecules. The high mobility of the dye molecules in PMMA also manifests itself in the finite value of the parameter k. Certainly the field induced dichroism is easily seen at the first harmonic of the applied field (k=–0.5, minus means that absorbance reduces with the field applied due to a deviation of chromophores to the z-axis) and a little less pronounced at the second harmonic (k=–0.1) just because of a certain inertia of the process.

Films Electrically Poled on Cooling. DEANAB in PMMA. Figure 5a shows the absorbance spectrum for the sample of a 0.5% solution of DEANAB in PMMA

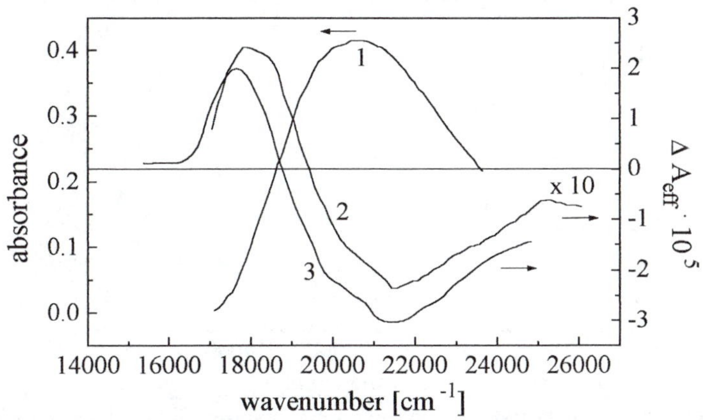

Figure 4. DEANAB in PMMA, isotropic 0.2% solution, film thickness 15.2µm: (1) Absorbance spectrum; (2) quadratic electroabsorption at the second harmonic, U_{eff}=120V (the scale should be reduced 10 times); (3) quadratic electroabsorption at the first harmonic, U_{eff}=117.5V, U_0=150V.

before (1) and after (2) electric field poling. The sample was prepared by squeezing a dye-polymer melt between two ITO covered glasses. A considerable difference in the maximum optical density for the two spectra in Figure 5 originates presumedly from a partial segregation of the dye that is from its transition to a solid state on heating. The thickness of the sample (15.5μm) was shown to be constant. Due to a very small field induced orientation of molecular dipoles the considerably large difference in absorbances cannot be accounted for by the field induced dichroism.

Figure 5b shows the quadratic electroabsorption spectrum of the sample taken before poling at the second harmonic of the applied voltage (U_{eff}=119V, curve 1), and the linear electroabsorption spectrum at the first harmonic of the applied a.c. voltage (U_{eff}=115V with the poling d.c. field switched off) taken after poling the sample in a cooling process with a d.c voltage of 150V applied (curve 2). The magnitude of the field induced absorption measured at the first harmonic of the applied a.c. voltage was negligible before poling (no quasilinear effect).

With the local field assumed to be of the Lorentz form (with dielectric constant of PMMA at 500Hz equal to 2.5) the fitting parameters found from the quadratic effect

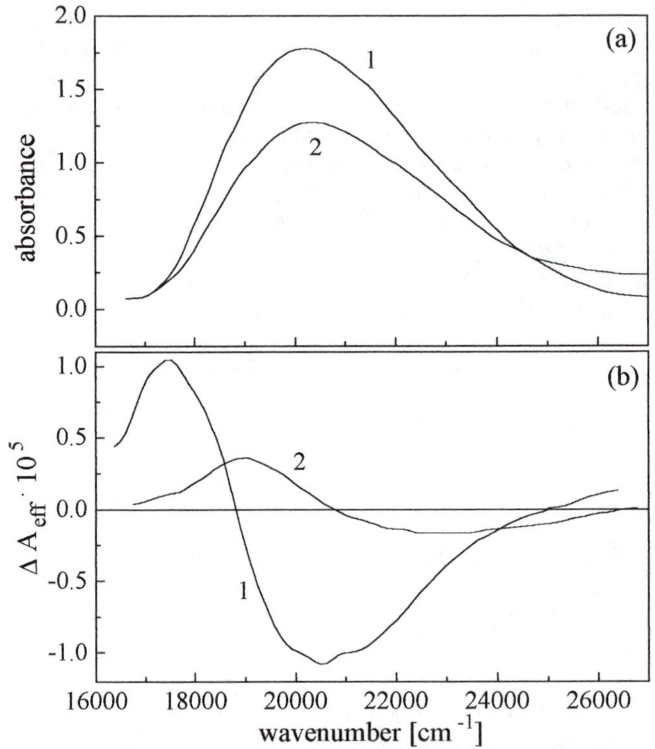

Figure 5. DEANAB in PMMA, isotropic 0.5% solution, film thickness 15.5μm: a) Absorbance spectrum before (1) and after (2) electric field poling; b) Quadratic electroabsorption spectrum taken before poling at the second harmonic of the applied voltage U_{eff}=119V (1) and the linear electroabsorption spectrum at the first harmonic of the applied a.c. voltage U_{eff}=115V taken after poling (2).

are $\Delta\mu$=8.3D, $\Delta\alpha$=46Å^3. The fitting of curve 2 shows that now the contribution of the second derivative term is negligibly small. From equation 13 the value of the static polar order parameter S_1 induced by the poling field is calculated to be of 1.6×10^{-4}, which is much smaller than S_1 to be induced by the poling field E_{ext}=7.5×10^4V/cm for the Boltzmann distribution of dipoles. The reason for this lies in a softness of the PMMA matrix discussed above, although a depolarizing role of the free ions, which (being mobile in a plastic state) screen the external field and reduce the polar order, may also be important.

Side Chain Polymer Poled by a Corona Discharge. The absorbance (curve 1) and electroabsorption (curves 2, 3) spectra of a 0.6μm thick film of the stilbene side chain polymer E/MA-ANS are shown in Figure 6. The spectral form of the quadratic Stark effect measured at the second harmonic of the a.c. applied voltage U_{eff}=20V, f=500Hz (curve 2) is mainly determined by the term proportional to the second derivative of the absorption spectrum. From the corresponding fitting a $\Delta\mu$ value of 8D has been calculated for the stilbene chromophore. This value is close to that anticipated from its molecular structure. The quadratic electroabsorption spectrum measured at the first harmonic of the applied field gave the same $\Delta\mu$ value. It means that E/MA-ANS may be considered as a rigid ensemble of chromophores. The spectrum of the linear Stark effect (curve 3) measured at the first harmonic of the applied voltage (U_{eff}=9.0V) two weeks after corona poling is mainly determined by the contribution proportional to the first derivative to absorption. Thus, with the previously found value of $\Delta\mu$=8D, the parameter of the field induced (static) polar order may be calculated. Its magnitude S_1=0.02 calculated from equation 11 is much higher than the value found above for the order parameter induced by the contact poling of PMMA based materials and comparable with the value (about 0.1 for μ_g=7D) anticipated from equation 15 for a local field strength of the order of 10^6V/cm typical for the corona discharge. The reason for a somewhat smaller value of S_1 is the relaxation of the polar order parameter observed also in the measurements of the time dependence of the SHG intensity, see below.

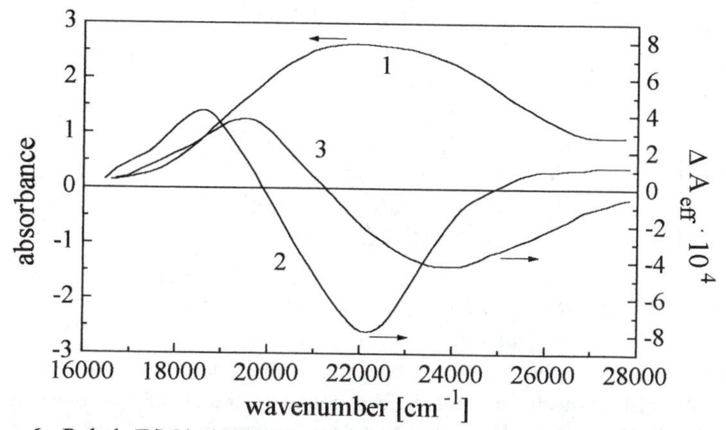

Figure 6. Poled E/MA-ANS sample: Absorbance (1) and electroabsorption spectra at the second (2, U_{eff}=20V) and the first harmonic (3, U_{eff}=9.0V).

Photoassisted Poled Systems. We made electroabsorption measurements on several systems electrically poled at room temperature with simultaneous irradiation of a film by intense light absorbed in the longest wavelength band of chromophores: Here we present only one example, namely the results obtained for the guest-host system DEANAB in PC, the other results will be published elsewhere. The wavenumber dependence of the absorbance of a 0.9μm thick DEANAB/PC film is plotted in Figure 7 (curve 1). Curve 2 shows the electroabsorption spectrum of the same sample taken before poling at the second harmonic of the applied voltage. With the local field assumed to be of the Lorentz form and dielectric constant of PC at 500Hz equal to 2.5 the fitting parameters found are $\Delta\mu$=10D and $\Delta\alpha$=54Å^3.

Before poling the magnitude of the field induced absorption measured in the absence of an external d.c. field at the first harmonic of the applied a.c. voltage was negligible. After the photoassisted poling of the sample with a d.c voltage of 20.5V the linear Stark effect appeared, which could be easily measured at the first harmonic of the applied a.c. voltage with the poling d.c. field switched off. The electroabsorption spectrum of the poled sample taken the next day after the poling is also shown in Figure 7 (curve 3). The fitting with equation 11 shows that now the contribution of the second derivative term is small compared with the first derivative. The static polar order parameter induced by the poling field calculated from the electroabsorption measurements is S_1=0.007 which is 6 times less than the value of 0.04 calculated with equation 15 from the Boltzmann distribution of the dipoles with the moment μ_g=7D in the local poling field of 33V/μm.

The light and field induced orientation of DEANAB molecules relaxes with time. The relaxation of the linear Stark effect has been measured by electroabsorption at wavelength of 470nm and two characteristic relaxation times were found: the first stage of relaxation with a time constant about 20min. Then the linear electroabsorption signal is leveled off and the second characteristic time estimated is about 280 hours.

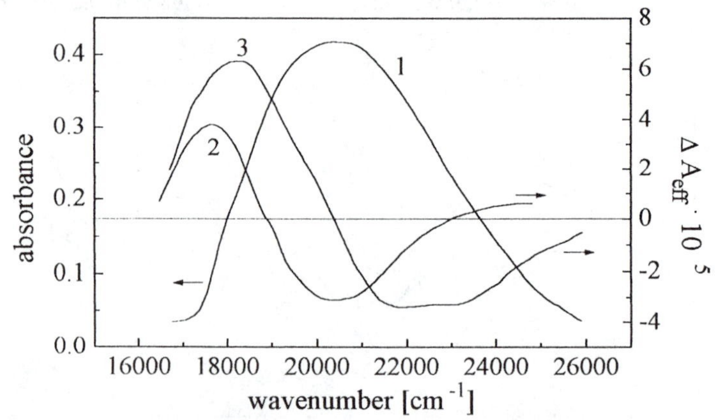

Figure 7. DEANAB/PC system, 0.9μm thick film: Absorbance spectrum (1), electroabsorption spectrum taken before poling at the second harmonic of the applied voltage U_{eff}=20.3V (2) and the electroabsorption spectrum after photo-assisted poling of the same film measured at the first harmonic of the applied voltage U_{eff}=20.5V (3).

For stilbene derivatives the polar order induced by photopoling is even less than for azobenzene chromophores. The latter has been attributed to an easier *trans-cis-trans* isomerisation of the azobenzene moiety.

Relationship between the Microscopic and Macroscopic Optical Tensor. Having the polar order parameter of the chromophores determined, the absolute value for the second order nonlinear susceptibility may be calculated on the basis of a microscopic approach . We have carried out this procedure for the corona poled side chain stilbene polymer E/MA-ANS. The results are to be compared with the experimental ones. The measurements of the time dependence of the SHG intensity show that the relaxation of the second order susceptibility takes place just after the poling process (Table I).

The electroabsorption measurements on the corona poled polymer (Figure 6) were carried out twice, two weeks and a half a year after the poling procedure and no change has been noticed in the spectra. This is consistent with the data of Table I, showing that the relaxation almost finished in a few days.

Table I. Time Dependence of the Second Order Susceptibility after Poling

Time after poling	10 min	17 min	26 min	20 h	128 d
$\chi^{(2)}_{zzz}$ / (pm/V)	20.3 ± 3	18.7 ± 3	18.7 ± 3	16.2 ± 3	8.8 ± 1.5

For the calculation of the macroscopic nonlinear optical tensor $\chi^{(2)}$ nonlinear optical chromophores are considered to be linear, i. e. only one element of the molecular hyperpolarizabilty tensor β is taken into account (*22*):

$$\chi^{(2)}_{zzz} = N\beta_{zzz} f^{2\Omega} f^{\Omega} f^{\Omega} \langle \cos^3\theta \rangle \tag{20}$$

$$\chi^{(2)}_{zxx} = N\beta_{zzz} f^{2\Omega} f^{\Omega} f^{\Omega} \langle \cos\theta \sin^2\theta \cos^2\varphi \rangle = N\beta_{zzz} f^{2\Omega} f^{\Omega} f^{\Omega} \tfrac{1}{2} \langle \cos\theta - \cos^3\theta \rangle \tag{21}$$

(Here N = number of dipoles per unit volume, β_{zzz}= main tensor element of the molecular hyperpolarizability, $f^{2\Omega}$, f^{Ω} are local field correction factors at the optical frequencies 2Ω and Ω, respectively. For noninteracting dipoles in thermodynamic equilibrium and $\alpha << 1$ it follows (*23*):

$$\langle \cos\theta \rangle = \coth\alpha - \frac{1}{\alpha} \approx \frac{\alpha}{3}, \tag{22}$$

$$\langle \cos^3\theta \rangle = \left(1 + \frac{6}{\alpha^2}\right)\coth\alpha - \left(\frac{3}{\alpha} + \frac{6}{\alpha^2}\right) \approx \frac{\alpha}{5}, \tag{23}$$

where $\alpha = f^0(\mu_g E_p/kT)$, μ_g = ground state dipole moment, E_p = poling field, k = Boltzmann constant, T = temperature, f^0 = local field correction factor for the poling field.

Therefore $< \cos^3\theta > \approx \frac{3}{5} < \cos\theta >$ and $\chi^{(2)}_{zzz}$ may be calculated from molecular properties and the value $< \cos\theta >$ obtained from the polar order parameter measured by the Stark spectroscopy. Using N=7.6×10^{20}cm^{-3} (calculated from molar weight of the total monomer unit M=717g/mole and the density d=1g/cm^3), β_{zzz}=274esu (the value for dimethylaminonitrostilbene, which is a good approximation for our chromophore (*24*)), Lorentz expressions for the local field correction factors (*18*) $f^{2\Omega} = (n_{2\Omega}^2 + 2)/3 \approx 1.4$, $f^{\Omega} = (n_{\Omega}^2 + 2)/3 \approx 1.4$ and $< \cos\theta >= 0.05$ we find $\chi_{zzz}^{(2)}$=8.5 pm/V in good agreement (within error range) with the measured values for $\chi_{zzz}^{(2)}$ (Table I). Thus a combination of SHG and Stark spectroscopy may be a powerful method for evaluation of the molecular hyperpolarizability of NLO chromophores dispersed in a polymer matrix.

Conclusion

A model for the electric field induced changes in the absorbance of a cylindrically symmetric polar molecular ensembles has been developed. Then the electroabsorption (Stark spectroscopy) technique was applied to calculate the dipole moment difference $\Delta\mu$ between ground and excited state of the chromophores for the relevant optical transition from the quadratic-in-field effect. Using this parameter the linear-in-field electroabsorption was applied to study the polar order parameter induced by thermal poling (either with the electrodes attached or a corona discharge) and photoassisted poling in an electric d.c. field of both dilute polymer solutions of azobenzene and stilbene dyes in PMMA and PC and a side chain polymer with stilbene chromophores as pending groups. For the thermally poled guest-host systems DEANAB in PMMA the polar order parameter found was small compared to the values anticipated from the Boltzmann distribution of dipoles in the electric field, which may be accounted for by screening the poling field by mobile ions. On the contrary, the order parameter of DEANAB in a photoelectrically poled DEANAB/PC system was shown to be much higher (0.007). For the corona poled side chain stilbene polymer, with the $< \cos\theta >$ value measured using the linear Stark effect, the second order nonlinear susceptibility has been calculated, which agrees with the same value measured directly by the SHG technique.

Acknowledgments

The authors are grateful to S. Saal, S. G. Yudin and N. N. Petukhova for technical help. Financial support by the Volkswagen-Stiftung (Germany), the International Science Foundation (grant No. M35000), and INTAS (Brussels, grant No. 93-1700) is gratefully acknowledged.

Literature Cited

1. see, e. g., *Nonlinear Optical Properties of Organic Materials VI*; Möhlmann, G. R., Ed.; Proc. SPIE 2025; The Society of Photo-Optical Instrumentation Engineers: Bellingham, WA, 1993.
2. Singer, K. D.; Lalama, S. J.; Sohn, J. E.; Small, R. D. In *Nonlinear Optical Properties of Organic Molecules and Crystals*; Chemla, D. S.; Zyss, J., Eds.; Academic Press: New York, NY, 1987; Vol. 1; Chap. II-8.
3. Williams, D. J. In *Nonlinear Optical Properties of Organic Molecules and Crystals*; Chemla, D. S.; Zyss, J., Eds.; Academic Press: New York, NY, 1987; Vol. 1; Chap. II-7.

4. Havinga, E. E.; van Pelt, P. *Ber. Bunsenges. Phys. Chem.* **1979**, *83*, 816.
5. Singer, K. D.; Sohn, J. E.; Lalama, S. J. *Appl. Phys. Lett.* **1986**, *49*, 248.
6. Singer, K. D.; Kuzyk, M. G.; Holland, W. R.; Sohn, J. E.; Lalama, S. J.; Comizzoli, R. B.; Kats, H. E.; Schilling, M. L. *Appl. Phys. Lett.* **1988**, *53*, 1800.
7. Sekkat, Z.; Dumont, M. *Appl. Phys.* **1992**, *B54*, 486.
8. Dumond, M.; Sekkat, Z. In *Nonconducting Photopolymers and Applications*; R. A. Lessard, Ed.;. Proc. SPIE 1774; The Society of Photo-Optical Instrumentation Engineers: Bellingham, WA, 1992; p. 188.
9. Singer, K. D.; Kuzyk, M. G.; Sohn, J. E. *J. Opt. Soc. Am.* **1987**, *B4*, 968.
10. Blinov, L. M. *Sov. Sci. Rev. (Phys. Ser.)* **1989**, *12 pt. 1*, 3.
11. Blinov, L. M.; Palto, S. P.; Yudin, S. G. *Mol. Mat.* **1992**, *1*, 183.
12. Barnik, M. I.; Blinov, L. M.; Palto, S. P.; Tevosov, A. A.; Weyrauch, Th. *Mol. Mat.* **1994**, *3*, 319.
13. Liptay, W. In *Modern Quantum Chemistry v. 3*; Sinanoglu, O., Ed.; Academic Press: New York, NY, 1965; p. 45.
14. Liptay, W. In *Adv. in Electronic Excitation and Relaxation*; Lim, E. C., Ed.; Excited States Vol. 1; Academic Press: New York, NY, 1974; p. 129.
15. Maker, P. D.; Terhune, R. W.; Nisenoff, M.; Savage, C. M. *Phys. Rev. Lett.* **1962**, *8*, 21.
16. Jerphagnon, J.; Kurtz, S. K. *J. Appl. Phys.* **1970**, *41*, 1667.
17. Singer, K. D.; Kuzyk, M. G.; Sohn, J. E. *J. Opt. Soc. Am.* **1987**, *B 4*, 968.
18. Singer, K. D.; Kuzyk, M. G.; Sohn, J. E. In *Nonlinear Optical and Electroactive Polymers*; Prasad, P. N.; Ulrich, D. R., Eds; Plenum Press: New York, NY, 1988; pp. 189-204.
19. Hampsch, H.; Yang, J.; Wong, G. K.; Torkelson, J. M. *Macromolecules* **1990**, *23*, 3648.
20. Shchapov, A. N.; Kornilov, A. I.; Basayev, P. M.; Chernyakovsky, F. P. *Zh. Fiz. Khim.* **1979**, *53*, 2568.
21. Hampsch, H.; Yang, J.; Wong, G. K.; Torkelson, J. M. *Macromolecules* **1990**, *23*, 3640.
22. Cross, G. H. In *Principles and Applications of Nonlinear Optical Materials*; Munn, R. W.; Ironside, C. N., Eds.; Blackie Academic & Professional: London, 1993; pp. 189-225.
23. Charney, E.; Yakamoto, K. *J. Am. Chem. Soc.*, **1972**, *94*, 8963.
24. Möhlmann, G. R.; van der Horst, C. P. J. M.; Huijts, R. A.; Wreesmann, C. T. J. In *Nonlinear Optical Properties of Organic Materials*; Khanarian, G., Ed.; Proc. SPIE 971; The Society of Photo-Optical Instrumentation Engineers: Bellingham, WA, 1988; p. 252.

RECEIVED April 10, 1995

Chapter 22

Pyroelectrical Investigation of Nonlinear Optical Polymers with Uniform or Patterned Dipole Orientation

S. Bauer, S. Bauer-Gogonea, Ş. Ylmaz, W. Wirges, and R. Gerhard-Multhaupt

Heinrich-Hertz-Institut für Nachrichtentechnik, Einsteinufer 37, D-10587 Berlin, Germany

In this survey, poled nonlinear optical (NLO) polymers are discussed as polymer electrets with molecular dipoles. A range of old and new poling techniques are briefly introduced. The pyroelectricity of poled NLO polymers can be exploited for analyzing the relaxation of the oriented chromophore dipoles under isothermal or non-isothermal conditions and for probing the distribution of the dipole orientation in the plane and across the thickness of poled films. The available experimental techniques and some typical results are briefly described. Advantages of pyroelectrical measurement techniques are their compactness, versatility, and ease of use. In view of possible waveguide applications of NLO polymers, two-layer systems with inverted-$\chi^{(2)}$ structures were prepared from one single NLO side-chain polymer by means of the combination of thermally assisted poling above the glass transition and photo-induced poling under reversed field below the glass transition as well as from two polymers with different glass-transition temperatures by thermally assisted poling at two different temperatures. The thermal stabilities of the resulting dipole-orientation patterns were studied by means of pyroelectric thermal analysis.

Second-order nonlinear optical (NLO) polymers are typical model systems of molecular dipole electrets. They consist of a glassy amorphous matrix with dispersed NLO dipole molecules [1]. Molecular-design approaches include guest-host systems in which the NLO dipoles are simply dopants dispersed within the host [2], side-chain systems with NLO dipoles chemically attached to the polymer backbone via flexible spacer units [3], main-chain polymers with NLO dipoles included into the main polymer chain [4], and cross-linkable polymers in which the NLO dipoles are linked to the polymer molecules by one or more chemical bonds during the poling process [5].

Poling, e.g. the permanent orientation of the NLO dipoles, is required for inducing the desired macroscopic NLO properties. The classical thermally assisted poling (TAP) technique (with electric fields from charge carriers in biased electrodes [2] or

0097–6156/95/0601–0304$12.00/0

from specifically deposited surface (corona) [6] or bulk (electron-beam) [7] charges) is in fact well known from the formation of other dipolar polymer electrets [8] such as e.g. ferroelectric polymers. In additon, several new poling techniques were recently introduced; they include photothermal poling (PTP) [9] with local heating by a focused light beam, photo-induced poling (PIP) [10] exploiting the photo-isomerism of the molecular dipoles, and suitable combinations of poling steps such as e.g. TAP and PIP [11].

Poling induces various effects in NLO polymers; they include piezo- and pyroelectricity (usually of concern in electret research), birefringence, linear electro-optic (Pockels) effect and any other second-order nonlinearity that is possible under the symmetry constraints of the respective tensor (such as e.g. second-harmonic generation (SHG)). Poled polymers lose their functionality as a consequence of dipole-relaxation processes. At least in principle, any property related to the dipolar order can be used for monitoring the stabilty of the dipole orientation. Dipole-relaxation processes were already studied by means of piezoelectrical [12] and pyroelectrical [13], birefringence [14], electro-optical [13], and SHG measurements [15].

A synergy of electret research (poling and dipole relaxation) and nonlinear optics (resulting optical nonlinearities and conceivable photonics applications) should benefit the further development of NLO polymers. In the following, NLO polymers are considered from the point of view of electret research, with a discussion of classical and new poling techniques as well as electrical investigations on the resulting dipole orientation. In particular, it is shown that pyroelectricity offers the most convenient way of investigating dipole-relaxation processes in compact, versatile, and easy-to-use experimental setups. In addition, the same setup can be employed for scanning the dipole polarization along all three space coordinates of a poled sample.

Poling of NLO Polymers

NLO polymers are most often poled by applying a sufficiently high voltage to evaporated electrodes at a temperature above the respective glass-transition temperature [2]. The oriented dipoles are stabilized by subsequent cooling under the applied electric field [2]. Poling via surface charging by means of a corona discharge is especially suited for the not always perfect samples from newly developed polymer films [6]. A well-designed corona setup includes a biased control grid and a rear electrode, which is grounded via a sensitive amperemeter [16]. While the grid (1) limits the surface potential of the polymer and (2) makes the surface-charge distribution more uniform, poling-current measurements allow for the determination of poling fields in a calibrated setup [16].

Some waveguide applications, such as phase-matched (PM) second-harmonic generation (SHG), require step-like profiles of the dipole orientation across the waveguide thickness. Selective poling of the lower half of a polymer film can be realized by electron-beam poling [7] or by photo-induced poling [11], in which the penetration depth of the required pump light is limited by absorption. Step-like dipole-orientation patterns can be realized in one polymer film by combining a TAP and a PIP process, while the electric field is reversed in between (Fig. 1 a) [11], or in a double-layer of NLO polymers with different glass-transition temperatures by combining two TAP processes at different temperatures (Fig. 1 b). One immediate concern with these combined poling steps is a suitable process control (see below).

For quasi-phase-matched (QPM) SHG, periodical poling patterns are required; they can be generated in a straightforward manner by use of a periodic electrode grating [17]. However, rectangular up/down patterns cannot be achieved in this way. Here, photothermal poling as shown in Fig. 2 [9] may be a more convenient technique. Table I summarizes the essential features and possible simultaneous combinations of poling techniques employed with NLO polymer electrets.

Table I. Poling Techniques for Nonlinear Optical Polymer Electrets

Poling Technique	*Orientation of Dipoles by*	*Mobility of Molecular Dipoles Induced by*	*Special Features and Advantages*
Thermal	Electric Field between Electrodes	External Heating to Temperatures $T > T_G$	Use of Device Electrodes for Poling
Corona	Electric Field from Surface Charges	External or Local Heating to $T > T_G$	Poling in Spite of Film Defects
Electron Beam	Electric Field from Bulk Charges	External or Local Heating to $T > T_G$	Selective Poling across Thickness
Photothermal	Electric Field from Various Sources	Local Heating with Light to $T > T_G$	Patterned Poling with Continuous Electrodes
Photo-Induced	Electric Field from Various Sources	*trans-cis* Isomerization of Chromophores	Selective Poling at Room Temperature

Pyroelectricity in Poled Polymers

The pyroelectric response of poled polymers is characterized by the so-called experimental pyroelectric coefficient, which is defined as the temperature derivative of the charge Q induced on the sample electrodes of area A upon heating (or cooling): $p_{exp} = (1/A)dQ/dT$. The pyroelectric effect of amorphous polymers results (1) from the dipole libration connected with thermal motion and (2) from the decreasing dipole density which is a consequence of thermal expansion. In poled polymers, dipole libration contributes only little to the overall pyroelectricity as compared to thermal expansion. Therefore, the experimental pyroelectric coefficient is directly related to the frozen-in polarization P [18, 19]:

$$p_{exp} = \alpha_x \frac{\epsilon_\infty + 2}{3} P, \tag{1}$$

where α_x and ϵ_∞ are the relative temperature coefficient of thermal expansion and the unrelaxed relative dielectric permittivity of the sample material, respectively. The pyroelectrical investigation of poled polymers is based on this simple relation.

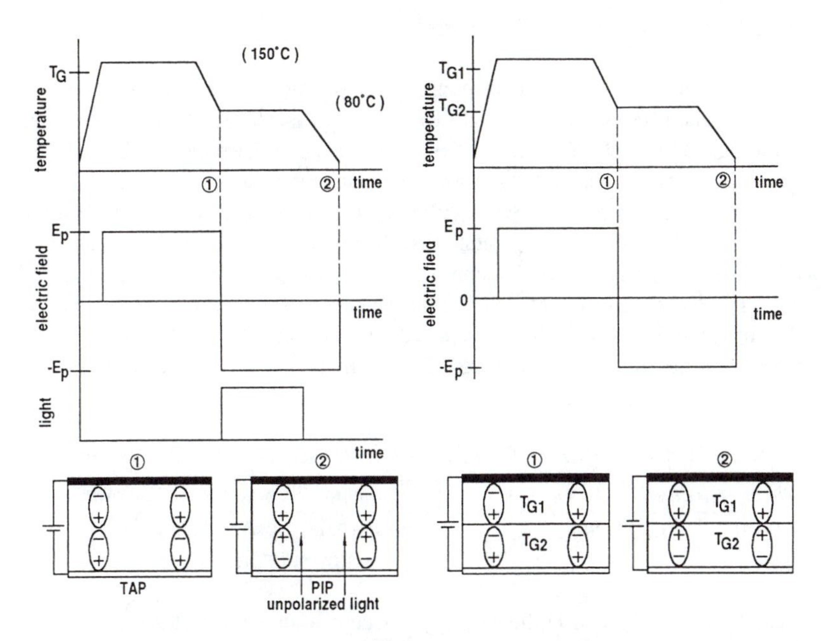

Figure 1. Preparation of inverted $\chi^{(2)}$-structures by combining a "hot" thermally assisted poling (TAP) and a "cold"photo-induced poling PIP process (left) and in a double layer of two NLO polymers with different glass-transition temperatures by means of two TAP processes with opposite poling fields (right).

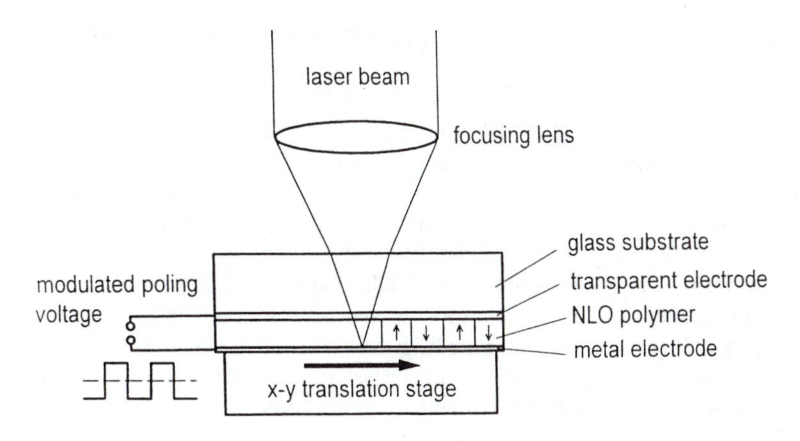

Figure 2. Experimental setup for photothermal poling with a fixed light source, a switchable power supply, and a movable sample holder; the polymer sample is supported by a glass substrate with a transparent ITO electrode on the top and coated with an aluminium electrode at the bottom.

Pyroelectrical Measurement Techniques

In order to obtain its pyroelectric response, a sample must be thermally excited. This can be conveniently achieved with light whose energy is converted into heat by means of absorption. However, other excitations such as heating the polymer via convection [19] are also possible. Heating by absorption forms the basis of the so-called photo-pyroelectrical absorption spectroscopy (PPAS) [20, 21] for measuring weak absorption on thin films. The resulting temperature distribution inside the sample depends on the specific form of the thermal excitation and can be used for determining its thermal parameters. Pyroelectrical microcalorimetry which is based on this principle has been used to measure the thermal diffusivities [22] and the specific heat [23] of films with thicknesses in the μm or sub-μm range. Its sensitivity is very high, as a sample mass of only a few μg is sufficient.

Table II. Pyroelectrical Techniques for Investigating NLO Polymers

Measured Physical Properties	*Experimental Technique (and Its Acronym)*
Absorption Coefficients	Photo-Pyroelectric Absorption Spectroscopy (PPAS)
Specific Heat, Thermal Diffusivity	Pyroelectric Microcalorimetry (PMC)
Dipole Relaxation	Pyroelectric Thermal Analysis (PTA)
Dipole-Orientation Distribution in the Film Plane	Scanning Pyroelectric Microscopy (SPM)
Dipole-Orientation Profile across the Film Thickness	Pyroelectric Depth Profiling (PDP)

The essential feature of pyroelectrical measurements is the direct electrostatic relation between dipole polarization and generated signal. It forms the basis for the recently introduced pyroelectrical thermal analysis [13] of dipole-relaxation processes in poled polymers as well as for the various microscopical probing techniques such as the scanning pyroelectrical microscope (SPM) of dipole-orientation patterns in the film plane [24] or the thermal-wave depth profiling of dipole-orientation profiles across the film thickness [25]. Since the pyroelectric signal is a linear function of the polarization P, not only the magnitude, but also the direction (up or down) of the dipole orientation can be obtained. The physical quantities that can be measured with pyroelectrical techniques as well as the relevant methods are summarized in Table II.

Sample Materials and Preparation

The following experiments were performed on the NLO polymers schematically shown in Fig. 3. Fig. 3 (left) depicts a styrene-maleic anhydride copolymer with chemically attached side groups of the azo dye Disperse Red 1 (DR 1) (glass-transition temperature of 140°C, 60 % weight DR 1, electro-optic coefficient of 15 pm/V at a wavelength of 1.5 μm with a poling field of around 100 V/μm); details of its chemical synthesis can be found in [26]. Fig. 3 (right) shows the chemical structure of the cross-linkable polymer Red-Acid Magly, where thermal cross-linking is possible between a carboxylic-acid functional group (COOH) located on the nonlinear chromophore and an epoxy side group of the polymer, while a chemical reaction between two chromophores is not possible (Liang, J.; Levenson, R.; Rossier, C.; Toussaere, E.; Zyss, J.; Rousseau, A.; Boutevin, B.; Foll, F.; Bose, D. *J. de Physique*, in press.). For the following experiments, approximately 1--5μm thick films were prepared from suitable solutions of all three polymers by means of spin-coating onto ITO-coated glass substrates. After vacuum evaporation of a top aluminium or gold electrode, the polymer films were poled under an electric field of around 50 V/μm.

Experimental Setup and Typical Results

Experimental Setup. The compact and easy-to-use experimental arrangement for pyroelectrical measurements on poled polymers is schematically shown in Fig. 4. It consists of a laser-diode assembly including all electronics necessary for intensity modulation with a waveform generator, the electroded polymer sample and a lock-in amplifier for measuring the pyroelectric signals. Further details are found in previous publications [13, 25].

Isothermal Relaxation of the Dipole Orientation. As the pyroelectric signal is directly proportional to the electric polarization, pyroelectrical measurements are ideally suited for the investigation of dipole-relaxation processes. In a recent theoretical and experimental study, it was demonstrated that the relaxation behaviors of the electric polarization and of the pyroelectric and electro-optic responses are almost identical, so that one experimental technique is usually sufficient for characterizing polymer electrets (Bauer, S.; Ren, W.; Yilmaz, S.; Wirges, W.; Gerhard-Multhaupt, R. *Nonlinear Optics*, in press.). Fig. 5 shows the isothermal decay of the pyroelectric response --- and thus also the dipole orientation --- in the side-chain polymer of Fig. 3 (left). Fitting curves according to the stretched-exponential or Kohlrausch-Williams-Watts function

$$\Phi(t) = \exp(-\frac{t}{\tau(T)})^{\beta} \tag{2}$$

are also included in Fig. 5. In the equation, $\tau(T)$ is the temperature dependent mean relaxation time and β is the stretching parameter. The depicted isothermal decay curves at three different temperatures can be reasonably well fitted with stretched-exponential functions which have the same stretching parameter β [27].

A different behavior is found for the cross-linkable polymer of Fig. 3 (right). Fig. 6 shows isothermal decay measurements at 90°C after different times of thermal cross-linking at 130°C. From Fig. 6, the enhanced stability after extended cross-linking is obvious. The experiments clearly indicate the presence of two types of dipoles in the cross-linked polymer: Already cross-linked (and thus thermally stable) dipoles and not yet cross-linked (and thus thermally unstable) dipoles [27]. The included fitting curves thus consist of superpositions of two relaxation processes according to

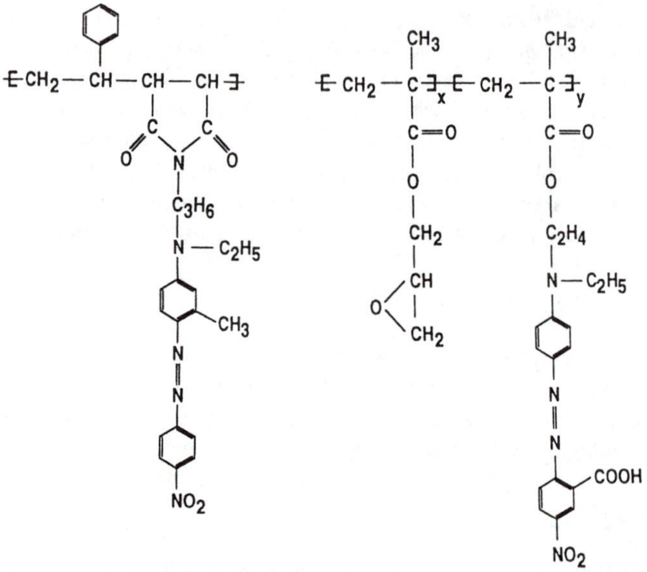

Figure 3. Chemical structures of the side-chain (left) and the cross-linkable (right) polymer.

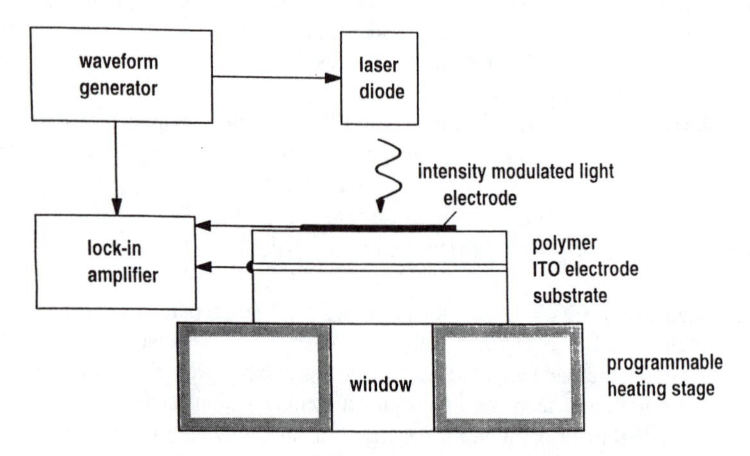

Figure 4. Experimental setup for pyroelectrical investigations.

$$\Phi(t) = \Phi_1 \exp[-(t/\tau_1)^{\beta_1}] + \Phi_2 \exp[-(t/\tau_2)^{\beta_2}], \tag{3}$$

where Φ_1 and Φ_2 are the respective fractions of not cross-linked and cross-linked dipoles, and τ_1 and τ_2 are the respective mean relaxation times. For the fast relaxation process of the not cross-linked dipoles, it is possible to derive the parameters τ_1 and β_1, while for the slow relaxation process, the time τ_2 is much larger than the time window of our experiment. The stretching parameter and the mean relaxation time for the fast, not cross-linked dipoles are comparable to the relaxation time for a similar DR1/PMMA guest-host polymer, which is to be expected because of their similar structural configuration.

Relaxation Behavior after Physical Aging. The thermal stability of the dipole orientation in glassy polymers can be strongly enhanced by aging under an applied electric field at temperatures below the respective glass-transition temperature. Fig. 7 shows isothermal relaxations of the pyroelectric response after different times of aging. Films of the side-chain polymer according to Fig. 3 (left) were poled under an electric field of approximately 75V/μm at 155°C for 10 minutes and subsequently quenched to 130°C within a few seconds. Immediately after quenching, a relatively steep decrease of the dipole orientation was observed (× × × in Fig. 7). As seen in Fig. 7, isothermal aging under the poling field for time periods of 10 (○ ○ ○), 100 (□ □ □), 1000 (△ △ △), and 4000 (◇ ◇ ◇) minutes yielded more and more stable dipole orientations. It is interesting to note that the unaged sample (× × ×) exhibits better stability at long times (beyond 100 minutes) than the sample aged for 10 minutes. Such a crossover with the isothermal relaxation of the unaged sample can also be expected for the longer aged samples at correspondingly later times (not shown here). These observations can be explained by the narrowing of the relaxation-time distributions during aging, which occurs not only at short relaxation times (where it improves the stability), but also at very long times. Work is in progress in order to further clarify this phenomenon which is of considerable practical interest.

Pyroelectric Thermal Analysis (PTA). While it is rather time-consuming to collect the necessary relaxation data in isothermal experiments, non-isothermal techniques allow for a relatively fast investigation of the relevant dipole-relaxation processes. During PTA, the polymer is heated at a constant rate, while the pyroelectric response is recorded. As shown in detail in an earlier publication [11, 13], the pyroelectric response during PTA is --- to a good approximation --- proportional to the frozen-in polarization given by

$$P(T) = P_0 \exp[-(h \int_{T_0}^{T} \frac{d\tilde{T}}{\tau(\tilde{T})})^{\beta}], \tag{4}$$

where P_0 is the frozen-in polarization at temperature T_0 and h is the inverse heating rate. Fig. 8 shows a comparison of the PTA responses after different poling procedures (PIP at 40 and 80°C and TAP) together with theoretical fitting curves according to eq.(4). All the experiments were performed with a heating rate of 5°C/min. From Fig. 8, it is obvious that PIP yields a lower thermal stability of the dipole orientation than TAP [11]. PTA responses similar to that of 40°C PIP-processed samples were

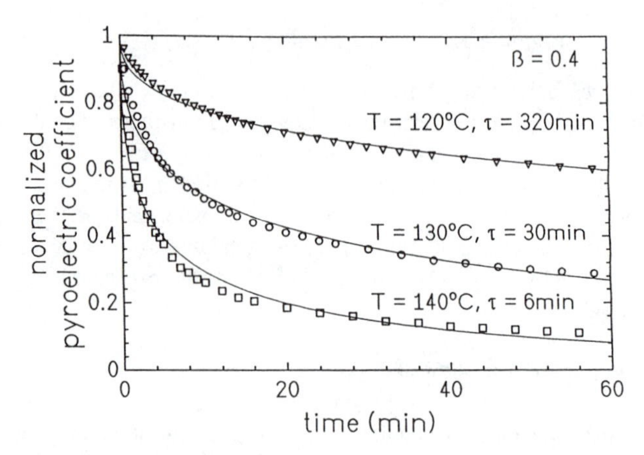

Figure 5. Isothermal relaxation of the pyroelectric response, together with fitting curves and parameters based on the stretched-exponential function.

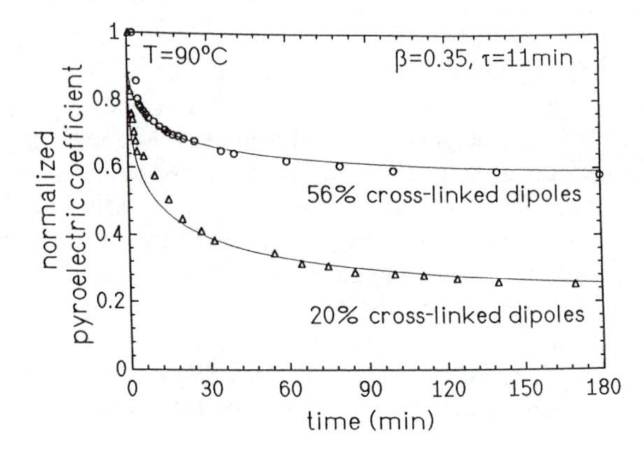

Figure 6. Isothermal relaxation at 90°C of the pyroelectric response after different times of cross-linking at 130°, together with fitting curves and parameters as above.

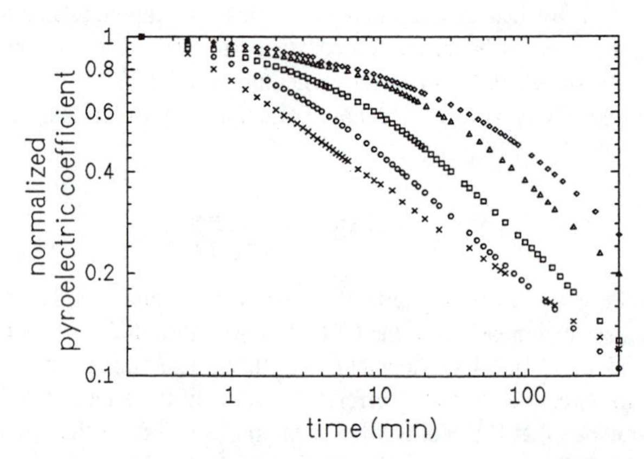

Figure 7. Isothermal relaxation at 130°C of the pyroelectric response after different times of aging: Unaged sample (× × ×) and samples aged for 10 (○ ○ ○), 100 (□ □ □), 1000 (△ △ △), and 4000 (◇ ◇ ◇) minutes under the poling field.

measured for photothermally poled samples. The reason for the limited stability of photothermal poling is the rapid cooling of the polymer below its glass-transition temperature, which freezes a large amount of free volume, while the limited stability of photo-induced poled samples is caused by the formation of additional free volume during the photo-isomerization process [11]. Fig. 9 shows the PTA response of an inverted $\chi^{(2)}$-structure, prepared according to the poling technique schematically shown in Fig. 1 (left) [11], and the PTA response of an inverted $\chi^{(2)}$ structure, prepared in a double layer of NLO polymers with different glass-transition temperatures. The theoretical curves in Fig. 9 were calculated according to

$$p(T) \propto \frac{P_1 d_1}{d_1 + d_2} \exp[-(h \int_{T_0}^{T} \frac{d\tilde{T}}{\tau_1(\tilde{T})})^{\beta}] + \frac{P_1 d_2}{d_2 + d_2} \exp[-(h \int_{T_0}^{T} \frac{d\tilde{T}}{\tau_2(\tilde{T})})^{\beta}, \quad (5)$$

where P_1 and P_2 are the frozen-in polarizations, τ_1 and τ_2 the temperature-dependent mean relaxation times and β_1 and β_2 the stretching parameters of the two polymers, respectively. For $\tau_1(T)$ and $\tau_2(T)$, Arrhenius laws are assumed. It is instructive to look at the parameter $\xi = -(P_1 d_1)/(P_2 d_2)$. For equal thicknesses ($d_1 = d_2$), $\xi = 1$ would indicate a perfect $\chi^{(2)}$-inverted structure. The fitting curves shown in Fig. 9 were obtained with $\xi = 1.136$ and $\xi = 0.936$, respectively; this shows that nearly perfect $\chi^{(2)}$-inverted structures were achieved.

Pyroelectric Profiling of Dipole-Orientation Patterns. While the PTA technique provides an intuitive picture of the underlying processes and allows for an analysis of $\chi^{(2)}$-inverted structures, it possesses the drawback of being destructive. However, as shown in detail elsewhere [25], pyroelectric measurements can also be employed for scanning dipole-orientation patterns across the thickness of polymer films. Fig. 10 shows the frequency-dependent pyroelectric response of a $\chi^{(2)}$-inverted structure, which had been prepared according to the scheme shown in Fig. 1. The included theoretical curves were calculated without any free parameters and thus demonstrate the feasibility of nondestructively probing $\chi^{(2)}$-inverted structures.

Conclusions

Several poling techniques for the global or patterned orientation of the molecular dipoles in nonlinear optical polymer electrets as well as various pyroelectrical measuring techniques for the investigation of dipole-relaxation processes and dipole-orientation patterns in poled polymer films were described. The proposed family of pyroelectrical methods represents an inexpensive means for the routine analysis of poled polymers during and after their preparation and for their optimization with respect to several attractive applications. In addition, two techniques for the preparation of inverted-$\chi^{(2)}$ double-layer structures were discussed. The resulting dipole-orientation patterns were investigated by means of pyroelectric thermal analysis (PTA). After the present proof of concept for the proposed techniques, it is now necessary to optimize the polymer materials, the preparation technologies, and the poling processes in order to obtain polymer samples with more stable and precisely controlled dipole-orientation patterns along all three space coordinates.

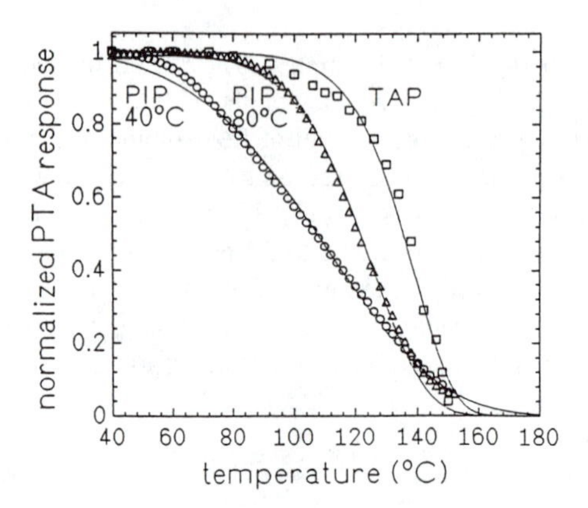

Figure 8. Pyroelectrical thermal analysis (PTA) after PIP at 40 and at 80°C and thermally assisted poling (TAP) at 155°C. The solid lines represent fitting curves.

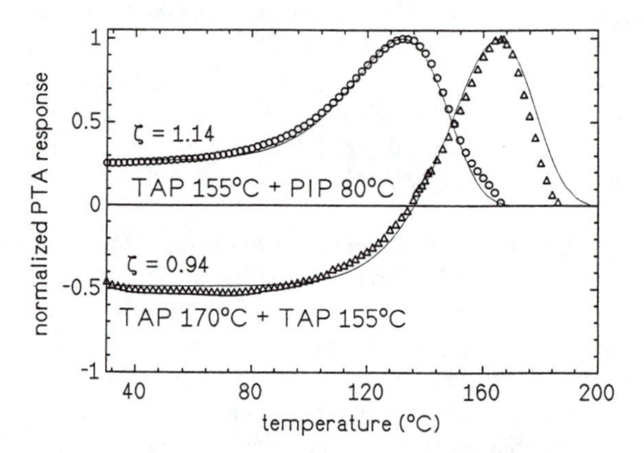

Figure 9. Pyroelectrical thermal analysis (PTA) results after the combination of PIP and TAP as well as for a double-layer system of two NLO polymers with different glass-transition temperatures.

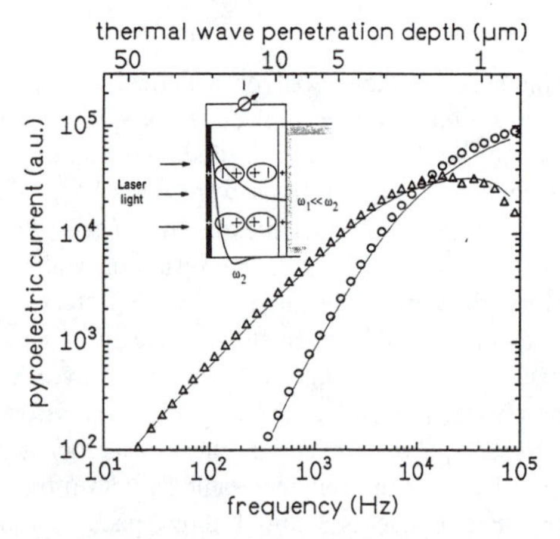

Figure 10. Frequency-dependent pyroelectric response of an inverted-$\chi^{(2)}$ double-layer structure together with its theoretical simulation.

Acknowledgments. The authors are indebted to Drs. M. Ahlheim, M. Staehelin, F. Lehr and B. Zysset (Sandoz Optoelectronics Huningue, France) as well as to Drs. J. Liang and J. Zyss (CNET Bagneux, France) for novel sample materials and stimulating discussions. Thanks are also due to the German Federal Minister for Research and Technology (BMFT) for financial support within the research project 03M4083E and under the HSP II program.

Literature Cited

[1] *Polymers for Lightwave and Integrated Optics;* Hornak, L. A., Ed.; Optical Engineering 32; Marcel Dekker: New York, New York, U.S.A., 1992.
[2] Singer, K. D.; Sohn, J. E.; Lalama, S. J. *Appl. Phys. Lett.* **1986**, *49*, 248--250.
[3] Eich, M.; Sen, A.; Looser, H.; Bjorklund, G. C.; Swalen, J. D.; Twieg, R.; Willson, C. G. *J. Appl. Phys.* **1989**,Ê*66*, 2559--2567.
[4] Teraoka, J.; Jungbauer, D.; Reck, B.; Yoon, D. Y.; Twieg, R.; Willson, C. G. *J. Appl. Phys.* **1991**, *69*, 2568-2576.
[5] Eich, M.; Reck, B.; Yoon, D. Y.; Willson, C. G.; Bjorklund, G. C. *J. Appl. Phys.* **1989**, *66*, 3241--3247.
[6] Singer, K. D.; Kuzyk, M. G.; Holland, W. R.; Sohn, J. E.; Lalama, S. J.; Comizzoli, R. B.; Katz, H. E.; Schilling, M. L. *Appl. Phys. Lett.* **1988**, *53*, 1800--1802.
[7] Yang, G. M.; Bauer-Gogonea, S.; Sessler, G. M.; Bauer, S.; Ren, W.; Wirges, W.; Gerhard-Multhaupt, R. *Appl. Phys. Lett.* **1994**, *64*, 22--24.
[8] *Electrets*, 2nd ed.; Sessler, G. M., Ed.; Topics in Applied Physics 33; Springer: Heidelberg, Baden-Württemberg, Germany, 1987.
[9] Yilmaz, S.; Bauer, S.; Gerhard-Multhaupt, R. *Appl. Phys. Lett.* **1994**, *64*, 2770--2772.
[10] Sekkat, Z.; Dumont, M. *Appl. Phys. B* **1992**, *54*, 486-489.
[11] Bauer-Gogonea, S.; Bauer, S.; Wirges, W.; Gerhard- Multhaupt, R. *J. Appl. Phys.* **1994**, *76*, 2627--2635.
[12] Gerhard-Multhaupt, R.; Yilmaz, S.; Bauer, S.; Molzow, W.-D.; Wirges, W.; Das-Gupta, D. K. In *1994 Annual Report, IEEE Conference on Electrical Insulation and Dielectric Phenomena (CEIDP);* IEEE: Piscataway, New Jersey, U.S.A., 1994; pp 743--748.
[13] Bauer, S.; Ren, W.; Yilmaz, S.; Wirges, W.; Molzow, W.-D.; Gerhard-Multhaupt, R.; Oertel, U.; Hänel, B.; Häussler, L.; Komber, H.; Lunkwitz, K. *Appl. Phys. Lett.* **1993**, *63*, 2018--2020.
[14] Graf, H. M.; Zobel, O.; East, A. J.; Haarer, D. *J. Appl. Phys.* **1994**, *75*, 3335--3339.
[15] Köhler, W.; Robello, D. R.; Dao, P. T.; Willand, C. S.; Williams, D. J. *J. Chem. Phys.* **1990**, *93*, 9157--9166.
[16] Ren, W.; Bauer, S.; Yilmaz, S.; Gerhard-Multhaupt, R. *J. Appl. Phys.* **1994**, *75*, 7211--7219.
[17] Khanarian, G.; Norwood, R. A.; Haas, D.; Feuer, B.; Karim, D. *Appl. Phys. Lett.* **1990**, *57*, 97--99.
[18] Mopsik, F. I.; Broadhurst, M. G. *J. Appl. Phys.* **1975**, *46*, 4204--4208.

[19] Goldberg, H. A.; East, A. J.; Kalnin, I. L.; Johnson, R. E.; Man, H. T.; Keosian, R. A.; Karim, D. In *Materials Research Society Symposium;* MRS Symposium Proceedings 175; MRS: New York, New York, U.S.A., 1990; pp 113--128.
[20] Coufal, H. *Appl. Phys. Lett.* **1984**, *44*, 59--61.
[21] Mandelis, A. *Chem. Phys. Lett.* **1984**, *108*, 388-392 .
[22] Coufal, H.; Hefferle, B. *Appl. Phys. A* **1985**, *38*, 213--219.
[23] Bauer, S.; Ploss, B. *IEEE Trans. Electr. Insul.* **1992**, *27*, 861--866.
[24] Yilmaz, S.; Bauer, S.; Wirges, W.; Gerhard-Multhaupt, R. *Appl. Phys. Lett.* **1993**, *63*, 1724--1726.
[25] Bauer, S. *J. Appl. Phys.* **1994**, *75*, 5306--5315.
[26] Ahlheim, M.; Lehr, F. *Macromol. Chem. Phys.* **1994**, *195*, 361--373.
[27] Bauer, S.; Ren, W.; Bauer-Gogonea, S.; Gerhard-Multhaupt, R.; Liang, J.; Zyss, J.; Ahlheim, M.; Stähelin, M.; Zysset, B. In *Proceedings 8th International Symposium on Electrets (ISE8);* Lewiner, J.; Alquié, C.; Morisseau, D., Eds.; IEEE: Piscataway, New Jersey, U.S.A., 1994; pp 800--805.

RECEIVED February 14, 1995

Long-Term Thermal, Photo, and Oxidative Stability

Chapter 23

Quantitative, Rational Predictions of the Long-Term Temporal Decay Properties of Second-Order Nonlinear Optical Polymers from the Analysis of Relaxation Dynamics

Ali Dhinojwala[1], Jacob C. Hooker[2], and John M. Torkelson[1,2]

[1]Department of Chemical Engineering and [2]Department of Materials Science and Engineering, Northwestern University, Evanston, IL 60208

An accelerated method of testing has been developed which allows quantitative, rational prediction of the temporal decay properties of second-order nonlinear optical (NLO) polymers under conditions of technological interest but which are experimentally inaccessible. This is accomplished by using a novel delay-trigger technique to monitor the reorientation of NLO chromophores doped and labeled in amorphous polymers over 12 decades in time (10^{-6} to 10^{6} sec). By fitting the time-dependent orientational component of the second-order macroscopic susceptibility, $\chi^{(2)}$, to the Kohlrausch-Williams-Watts expression, an average reorientation relaxation time, $<\tau>$, is determined in both the glassy and rubbery states. The reorientation dynamics of Disperse Red 1 (DR1) above T_g, is coupled to the α-relaxation of polystyrene (PS) and poly(isobutyl methacrylate) (PIBMA). The temporal stability of $\chi^{(2)}$ can be enhanced by physically aging the sample or by covalent attachment of the chromophore. However, increases in $<\tau>$ at a given temperature for the methacrylate-based copolymers with increasing functionalization is due to the increasing T_g of the copolymer with greater chromophore content. Finally, the issue of poling method and conditions are considered in terms of correctly interpreting the temporal behavior of SHG properties in NLO polymers.

The subject of the temporal stability of second-order nonlinear optical (NLO) properties of poled polymers has received substantial attention over the past half dozen years [1-10]. Much of this interest has been driven by the need to assure the reliability of potential technological devices over periods of years at specified use temperatures. Unfortunately, the actual time scales of interest make exact experimental simulations too time consuming for practice. This has left the question

0097–6156/95/0601–0318$12.00/0

of how to perform accelerated tests which allow rational predictions of the second-order NLO properties of poled polymers under use conditions.

Recently, we developed [7-11] a delay-trigger approach which allows us to measure the decay of second harmonic generation (SHG) intensity (after turning off the dc-poling-field) or the onset of SHG intensity (after turning on the dc-poling-field) on time scales of microseconds to seconds. Using this approach in conjunction with normal time-domain measurements of SHG intensity decay or onset (from 10 to 10^6 sec), the reorientation dynamics of NLO chromophores doped in or covalently attached to amorphous polymers may be followed over approximately twelve orders of magnitude in time. This has allowed for exact determinations of the degree of coupling of NLO-chromophore reorientation dynamics to polymer α-relaxation dynamics, and the effects of physical aging and polymer structure on those dynamics.

Given this background, it is then possible to make determinations of interest both scientifically and technologically. For example, this approach allows the determination of the distribution of NLO-chromophore reorientation relaxation times and thereby α-relaxation times in the polymer as functions of temperature, physical aging, chromophore content, etc. Using the knowledge gained on the temperature dependence of the distribution of relaxation times and average relaxation times, it is also possible to make rational predictions of the temporal stability of second-order NLO properties under conditions which are experimentally inaccessible. Presented in this manuscript is a description of the approach used to characterize the second-order NLO behavior of poled polymers, both with NLO chromophores doped or covalently attached, and a description of how this information can be used to quantify the relaxation behavior of the polymers and thereby allow rational prediction of long-term decay of SHG properties. Warnings of how mischaracterizations of these properties can be made will also be given.

Second Harmonic Generation Applied to Relaxation Dynamics

A complete description of our approach has been detailed in ref. 8. A brief explanation is given below. Noncentrosymmetry of NLO-chromophore orientation necessary for SHG can be achieved by applying a dc-poling-field, E_z. For a steady-state situation the SHG intensity, I(2ω), is given as [12]

$$[\ I(2\omega)\ ]^{0.5} \propto \chi^{(2)}{}_{zzz} \propto N\, E_z \left[\gamma + \frac{\mu\beta_{zzz}}{5kT}\right] \tag{1}$$

where $\chi^{(2)}{}_{zzz}$ is the second-order macroscopic susceptibility ($\chi^{(2)}{}_{zzz}$ will be abbreviated as $\chi^{(2)}$), z is the direction of the incident polarization of the fundamental beam and the direction of the dc field, and N is the number density of chromophores. γ is due to an electric-field-induced third-order effect which appears and disappears instantaneously upon application and removal of the dc field. (The repeated instantaneous appearance and disappearance of this electric-field-induced effect in an already poled polymer sample in the glassy state can be observed upon alternating the removal and application of an electric field.) $NE_z[\mu\beta/(5kT)]$, where μ is the dipole moment, β is the microscopic susceptibility, k is the Boltzmann constant, and T is absolute temperature, is due to chromophore orientation.

To demonstrate a quantitative relationship of the temporal decay of $\chi^{(2)}$, and thereby chromophore orientation, to polymer dynamics, it is necessary to account for γ. We recently reported a method [11] which can determine the contribution of γ to the overall value of $\chi^{(2)}$ (with the dc field applied). After determining this relative contribution, here designated "y", the decay of $\chi^{(2)}_{NO}$ ($\chi^{(2)}$ normalized with respect to its orientation component only at zero time) can be written as

$$\chi^{(2)}_{NO} = \frac{\chi^{(2)}_{N}}{(1-y)} \tag{2}$$

where $\chi^{(2)}_{N}$ is the ratio of $\chi^{(2)}$ normalized to the value of $\chi^{(2)}$ at zero time that accounts for the combined contributions of the orientation component and the induced effect, γ. (It should be noted that $\chi^{(2)}_{N} = \chi^{(2)}(t)/\chi^{(2)}(0)$ where the denominator, $\chi^{(2)}(0)$, is the sum of the orientation and electronic contributions at t = 0, hence, accounting for both effects. In contrast, $\chi^{(2)}_{NO}$ is the relative value of the orientation component of $\chi^{(2)}$ at some time t > 0 divided *only* by the relative value of the orientation component of $\chi^{(2)}$ at t = 0. Thus, $\chi^{(2)}_{N}$ will always be less than 1 given $\gamma > 0$, whereas $\chi^{(2)}_{NO}$ can have a value of 1 at sufficiently short times and low enough temperatures.) Once the orientation component of $\chi^{(2)}_{N}$ has been determined, the temporal decay of $\chi^{(2)}_{NO}$ at a given temperature can be related to the Kohlrausch-Williams-Watts (KWW) [13] equation:

$$\chi^{(2)}_{NO} = \exp\left(-\left(\frac{t}{\tau}\right)^{\beta_w}\right) \tag{3}$$

where τ and β_w are KWW parameters. (β_w indicates the breadth of the distribution of relaxation times. When $\beta_w = 1$, there is a single relaxation time; as β_w decreases toward 0, the distribution of relaxation times becomes increasingly broad.) A comparison of polymer dynamics to the decay of $\chi^{(2)}_{NO}$ at a variety of temperatures may be achieved by defining an average reorientation relaxation time constant, $\langle\tau\rangle$:

$$\langle\tau\rangle = \int_0^\infty \exp\left(-\left(\frac{t}{\tau}\right)^{\beta_w}\right) dt = \frac{\tau\Gamma\left(\frac{1}{\beta_w}\right)}{\beta_w} \tag{4}$$

where Γ is the gamma function.

Experimental

The NLO polymers, both with chromophores doped and covalently attached, were prepared using approaches described in refs. [7-10]. Polymers which underwent in-plane poling were dissolved in spectroscopic grade chloroform and spin coated onto a quartz substrate patterned with planar chrome electrodes (800 μm gap) by standard photolithographic techniques. The films (5-10 μm thick) were dried below T_g for 24 hrs and above T_g for 12 hrs under vacuum. Ref. 8 provides details on the optical equipment, dc fields, and delay-trigger measurements employed in these studies. Refs.

8 and 9 indicate the conditions used for dielectric relaxation measurements. For the few cases described here involving corona poling of samples, procedures described in ref. 1b were followed. (Unless it is specifically pointed out in the manuscript that the samples were poled by corona techniques, samples were poled using in-plane poling.)

Results and Discussion

Figure 1 shows the SHG decay mode measurements for polystyrene (PS) + 2 wt% Disperse Red 1 (DR1) system at several temperatures in the glassy state. (See Table I for the glass transition temperatures, T_g, of the polymers reported in this study.) The decay mode data are represented as $\chi^{(2)}_N$, which is $\chi^{(2)}$ normalized to its value just prior to switching off the dc field. Within 5 μs of the removal of the dc field, $\chi^{(2)}_N$ has decreased to a value of about 0.83 for all temperatures deep in the glassy state, signifying a 17% decrease in $\chi^{(2)}_N$ as compared to that before removal of the dc field. This decrease is associated with the electric-field-induced third-order effect [11,12] which must be accounted for if an accurate determination is to be made of the temporal decay of the orientation component of SHG properties.

Figure 2 shows the decay mode dynamics associated with the orientation component of $\chi^{(2)}_N$, here designated $\chi^{(2)}_{NO}$, for selected temperatures above and below T_g. $\chi^{(2)}_{NO}$ has been determined at each temperature from data such as those in Figure 1 using equation (2) with y = 0.17. Noteworthy is the tremendous breadth of relaxation behavior below T_g as compared to above T_g. The solid curves in Figure 2 correspond to best fits using equation (3). Values of β_w exhibit little temperature dependence below T_g, with a typical value being 0.21. However, above T_g, values of β_w exhibit a dramatic temperature dependence, increasing from 0.24 at 95°C to 0.51 at 128°C. The increase in β_w with temperature in the rubbery state indicates that the relaxation dynamics associated with reorientation of DR1 in PS is not thermorheologically simple [14], i.e., time-temperature superposition does not hold strictly over this range in temperature.

If thermorheological simplicity were to hold in the rubbery state, this would imply that the breadth of the distribution of relaxation times would be unchanged with temperature; instead, they would only shift to shorter times with increasing temperature. The distribution of reorientational relaxations for several temperatures above and below T_g may be determined using the fits to equation (3) and the method of Emri and Tschoegl [15]. Figure 3 illustrates the tremendous breadth of the distribution of reorientational relaxation times in the glassy state, over twelve orders of magnitude. Above T_g the distribution narrows but is nevertheless substantial. The breadth of the distribution of relaxation times in both the rubbery and glassy states makes clear the inadequacy of fitting SHG decay data by a simple biexponential function and the necessity to measure relaxation over many orders of magnitude in time.

The ability of the delay-trigger approach [7-11] to allow for characterization of NLO chromophore reorientation dynamics on time scales from microseconds to seconds is crucial from several standpoints. First, as most if not all of the reorientational relaxation in the rubbery state is complete within the first second after switching off the dc field, the delay-trigger approach is necessary if any characterization is to be done in the rubbery state. (Characterization in the rubbery

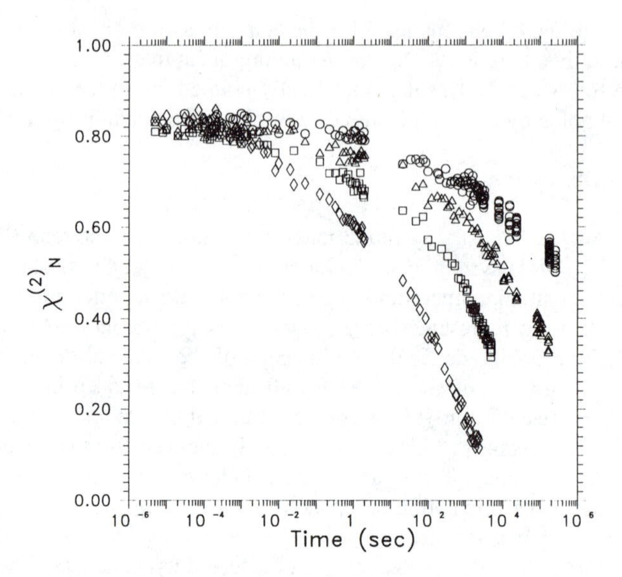

Figure 1. Decay data in PS + 2 wt% DR1 at 28 (○), 47 (▵), 71 (□), 91°C (◊). $\chi^{(2)}_N$ is the susceptibility normalized with respect to the value at t = 0 before switching off the dc-poling-field.

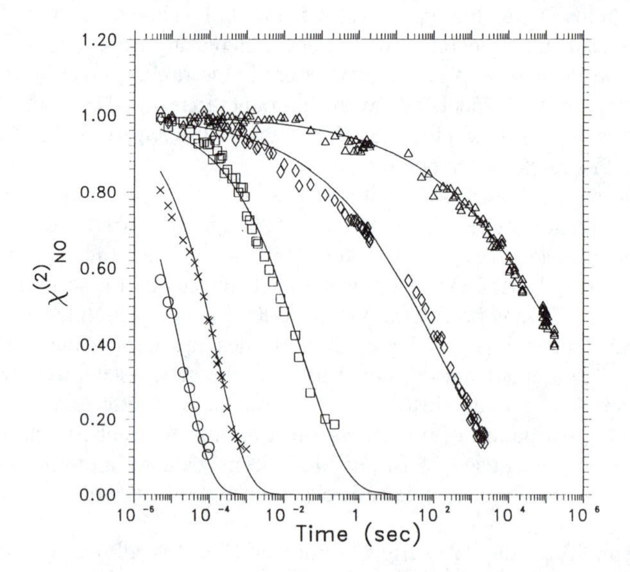

Figure 2. Decay data in PS + 2 wt% DR1 at 47 (▵), 91 (◊), 105 (□), 119 (x), and 128°C (○). $\chi^{(2)}_{NO}$ is the orientation component of $\chi^{(2)}_N$ .The solid curves are fits to equation (3).

state is useful in order to determine the degree to which the NLO chromophore reorientation is coupled to the α-relaxation processes of the polymer itself.) Second, in the glassy state it is impossible to characterize the full breadth of the distribution of relaxation times if time scales on the microsecond to second time scale are excluded. The characterization of the full distribution of relaxation times in the glassy state is required if a rational, quantitative prediction of the temporal decay behavior of SHG of poled polymers is to be made under potentially useful technological conditions.

It is important to note that if an accurate characterization of the temporal decay is not achieved, the temperature dependence of the average reorientational relaxation time, $\langle\tau\rangle$, in the glassy state can be significantly understated and the viability of these polymeric materials for technological application misunderstood. For example, if the data reported for PS + 2 wt% DR1 system of the type given in Figures 1 and 2 were truncated so that results were available only at time scales longer than 100 to 200 sec after removal of the dc field and fits were done using the point at 100 to 200 sec as $\chi^{(2)}_{NO} = 1$, as done in some recent studies [16,17], a much smaller temperature dependence of $\langle\tau\rangle$ would be calculated fitting data either with a KWW equation or double exponential fit. (See Figure 4.) The complete experimental data sets in the glassy state, with β_w approximately equal to 0.21 independent of temperature, yield $\langle\tau\rangle$ = 200 to 300 sec at T_g and an apparent Arrhenius temperature dependence below T_g with an apparent activation energy of 45-50 kcal/mol. However, if experimental data were accessed only at 100 to 200 sec and later, the much narrower range of relaxation times being accessed results in an artificially large β_w value (implying a narrower distribution of relaxation times) of 0.35 in the glassy state and a smaller apparent activation energy. (See Figure 4.)

The accurate determination of the temperature dependencies of $\langle\tau\rangle$ and β_w in the glassy state has important technological implications. For example, if application requires no more than a 5% loss in $\chi^{(2)}_{NO}$ over a four-year period, then a device involving the PS + 2 wt% DR1 system could require $\langle\tau\rangle$ greater than 10^{15} sec. This calculation employs a constant value of $\beta_w = 0.21$ in the glassy state and equations (3) (from which τ is determined) and (4) (from which $\langle\tau\rangle$ is determined). Using an apparent activation energy of 45 kcal/mol, a 5% loss in $\chi^{(2)}_{NO}$ over four years is predicted to be achieved at T_g-114°C (or T ≈ - 20°C for the PS systems used here). Some technological applications have maximum use temperatures of 80°C [18]; this implies that if another NLO polymer system with a T_g 100°C higher than that of the PS system studied here is found to possess similar β_w values and apparent activation energies below T_g, then it may be expected to have sufficient temporal stability to meet the application requirements. This suggests that polymers with T_g's of 200°C may prove to be viable candidates for certain second-order NLO applications. However, if instead of using data obtained from μsec onward we employed only data 200 sec after turning off the dc field and longer, represented by the diamonds in Figure 4, we would predict an operating requirement of T_g-125°C (as opposed to T_g-114°C using the more complete data set), thereby incorrectly eliminating certain polymers from consideration. The double exponential fit leads to even more dire predictions. Thus, it is clear that accurate characterization of second-order NLO decay dynamics is important in predicting technological utility; in fact, if the current comparison has generality, then the viability of most materials studied in the past has been understated due to characterization over an inadequate range of relaxation times.

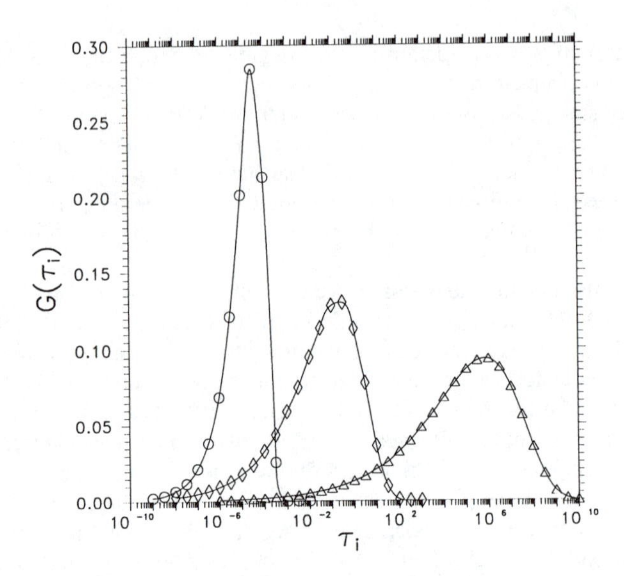

Figure 3. The distribution of reorientation relaxation times for PS + 2 wt% DR1 at 47 (△), 91 (◊), and 128°C (○).

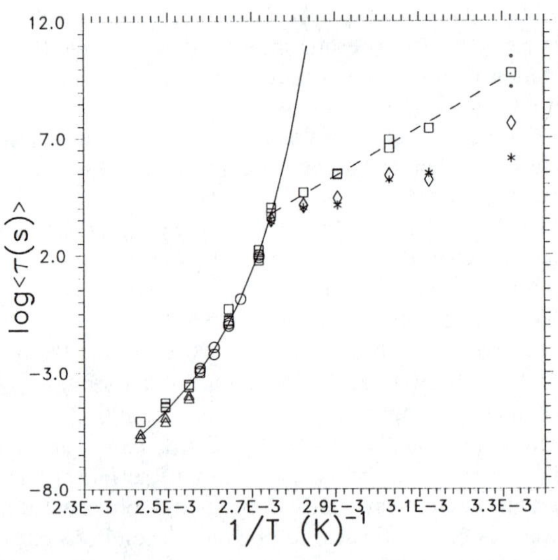

Figure 4. Temperature dependence of the average reorientation relaxation time constant, $<\tau>$, above and below T_g in PS + 2 wt% DR1 poling onset (○) and decay (□) data, and dielectric relaxation measurements (△). Data without error bars are equal to or smaller than the symbol size. Solid curve corresponds to a fit using the WLF equation. Dashed line corresponds to a fit at $T < T_g$ using an Arrhenius dependence and E_a = 45-50 kcal/mol. Data given by (◊) and (*) employ KWW and biexponential fits, respectively, taking data in Figure 1 and renormalizing with the value of $\chi^{(2)}_{NO}$ at t = 200 sec as being 1.0.

The data provided in Figure 4 also allow for determination of the coupling between NLO chromophore reorientation dynamics and polymer segment dynamics. The fact that $<\tau>$ has a value of 200 to 300 sec at T_g strongly suggests that the reorientation dynamics is coupled to the polymer α-relaxation process, i.e., the cooperative segmental motion associated with T_g. Previous studies have indicated that at T_g the characteristic relaxation time for the α-relaxation process in polymeric as well as low molecular weight glass-formers is 1 to 5 min. [19] Providing further evidence of the coupling to the polymer α-relaxation is the fact that $<\tau>$ shows different temperature behaviors above and below T_g, with the temperature dependence in the rubbery state able to be fit (solid curve in Figure 4) to the Williams-Landel-Ferry (WLF) equation [14]:

$$\log\left(\frac{<\tau>}{<\tau>_{T_g}}\right) = \frac{-C_1(T - T_g)}{C_2 + T - T_g} \qquad (5)$$

where T is absolute temperature and C_1 and C_2 are WLF parameters. Even more convincing is the fact that the parameters $C_1 = 15$ and $C_2 = 40$ K obtained from the fit are in good agreement with expectations for an α-relaxation process [14]. (It should be noted that over a limited-temperature range, data following an Arrhenius-type model may be fit to an expression of the form of equation (5) but often with physically unreasonable parameters, so the actual values determined for C_1 and C_2 are important in drawing conclusions.) Finally, we note that a comparison of the temperature dependence of our rubbery-state $<\tau>$ data with that found for the α-relaxation process in PS by a variety of techniques, including NMR [20], photon correlation spectroscopy [21], dielectric relaxation [22,23] and mechanical relaxation [24], exhibits excellent agreement. (See ref. 9 for more detail.) The combination of these results provides convincing evidence of the coupling of the chromophore reorientation process to the cooperative segmental motion associated with the α-relaxation in PS. Similar conclusions have been drawn regarding DR1 and similarly sized chromophores doped in methacrylate-based polymers [7,8]. (It should be noted that chromophores of sufficiently small size, such as N,N-dimethyl-p-nitroaniline, can exhibit significant decoupling from the α-relaxation process in these polymers. [25,26])

Besides extending reorientation relaxation times by going deeper into the glassy state, it is also possible to improve stability of SHG properties by physically aging the system below T_g. Physical aging is a process by which excess quantities such as specific volume or enthalpy relax towards equilibrium upon isothermal annealing. The loss in specific volume results in a decrease in local mobility in the system. [27] Figure 5 illustrates the effects of physically aging the polymer for 72 hrs at 57°C, with the dc field on, before measuring the decay dynamics. Within experimental error, there is no change in the breadth of the relaxation distribution between the physically aged and quenched samples; instead, for the physically aged samples the distribution of relaxation times simply shifts by nearly 2 orders of magnitude to longer times. (This is consistent with previous observations involving stress relaxation measurements where physical aging shifts the relaxation spectrum to longer times. [28]) For PS + 2 wt% DR1 samples which were aged 70-72 hours at temperatures ranging from 57° to 86°C, the shift in relaxation processes to longer times indicates that technological viability is possible at temperatures 15 to 20°C higher than those for samples that were quenched rather than physically aged. Similar conclusions on the effects of physical aging have also been drawn for methacrylate-based polymers. [10,25,26]

An additional option which has been employed for extending the reorientation relaxation to longer times is to covalently attach the chromophore to the polymer. Figure 6 compares $<\tau>$ values, calculated from KWW fits of decay- and onset-mode data, for doped polymer and covalently attached copolymer systems over a broad temperature range. Here the polymer is poly(isobutyl methacrylate) (PIBMA) which can be made as a copolymer with a methacrylate monomer functionalized with a DR1 chromophore, which we refer to as IBMA-MADR1 [10]. See Figure 6 for the copolymer studied. This figure illustrates a major increase in $<\tau>$ over all temperatures studied with increasing DR1 content in the copolymer.

The underlying reason for the increasing relaxation times observed with increasing chromophore-content copolymers had in the past been attributed to a variety of factors. However, by recasting the data from Figure 6 in a scaled form using the reduced variable T_g/T (see Table I for T_g values), we observe in Figure 7 that there is substantial similarity in the temperature dependence of $<\tau>$ for all copolymer compositions and the doped sample. That the temperature dependence is little affected by copolymer composition (up to 60 mol% DR1-containing chromophore content) on this rescaled plot proves several points. First, the degree of cooperativity involved in the reorientation relaxation process is not strongly affected by composition in these copolymers. This may be expected given that several methacrylate-based systems have shown exactly the same temperature dependence of $<\tau>$ both above and below T_g when compared on a T_g/T basis [8]. Second, the reorientation relaxation is coupled to the polymer α-relaxation in both the doped and copolymer systems as evidenced by the WLF dependence of $<\tau>$ in the rubbery state, the roughly 100-200 sec value of $<\tau>$ at T_g and the deviation of the temperature dependence in the nonequilibrium glassy state from the equilibrium-based WLF equation. The description given by Goodson et al. [29] which attributes the increase in the relaxation time with increasing chromophore concentration to the "orientational pair correlation associated with intermolecular interactions" (dipole-dipole) is unnecessary to explain the temporal decay behavior of NLO properties in the copolymer systems studied here. (Future study will be needed to investigate whether copolymerization of the labeled monomer employed in the present study with styrene may result in a copolymer-composition effect on the temperature dependence of $<\tau>$ when compared on a T_g/T basis. As doped polystyrene and methacrylate-based polymers exhibit different temperature dependencies in the rubbery state [9], such an effect may be anticipated.) Ultimately, the enhancement of temporal stability in these labeled systems can be attributed to the increasing T_g of the copolymer with increasing chromophore content (see Table I).

Although the $<\tau>$ values at equivalent T_g/T values are comparable for the doped and functionalized systems, it has been shown [10] that the glassy state β_w's are approximately 0.25 in the doped PIBMA systems at T_g-25°C < T < T_g, while β_w = 0.40-0.50 for the side-chain functionalized systems and at most decrease slightly with decreasing temperature below T_g-25°C. As a result of the higher glassy-state β_w

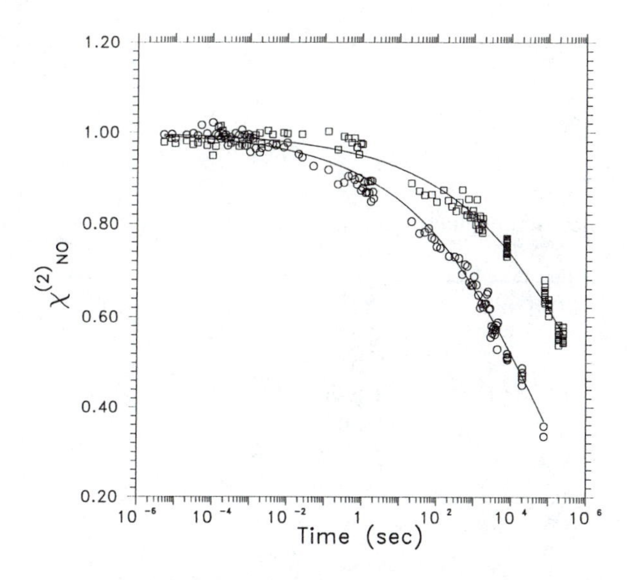

Figure 5. SHG decay data at 57°C in PS + 2 wt% DR1: quenched glass (○) and physically aged 72 hrs with poling field applied at 57°C (□). $\chi^{(2)}_{NO}$ is the normalized orientation component of $\chi^{(2)}_{N}$. The solid curves are fits to the KWW equation.

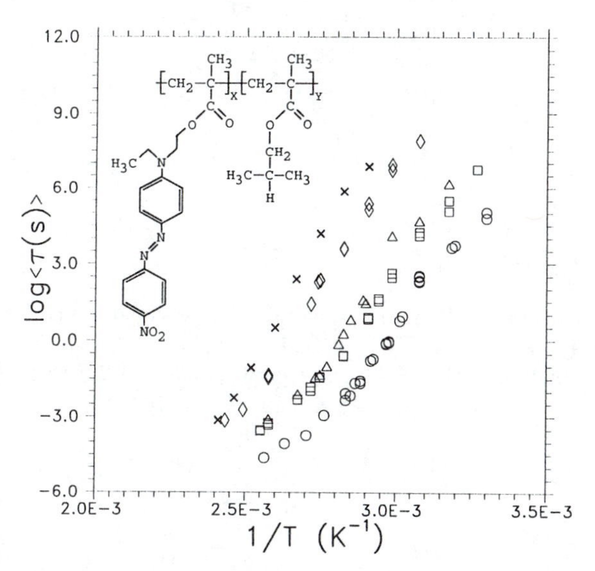

Figure 6. Temperature dependence of <τ> above and below T_g obtained from SHG decay and onset measurements: (○) PIBMA + 2 wt% DR1; (□) 1 mol%, (△) 6 mol%, (◊) 36 mol%, and (×) 60 mol% DR1 content in IBMA-MADR1 copolymer.

Table I. Polymer-chromophore systems used in this study and their corresponding T_g's.

Polymer	% Chromophore	T_g (onset), °C
PS	2 wt% DR1	94
PIBMA	2 wt% DR1	53
IBMA-MADR1	1 mol% DR1	67
	6 mol% DR1	73
	36 mol% DR1	93
	60 mol% DR1	102

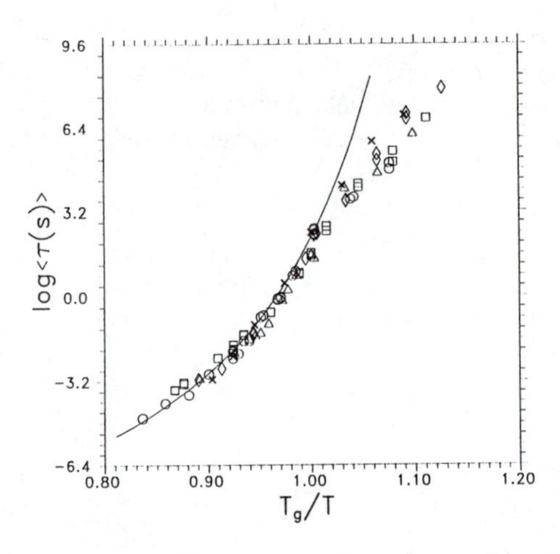

Figure 7. Values of <τ> from SHG measurements scaled using reduced variable T_g/T: (○) PIBMA + 2 wt% DR1; (□) 1 mol%, (△) 6 mol%, (◊) 36 mol%, and (×) 60 mol% DR1 content in copolymer. Solid curve corresponds to WLF fit to doped PIBMA + 2 wt% DR1 with C_1 = 12, C_2 = 45 K, and $\langle\tau\rangle_{T_g}$ = 140 sec.

values, i.e., narrower distribution of relaxation times, the functionalized systems do afford some advantage over their doped counterparts with regards to technological applications. Given the same device performance requirements outlined previously, a 5% loss in $\chi^{(2)}_{NO}$ over a four-year period would require $<\tau> = 2\text{-}3 \times 10^{11}$ sec with $\beta_w = 0.45$; in contrast, the $\beta_w = 0.25$ associated with a device made from a doped system could require $<\tau> \geq 10^{14}$ sec. Assuming an activation energy of 45-50 kcal/mol in the glassy state for both systems, the upper limiting use temperature for the copolymer would be T_g-85°C while for the doped case it would be T_g-104°C. Thus, in conjunction with the higher T_g's of the copolymers, the narrower distribution of relaxation times of the side-chain functionalized systems in this study enhances the temporal stability of the SHG properties as compared to that found in doped systems.

Finally, some warning must be given regarding how certain approaches used for poling NLO polymers may result in significant errors in the quantitative interpretation (and thereby in the predictions) of the temporal decay of SHG properties. All work described previously in this manuscript was performed under in-plane poling conditions in which charge injection effects were negligible. (This was concluded due to the linear increase in the measured current [on the order of tens of nanoamps] as function of electric field, results comparable to dieletric measurements, and the evolution of a distinct additional contribution to $\chi^{(2)}$ under harsher poling conditions.) A number of studies [1,4,16,26,29-31] have shown that the SHG signal and temporal behavior depend significantly on poling methods and conditions. In particular, significant surface charge and charge injection effects are present when corona poling methods are used with the magnitude of the effects depending on temperature, humidity, substrate, magnitude of the poling field, poling field bias, poling time, etc. [1,31] However, it should be noted that charge injection effects can be considerably reduced if corona poling conditions and geometry are optimized [31,32] although surface charges are still present.

As an example of how erroneous interpretations may be drawn when corona poling is employed, Figure 8 compares the SHG decay dynamics at 40°C for an IBMA-MADR1 copolymer functionalized with 6 mol% DR1 (T_g = 73°C). This polymer sample was corona or in-plane poled at T_g+10°C (for 60 sec in Figure 8a and 15 min in Figure 8b) in air before quenching to the decay measurement temperature. During corona poling, -3 kV (considered to be a relatively mild corona poling voltage [30]) was applied with a sharp tungsten needle 1 cm from the polymer film surface (reaching a steady-state value *only* for the sample poled 15 min). In contrast, the in-plane poling involved applying a 15 kV/cm field (an order of magnitude smaller than poling fields used in other studies [4,16,29]) with the SHG signal reaching a steady-state value (when poled 60 sec or 15 min) prior to quenching to 40°C. Figure 8 clearly demonstrates very significant differences in the decays of $\chi^{(2)}_N$ of the corona and in-plane poled samples. As expected for the in-plane poled sample, $\chi^{(2)}_N$ drops instantaneously by nearly 10% upon removal of the dc field due to the loss of the electric-field-induced third order effect [11] (γ in equation (1)), with further gradual decline in $\chi^{(2)}_N$ at longer times due solely to the reorientation dynamics of the DR1 chromophore. In contrast, the corona poled sample does not exhibit this instantaneous drop upon turning off the corona voltage source and instead exhibits only a gradual decrease in $\chi^{(2)}_N$ with time.

The absence of the instantaneous drop off in $\chi^{(2)}_N$ in the corona poled sample

is evidence of the fact that after turning off the corona voltage source, surface charge [1,31] and injected charge remain in the sample, meaning that the polymer sample retains some effective electric field. Over time, surface and injected charges decay, with the effective decay times ranging up to weeks depending on a host of poling and decay conditions [31]. The fact that some effective electric field is retained in the polymer sample implies that the reorientation dynamics of the chromophore also slow down as compared to a sample which does not retain an effective electric field, as with the in-plane poled sample. However, once the surface and injected charge effects in the corona poled sample decay completely, it is expected that the reorientation dynamics of the NLO chromophores would be similar to those in the in-plane poled samples.

Hence, erroneous predictions of long term SHG decay behavior must result from corona poled samples if the predictions are made by extrapolating short term behavior which reflects a combination of the reorientation dynamics of the NLO chromophores (modified by the presence of a decay in the effective electric field due to surface charge and injected charge) and slow decay of the electric-field-induced third-order NLO effect (associated with γ in equation (1)). Furthermore, as technological application of second-order NLO polymers is anticipated to employ in-plane poling rather than corona poling [18,34], it is clear that any predictions of the decay behavior of the SHG polymers should be made from in-plane poled samples prepared under conditions where no measurable effects of charge injection exist. (We note that such effects may even be observed in some in-plane poled samples depending on substrate, electrode-type and especially on the poling strength, time, and temperature, examples of which are given elsewhere [26,33].)

Literature Cited

1. (a) Hampsch, H.L.; Yang, J.; Wong, G.K.; Torkelson, J.M. Macromolecules 1988, 21, 526; (b) Hampsch, H.L.; Yang, J.; Wong, G.K.; Torkelson, J.M. Macromolecules 1990, 23, 3640.
2. Singer, K.D.; King, L.A. J. Appl. Phys. 1991, 70, 3251.
3. Lindsey, G.A.; Henry, R.A.; Hoover, J.M.; Knoesen, A.; Mortazavi, M. Macromolecules 1992, 25, 4888.
4. Koehler, W.; Robello, D.R.; Dao, P.T.; Willand, C.S.; Williams, D.J. J.Chem. Phys. 1990, 93, 9157.
5. Eich, M.; Looser, H.; Yoon, D.Y.; Twieg, R.; Bjorklund, G.; Baumert, J.C. J. Opt. Soc. Am. B 1989, 6, 1590.
6. Man, H.-T.; Yoon, H.N. Adv. Mater. 1992, 4, 159.
7. Dhinojwala, A.; Wong, G.K.; Torkelson, J.M. Macromolecules 1992, 25, 7395.
8. Dhinojwala, A.; Wong, G.K.; Torkelson, J.M. Macromolecules 1993, 26, 5943.
9. Dhinojwala, A.; Wong, G.K.; Torkelson, J.M. J. Chem. Phys. 1994, 100, 6046.
10. Dhinojwala, A.; Hooker, J.C.; Torkelson, J.M. J. Non-Cryst. Solids, 1994, 172-174, 286.

11. Dhinojwala, A.; Wong, G.K.; Torkelson, J.M. J. Opt. Soc. Am. B, 1994, 11, 1549.
12. Levine, B.F.; Bethea, C.G. J. Chem. Phys. 1975, 63, 2666.
13. (a) Kohlrausch, R. Ann. Phys. (Leipzig) 1847, 12, 393; (b) Williams, G.; Watts, D.C. Trans. Faraday Soc. 1970, 66, 80.
14. Ferry, J.D. Viscoelastic Properties of Polymers; Wiley: London, 1980.
15. Emri, I.; Tschoegl, N.W. Rheol. Acta 1993, 32, 311.
16. Stahelin, M.; Walsh, C.A.; Burland, D.M.; Miller, R.D.; Twieg, R.J.; Volksen, W. J. Appl. Phys. 1993, 73, 8471.
17. Walsh, C.A.; Burland, D.M.; Lee, V.Y.; Miller, R.D.; Smith, B.A.; Twieg, R.J.; Wolksen, W. Macromolecules 1993, 26, 3720.
18. Lytel, R. presentation, Fall 1994 National ACS Meeting, Washington, D.C.
19. Angell, C.A. J. Non-Cryst. Solids 1991, 131-133, 13.
20. Pschorn, U.; Rossler, E.; Kaufmann, S.; Sillescu, H.; Spiess, H.W. Macromolecules 1991, 24, 398.
21. Patterson, G.D.; Lindsey, C.P. J. Chem. Phys. 1979, 70, 643.
22. Saito, S.; Nakajima, T. J. Appl. Polym. Sci. 1959, 4, 93.
23. Kastner, S.; Schlosser, E.; Pohl, G. Kolloid Z. 1963, 192, 21.
24. Plazek, D.J. J. Phys. Chem. 1965, 69, 3480.
25. Dhinojwala, A.; Torkelson, J.M. manuscript in preparation.
26. Dhinojwala, A. Ph.D. thesis, Northwestern University, 1994.
27. Royal, J.S.; Torkelson, J.M. Macromolecules 1993, 26, 5331.
28 Matsuoka, S. Relaxation Phenomena in Polymers; Hanser: New York, 1992.
29. Goodson III, T.; Gong, S.S.; Wang, C.H. Macromolecules 1994, 27, 4278.
30. Guan, H.W.; Wang, C.H.; Gu, S.H. J. Chem Phys. 1994, 100, 8454.
31. Hampsch, H.L.; Torkelson, J.M.; Bethke, S.J.; Grubb, S.G. J. Appl. Phys. 1990, 67, 1037.
32. Giacometti, J.A.; DeReggi, A.S. J. Appl. Phys. 1990, 74, 3357.
33. Dhinojwala, A.; Hooker, J.C.; Hamilton, K.; Torkelson, J.M. manuscript in preparation.
34. Burland, D.M. presentation, Fall 1994 National AIChE Meeting, San Francisco, CA.

RECEIVED January 30, 1995

Chapter 24

Optical Damage in Nonlinear Optical Polymer Waveguides

K. Song, M. Mortazavi, and H. Yoons[1]

Hoechst Celanese Research Division, 86 Morris Avenue, Summit, NJ 07901

Intensity of transmitted light through NLO polymer waveguides significantly decreased during continuous operation with a low-power near infrared laser. Results of an investigation of the optical damage of the polymer waveguides are presented for several NLO polymers. The damage was found to originate from a reduction of the index of refraction of the NLO polymers in the waveguides and subsequent loss of optical beam confinement in the waveguides. The decrease of the refractive index of the polymers is a result of photo-degradation of NLO chromophores through an interaction with the near infrared light and dissolved oxygen molecules in the polymers.

Recently, nonlinear optics (NLO) has become a subject of intense research interest because developments in the field of nonlinear optics hold promise for important applications in optical information processing, telecommunications, and integrated optics.[1,2] Organic NLO polymers have emerged as a rapidly developing research area since they possess a number of important characteristics for NLO device applications such as large nonresonant susceptibility, fast response time, and low dielectric constant. However, for successful implementation of device applications, the NLO polymers must satisfy additional requirements such as waveguide processability, good optical transparency, acceptable mechanical and thermal stability and photostability. This paper is a second report of our continuing investigation on the optical stability of NLO polymers in relation to waveguide device applications.

Much research has focused on elucidation of the nature of high-power laser induced damage in optical materials.[3] Optical damage behavior of the inorganic crystals has been extensively studied and is now recognized as due to photo-refractive index change upon laser irradiation.[4] Studies of optical damage in $LiNbO_3$ waveguides have established a basic understanding of

[1]Corresponding author

0097–6156/95/0601–0333$12.00/0

damage mechanisms, wavelength sensitivity, and power density thresholds for the inorganic materials.[5] Previous studies of optical damage in organic polymers largely dealt with single and multiple-shot damage behavior of plastics when exposed to high-power lasers.[6-9] The polymeric materials typically exhibit optical damage thresholds in the range of GW/cm^2 under these conditions and the damage is mainly due to the presence of impurities or absorbing centers in the materials. Scarce attention has been expended on the ability of NLO polymers to withstand continuous exposure to low-power near infrared (IR) light and photochemical stability of the polymers in extended waveguide operation.

We recently reported that NLO polymer waveguides easily suffer optical damage and the intensity of transmitted light through the waveguides quickly decreases during a continuous operation with an 1.3 μm diode pumped Nd:YAG laser.[10, 11] It was surprising to observe such rapid optical deterioration of the waveguides under a low-power near IR light at a wavelength far from the optical absorption wavelengths of the polymers. It further raises a challenging question on how to optimize the design of NLO polymers for optical communication applications since the prevailing wavelength of optical communication is 1.3 μm. Here we report on the dependence of the optical damage of NLO polymer waveguides on the operating parameters and molecular structure of the polymers. Also, a molecular mechanism for the photodegradation of NLO polymers at near IR wavelengths is proposed and directions for designing photochemically stable NLO dyes are suggested.

Experimental

Optical damage of waveguides of a series of side chain NLO polymers based on a poly-(methyl methacrylate) (PMMA) chain was investigated in the present study. Chemical structures of the NLO polymers used are shown Figure 1. Single mode inverted rib waveguides were prepared from these polymers by spin-coating followed by photobleaching. The vertical structure of the waveguides were first prepared by spin-coating a polymer film on silicon wafers. The wafers had a 2 μm thick SiO_2 top layer which was used as the lower cladding layer of the waveguides. The spin-coating solutions were prepared by dissolving the polymers in cyclohexanone to a concentration of 20 wt. % and were filtered through 0.2 μm pore Teflon filters to remove particulate impurities. The solutions were then spun on the silicon wafers and the resulting films were dried at about 160 ^oC which was about 20-30 ^oC above the glass transition temperature of the polymers. The drying condition was chosen to minimize build-up of shrinkage stresses associated with evaporation of solvent. The films typically had a thickness of about 5 μ m.

The lateral structure of the waveguides was defined using a photobleaching technique.[12] Photobleaching of polymers in the regions immediately outside the guiding region with visible light lowers the refractive index in the exposed region and thus defines the lateral structure of the waveguides. In this technique, the polymer films were bleached by exposing them to a broad band high pressure Xenon lamp through a mask which was placed in physical contact with the films. The lateral dimension of the waveguides were typically 5 μm. Detailed results of an investigation of the kinetics and the mechanisms of photobleaching of NLO polymers will be reported elsewhere.[13] Waveguide samples of polymers that do not photobleach were prepared by ablating away the polymers in the immediately adjacent lateral regions to the guiding region with a NaF excimer laser (190 nm wavelength).

The rate of optical damage of the waveguides was determined by monitoring the throughput intensity of 1.3 μm light as a function of exposure time. The light source was an 1.3 μm diode pumped Nd:YAG laser with the maximum power of 175 mW. The beam was coupled into a cleaved end face of the waveguides of which typical cross-sectional dimensions were 5 x 5 μm and length 3 cm. The output beam was imaged onto an infrared detector and the beam intensity was monitored as a function of time. For experiments in which stability of optical coupling was critical for an extended period of time, optical fibers were pigtailed at both ends of the waveguide using an epoxy adhesive. A more detailed description of the test equipment and method was given in an earlier report.[11]

UV-Vis and near infrared absorption spectra of the NLO polymers were obtained using a Perkin-Elmer Lambda 9 spectrophotometer. Thin polymer films (0.1 μm thick) deposited on a quartz plate were used for the spectroscopic study.

Results and Discussion

A significant decrease of the beam intensity with time is observed for waveguides prepared from a series of methyl methacrylate (MMA) copolymers containing 4-dimethylamino-4'-nitrostilbene (DANS) as a side chain NLO moiety. Figure 2 shows the decay of output intensity of 1.3 μm light transmitted through the waveguides relative to the initial output intensity. The relative transmission loss is given in the dB units in which a decrease of 3 dB is equivalent to the 50 % decrease of the transmitted beam intensity compared to the initial transmittance. The waveguide prepared from the copolymer of 5/95 DANS/MMA composition loses more than half of its initial intensity within two days when operated with 2 mW of 1.3 μm laser in an atmospheric environment. As the concentration of DANS chromophores in the copolymers increases, the rate of optical decay of the waveguide increases, strongly indicating that the optical damage is closely related to the NLO chromophore.

Changes in shape of the beam exiting the waveguides in the course of the near IR operation suggest that the optical transmission loss is predominantly due to a loss of beam confinement in the waveguides resulting from lowering of the refractive index of the polymer in the guiding region.[11] Figure 3 shows time-lapsed photographs of the output beam profile as a function of time. Initially, the beam profile is circular, as expected from a square cross-sectional shape of the waveguide. With further exposure to 1.3 μm laser light, the output image becomes first oval-shaped, and then completely dispersed in the lateral direction. These time lapsed photographs of the output image clearly exhibit the loss of beam confinement in the waveguide during exposure to the near IR light. The loss of beam confinement is a result of a decrease in refractive index differences between polymers in the guiding and the lateral cladding regions. As the polymer in the guiding region is continuously exposed to the laser light, its refractive index decreases and approaches to that of the polymer in the cladding region. Such a loss of light confinement was also observed in optical microscopic studies of damaged waveguides.[11]

Atmospheric oxygen plays a critical role in the optical damage of the NLO polymer waveguides at 1.3 μm.[10] The effects of atmospheric oxygen on the optical damage of the waveguides of the DANS/MMA copolymer were determined by measuring a change of output intensity through the waveguides in both air and argon environments. A waveguide was placed in a sealed metallic box which was continuously purged with argon gas in order to keep an oxygen-free environment. Input and output fibers were pigtailed to the waveguide prior to complete sealing. The optical damage rate is significantly reduced when the waveguide is operated in an argon atmosphere. The rate of the output intensity deterioration was found to be less than 0.25 dB/hr in the oxygen-free environment whereas 3 dB/hr was founded for the waveguides in an air atmosphere. Another evidence for participation of oxygen in the deterioration of the waveguides was found from a study of waveguide with multilayer structure. When a guiding layer of the waveguide is overcoated with another layer, the optical damage rate becomes much slower compared to that in waveguides without an overcoated layer. The overcoated layer acts as a barrier against diffusion of ambient oxygen and prevents oxygen from reaching the waveguide layer, resulting in reduction of the rate of optical damage.

Waveguides prepared from PMMA homopolymer (prepared by excimer laser ablation) did not show a decrease of the output intensity of 1.3 μm light transmitted through the waveguides even in the ambient environment. The result strongly suggests that the source of the optical damage is the NLO chromophores. The optical damage in NLO polymer waveguides apparently originates from degradation of a portion of the NLO chromophores through an interaction with the 1.3 μm light and atmospheric oxygen. To understand the relationship between the chromophore molecular structure and the optical damage characteristics, a series of NLO polymers containing different chromophores were investigated. New NLO polymers of which chemical structure is similar to DANS/MMA (see Figure 1) were prepared for this purpose. Instead of DANS, 4-methoxy-4'-nitrostilbene (ONS) was incorporated in PMMA as a side chain NLO chromophore. The chemical structure of ONS/MMA copolymer is the same as DANS/MMA copolymer except that nitrogen in the amine group of DANS has been replaced with

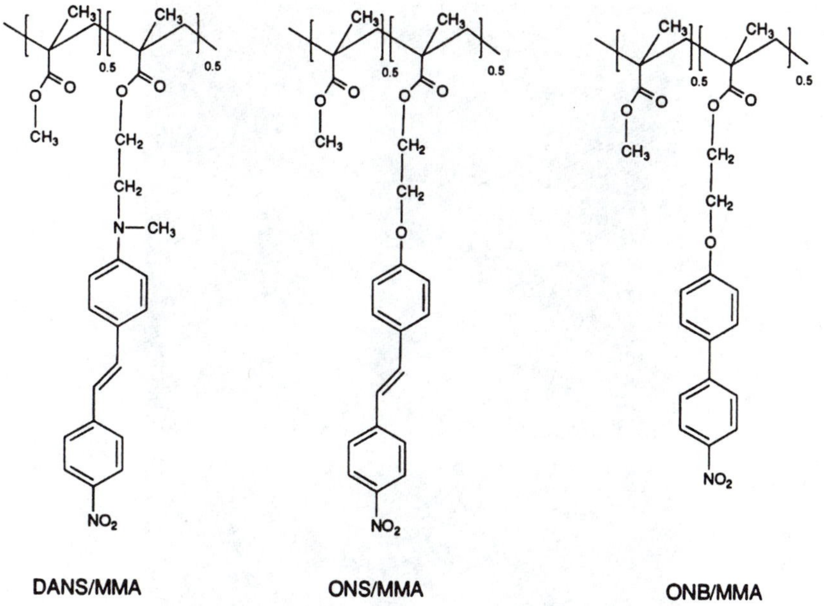

DANS/MMA ONS/MMA ONB/MMA

Figure 1. Chemical Structure of NLO polymers.

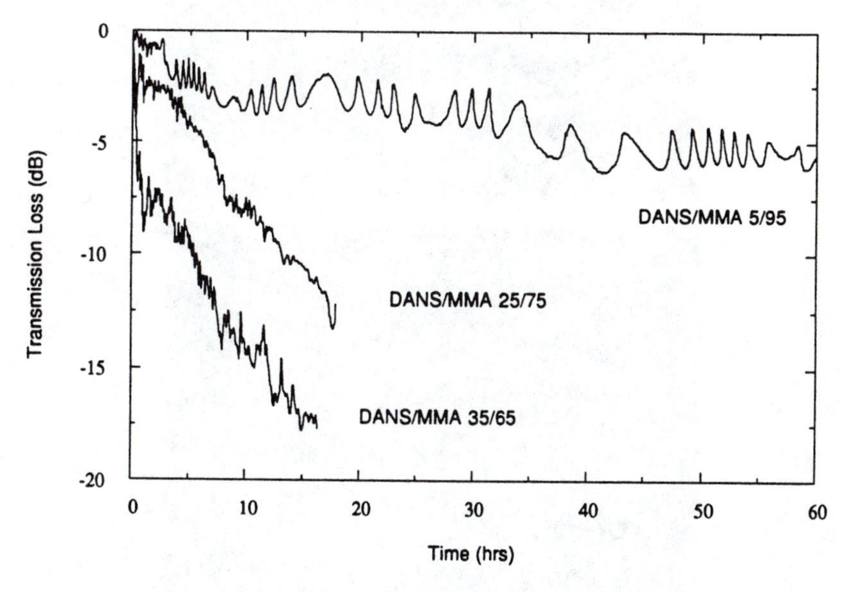

Figure 2. Optical transmission changes of 1.3 μm light for waveguides prepared from DANS/MMA copolymers with different comonomer compositions.

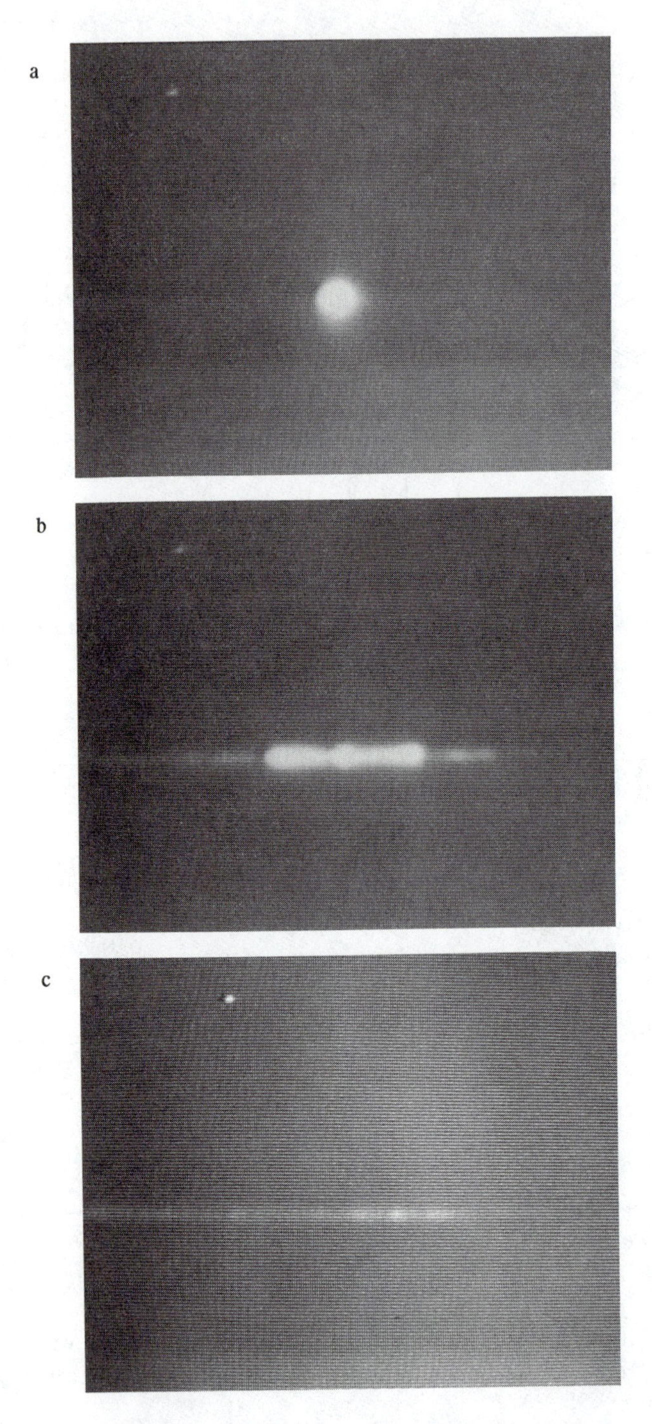

Figure 3. Photographs of output beam profile of transmitted light through the NLO polymer waveguide after (a) immediately, (b) 2 hr, and (c) 16 hr of operation with 1.3 μ m mm laser light.

oxygen. Another NLO chromophore examined, 4-methoxy-4'-nitro-biphenyl (ONB), has a structure similar to ONS except that the stilbene double bond is missing in ONB.

Strong dependence of the optical decay on the molecular structure of NLO chromophores was found. Interestingly, the waveguide from ONB/MMA copolymer did not exhibit any optical damage at 1.3 μm whereas the decay behavior of the waveguide of ONS/MMA copolymer was similar to that of DANS/MMA polymers albeit in a lesser extent. Figure 4 shows the decrease of the transmitted intensity through waveguides for NLO copolymers with DANS, ONS, and ONB chromophores. Copolymers of DANS and ONS lead to a continuous decrease of output intensity while a waveguide from ONB/MMA copolymer maintains the initial output intensity constant.

The above observations clearly demonstrate that the optical decay of the polymer waveguides is related to the NLO chromophore in the polymers. We examined a number of possible mechanisms for photochemical interactions of the NLO chromophores that may result in reduction of the refractive index of the polymers: (1) The long wavelength tail of the primary absorption peak extends into the 1.3 μm region and the optical absorption leads to a photochemical degradation of the chromophores much as in photobleaching with visible light; (2) Two- or multi-photon absorption of 1.3 μm light by the polymer falls into the primary absorption band and results in photochemical degradation; (3) Absorption of the near IR light leads to a thermal degradation of the chromophores; (4) There are unspecified photochemical mechanisms related to absorption of the near IR light.

Available spectroscopic data of the polymers do not support the decay mechanism related to photochemcial degradation resulting from the primary absorption at 1.3 μm by the chromophores. The first law of photochemistry states that only the light which is absorbed by a molecule can be effective in producing a photochemical change of the molecule. All NLO chromophores studied in this study have an absorption band in the 350-450 nm region with the half width at half maximum absorption of about 100 nm and it is highly unlikely that the absorption tail would extend to 1.3 μm region. The 50/50 DANS/MMA copolymer has the primary absorption band centered at 430 nm (Figure 5) which is far from the laser radiation wavelength of 1.3 μm. It is thus difficult to imagine the absorption of the 1.3 μm light by any resonant effects of the chromophore with light. Even the tail of the absorption band does not extend into the near IR region and becomes almost negligible at 1.3 μm which is the minimum point of the absorption spectrum (Figure 6). Furthermore, the electronic absorption bands of ONB/MMA and ONS/MMA copolymers centered at 350 nm and 370 nm, respectively, are very close to each other (Figure 5) and their optical decay behavior were markedly different. The difference in the absorption peak positions between the two copolymers is too small to account for the difference in the optical damage behavior of the polymer waveguides. Such a small shift in the absorption peak in the UV regime may not seem to bring about a significant change in the absorption tail at

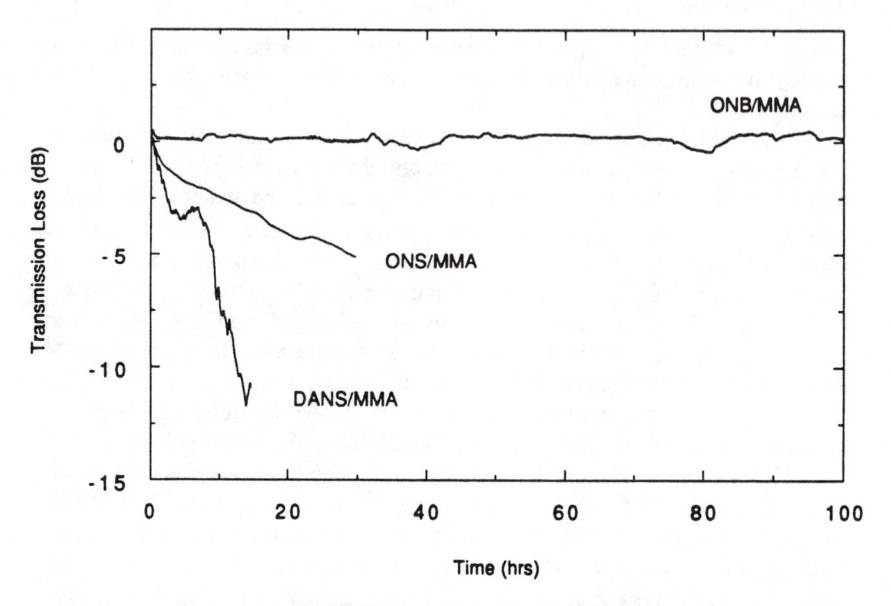

Figure 4. Optical transmission changes of 1.3 μm light for waveguides prepared from different NLO polymers.

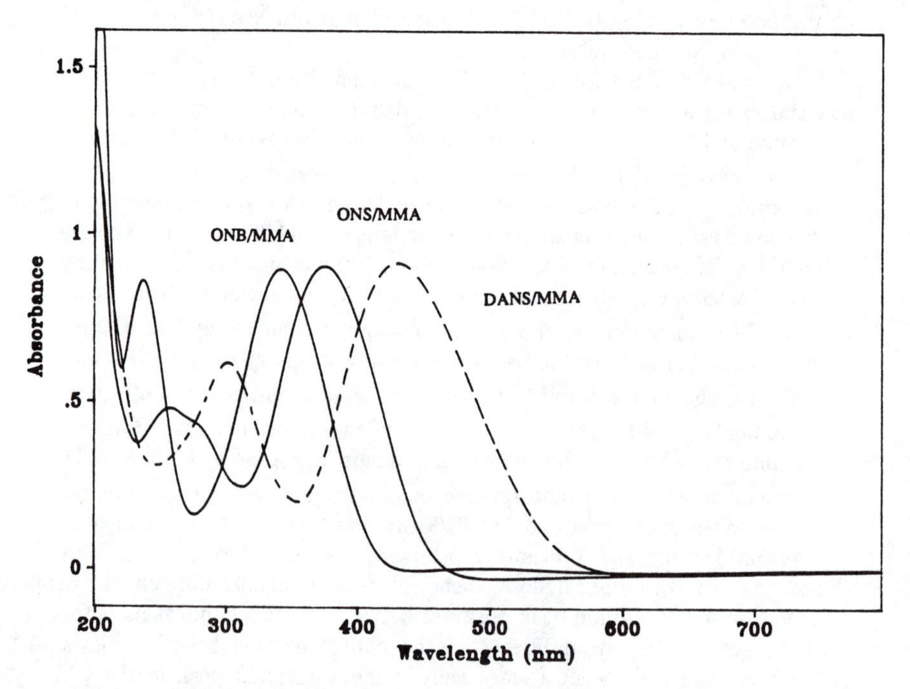

Figure 5. UV-Vis and near IR absorption spectra of thin films of the NLO polymers.

1.3 μm. In fact, the near IR absorption spectra of all the polymers, ONS/MMA, ONB/MMA and DANS/MMA copolymers are essentially the same.

We also ruled out the possibility of two- or multi-photon processes responsible for the optical degradation of the NLO chromophores. Investigation of the incident laser power dependence of the optical damage of the waveguides of DANS/MMA copolymers show that the rate was linear with the incident power.[14] In fact, the decay was observed even at very low incident laser power (below 1 mW) at 1.06 or 1.3 μm light within a few days.

A possibility of thermal degradation of the NLO polymers in the waveguides associated with temperature rise due to absorption of 1.3 μm light was also discarded. The near IR absorption spectrum of the 50/50 DANS/MMA copolymer (Figure 6) shows that the absorption at 1.3 μm corresponds to the minimum of the absorption spectrum in the near IR region. A comparison with the solution spectra shows that the finite absorption at 1.3 μm observed for the bulk sample shown in Figure 6 may be mostly from the Rayleigh scattering. A computer modeling study of the energy absorption and heat transfer in the NLO polymer waveguides showed that the temperature rise even with a 100 mW laser at 1.3 μm was less than 3 ^{o}C, too small to bring about thermal degradation of the chromophores or irreversible volume change (and associated refractive index change) in the guiding region of the waveguides.[14] Further, most absorption bands around 1.3 μm are due to the aliphatic C-H stretch overtones and its combination bands and therefore the near IR absorption behavior of the polymers samples were essentially identical while their optical decay behavior were markedly different. Comparison of the decay behavior at 1.3 and 1.5 μm (Figure 7) also disputes the thermal degradation mechanism. Absorption of the polymers at 1.5 μm is almost three times higher than that at 1.3 μm but no optical decay was observed at 1.5 μm. Through a similar reasoning, we also rejected the possibility that there may be unspecified photochemical reactions through the interaction between the chromophores and 1.3 μm light.

The oxygen effects on the optical damage of the NLO polymer waveguides and the dependence on the chemical structure of the chromophores strongly suggest that the photodegradation of the NLO polymers and resulting optical decay of the waveguides occur through a photosensitized oxygeneration reaction of olefinically unsaturated bonds. The photosensitized oxidation may proceed via formation of singlet oxygen by energy transfer from the excited states of chromophores.[15] In this model, the singlet oxygen may be generated in the NLO polymers by energy transfer from the excited chromophores to dissolved molecular oxygen, and they, in turn, react with the chromophores leading to a photo-degradation of the chromophore molecules. Such processes of photosensitized oxygenation of olefinic polymers involving singlet oxygen have been well documented.[15-18] Energy of the IR light may be absorbed by metastable excited intermediates which are generated by complexes of

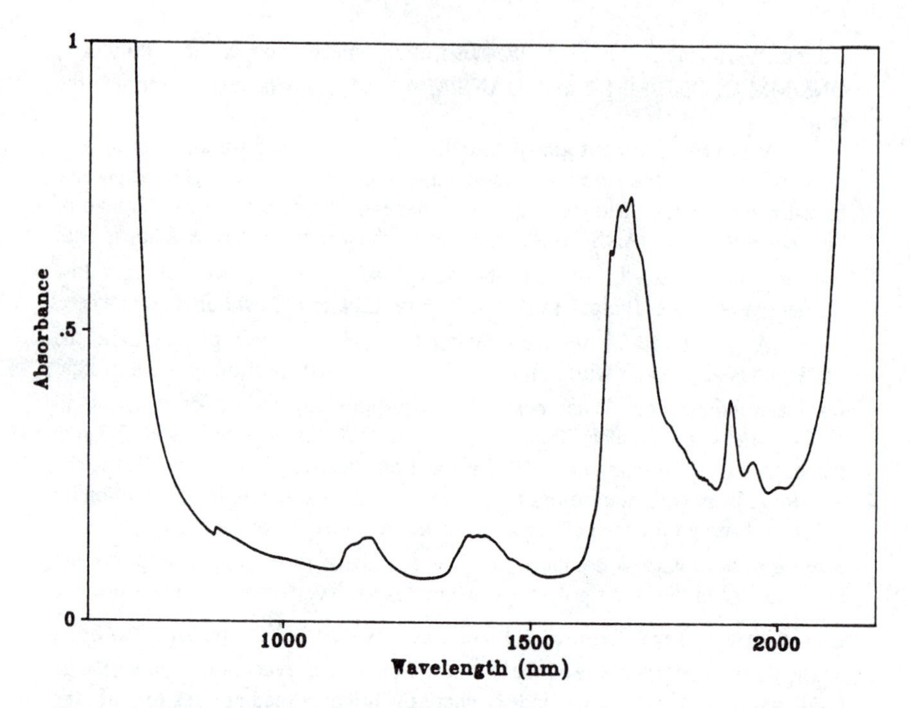

Figure 6. Near IR absorption spectrum of a 3 mm thick pellet of DANS/MMA 50/50 copolymer.

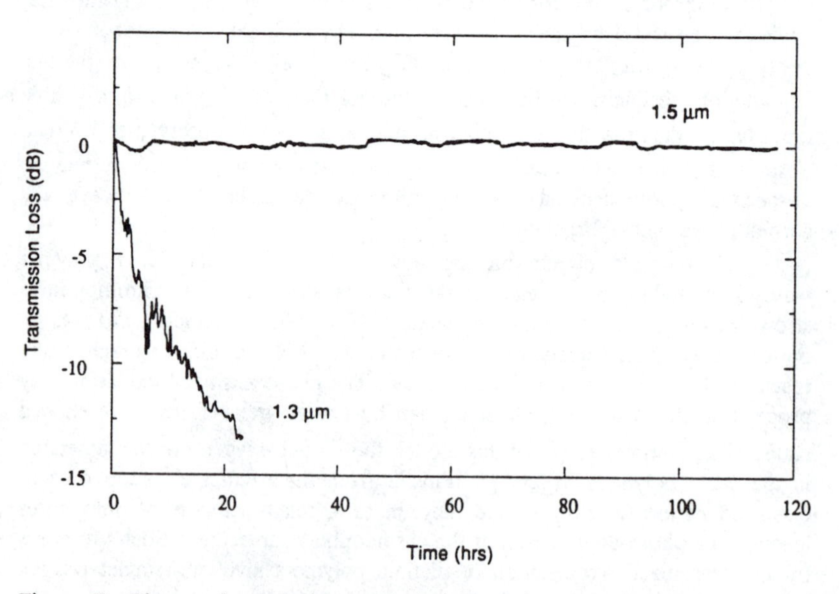

Figure 7. Changes in optical transmission through waveguides prepared from DANS/MMA 50/50 copolymer at 1.3 μ m and 1.5 μ m light.

chromophores with oxygen[19] or complexes of chromophore themselves.[9] Another probable source of the singlet oxygen is direct photochemical excitation of oxygen molecules absorbed in the NLO polymers by 1.3 μm laser light.[20,21] The wavelength of 1.3 μm light almost coincides the energy (1.27 μm) of the electronic transition from the triplet ground state oxygen to the lowest excited state singlet oxygen.[22] This implies that energy transfer from the near IR light to molecular oxygen is probable by the energy matching and thus generates excited singlet oxygen.

Another evidence that supports the photosensitized oxygenation mechanism is found in Figure 7 in which the wavelength dependence of waveguide optical damage is shown for 1.3 μm and 1.5 μm light. These two wavelengths are two primary wavelengths of the current optical communication networks. Waveguides from a 50/50 DANS/MMA copolymer operated without any performance decay over a week when operated with 30 mW of 1.5 μm diode pumped Nd:YAG laser light. The intrinsic absorption of the DANS/MMA copolymer at 1.5 μm is three times greater than at 1.3 μm and both wavelengths are far from the electronic absorption of the chromophore. However, Since the wavelength of 1.5 μm is of insufficient energy to excite oxygen molecules in the ground state to the singlet oxygen, molecular oxygen cannot be activated and thus the oxidation reaction does not occur at this wavelength. The superior photostability of the waveguides prepared from the ONB/MMA copolymer (see Figure 4) also supports the singlet oxygen mechanism. The ONB/MMA copolymer does not contain olefinic centers that are readily attacked by singlet oxygen. We conclude, based on thse evidences, that when waveguides prepared by DANS/MMA or ONS/MMA copolymers are operated at 1.3 μm light, the NLO copolymers experience a photosensitized singlet oxygen reaction at the stilbene C=C bond resulting in oxidation of the chromophore. This photo-oxidation reaction brings about the loss of conjugation in the NLO chromophores and reduces the oscillator strength of the primary absorption band. This results in a decrease of the index of refraction of the NLO polymer in the guiding region and, eventually, a loss of the beam confinement in the waveguide.

Conclusions

We observed for the first time the optical damage of NLO polymer waveguides by low-power near IR light. The damage manifests as continuous decrease of light transmission through the waveguides when operated with a laser of a wavelength near 1.3 μm. The damage was found to be loss of optical confinement in the waveguides resulting from a reduction of the refractive index of the NLO polymers through a photochemical interaction of the polymers, atmospheric oxygen and the near IR light.

Experimental data on the effects of operating variables on the damage suggest that the photochemical reaction centers are the NLO chromophores. The damage was not observed in waveguides fabricated with PMMA homopolymers while it intensifies in waveguides of DANS/MMA copolymer as the chromophore content increased. The damage behavior of NLO polymers varied with the chemical structure of the NLO chromophores; NLO chromophores that do not contain an aliphatic unsaturated bonds do not exhibit the damage with 1.3 μm. The photochemical reaction also involves atmospheric oxygen and the damage therefore was not observed when the waveguides were operated in Ar atmosphere. Additionally, in contrast to the behavior with 1.3 μm light, all the NLO polymer waveguides did not show any damage in operation with 1.5 μm light.

We conclude, based on the experimental findings, that the optical damage of the NLO polymer waveguides is due to a photosensitized oxidation of the NLO chromophores. The 1.3 μm light generates excite molecular oxygen to produce singlet oxygen molecules which readily react with the olefinic unsaturated bonds in the chromophores. The resulting loss in electronic conjugation reduces the refractive index of the polymer in the waveguiding region and thus lowers the optical confinement property of the waveguides.

For practical application of NLO polymer technology in telecommunication, the polymers must exhibit not only high NLO activity but acceptable optical stability in the operating wavelengths. The results of the present study suggest directions for molecular design of NLO chromophores with good optical stability. More detailed studies of the effects of the molecular structure of NLO chromophore, the optical wavelengths and the beam intensity on the optical damage of NLO polymer waveguides have been carried out to establish the damage mechanism(s) and molecular design principle of the optically stable NLO chromophores. The results will be reported elsewhere[23].

Acknowledgment

We gratefully acknowledge the suggestions and support of Drs. O. Althoff and C. C. Teng of Hoechst Celanese Research Division.

References

1. Williams, D. J. (Ed.); "*Nonlinear Optical Properties of Organic and Polymeric Materials*", ACS Symp. Ser. 233, Washington D.C., **1983**.
2. Tripathy, S.; Cavicchi, E.; Kumar, J.; Kumar, R. *Chemtech.* **1989**, 620.
3. A. J. Glass, A. J.; Guenther, A. H. *Appl. Opt.* **1977**, 16 (5), 1214.
4. Furukawa, Y.; Sato, M.; Kitamura, K.; Yajima, Y.; Minakata, M. *J. Appl. Phys.* **1992**, 72, 3250.
5. (a) Tangonan, G. R.; Barnoski, M. K.; Lotspeich, J. F.; Lee, A. *Appl. Phys. Lett.* **1977** 30, 238, (b) Jackel, J.; Glass, A. M.; Peterson, G. E.; Rice, C. E.; Olson, D. H.; Veselka, J. J. *J. Appl. Phys.* **1984**, 55, 269, (c) Wen, J. K.; Wang, Y. S.; Tang, Y. S.; Wang, H. F. *Appl. Phys. Lett.* **1988** 53, 260.
6. O'Connell, R.; Saito, T. *Optical Engineering* **1983**, 22, 393.

7. Chirila, T.; Constable, J.; VanSaarloos, P.; Barrett, G. *Biomaterials* **1990**, 11, 305.
8. O'Connell, R.; Deaton, T.; Saito, T. *Appl. Opt.* **1984**, 23(5), 682.
9. Dyumaev, K.; Manenkov, A.; Maslyukov, A.; Matyushin, G.; Nechitailo, S; Prokhorov, A. *J. Opt. Soc. Am. B* **1992**, 9(1), 143.
10. Song, K.; Mortazavi, M.; Chen, R.; Rafalko, J.; Goldberg, H.; Yoon, H. *Bull. Am. Phys. Soc.* **1993**, Vol. 38, 190.
11. Mortazavi, M.; Yoon, H.; Teng, C. *J. Appl. Phys.*, **1993** 74(8), 4871.
12. Diemeer, M.; Suyten, F.; Trommel, S.; McDonach, A.; Copeland, J.; Jenneskens, L.; Horsthuis, W. *Elect. Lett.* **1990**, 26, 380.
13. Song, K.; M.; Man, H., I. McCulloch, Yoon. H.; *J. Appl. Polym. Sci.* **1994**, 53, 665
14. Yoon, H.; Goldberg, H.; Mortazavi, M.; Song, K. HCC Internal Research Report.
15. Kearns, D.; *Chem. Reviews* **1971**, 71 (4), 395.
16. Jensen, J.; Kops, J.; *J. Polym. Sci. Polym. Chem. Ed.* **1980**, 18, 2737.
17. Trozzolo, A; Winslow, F. *Macromolecules* **1968**, 1, 98.
18. Ranby B.; Rabek, J. (Ed.), "*Singlet Oxygen*", John Wiley & Sons, **1980**.
19. Buchachenko, A. *Russ. Chem. Rev.* 1985, 54 (2), 117.
20. Evans, D. *Chem. Comm.* **1969**, 367.
21. Nathan, R.; Adelman, A. *J. Chem. Soc. Chem. Comm.* **1974**, 674.
22. Zweig, A.; Henderson, W. *J. Polym. Sci. Polym. Chem. Ed.* **1975**, 13, 717.
23. Mortazavi, M.; Song, K.; Yoon, H. in preparation.

RECEIVED February 14, 1995

Chapter 25

Polyimides for Electrooptic Applications

P. Kaatz[1,3], P. Prêtre[1,4], U. Meier[1], P. Günter[1], B. Zysset[2], M. Anlheim[2], M. Stähelin[2], and F. Lehr[2]

[1]Nonlinear Optics Laboratory, Institute of Quantum Electronics, Swiss Federal Institute of Technology, ETH Hönggerberg, CH–8093 Zürich, Switzerland
[2]SANDOZ Optoelectronics Research, SANDOZ Huningue SA, F–68330 Huningue, France

New modified polyimide polymers with pendant side group nonlinear optical (NLO) azo chromophores and moderate to high glass transition temperatures (140 °C < T_g < 190 °C) have been prepared. Corona poled films of these polymers possess large nonlinear optical susceptibilities of d_{31} = 23 pm / V and electro-optic (EO) coefficients of r_{13} = 6.5 pm / V at a wavelength of λ = 1.3 μm. The structural properties (glass transition, molecular weight, chromophore density) and optical properties (refractive index, optical nonlinearity) of these polyimides can easily be varied to fulfill the requirements of potential electro-optic devices. Due to the relatively high glass transition temperatures of these polymers, long-term stability of the optical nonlinearity of typically one to hundreds of years at operating temperatures of 80-100 °C is predicted from accelerated time-temperature measurements.

Novel amorphous nonlinear optical (NLO) polymers have been synthesized for applications as electro-optic materials *(1)*. High glass transition temperatures, good temporal stability, good processability and the ease of modification of the NLO side chain chromophores are the most significant aspects of these new polymers. The dispersion of the linear and nonlinear optical properties of these polymers has been determined.

Relaxation processes in nonlinear optical polymers are of considerable interest for obtaining a better understanding of the long-term stability of potential devices fabricated from these materials. Extensive relaxation measurements as a function of temperature have been performed above and below the glass transition, which plays the key role for the understanding of the orientational stability of poled amorphous polymers. It is found that the time-temperature relaxation behaviour of the NLO chromophores in a variety of polymer systems can be understood in terms of a phenomenological description of the glass transition with the aid of a scaling relation.

[3]Current address: Department of Physics, University of Nevada, Las Vegas, NV 89154
[4]Corresponding author

0097–6156/95/0601–0346$12.00/0

Synthesis of NLO Polymers

The polymers described in this article were prepared by polymer analogous reaction of styrene-maleic-anhydride copolymers with aminoalkyl functionalized azo chromophores *(2)*.

NH NH_2 R_2 N N O O O + NO_2 Cl C≡N N N N_2+ R_2 N NH_2 O O NO_2 a) m O N NH_2 b) O m HO NH m O N O N $(CH_2)_n$ NH_2 N $(CH_2)_n$ O N $(CH_2)_n$ R_1 R_1 R_1 N=N N=N N=N R_2 R_2 R_2 NO_2 NO_2 NO_2

In such NLO polymers, a precursor polymer containing reactive functional groups for the reaction with the NLO active unit is first prepared. For the synthesis of the precursor polymers, maleic anhydride was chosen as comonomer because of its high reactivity towards various vinyl monomers leading to preferably alternating copolymers. This offers the possibility of the synthesis of tailored NLO polymers. These precursor polymers were treated with aminoalkyl functionalized azo chromophores leading to polyamic acids, which were cyclized in a one-pot procedure to the corresponding polyimides with N,N-dimethylformamide (DMF) or N-methyl-2-pyrrolidone (NMP) as solvents .

Three polymers denoted by A-095.11, A-097.07, and A-148.02 with the chromophores and glass transition temperatures T_g indicated in Table I were chosen for the detailed measurements described in this work.

Table I. Azo Dye Substitution Patterns and Glass Transition Temperatures

Polymer	n	R_1	R_2	T_g [°C]
A-095.11	3	CH_3	H	137
A-097.07	3	CH_3	Cl	149
A-148.02	2	H	H	172

Glass transition temperatures for the C2 and C3 spacer polymers as a function of dye concentration are drawn in Figure 1 together with the abbreviations for the three polymers used in this work.

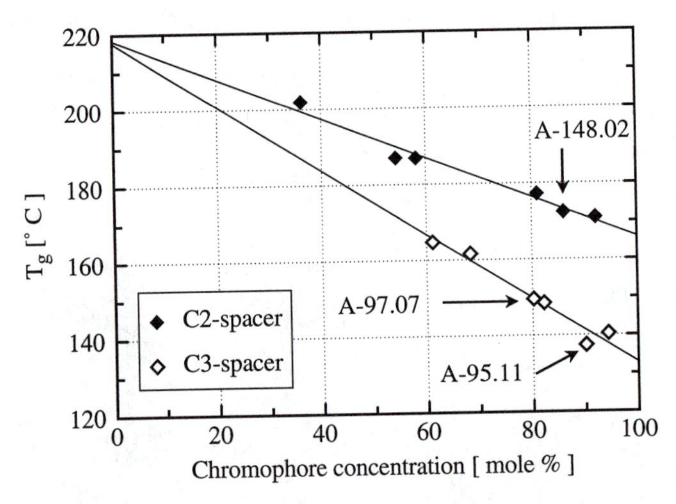

Figure 1. Glass transition temperatures as a function of chromophore content.

Thin Film Preparation

Thin films for optical and nonlinear optical measurements were spin cast from solutions of cyclopentanone/NMP (5:1) of varying polymer concentrations (typically 20 wt %). Films were backed on a hot plate at 200 °C for 1 h reducing the temperature to 170 °C for the rest of 1 day to prevent possible chromophore degradation.

Optical and Nonlinear Optical Properties

Refractive Indices. There exist several methods for the determination of refractive indices of spin cast polymer films. Quick, simple, and capable of reasonable accuracy is transmission spectroscopy *(3)*. The difference between the refractive indices of the polymer and the substrate (preferably fused silica) produces a modulation of the transmission properties due to optical interference in the thin ($\approx$ 1 μm) polymeric films. With the knowledge of the dispersion of the refractive indices of the substrate, the real part of the polymer's index can be fitted to a Sellmeier-type dispersion formula by the following equation

$$n_f^2(\lambda) - 1 = \frac{q}{1/\lambda_o^2 - 1/\lambda^2} + A \qquad (1)$$

where λ_o, the absorption wavelength of the dominant oscillator, was obtained directly from measured absorption curves. The parameter *A* was kept constant as it describes the contribution from the polymer backbone to the refractive index, assuming that the dispersion from the chromophores is approximately described by a single oscillator term. Figure 2 is an example of a transmission spectrum showing a characteristic modulation of the transmissivity due to interferences in the thin film ($\approx$ 1 μm). (Table II)

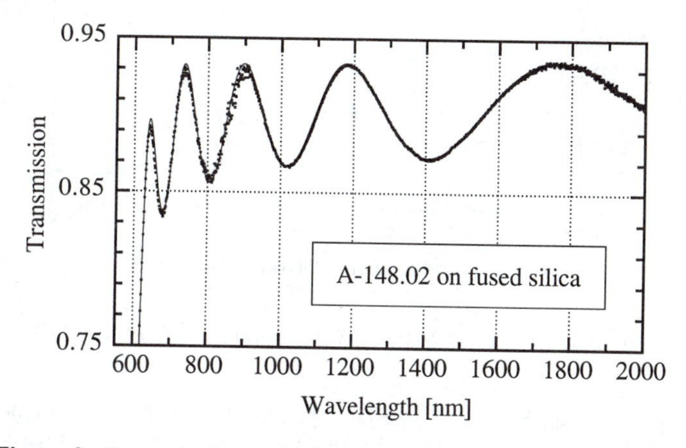

Figure 2. Transmission spectrum of a 1060 nm thick A-148.02 film.

Table II. Sellmeier Parameters and Refractive Indices at Selected Wavelengths

Polymer	A-095.11	A-097.07	A-148.02
λ_o [nm]	491	510	470
q [10^{-6} nm^{-2}]	1.874	1.809	2.074
A	1.234	1.234	1.234
n_f (633 nm)	1.84±0.02	1.89±0.02	1.80±0.02
n_f (1313 nm)	1.65±0.01	1.67±0.01	1.66±0.01

Nonlinear Optical Measurements. Nonlinear optical measurements were performed using a standard Maker-fringe technique *(4)*. Several films of each polymer (thickness 500 - 2200 nm) on ITO-coated substrates were poled for at least half an hour near the glass transition with a negative and positive corona discharge at voltages between 8 and 15 kV. The distance between the corona needle and the polymer film was approximately 1 - 2 cm. The second harmonic signals (SHG) from the polymer films were compared to a quartz reference (d_{11} = 0.4 pm / V).

The theoretical curves (2-level model) for the dispersion of the $\boldsymbol{d}$ coefficient given by

$$d_{ijk}(-2\omega,\omega,\omega) \propto f_i^{2\omega} f_j^{\omega} f_k^{\omega} \cdot \frac{\omega_0^2}{(\omega_0^2-\omega^2)(\omega_0^2-4\omega^2)}, \tag{2}$$

where ω_0 is the resonant frequency corresponding to λ_0 , are shown as solid lines in Figure 3 using the Lorentz approximation for the local field factors.

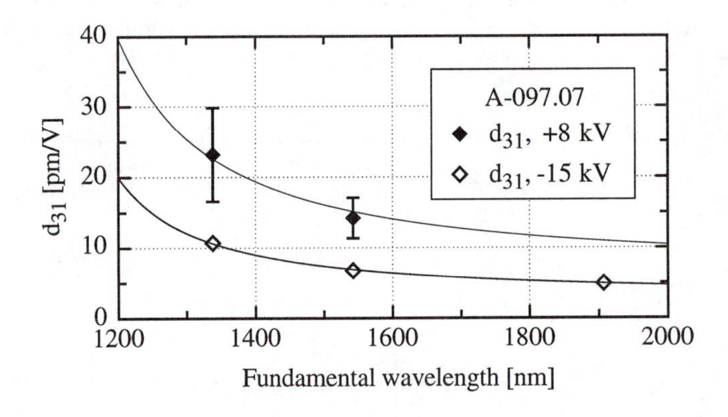

Figure 3. Dispersion of the nonlinear optical susceptibility d_{31}.

Electro-optic Measurements. SHG experiments can be used to deduce the electronic part of the electro-optic (EO) effect r^e and provide a measure of the expected EO-response of a material since for organics the effect is assumed to be mainly of electronic origin. Consequently, the electro-optic and nonlinear optical coefficients are directly related by*(5)*:

$$r_{ijk}^{e;-\omega,\omega,0} = \frac{-4}{(n_i^{\omega} n_j^{\omega})^2} \cdot \frac{f_i^{\omega} f_j^{\omega} f_k^{0}}{f_k^{2\omega'} f_i^{\omega'} f_j^{\omega'}} \cdot \frac{(3\omega_0^2-\omega^2)\cdot(\omega_0^2-\omega'^2)^2\cdot(\omega_0^2-4\omega'^2)}{(3\omega_0^2-\omega'^2)\cdot(\omega_0^2-\omega^2)^2\cdot\omega_0^2} d_{kij}^{-2\omega',\omega',\omega'}$$

The unclamped electro-optic coefficients r^T were measured at λ = 1313 nm either in the ellipsometric configuration *(6)* or by the interferometric technique *(5)* at a modulation frequency of 1 kHz. The sampled films on ITO-coated substrates with a 100 nm thick top gold electrode were poled at T_g for 35 minutes at poling fields between 120 and 150 V/µm. It is unclear whether the low value of r^T for A-097.07 is due to different poling conditions or to a inherent difference in the measurement technique. (Table III)

Table III. Electro-Optic Coefficients r_{13} at the Wavelength λ = 1313 nm

Polymer	r_{13}^{e} [pm/V]	r_{13}^{T} [pm/V]
A–095.11[1)]	6.0 ± 0.6	6.7 ± 0.4 (E_p = 150 V/μm)
A-097.07[2)]	5.4 ± 1.8	2.3 ± 0.2 (E_p = 120 V/μm)
A-148.02[1)]	5.5 ± 1.3	6.5 ± 0.3 (E_p = 150 V/μm)

[1)] ellipsometric / [2)] interferometric measurement

Relaxation Processes

Relaxation processes in NLO polymers are strongly coupled with aspects of the glass transition. The relaxation mechanisms of the side-chain chromophores in these polymers have been investigated above and below the glass transition by second-harmonic decay (SHG), dielectric relaxation and differential scanning calorimetry (DSC) measurements.

Differential Scanning Calorimetry (DSC) Measurements. The structural relaxation measured by DSC was modeled in terms of the fictive temperature concept, T_f, as proposed by Tool and calculated from the procedure of Narayanaswamy *(7)* and as further developed by Moynihan *(8)* and Hodge *(9)*. In this picture, T_f would be the equilibrium temperature for a polymer in the glassy state. Therefore, T_f equals T above T_g and reaches the final value T_g in simple cooling processes. The structural relaxation time was assumed to have a temperature dependence that is described by an Adam-Gibbs expression *(10)* as subsequently developed by Hodge *(11)*:

$$\tau(T) = A \exp\left(\frac{B}{T(1 - T_2 / T_f)}\right), \tag{4}$$

where A and B are constants. This expression for the relaxation time reproduces the Vogel-Fulcher / Arrhenius behaviour above / below T_g . It also implies the existence of a thermodynamic transition temperature, T_2, below the glass transition, at which the configurational entropy of the relaxing molecular segments vanishes.

DSC measurements were done with a commercial Perkin Elmer DSC-2C apparatus. The samples were prepared as pressed powder capsules of about 20 - 30 mg. Figure 4 shows the DSC traces of polymer A-095.11 and the corresponding theoretical fits (see also below). A typical measurement was accomplished as follows: the polymer sample was first heated to an initial temperature of about T_g + 50 K for approximately 15 min. Subsequent cooling was done at a constant rate to at least 50 K below T_g, and then further reheated at a constant rate to the initial temperature. Data was collected on the heating portion of the cycle. DSC measurements of the unsubstituted and side-chain polymers indicate that optimal T_2 values are obtained at temperatures of about 175 - 200 K. The cooling and heating rates were at ± 20 K / min.

As the main result of these measurements, we conclude that the preexponential time factor A can be replaced by the relation

$$A = \tau_g \exp\left(\frac{-B}{T_g - T_2}\right), \tag{5}$$

with τ_g the relaxation time at T_g.

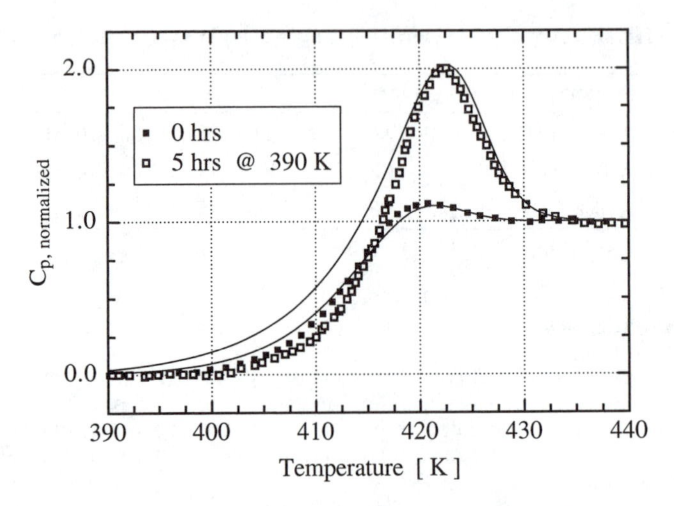

Figure 4. DSC-traces of A-095.11 without and with sub-T_g annealing

Dielectric Measurements. A Hewlett-Packard HP 4129A LF Impedance Analyzer with a frequency range of 5 Hz to 13 MHz was used for dielectric measurements to investigate the relaxation of the side-chain chromophores above the glass transition temperature. Films were spin-cast onto glass substrates with evaporated gold electrodes. The impedance measurements were done according to the "lumped circuit" method in either a parallel or series configuration. Figure 5 shows the relaxation times τ above T_g.

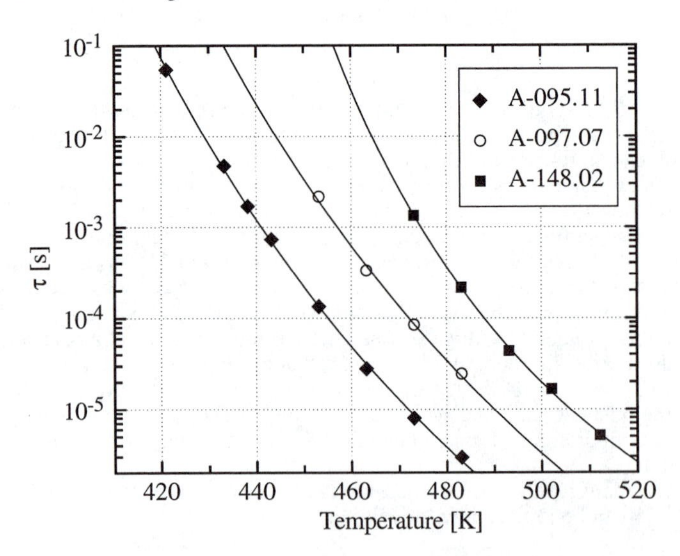

Figure 5. Dielectric relaxation times above T_g

The dielectric loss was fitted to the Fourier transform of the Kohlraus-Williams-Watts (KWW) stretched exponential function *(12)* (see equation 6). All of the dielectric loss spectral measurements show a large rise at low frequencies that is most likely due to the DC conductivity of the polymer. This DC-background can be eliminated by fitting the dielectric modulus $M(\omega)$ (or inverse dielectric constant) rather than the dielectric constant $\varepsilon(\omega)$ itself *(13)*. The Vogel-Fulcher behaviour is well demonstrated with fitted T_2 values slightly higher than expected from DSC measurements.

Second Harmonic Decay (SHG) Measurements. Relaxation of the side-chain chromophores below the glass transition was investigated by the decay of the SHG signal from corona poled films. The time dependence of the SHG decay was usually found to be well represented by a "stretched exponential",

$$d(t) = d_0 \cdot \exp(-(t / \tau)^b), \tag{6}$$

where $0 < b \leq 1$ is a measure of the nonexponentiality. The stretching coefficient b was found to be approximately linear over the temperature range of the measurements (T_g - 80 K < T < T_g - 20 K) with values in the range 0.2 - 0.4, with broader distributions at lower temperatures. Figure 6 shows examples of decay curves for polymer A–148.02 at two different decay temperatures in a semilogarithmic plot which is the most appropriate representation for this kind of relaxation.

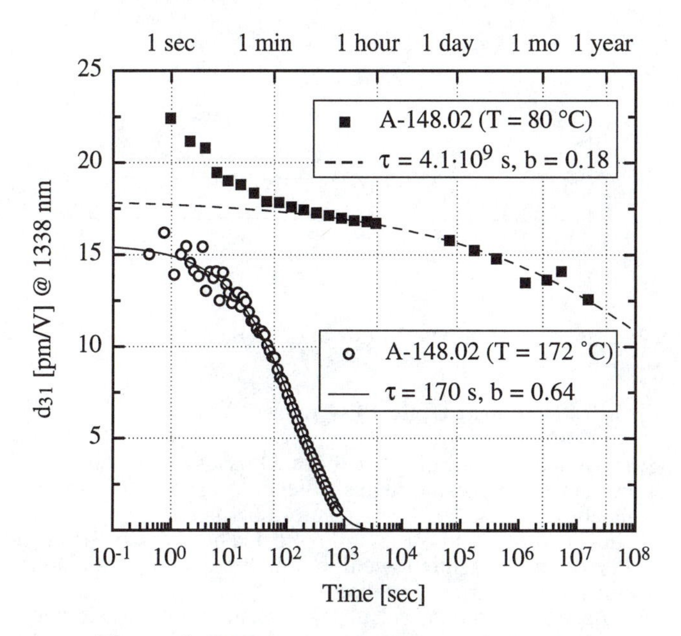

Figure 6. SHG decay at temperatures below T_g

All of the decay experiments exhibit a fast initial decay amounting to about 20 - 30% of the total initial SHG signal during the first 1 - 10 minutes. We attribute this decay to the release of charge carriers trapped in the film after the corona voltage has been turned off. In most cases the contribution from this fast decay has been isolated by introducing in the above equation an extra fit parameter t_0 which provides a time offset. It is evident from the fit which points at the beginning of the decay should be eliminated from subsequent data analysis.

For temperatures lower than about 30 K below T_g, the relaxation shows Arrhenius type behaviour with activation energies of about 220 - 250 kJ / mole. Inserting equation 5 in 4 reveals that the logarithm of the relaxation time associated with the α process or main chain segmental motion should be linear with $(T_g - T) / T$ with a slope given by $B / (T_g - T_2)$. The SHG data for the side-chain polymers are plotted in Figure 7 along with the results of Stähelin et al. for guest-host polymer matrices *(14)*. The SHG results indicate that the relaxation time of the NLO chromophores of side chain polymers correlate better with this main chain segmental motion than do guest-host NLO polymer matrices.

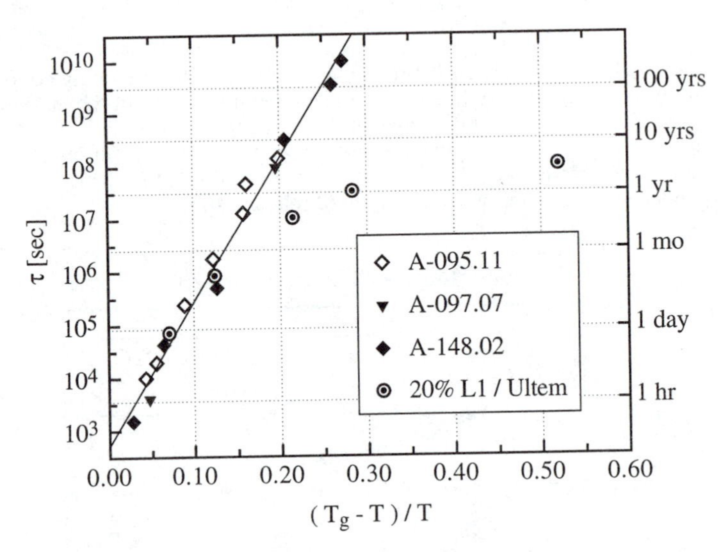

Figure 7. SHG decay times with $(T_g - T) / T$ as the relevant scaling parameter

Physical Ageing of Nonlinear Optical Polymers

Physical ageing or sub-T_g annealing is expected to provide an additional increase in relaxation times since the polymer tends to relax to the equilibrium or stable state during the annealing process. A DSC trace of polymer A-095.11 for an annealing time of 5.5 hours at 390 K can be seen in Figure 4. The overshoot is due to the rapid approach of the polymer structure to equilibrium during reheating after being highly relaxed in the glassy state during annealing.

Near the glass transition, the calculated relaxation times are in good agreement with measured relaxation times. The program predicts that annealing the polymers at ≈ 15 K below T_g increases the enthalpic relaxation time below T_g by about a factor of five. Preliminary measurements by SHG relaxation indicate that the chromophore orientation can be stabilized by about the same amount using this annealing process.

Conclusions

The polyimide side-chain NLO polymers reported in this work have the useful feature of allowing the simple attachment of a wide variety of chromophores. With a typical substitution using azo chromophores as the NLO active moiety, we measure similar NLO and EO coefficients as those measured in more conventional polymeric systems *(15)* with however, substantially increased glass transition temperatures. Phenomenological theories of the glass transition allow us to model some aspects of the chromophore relaxation. These theories predict an Vogel-Fulcher type temperature dependence *above* the glass transition. Dielectric relaxation measurements in the range of $T_g + 10$ to $T_g + 70$ K confirm this prediction. *Below* the glass transition, in the range of about T_g to T_g - 50 K, relaxation of NLO chromophores appears to be primarily influenced by large scale molecular motions of the polymeric chains. Consequently, relaxation in several types of polymer systems seems to scale well with regard to the normalized temperature $(T_g - T) / T$. Experimentally we find that the orientational stability of covalently bound side-chain NLO chromophores is significantly improved as compared to guest-host type polymer systems having the same glass transition temperature. A further increase in relaxation times of the order of five can be expected with a sub-T_g annealing process at temperatures 15 K below T_g.

Acknowledgements

We thank U. Stalder, R. Iseli, and G. Recoque for their help with the experiments. J. Söchtig of the Paul Scherrer Institute - Zürich provided helpful assistance with the electrode poling of the polymer films. We also thank H. Scherrer of ETH Zürich for the deposition of the gold electrodes used for the dielectric measurements.

Literature Cited

(1) Burland, D. M.; Miller, R. D.; C.A.Walsh, *Chem. Rev.* **1994**, 94, 31-75
(2) Ahlheim, M.; Lehr, F., *Makromol. Chem* **1994**, *195*, 361 - 373.
(3) Manificier, J. C.; Gasiot, J.; Fillard, J. P., *J. Phys. E: Sci. Instrum.* **1976**, *9*, 1002-1004.
(4) Jerphagnon, J.; Kurtz, S. K., *Journal of Applied Physics* **1970**, *41*, 1667.
(5) Bosshard, C.; Sutter, K.; Schlesser, R.; Günter, P., *J. Opt. Soc. Am.* **1993**, *B10*, 867.
(6) Teng, C. C.; Man, T. H., *Appl. Phys. Lett.* **1990**, *56*, 1734-1736.
(7) Narayanaswamy, O. S., *J. Am. Cer. Soc.* **1971**, *54*, 471.
(8) Moynihan, C. T.; Crichton, S. N.; Opalka, S. M., *J. Non-Cryst. Solids* **1991**, *131-133*, 420.
(9) Hodge, I. M., *J. Non-Cryst. Solids* **1991**, *131-133*, 435.
(10) Adam, G.; Gibbs, J. H., *J. Chem. Phys.* **1965**, *43*, 139 - 145.
(11) Hodge, I. M., *Macromolecules* **1986**, *19*, 936.
(12) Williams, G.; Watts, D. C.; S.B., D.; North, A. M., *Trans. Far. Soc.* **1971**, *67*, 1323.
(13) Moynihan, C. T.; Boesch, L. P.; Laberge, N. L., *Phys. Chem. Glass.* **1973**, *14*, 122-125.
(14) Stähelin, M.; Burland, D. M.; Ebert, M.; Miller, R. D.; Smith, B. A.; Twieg, R. J.; Volksen, W.; C.A.Walsh, *Appl. Phys. Lett.* **1992**, *61*, 1626.
(15) Nahata, A.; Shan, J.; Yardley, J. T.; Wu, C., *J. Opt. Soc. Am.* **1993**, *B10*, 1553-1564.

RECEIVED January 30, 1995

Chapter 26

Electrooptic Cardo-Type Polymers with High Glass-Transition Temperatures

Chengjiu Wu, Charles Knapp, Victor Lu, Michael J. McFarland, Ajay Nahata, Jianhui Shan, and James T. Yardley

AlliedSignal, Inc., Polymer Science Laboratory, Research and Technology, Morristown, NJ 07960

We report the electro-optic and thermal stability properties of a new class of novel thermostable fluorene-based cardo-type polymers. The electro-optic coefficients found for these materials are comparable to those obtained from many other reported side-chain polymers. Isothermal decay measurements for both the charge transfer absorption bands and for the electro-optic responses demonstrate that the thermal degradation of electro-optic response in these materials is dominated by chemical degradation and not by depoling, at least over a large range of elevated temperatures. One of these materials for example is characterized by a measured 1/e decay time of over 40 days at 190°C with extrapolated lifetimes of over 1 year at 175°C.

There is a profound need for electro-optic materials which can satisfy the needs for integrated optical waveguide devices suitable for use in high speed data transmission and manipulation. Such materials must exhibit large intrinsic electro-optic response and must be compatible with traditional electronic processing techniques. Polymeric materials are extremely attractive for this purpose since materials have been demonstrated with significant response and since polymers can be chemically tailored to optimize optical properties and can be processed using conventional techniques compatible with semiconductor devices. In particular, polymers containing chromophores which are characterized by large hyperpolarizabilities have been "poled" through application of electric fields at temperatures near the glass transition to display a non-centrosymmetric structure exhibiting significant second-order non-linearities (1). However, real-world devices must also withstand a wide range of environmental conditions. Even at temperatures below the glass transition temperature, T_g, many polymers reported to date will exhibit reduced electro-optic response due to thermal degradation and thermodynamic relaxation of the dipolar alignment as a result of

0097–6156/95/0601–0356$12.00/0

processing at elevated temperatures or of long-term operation under normal operating conditions. Approaches directed toward the achievement of improved poling stability include the synthesis of main chain polymers with reduced rotational mobility (2), the creation of thermally (3) or optically (4) crosslinked chromophores within the polymer matrix, or the creation of polyimide-based guest-host systems (5). Our earlier work has demonstrated that, for electro-optic polymers with a low T_g such as the polymethacrylate with the disperse red dye-1 as pendant ester group, the loss of EO response at elevated temperature is mainly caused by reorientation of the aligned EO chromophores (depoling) (6,7) . Our approach thus seeks to incorporate the EO chromophores into a thermally stable polymer backbone. We have described EO polymers made by incorporating Disperse Red-1 type chromophores onto high T_g polymers such as polyvinyl phenol or styrene-maleimide alternate copolymer (6). In this report, we describe the electro-optic response and the observed thermal stability for a new family of EO polymers which is derived from the fluorene-based cardo-polymers (7). These polymers are based on a structure which has been shown to provide a base for engineering thermoplastic polymers characterized by high T_g and excellent thermal stability (8-10). In addition, these polymers can be processed into electro-optic material in a manner identical to previously-described side-chain polymers. Therefore, they do not require any unusual poling procedures and do not exhibit the brittleness associated with many crosslinked polymers.

2. NEW CARDO-TYPE ELECTRO-OPTIC POLYMERS

Cardo-polymers are well known for their high T_g and outstanding thermal stability (8). The fluorene-based cardo-polymer is shown in the following structure:

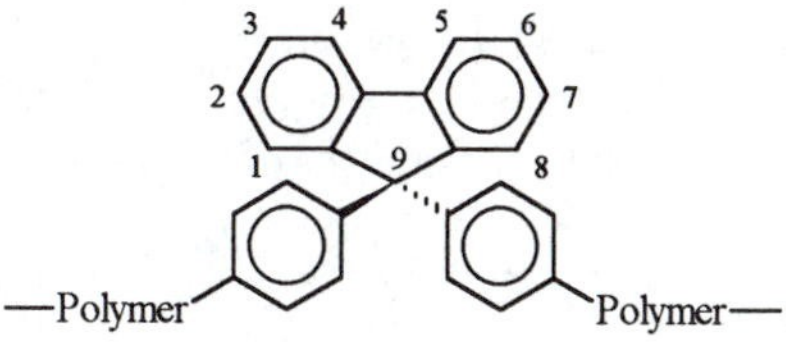

3 4 5 6
2 7
1 9 8
—Polymer Polymer—

Fluorene-based Cardo-polymer

D A
—Polymer Polymer—

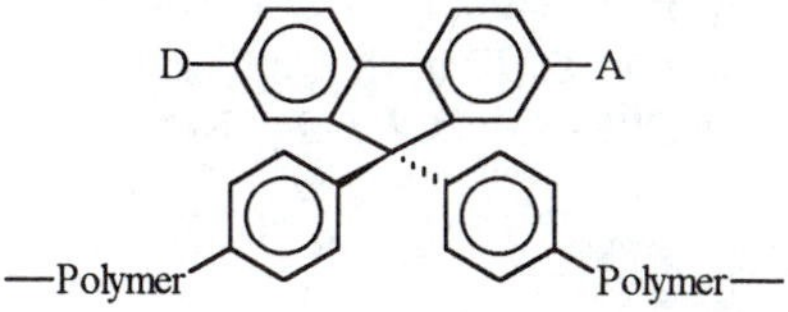

EO active Fluorene-based Cardo-polymer

The decomposition temperatures for such fluorene-based cardo-polymers are usually ≥ 450°C and glass transition temperatures greater than 200°C are easily attainable. We have demonstrated that certain cardo-polymers with high T_g are useful polymer hosts for EO systems. Due to the spiro nature at the 9-position, the planar biphenyl in the fluorenyl system is electronically isolated from the two 9-phenyls which are used for further polymerization. In order to create an EO material, we have attached at the 2- and the 7-positions of the fluorenyl ring an electron donating and an electron accepting substituent respectively.

We report here the synthesis and properties of several new fluorene-based electro-optically active thermoplastic cardo-polymers, as shown in the Figure 1. The electron accepting group in all cases is a nitro. In one type of polyesters, **BPE-n** where n is the average number of CH_2 linkages per repeating unit, and in a polycarbonate **BPC,** the electron donating group is a dimethylamino. In another type of polyesters

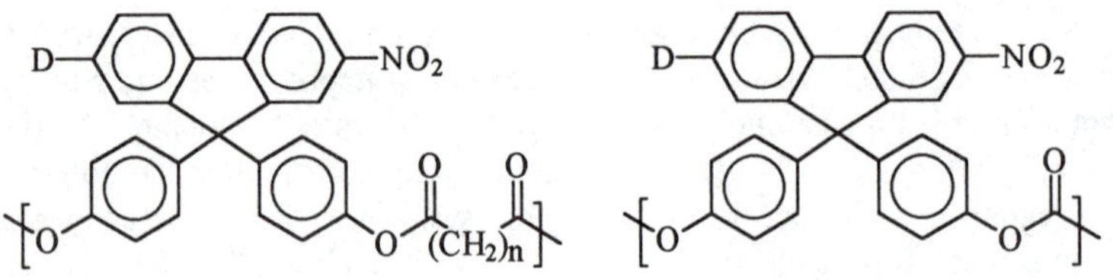

Polyester
D = $-N(CH_3)_2$, **BPE-n**;
$-OCH_3$, **OPE-n**

Polycarbonate
D = $-N(CH_3)_2$, **BPC**;
$-OCH_3$, **OPC**

Figure 1. New electro-optically active cardo-polymers

(**OPE-n**) and another polycarbonate (**OPC**), the electron donor is a methoxy group.

The synthesis of the monomers and polymers is depicted in Figure 2. The commercial available 2,7-dinitrofluorenone, **DNF**, was first converted to its ketal, **DNFK**. An aromatic nucleophilic substitution then replace one of the nitro groups on the fluorenyl ring with a electron donating group to produce **DANFK** when D is a dimethylamino or **MNFK** when D is a methoxy group. The monomer, **BISMAN** or **BISMON**, were prepared by treating **DANFK** or **MNFK** with phenol in the presence of an acid. By the same methodology, other monomers in which the polymerizing groups are amines or glycidyl ethers have also been prepared. The detailed preparation of these monomers will be reported elsewhere (11). For polycondensation we have used an active ester method (12, 13). The 2- or 4-nitrophenyl ester of the diacids or phosgene reacted with the fluorenyl bisphenol in a homogeneous solution of methylene chloride at room temperature. The advantages of this method include smooth reaction and no contamination of the troublesome ionic species usually encountered in condensation reaction. These polymers are amorphous, highly soluble in organic solvent such as DMF or diglyme, and film-forming. The glass transition temperatures (T_g) of the **BPE-n** and **OPE-n** polyesters are plotted against the number of the CH_2

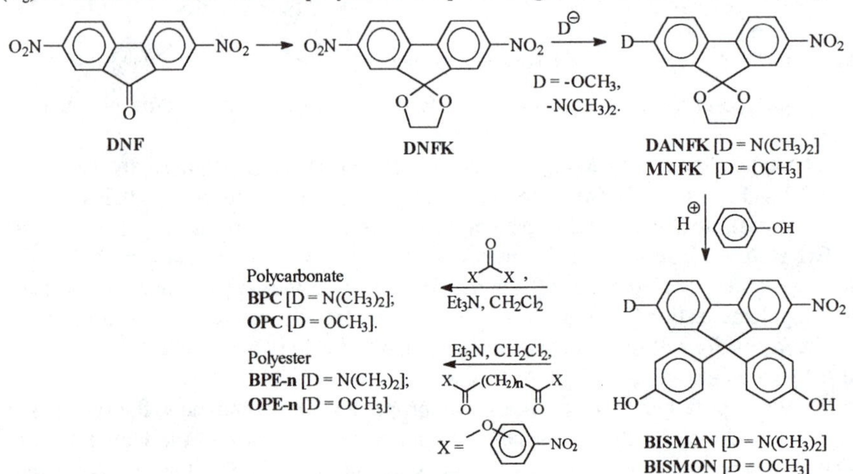

Figure 2. General scheme of synthesis

carbons in the diacid comonomer, as shown in Figure 3. For comparison, the T_g of the **OPC** and **BPC** polycarbonates are also shown despite the slight differences in the relating structures. The T_g ranges from 150°C for **BPE-4** to 278°C for **BPC** and 294°C for the **OPC**. As was proved in the **BPE-n** and the **OPE-n** case, T_g of the polyesters can be controlled by the composition of the diacid comonomers. Here we will be primarily concerned with **BPE-2.5** built from equal amounts of $(CH_2)_2$ groups and $(CH_2)_3$ groups for which T_g is 205°C.

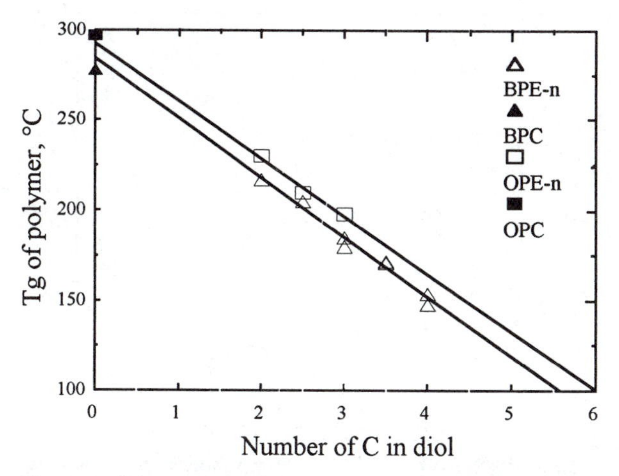

Figure 3. Glass transition temperatures of **BPE-n** and **OPE-n** polyesters as a function of numbers of CH_2 in diacids. Tg's of **BPC** and **OPC** polycarbonates are also shown.

3. ELECTRO-OPTIC PROPERTIES

The electro-optic characterization of the **BPE-2.5** was performed using apparatus and methodology as described earlier (14). Films of thickness 0.5 μm and 1.0 μm were prepared on quartz substrates fitted with photolithographically defined coplanar aluminum electrodes (2000 A thick) separated by a 25 μm gap. The samples were vacuum-poled at the glass transition temperature (T_g = 205°C) in a contact poling apparatus. During the poling process, the sample temperature was held at T_g with the field applied for a period of 10 minutes before being cooled to room temperature. Samples poled for as little as 2 minutes and as long as 1 hour at T_g showed negligible variation in the observed electro-optic coefficients. The characterization was performed 12 to 24 hours after poling to ensure dissipation of any possible surface injected charge.

Figure 4 shows the observed electro-optic response, ($n_e^3r_{33}$ - $n_o^3r_{13}$), measured at 810 nm as a function of poling field where for the $C_{\infty v}$ symmetry of our poled polymer system n_o and n_e are the ordinary and extraordinary refractive indexes and r_{33} and r_{13} the unique electro-optic coefficients (15). The response is linear at low fields, but becomes increasingly sublinear at higher fields. We have also measured the electro-optic response at wavelengths of 632.8 nm, 810 nm, and 1305 nm for a sample poled at 0.5 MV/cm. Using the observed combination of electro-optic coefficients ($n_e^3r_{33}$-

$n_o^3 r_{13}$), measured refractive index data, and the assumptions that $n_e \sim n_o$ and $r_{33} \sim 3r_{33}$ (presumed to be valid for this poling field), we have calculated r_{33} at these wavelengths to be 4.6, 2.7, and 1.8 pm/V respectively as described previously (7). In order to extrapolate the data to other wavelengths, we further assume that this polymer can be modeled by an independent response two-level model (16) and that the dominant tensor component of the molecular hyperpolarizability, β_{IJK}, is β_{ZZZ}. Using the appropriate local field factors and experimental poling parameters and the fit described above, we

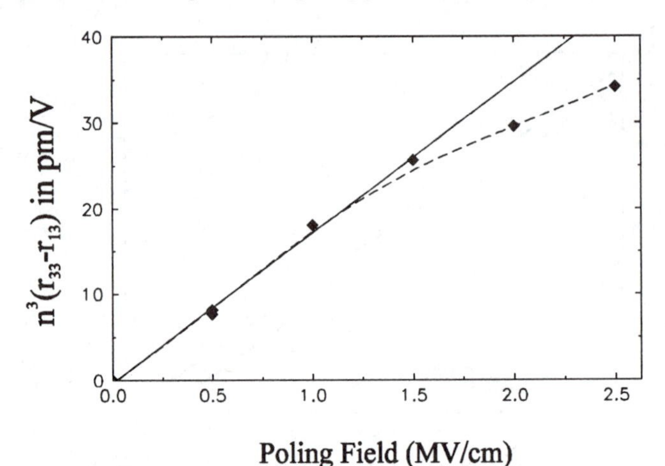

Figure 4. Electro-optic response for **BPE-2.5** as a function of poling field measured at 810 nm wavelength.

find that $\mu\beta^0{}_{ZZZ} = 276 \times 10^{-30}$ Debye cm^5/esu where μ is the dipole moment and $\beta^0{}_{ZZZ}$ is the hyperpolarizability at infinite wavelength. If the non-linearity seen in Figure 4 arises from saturation of the orientation, we may use this independent response model as described previously (17) to extract approximate values for μ and for $\beta^0{}_{ZZZ}$. If we neglect the variation in the refractive indexes due to the poling of the polymer, we find that the best fit to the data yields a dipole moment of 16 Debye. The corresponding $\beta^0{}_{ZZZ}$ is 17.3×10^{-30} cm^5/esu. These results show that the **BISMAN** chromophore is quite comparable to other chromophores commonly used in poled polymeric electro-optic materials and illustrate the utility of these polymer systems for electro-optic systems. The other **BPE-n** polyesters, as well as the **BPC** polycarbonate, all have electro-optic responses close to the **BPE-2.5**. The **OPE** polyesters and the **OPC** polycarbonate, in which the electron donor has changed to a methoxy group, have lower electro-optic responses. For example, the r_{33} of **OPC** at 810nm poled at 0.5 MV/cm is 1.0 pm/V.

4. CHEMICAL THERMAL STABILITY

The overall stability of an electro-optic polymer system is dependent upon many things, but in general will be limited by at least the intrinsic thermal stability of the chromophore which provides the electro-optic response. Thus a study of the thermal degradation of the basic chromophore is critical to understanding the potential stability of the overall system. We have examined this stability by studying changes in the ultraviolet absorption of thin polymer films as a function of time under isothermal conditions. For example, Figure 5 shows the ultraviolet absorption spectrum of a thin film of **BPE-2.5** polymer as a function of time at 180°C in air. The primary charge transfer absorption band with peak at 438 nm attributed to the substituted flourene chromophore **BISMAN** is seen to decay in time while other absorptions gradually appear. There is a marked blue-shift in the peak of the main charge transfer absorption band with increasing time. The existence of an isospestic point at 393nm is indicative of

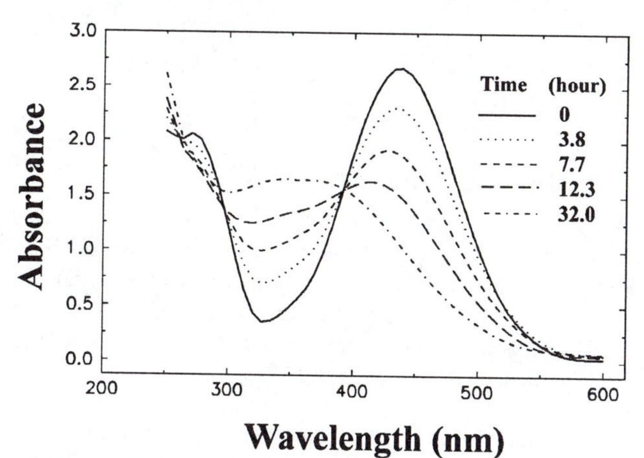

Figure 5. Observed ultraviolet absorbance as a function of wavelength for isothermal exposure of thin film of **BPE-2.5** at 180°C for given exposure times.

a simple thermal decay of the initial species into a second species characterized by a different absorption spectrum (18). The figure also shows however, that this behavior is not seen at all wavelengths and thus such an assumption should be interpreted carefully. For this reason, we have chosen to study the time decay of absorbance at 438 nm only in the initial phase of degradation to minimize errors due to potential complex chemical degradation paths. Identical measurements were performed with samples prepared in a similar manner at several temperatures between 140°C and 210° C. Figure 6 shows, along with a linear fit, the apparent first order exponential decay rate constants for the change in absorbance at 438 nm for thermal degradation of **BPE-2.5** in air plotted as a function of inverse temperature over the range 140°C to 210°C. The linearity of such an Ahrrenius plot is suggestive of a simple first order thermal decay. Although our errors are high, the data are consistent with such a thermal degradation mechanism. In such cases the slope of the plot gives an apparent activation energy for the reaction, in this case 23 kcal/mole. Figure 6 also shows corresponding data for **BPC** polycarbonate in air. Despite the fact that the **BPE-2.5** and the **BPC** polymer have a T_g difference of 73°C, within our overall error the results are identical suggesting a common decay mechanism. Experiments for films maintained within a 1 atmosphere nitrogen atmosphere give decay rates about a factor of two lower, but with approximately the same activation energy.

When the electron donor group in the EO chromophore changed from a dimethylamino to a methoxy, the thermal stability is significantly improved. Figure 7

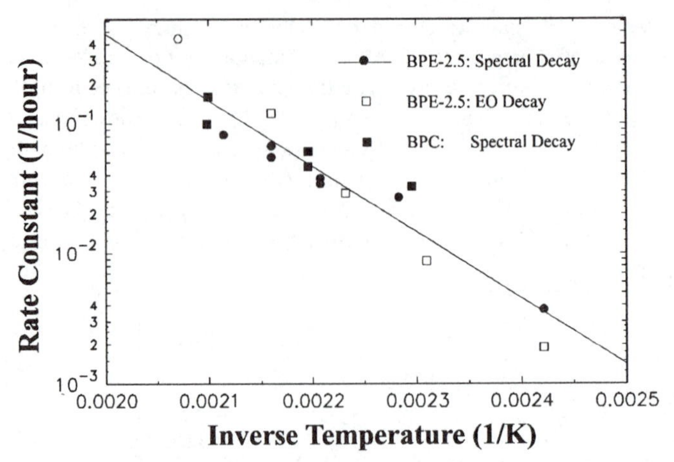

Figure 6. Rate constant for degradation of **BPE** polymers plotted on a semilogarithmic scale as a function of inverse temperature. Closed circles are for **BPE-2.5** polyester determined from decay of ultraviolet absorbance. Open squares are for the same material obtained from electro-optic decay measurements. Closed squares give degradation rates for **BPC** polycarbonate obtained from ultraviolet absorbance. The straight line corresponds to an activation energy of 23 kcal/mole.

shows similar spectral decay data for the **OPC** polycarbonate polymer film in air, but at a temperature of 240°C. Although an isospestic point is less clear for this polymer the qualitative behavior is similar. We have thus interpreted the decay of absorbance at the charge transfer peak (368 nm) in a like manner. The resulting Ahrrenius plot of the spectral decay, shown in Figure 8, is linear to within the experimental error.

The fit to the observed data in Figure 8 gives an activation energy of about 33 kcal/mole, although the uncertainty is large. For **OPC** polymer in 1 atmosphere of nitrogen, we observe a decrease in apparent decay rate of a factor of two over the same temperature range, very much like our observations for **BPE-2.5**.

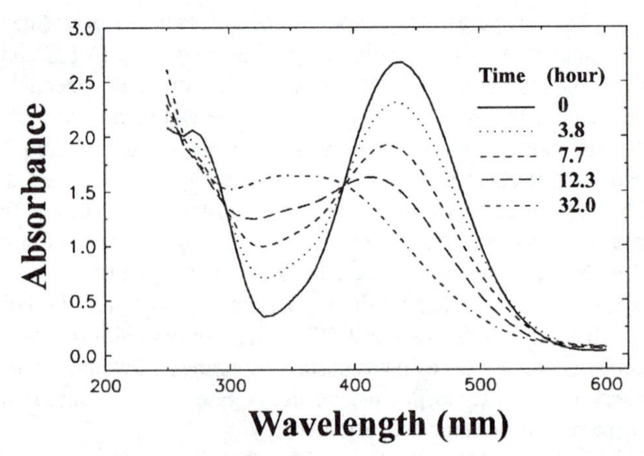

Figure 7. Observed ultraviolet absorbance as a function of wavelength for isothermal exposure of thin film of **OPC** polycarbonate at 240°C for given exposure times.

These experiments demonstrate clearly the unusually high thermal stability of the chromophores used in the **OPC** thermoplastic polymer system. They also demonstrate that the degradation follows first order (pseudo unimolecular) kinetics during at least the initial decay over the temperature ranges examined and is relatively independent of environment (atmosphere, polymer backbone, etc.).

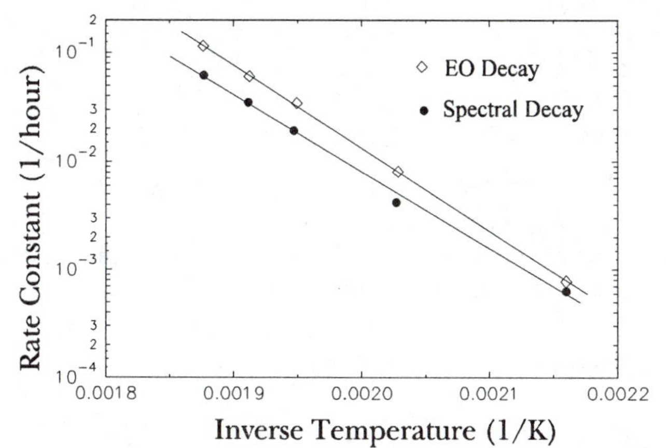

Figure 8. Rate constant for the spectral decay (chromophore degradation) and the EO decay of **OPC** polycarbonate plotted on a semilogarithmic scale as a function of inverse temperature. The straight lines of the spectral decay and of the EO decay correspond to activation energies of 33 and 35 kcal/mole, respectively.

5. ELECTRO-OPTIC THERMAL STABILITY

Thermal stability of chromophore does not assure stability of electro-optic response. We have therefore measured the isothermal decay of the apparent electro-optic response for polymers poled to 0.5 MV/cm as described above. The isothermal electro-optical decays of **BPE-2.5** and **OPC** polymers are shown in Figure 9 and 10, respectively. In these experiments the samples were maintained in an oven at controlled temperature under ambient conditions for the indicated time. The samples were withdrawn, cooled, and examined, then returned to the oven for additional exposure. The initial fast drop of EO response at a very short time period, which accounts for less than 5% of the total decay, is partially non-exponential, especially at higher temperatures. However at longer times the decay is quite exponential and the exponential component apparently governs the major fraction of the initial response. For **BPE-2.5** polymer, the first order decay rate constants inferred from the semilogarithmic plots of EO response versus time at long times are plotted in Figure 6 along with corresponding data from spectral decay measurements for **BPE-2.5** and for **BPC**. Within experimental error, the results are in excellent agreement over the temperature range employed and suggest clearly that the primary mechanism for degradation of electro-optic response for these polymers in air at least for temperatures in the range 140°C to 210°C is simple chemical decomposition of the chromophore which gives rise to the electro-optic response. At short times, additional loss of response is observable over this temperature range and this may possibly be due to

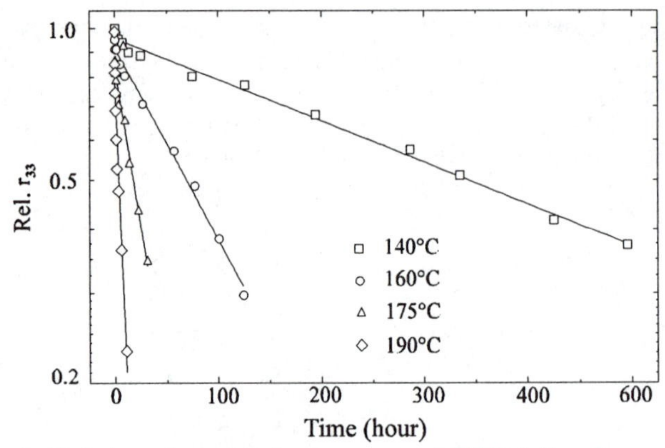

Figure 9. Relative electro-optic response of **BPE-2.5** polymer films observed at different temperatures as a function of wavelength of time.

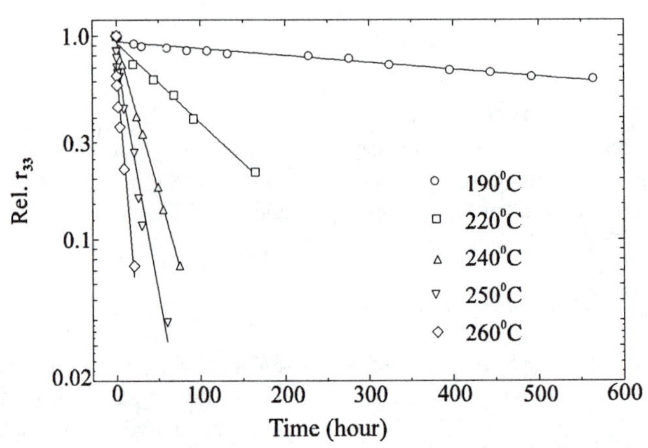

Figure 10. Relative electro-optic response of **OPC** polymer films observed at different temperatures as a function of wavelength of time.

some orientational relaxation. Unfortunately, our data for **BPE-2.5** are not sufficiently detailed to be able to separate out the thermal degradation so that additional mechanisms can be delineated.

The first order decay rate constants of EO response of **OPC** polycarbonate are plotted in Figure 8 along with corresponding data from spectral decay measurements for the same polymer. The EO rate constants are larger than those obtained from the spectral decay approximately by a factor of two, however, their regression lines have similar slope. The Arrhenius fit of the spectral decay and of the EO decay correspond to activation energies of 33 and 35 kcal/mole, respectively. This again demonstrates that the primary cause for the EO and the spectral decay of **OPC** polymer are probably stems from the same mechanism. Further work is continued to determine whether the rate difference of the two processes is caused by chromophore relaxation.

We have described earlier the study of **MA-1** side chain EO polymer (6,7). The

MA-1 is a side chain EO polymer in which the EO chromophore - disperse red dye #1 - is attached to the pendent ester group of a polymethacrylate molecule and the polymer has a T_g of 124°C. Figure 11 shows the spectral and EO decay rates of **MA-1** polymer measured by the same technique described above. The EO decay occurs well below the thermal degradation temperature as indicated by the spectral decay line. This example illustrates at lower T_g, the loss of EO response is mainly caused by reorientation of EO chromophores (depoling). With the increasing T_g of the polymer, as in the present case of **BPE-2.5** and **OPC**, the chemical stability becomes a dominating factor in determining the service life of a EO polymer. EO chromophores of better thermal stability is needed for a polymer having $T_g \geq 200°C$. Figure 12 shows on a semilogarithmic scale a plot of half lives ($t_{1/2}$) calculated from the assumed first order rate constants of EO decay of the **BPE-2.5**, **OPC**, and **MA-1** polymers. For a half life of a one year period, the corresponding temperatures of **MA-1**, **BPE-2.5**, and **OPC** polymers are 75°, 115°, and 175°C, respectively.

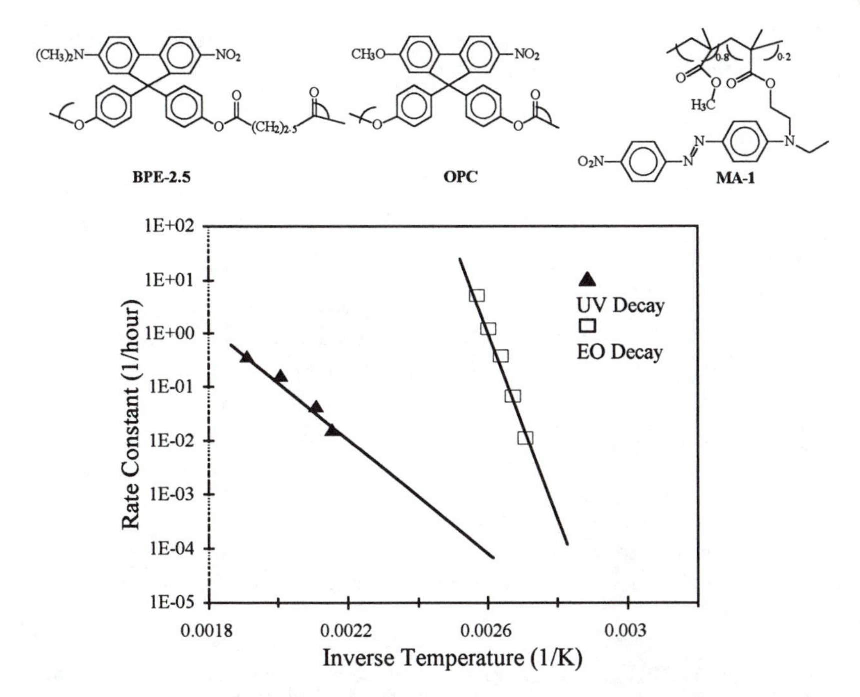

Figure 11. Rate constants for electro-optical decay and for UV decay of **MA-1** polymer (Disperse red 1 covalently bounded to polymethacrylate side chain) plotted on a semilogarithmic scale as a function of inverse temperature.

6. CONCLUSIONS

The experiments reported here have considerable technical significance. Figure 13 shows on a semilogarithmic scale a plot of first order rate constant for chemical decomposition as measured by spectral decay plotted as a function of temperature on

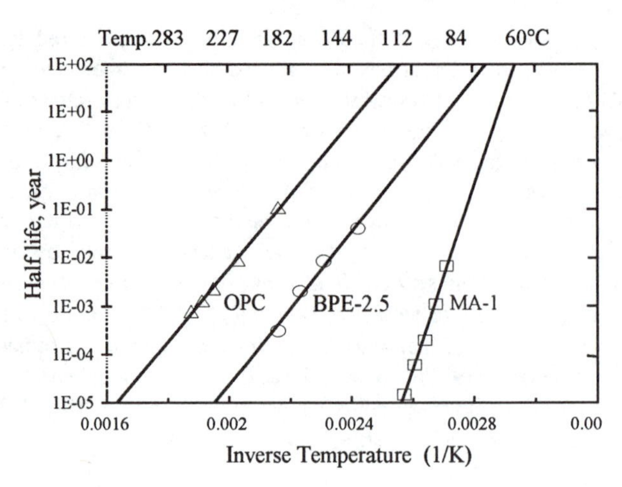

Figure 12. Half lives calculated from the assumed first order rate constants of electro-optical decay for the **MA-1**, **BPE-2.5**, and **OPC**.

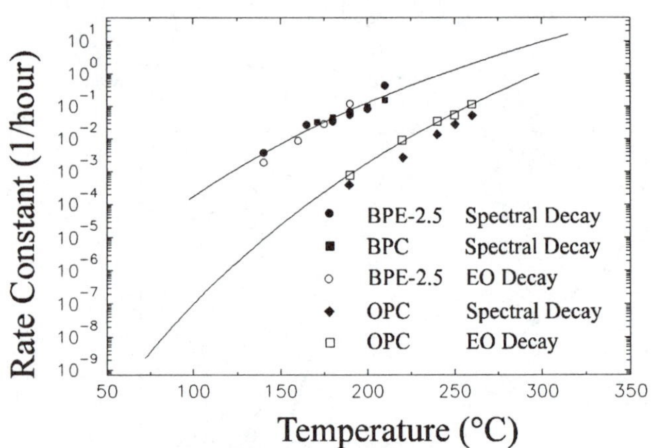

Figure 13. Rate constants for degradation of **BISMAN** and **BISMON** polymers plotted as a function of temperature (°C). Lines give extrapolations from Ahrrenius fits to the observed data.

an expanded scale for **BPE-2.5**, **BPC**, and **OPC**. Also shown in the plot are rates for electro-optic decay of **BPE 2.5** and **OPC**. Also shown are Ahrrenius extrapolations to both lower and higher temperatures. Since these extrapolations assume that the degradation follows the same kinetics over a wide temperature range, such extrapolations must of course be used with caution. However these data certainly give expectations that the chromophores used here give long term chemical stability for 1 year at temperaturesranging from over 115°C for the **BISMAN** chromophore to over 175°C for the **BISMON** chromophore. When this level of stability is contained within a thermoplastic polymer backbone of high thermal stability and extremely high controllable glass transition temperatures, the capability for creating highly stable electro-optic polymer systems is readily demonstrated.

The experiments reported here also possess considerable scientific interest. The exact nature of the thermal degradation of the chromophores studied here is not known. Apparent activation energies of approximately 23 kcal/mole are quite low for chemical systems as stable as these. This suggests that the actual chemical decay mechanism is more complex than simple first order decay. To the extent that this is true the real chemical stability of these systems at low temperatures (say below 100°C) may in fact be even greater than that extrapolated from the data presented here.

References

1. See for example, *Nonlinear Optical Properties of Organic and Polymeric Materials;* Williams, D. J. ed., ACS Symposium Series 455, American Chemical Society, Washington, 1991 or Burland, D. M.; Miller, R. D.; Walsh, C. A.; *Chem. Rev.* **1994**, *94*, 31.
2. Teraoka, I.; Jungbauer, D.; Reck, B.; Yoon, D. Y.; Twieg, R.; Willson, C. G. *J. Appl. Phys.* **1991**, *69*, 2568.
3 Eich, M.; Reck, B.; Yoon, D. Y.; Willson, C. G.; Bjorklund, G. C. *J. Appl. Phys.*, **1989**, *66*, 3241.
4. Mandal, B. K.; Kumar, J.; Huang, J. C.; Tripathy, S. K. *Makromol. Chem. Rapid Commun.*, **1991**, *12*, 63.
5. Wu, J. W.; Valley, J. F.; Ermer, S.; Binkley, E. S.; Kenney, J. T.; Lipscomb, G. F.; Lytel, R. *Appl. Phys. Lett.* **1991**, ***58***, 225.
6. Wu, C.; Beeson,.K. W.; Ferm, P.; Knapp, C.; McFarland, M. J.; Nahata, A.; Shan, J.; Yardley, J. T. *MRS Symposium Proceedings*, **1994**, *328*, 477.
7. Nahata, A.; Wu, C.; Knapp, C.; Lu, V.; Shan, J.; Yardley, J. T. *Appl. Phys. Lett.* **1994**, *64*, 3371.
8. Karshak, V. V.;Vinogradova, S. V.;Vigodskii, Y. S. *J. Macromol. Sci. Rev. Macromol. Chem.* **1974**, *C11*, 45.
9. Ueda, M.; Okada, K.; Imai, Y. *J. Poly. Sci. Chem. Ed.*, **1976**, *14*, 7665.
10. Morgan, P. W. *Macromolecules*, **1971**, *4*, 536.
11. Wu, C.;Knapp, C.; Nahata, A.; Shan, J.; Yardley, J. T. to be published.
12. Korshak, V. V.; Vasnev, V. A. *Vysokomol. Soedin. Ser. B*, **1982**, *24*, 198.
13. Iyer, V. S.; Sehra, J. C.; Sivaram, S. *Makromol. Chem. Rapid Commun.* **1993**, *14*, 174.
14. Nahata, A.; Wu, C.; Yardley, J. T. *IEEE Trans. on Instrumentation and Measurement*, **1992**, *41*, 128.
15. Yariv, A. and Yeh, P. *Optical Waves in Crystals,* John Wiley and Sons, New York, 1984.
16. Singer, K.D.; Kuzyk, M.G.; Sohn, J.E. *J. Opt. Soc. Am. B* **1987,** *4*, 968.
17. A. Nahata, A.; Shan, J.; Yardley, J. T.; Wu, C. *J. Opt. Soc. Am. B* **1993**, *10*, 1553.
18. See for example Moore, J. W.; Pearson, R. G. *Kinetics and Mechanism* John Wiley and Sons, New York, 1981, pp 48ff.

RECEIVED January 30, 1995

Chapter 27

Nonlinear Optical Activity Relaxation Energetics in a Polyether Thermoplastic Nonlinear Optical Polymer with a High Glass-Transition Temperature

Robert J. Gulotty[1], Claude D. Hall[1], Michael J. Mullins[1], Anthony P. Haag[1], Stephen E. Bales[1], Daniel R. Miller[1], Charles A. Berglund[1], Marabeth S. LaBarge[2], Andrew J. Pasztor, Jr.[2], Mark A. Chartier[1], and Kimberly A. Hazard[1]

[1]Dow Chemical Company, Central Research and Development, Midland, MI 48674
[2]Dow Chemical Company, Anaytical Sciences, Midland, MI 48667

A polyether polymer comprised of the 4-nitrophenyl hydrazone of bisphenol K and bisphenol A in a 50/50 comonomer ratio exhibits stable NLO activity at 100°C for years. In an effort to understand the kinetics and energetics of relaxation, isothermal aging studies were done at 150 °C, 175 °C and 200 °C. Williams-Watts stretched exponential and Arrhenius analysis were used to determine the activation energy and prefactors. In addition, polymer relaxations were studied by dynamical mechanical analysis and dielectric spectroscopy. The conclusion from this work is that the energetics of orientational relaxation are that of a sub-Tg relaxation, or beta relaxation, consistent with a crank shaft motion of the chromophore about the polymer axis. However, this relaxation only turns on at temperatures very near the polymer Tg, where enough free volume is developed for the motion to occur. We have termed this a Tg assisted beta relaxation.

A fundamental requirement of nonlinear optical (NLO) polymers for electro-optical (EO) applications is stability at elevated temperatures for long periods of time. For example, an EO modulator for telecommunication applications needs to be stable at -50 °C to 70 °C for many years. Similarly, an EO modulator in a desk top computer may be exposed to temperatures as high as 80 °C during use. It is generally asserted that stabilities at 100 °C for 10 years will be a benchmark for commercial applications.

NLO thermoplastics developed at The Dow Chemical Co. have been shown to have excellent thermal stability at 100 °C. For example, a corona poled sample of the 4-nitrophenyl hydrazone of bisphenol K/bisphenol A copolyether (NBPE, see structure below) has been shown to exhibit stable NLO activity at 100°C in air for a period of 3.8 years, with an estimated 1/e lifetime of 35 years, see Figure 1.

0097–6156/95/0601–0368$12.00/0

NBPE

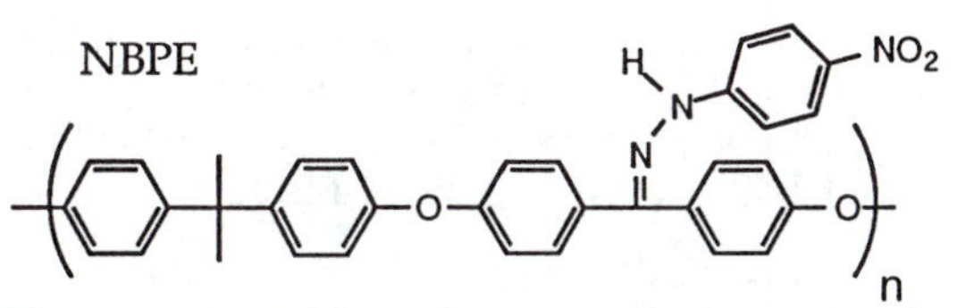

The purpose of this study was to obtain a more detailed understanding of the NLO stability and thermal degradative processes affecting the stability of the NBPE.

Experimental

Sample Preparation. The NBPE polymer was synthesized in a manner similar to that described elsewhere (1). Two batches were prepared. The first batch had a Tg of 207 °C by DSC and was used in the long term aging study at 100 °C. The second batch had a Tg of 219 °C by DSC and was used for the study of NLO relaxations at elevated temperatures.

TG/GC/MS Evolved Gas Analysis. The evolved gas analysis is a thermogravimetry - mass spectrometry (TG/MS) experiment performed simultaneously with a thermogravimetry - gas chromatography - mass spectrometry (TG/GC/MS) experiment. A custom built evolved gas analysis instrument was developed by interfacing a thermogravimetric analyzer (Cahn 131 TGA) with a gas chromatographic - mass spectrometer system (Fisions, Trio-1) with heated fused silica capillary transfer lines operated at 280 °C. The thermogravimetric analysis was performed by heating ca. 15 mg of sample contained in a platinum pan to approximately 315 °C in a helium atmosphere. A fraction of the evolving gases was sent to the quadruple mass spectrometer for real time detection of the volatiles during the thermogravimetric analysis (e.g. TG/MS). The mass spectrometer was operated in +EI mode with a scanning range from 12 to 800 amu. Simultaneous with the real time evolved gas analysis, a second small fraction of the evolving gases was directed to a cryotrap (at approximately -180 °C) over the entire experiment. After the thermogravimetric program was completed, these trapped volatiles were analyzed using sub-ambient GC/MS. This latter part of the experiment (e.g. TG/GC/MS) is critical in identifying which components evolved and in correctly interpreting the TG/MS results. Using the TG/GC/MS results, unique ions are selected for each component in order to assign their evolution profiles in the TG/MS experiment to the ion profiles. Using this approach, ions m/z=138, 112, and 49 were assigned to nitroaniline, chlorobenzene, and dichloromethane, respectively. Unfortunately, unique ions which would allow discrimination between the nitrobenzene and an unknown aromatic compound were not found. Therefore, these component evolution profiles are combined and assigned to ion m/z = 123.

TGA and DSC Analysis. Thermogravimetric (TGA) measurements of weight loss during heating in air were run on a Dupont Instruments Model 951 Thermogravimetric Analyzer . The Differential Scanning Calorimetry measurements were run on a Dupont Instruments Model 9900 Thermal Analyzer. The scan rates were 20 °C/min and the atmosphere was air.

Absorption Spectroscopy. Absorption spectra were recorded using a Shimadzu model UV-3101PC Spectrophotometer. For the isothermal aging data, wavelengths were chosen such that the absorbance values were less than 2 absorption units initially.

SHG Measurements. Measurement of the second-order nonlinear optical coefficient, d_{33}, was done by second harmonic generation (SHG) and Maker fringe analysis as described in (2). An excitation wavelength of 1064 nm was used. The parameter values of n_{ω} of 1.5677, $n_{2\omega}$ of 1.5929 and d_{11} quartz of 1.2×10^{-9} esu were assumed.

Dynamic Mechanical Data. Dynamic mechanical measurements were made in torsional shear using a Rheometrics RDS 2 mechanical spectrometer. Rectangular bars about 0.85 x 11.7 x 50 mm were cast from solution and measurements were made at a constant frequency of 1.0 rad/s. Temperature was controlled with a nitrogen gas flow. Strains were 0.2% or less, which was well within the linear viscoelastic regime.

Dielectric Spectroscopy. Dielectric measurements were made using a TA Instruments 2970 Dielectric Analyzer, attached to a 2100 Analyzer which controls the instrument, collects data and performs data analyses. The instrument was operated in the single surface mode, and was calibrated according to the instrument manufacturers' instructions. Thin films were fabricated on 1x1" ceramic single surface interdigitated electrode plates by spin-coating polymer/cyclopentanone solutions (20 %w/v). Measurements were made on unpoled samples from -150 to 230 °C with a temperature scan rate of 2 °C/min over the frequency range from 10^{-1} Hz to 10^{5} Hz at 1 volt applied.

Results and Discussion

NLO Activity. The NBPE polymer had an onset Tg of 219 °C measured by DSC. The NLO activity or second-order NLO coefficient (d_{33}) was measured for NBPE films parallel-plate poled at 211, 221 and 231 °C with an applied field of 50 V/μm. Poling near the Tg (221 °C) or above the Tg (231 °C) gave the same result, $d_{33} = 1.0 \times 10^{-8}$ esu. Poling at 211 °C (below the Tg) gave 3.7×10^{-9} esu. Corona poled films gave activities of 20×10^{-9} esu.

Thermal Stability. Figure 2 shows a TGA scan (20 °C/min) for the NBPE polymer in air. The sample loses 3% of the weight by 310 °C and shows an extrapolated onset of decomposition at 271 °C. A more detailed understanding of the origin of the weight loss was determined by measuring the decomposition fragments for a related polymer BHPF-NBPE (Tg = 265 °C),

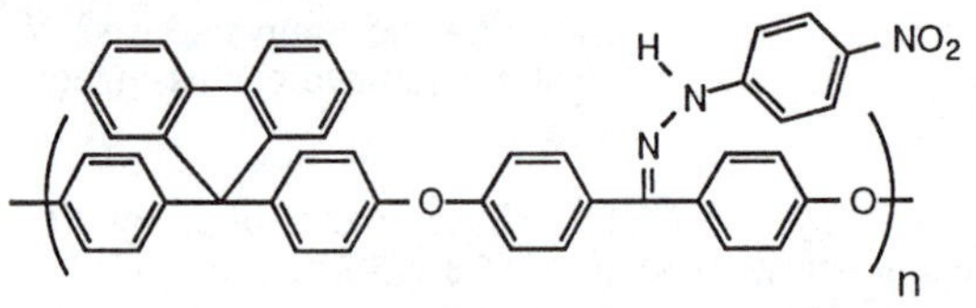

H N N NO$_2$ O O n

by TG/GC/MS.

In Figure 3 it is shown that BHPF-NBPE suffers a weight loss of approximately 5.6% upon heating to 315 °C and holding for ten minutes in an anaerobic (He) atmosphere. The weight loss is principally due to the evolution of nitroaniline, nitrobenzene, and an unidentified aromatic component. These major components are observed to evolve during the major weight loss step centered at approximately 30

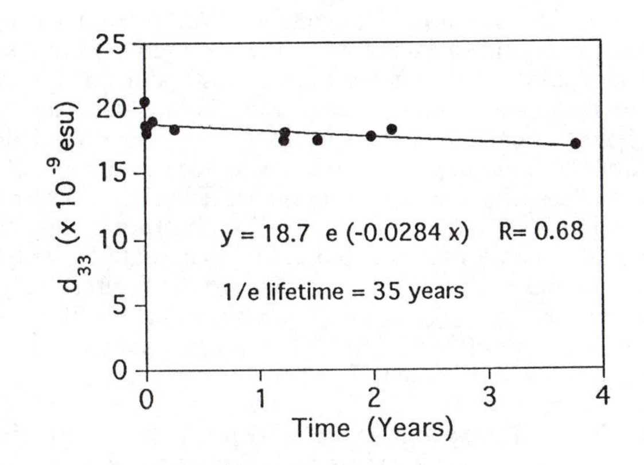

Figure 1. Stability of Corona Poled NBPE in Air at 100°C.

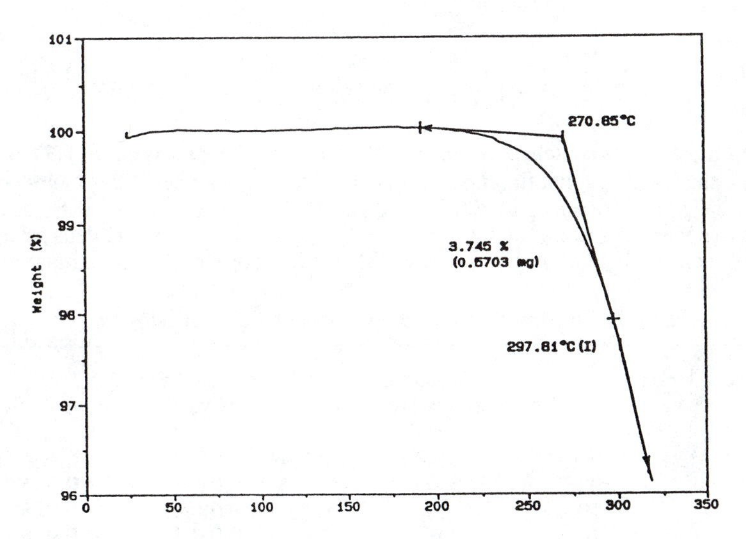

Figure 2. TGA Scan of NBPE in Air.

minutes and 315 °C (Figure 3). Dichloromethane and chlorobenzene (residual synthesis solvents) also evolved, but are not major contributors to the weight losses as they principally evolve over a region where only minor weight losses are observed. The identification of each of these components are based on the electron impact mass spectra obtained from the TG/GC/MS experiment.

Absorption Decay Kinetics and Energetics. We have shown by TG/GC/MS that the decomposition occurs by loss of the nitroaniline group from the chromophore. This loss also results in loss of the charge-transfer absorption band in the visible. For example, Figure 4 shows the change of absorbance before and after aging at 200 °C until no more change occurred in the spectrum. We have used the decrease in absorbance at 470 nm in film samples as a measure of the rate of thermal degradation in isothermal aging studies. Figure 5 shows the isothermal decay plots for samples poled at 221 °C and aged at 150, 175 and 200 °C. The absorption was monitored in an unpoled region of the parallel-plate poled films. The data is well fit to a single exponential rate law with the fit results given above the figure. Figure 6 shows an Arrhenius plot (labeled $k_{degradation}$) of the absorbance data for films poled at 221 °C, yielding an activation energy of 26 kcal/mol. The kinetics and activation energies were similar for samples poled at 211 °C and 231 °C.

Analysis of the NLO Relaxation Data. Figure 7 shows a typical isothermal decay of NLO activity for a film which was parallel-plate poled with a field of 50 V/μm at 221 °C and aged at 200 °C in air. The data was fit to equation 1, where the degradation rate constant ($k_{degradation}$) was determined by the absorbance decay of the same samples during the aging study.

$$d_{33}(t) = d_{33}(0) \cdot e^{-(k_{orientation} \cdot t)^{\beta_w}} \cdot e^{-(k_{degradation} \cdot t)} \quad (1)$$

The Williams-Watts stretched exponential fit (curve a, where β_w is varied) (3) is better than the single exponential fit (curve b, $\beta_w = 1$). The fit parameters for samples poled at 221 °C, aged at 150, 175 and 200 °C and fit using equation 1 in both stretched and single exponential form are listed in Table 1. Note that the 150 °C decay data was single exponential. The samples poled at 231 °C (data not shown) gave similar results.

Table 1. Fit parameters for orientational relaxation using Eq. 1.

T_{poling} (°C)	T_{aging} (°C)	$d_{33}(0)$ (10^{-9}esu)	$k_{orientation}$ (days^{-1})	β_w	$k_{degradation}$ (days^{-1})	r (correlation coefficient)
221	150	9.11	0.0105	0.836	0.0131	0.986
		9.02	0.0126	1	0.0131	0.985
	175	12.4	0.411	0.423	0.0831	0.995
		10.0	0.346	1	0.0831	0.968
	200	10.8	4.78	0.501	0.328	0.998
		9.38	4.19	1	0.328	0.972

Arrhenius Analysis. The rate constants for orientational decay determined above can be used to estimate the activation energy of orientational relaxation. Here we convert the rate constants measured in the stretched exponential fit (see Table 1) to an average value using equation 2 as in (4).

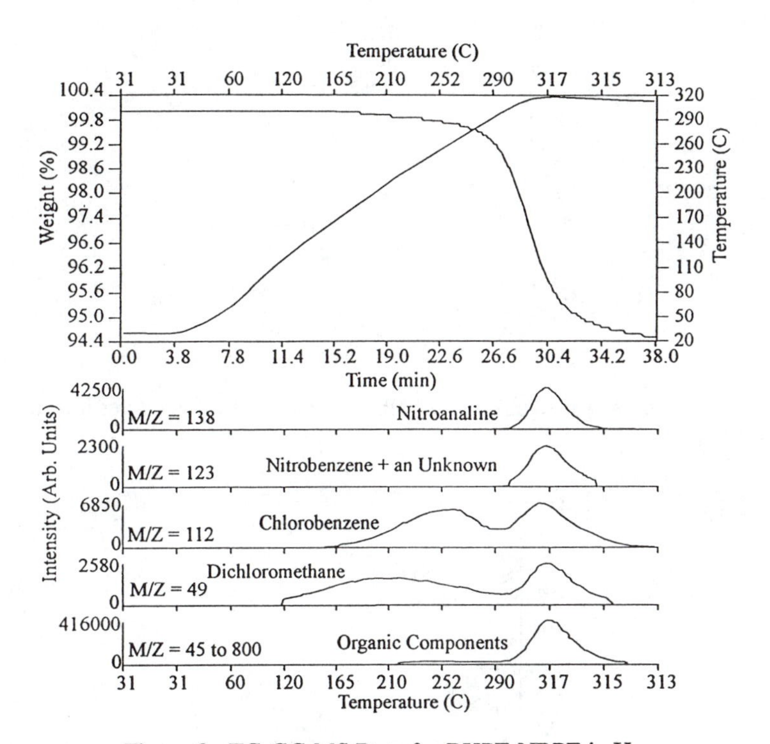

Figure 3. TG-GC-MS Data for BHPF-NBPE in He.

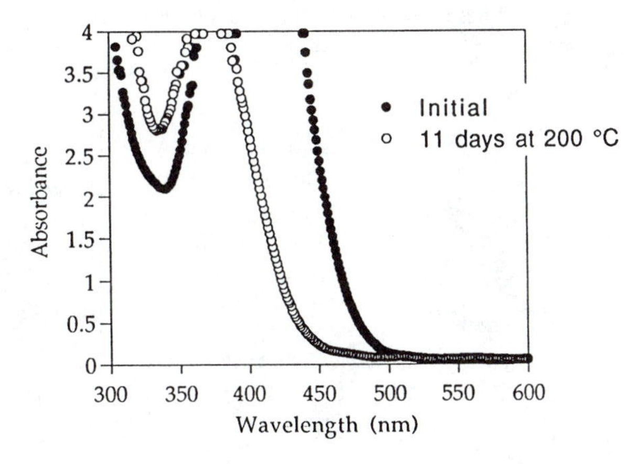

Figure 4. Absorbance of NBPE films with aging at 200°C.

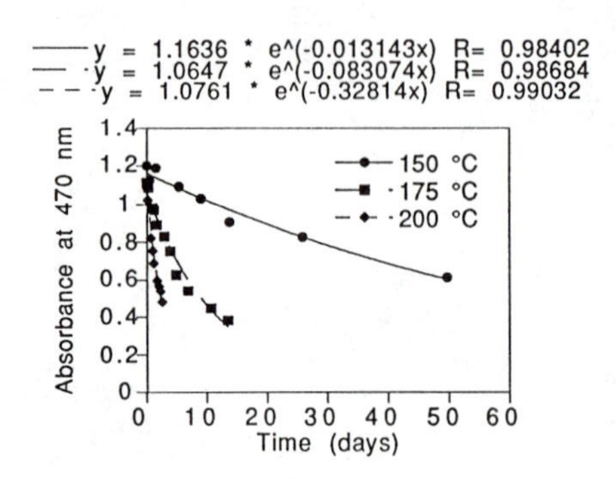

Figure 5. Absorbance decay of NBPE at elevated temperatures in Air.

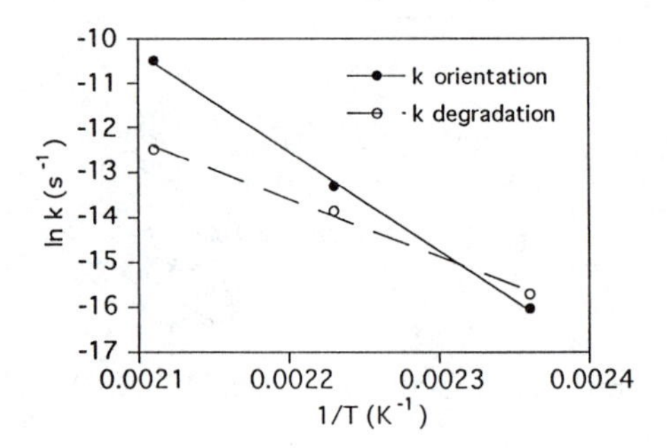

Figure 6. Arrhenius plot of isothermal aging data for NBPE.

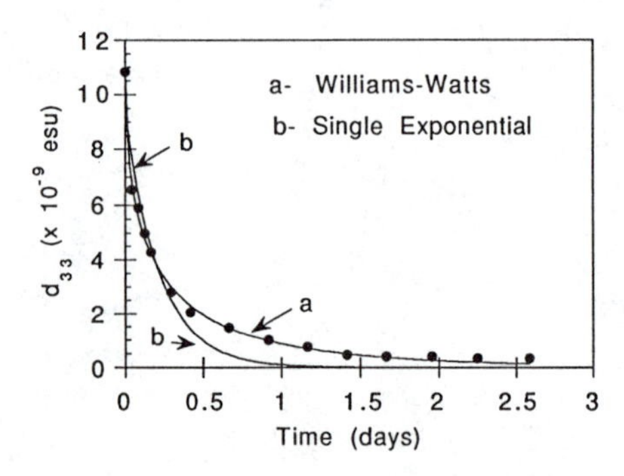

Figure 7. NLO Relaxation of NBPE at 200 °C in Air.

$$\langle 1 / k_{orientation} \rangle = \langle \tau \rangle = \tau \cdot \Gamma\left(\frac{1}{\beta_w}\right) \Big/ \beta_w \tag{2}$$

The rate constants at each depoling temperature are plotted in Figure 6 as an Arrhenius plot. Linear behavior is observed. The activation energies for orientational relaxation are higher than those of thermal degradation, see Table 2. Note that in our publication of these results as a Polymer Preprint (5), we did not use the average rate constants calculated from equation 2 for the Arrhenius analysis, the $k_{orientation}$ values in Table 1 were used directly. The activation energies for orientational relaxation decreased by about 6 kcal/mol by using the average rate constants.

Table 2. Summary of Arrhenius Activation Energies

T_{poling} (°C)	E_a orientation (kcal/mol)	E_a degradation (kcal/mol)
221	43.9	25.6
231	41.3	25.7

A comparison of the rates of orientational relaxation and degradative relaxation (see Table 1 or Figure 6) for aging at 175 °C and 200 °C shows that the NLO relaxation is dominated by orientational relaxation at these temperatures. However at 150 °C, the rates become comparable. At lower temperatures, the degradation of the chromophore should dominate the kinetics of NLO relaxation, (see Figure 6).

Dynamical Mechanical Analysis. The activation energy for the T_g process for a polymer similar to NBPE was found to be about 223 kcal/mol, which is typical for a polymer with a T_g over 200 °C (6). The activation energies derived from NLO measurements, on the order of 40 kcal/mol, are characteristic of sub-T_g beta relaxations which typically involve motions of short segments of polymer chain backbones. For example, the activation energy of the beta relaxation in bisphenol A polycarbonate is 46 kcal/mol (7). Usually beta relaxations give rise to maxima in G" and tan δ at temperatures below the T_g in isochronal dynamic mechanical data. The fact that no such maxima are observed for NBPE, as seen in Figure 8, suggests that a beta relaxation may exist, but is obscured in the dynamic mechanical data by the maxima in G" and tan δ due to the glass transition. The situation is similar for motion of the pendant chromophore independent of the backbone. That relaxation would be expected to be observed in the dynamic mechanical data, and if it occurs at high temperature could also be obscured by T_g maxima in G" and tan δ. Since there is no loss in NLO activity at low temperature, the chromophore reorientation obviously does not occur at temperatures well below T_g.

Dielectric Spectroscopy. Dielectric spectroscopy (DES) is a useful technique for studying the dynamic motions of dipolar components in polymer systems (8, 9). For example, the application of DES to study transitions and dielectric relaxations in polymethyl methacrylate (PMMA) has been thoroughly described (8). For NLO polymers with the large chromophore dipole moment and the molecular hyperpolarizability perpendicular to the copolymer backbone, DES has been particularly useful in determining the kinetics and energetics of orientational relaxation (10). Plots of log frequency versus dielectric loss factor (ε") at maximum temperature were used to determine the activation energy for molecular processes.

A measure of the energy of dipolar relaxation in the shoulder of the Tg peak was performed using DES. Figure 9 shows a plot of ε" versus temperature for various frequencies between 25 °C and 221 °C. Absolute values of ε" were artificially low due to small sample thicknesses.

To determine an Arrhenius activation energy for dielectric relaxation at Tg and below, an Arrhenius plot was constructed by plotting ln frequency versus reciprocal temperature at a constant ε" value of 0.02 (see Figure 10). Low frequencies (< 5 Hz) were excluded from the plot to minimize contributions from ionic conductivity to the apparent activation energy. From the slope of this plot, the activation energy was determined to be approximately 30 kcal/mole. A similar plot was constructed for a constant value of ε" = 0.1 (but is not shown) and gave the same activation energy. This value is consistent with the activation energy determined from SHG d_{33} orientation relaxation measurements in Table 2.

Estimated Lifetimes Predicted from Arrhenius Analysis. The above data can be used to estimate the expected stability of NLO activity of the NBPE polymer at lower temperatures. Figure 11 shows the orientation and relaxation rates expected with temperature for the samples poled at 221 °C. The Arrhenius data predicts a 1/e lifetime of 10 years at 100 °C due to degradation, with the orientational stability being much higher. However, the Arrhenius data (measured at 150- 200°C) underestimates the actual stability measured for the corona poled NBPE at 100 °C, where a 1/e lifetime of 35 years is estimated (Figure 1). The origin of this difference is unknown.

Comparison to Other Literature Data and WLF Analysis. There have been numerous studies of the relaxation of nonlinear optical activity in poled polymer films (4, 11-24). These studies have been of guest-host polymers (11-15) and of NLO polymers with covalently bonded chromophores (16-24). The studies of NLO relaxation in guest-host and side-chain type NLO polymers all show NLO relaxation onsets as one approaches the Tg. In all cases, where reported, the energetics of relaxation are consistent with the energetics of a beta relaxation, 20-70 kcal/mol, while the onset of this relaxation is correlated with the Tg. Our findings above are consistent with these data. To illustrate this, in Figure 12 we have plotted data from references (12, 13) with our data for NBPE poled at 221 °C. Similar slopes are observed in the relaxation data in the regions near the Tg and the onsets track the Tg onset as discussed above. The decision as to whether WLF theory (25) should be applied to the data is more complicated.

It is universally asserted that the WLF theory describes the alpha relaxation process above the Tg. However, the behavior below the Tg is less well characterized because of the time scales involved to get data over a large enough temperature range. It is also universally asserted that the beta relaxation process, when decoupled from the alpha transition, displays Arrhenius behavior. As stated above, the beta relaxation in the NLO polymer systems is correlated with the Tg. Various researchers have applied WLF theory to their NLO relaxation data below the Tg. We found that the Arrhenius plots of the NBPE data were linear over the temperature range of our study. If one applies WLF analysis to the relaxation data by plotting log τ versus 1/(Tg+50-T), the data is coarsely consistent with the similar (master) plot of IBM (12) for guest-host blends. However, using this analysis leads to predicted lifetimes of NLO relaxation which significantly underestimates the measured stability of the polymers at temperatures below the Tg. For example, WLF analysis by plotting ln $\langle\tau\rangle$ versus 1/(Tg+50-T) predicts orientational decay lifetimes of 2 years at 100 °C for the NBPE polymer. This clearly underestimates the stability observed in isothermal data at 100 °C for the NBPE polymer (Figure 1) where a lifetime of about 35 years, including polymer degradation is observed.

The data in Figure 12 for the L1/Ultem blends (12) shows relaxation data was measurable at temperatures as low as 30 °C. This suggests that in addition to beta

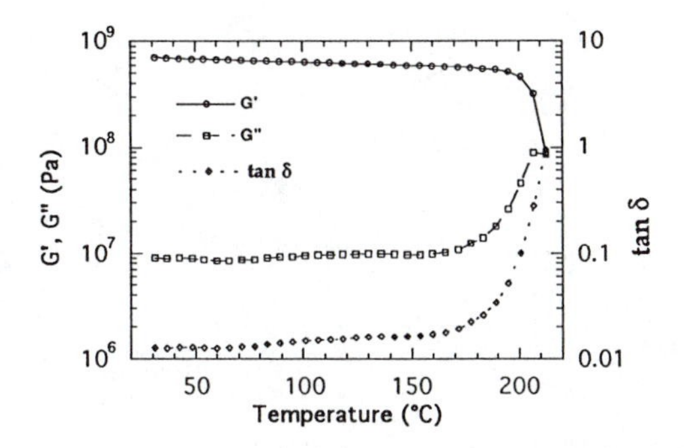

Figure 8. Dynamic mechanical data at 1.0 rad/s for NBPE.

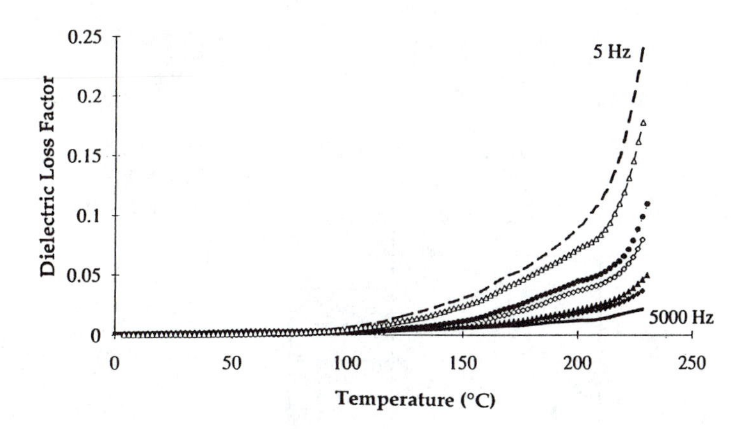

Figure 9. Dielectric loss of NBPE at elevated temperatures.

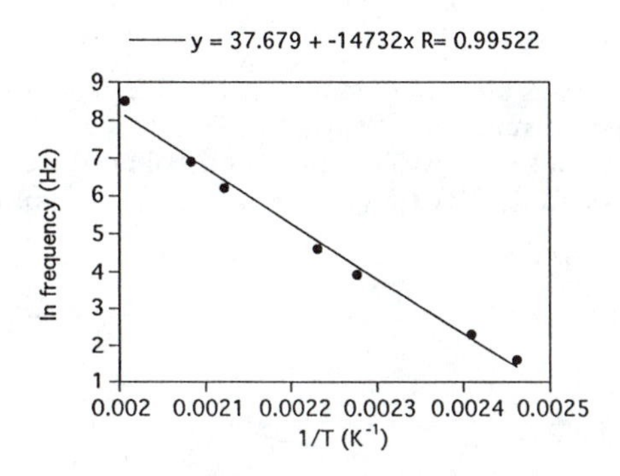

Figure 10. Arrhenius plot from dielectric data.

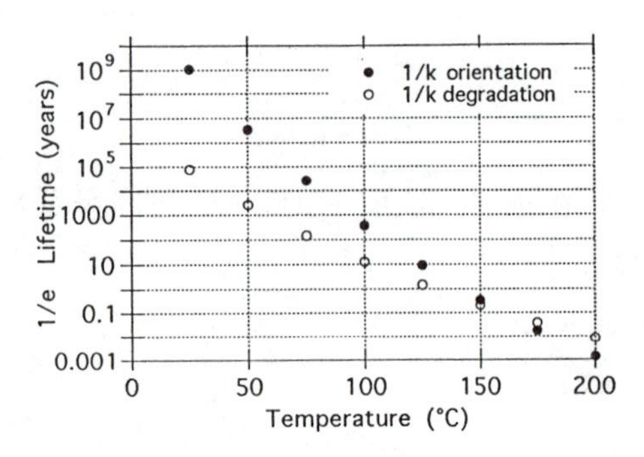

Figure 11. Predicted lifetime of NBPE from Arrhenius data.

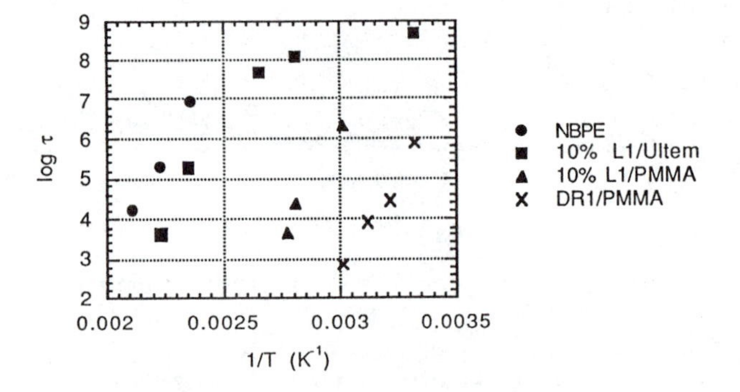

Figure 12. Arrhenius plots of literature and NBPE data. Reproduced with permission from reference 12. Copyright 1993 American Instutute of Physics. L1 (4,5-di(4-methoxyphenyl)-2-(4-nitrophenyl)-imidazole), Ultem (General Electric Co.), DR1 (Disperse Red 1), PMMA (poly(methyl methacrylate).

relaxation correlated to the Tg, there is a relaxation well below the Tg which can cause NLO relaxation in the L1/Ultem blends. If this is the case, the non-exponential Arrhenius plot should not be corrected for by use of the WLF theory.

Interpreting the NLO relaxation in guest-host and side-chain polymers as Tg assisted beta relaxations was also supported by work of Kohler (17) where TSDC (thermally stimulated depolarization currents) were used to compare a side-chain and main chain polymer system. In the side-chain system they found dipolar relaxation near the Tg which they termed an alpha peak; this dipolar relaxation would correspond to the Tg assisted beta relaxation. For the main chain system both an alpha peak (Tg assisted beta relaxation) and what was termed an alpha' peak at temperatures somewhat above the Tg were observed. The alpha' peak in the main-chain system corresponds to the loss of polar order of large backbone segments in the Tg process. This process would be expected to have an activation energy of 100-300 kcal/mol, consistent with an alpha transition as measured by DMA.

Conclusions

The activation energies for orientational relaxation (42 kcal/mol) in poled NBPE films are much lower than expected for a Tg (or alpha) process (typically 100-300 kcal/mol). The activation energy is more consistent with a sub-Tg relaxation (beta relaxation) process. However, dynamic mechanical data for a bar of this polymer, and dielectric data for films of the polymer show no sub-Tg relaxations between 0 °C and Tg. The NLO relaxation appears to originate from a Tg assisted beta relaxation in the shoulder of the Tg peak. The conclusion is consistent with the usual connection of beta relaxations with crankshaft motions of 1-3 monomer units. That is, at temperatures approaching the Tg the volume of the polymer increases enough to allow the crankshaft motions to occur. Full large scale motion of the polymer chain, as in a Tg process where 20-50 monomer units are involved, is not a requirement for NLO relaxation as the NLO chromophore is pendant to the backbone and not aligned along the backbone.

Arrhenius plots of the NBPE NLO relaxation data exhibit linear behavior, suggesting Arrhenius behavior. Estimation of NLO stability of the NBPE at temperatures of 100 °C using Arrhenius behavior gave better agreement with the 100 °C aging data than WLF analysis, which underestimates the stability observed.

Acknowledgments

The authors acknowledge partial funding of this research by the US Air Force at Wright-Patterson AFB, contract F33615-93-C-5359 and helpful contributions of R.A. Cipriano, J.J. Curphy, C.A. Langhoff, M.D. Newsham and S.P. Webb to these results.

Literature Cited

1. Bales, S.E.; Brennan, D.J.; Gulotty, R.J.; Haag, A.P.; Inbasekaran, M.N. *US Patent 5208299* **1993**.
2. Gulotty, R.J.; Langhoff, C.A.; Bales, S.E. *SPIE* **1990**, 1337, 258.
3. Williams, G. and Watts, D.C. *Trans. Faraday Soc.* **1970**, 66, 80.
4. Dhinojwala, A.; Wong, G.K.; Torkelson, J.M. *Macromolecules* **1993**, 26, 5943.
5. Gulotty, R.J.; Hall, C.D.; Mullins, M.J.; Haag, A.P.; Miller, D.A.; Berglund, C.A.; LaBarge, M.S.; Pasztor, A.J.; Chartier, M.A.; Hazard, K.A. *Polymer Preprints* **1994**, 35, 196.
6. Boyer, R.F.; in *Encyclopaedia of Polymer Science and Technology*, Supplement, Vol. II, Interscience; 1977, p. 745.
7. Varadarajan, K.; Boyer, R.F. *J. Polym. Sci.: Polym. Phys. Ed.* **1982**, 20, 141.
8. Aklonis, J.J.; MacKnight, W.J.; in *Introduction to Polymer Viscoelasticity*, Second Ed.; J. Wiley & Sons; 1983, pp. 207-208.

9. McCrum, N. G., Read, B. E., Williams, G.; in *Anelastic and Dielectric Effects in Polymeric Solids*, Dover Publications, 1991, p 525.
10. Jungbauer, D.; Teraoka, I.; Yoon, D. Y.; Reck, B.; Swalen, J. D.; Twier, R.; Willson, C. G.; *J. Appl. Phys.*, 1991, 69, 8011.
11. Staehelin, M.; Burland, D.M.; Ebert, M.; Miller, R.D.; Smith, B.A.; Tweig, R.J.; Volksen, W.; Walsh, C.A. *J. Appl. Phys.* **1992**, 61, 1626.
12. Staehelin, M.; Walsh, C.A.; Burland, D.M.; Miller, R.D.; Tweig, R.J.; Volksen, W. *J. Appl. Phys.* **1993**, 73, 8471.
13. Singer, K.D.; King, L.A. *J. Appl. Phys.* **1991**, 70, 3251.
14. Dhinojwala, A.; Wong, G.K.; Torkelson, J.M. *J. Chem. Phys.* **1994**, 100, 6046.
15. Hampsch, H.L.; Yang, J.; Wong., G.K.; Torkelson, J.M. *Macromolecules* **1990**, 23, 3648.
16. Kohler, W.; Robello, P.T.; Willand, C.S.; Williams, D.J. *Macromolecules* **1991**, 24, 3589.
17. Kohler, W.; Robello, D.R.; Dao, P.T.; Willand, C.S.; Williams, D.J. *J. Chem. Phys.* **1990**, 93, 9157.
18. Wang, H.; Jarnagin, R.C.; Samulski, E.T. *Macromolecules* **1994**, 27, 4705.
19. Teraoka, I.; Jungbauer, D.; Yoon, D.Y.; Reck, B.; Swalen, J.D.; Tweig, R.; Willson, C.G. *J. Appl. Phys.* **1991**, 69, 8011.
20. Jungbauer, D.; Teraoka, I.; Reck, B.; Yoon, D.Y.; Tweig, R.; Willson, C.G. *J. Appl. Phys.* **1991**, 69, 2568.
21. Man, H.T.; Yoon, H.N. *Adv. Mater.* **1992**, 4, 159.
22. Lindsay, G.A. *TRIP* **1993**, 1, 138.
23. Lindsay, G.A.; Henry, R.A.; Hoover, J.M.; Knoeson, A.; Mortazavi, M.A. *Macromolecules* **1992**, 25, 4888.
24. Chen, M.; Dalton, L.R.; Yu, L.P.; Shi, Y.Q.; Steier, W.H. *Macromolecules* **1992**, 25, 4032.
25. Williams, M.L.; Landel, R.F.; Ferry, J.D. *J. Am. Chem. Soc.* **1955**, 77, 3701.

RECEIVED January 30, 1995

Chapter 28

Guest–Host Cross-linked Polyimides for Integrated Optics

T. C. Kowalczyk[1], T. Z. Kosc[1], K. D. Singer[1], A. J. Beuhler[2,4], D. A. Wargowski[2], P. A. Cahill[3], C. H. Seager[3], and M. B. Meinhardt[3]

[1]Department of Physics, Case Western Reserve University, Cleveland, OH 44106–7079
[2]Amoco Chemical Co., Amoco Research Center, Naperville, IL 60566
[3]Sandia National Laboratories, Division 1811, Albuquerque, NM 87185–1407

We report on the optical and electrical characterization of aromatic, fluorinated, fully imidized, organic soluble, thermally and photochemically crosslinkable, guest-host polyimides for integrated optics. Refractive indices and optical losses were measured to evaluate the performance of these materials for passive applications. Materials were doped with two high temperature nonlinear optical chromophores, and poled during crosslinking to produce nonlinear optical materials. Measurements of electro-optic coefficient, macroscopic second order susceptibility, and conductivity were performed to assess these materials as potential candidates for active devices.

High speed optical interconnects and switches are being investigated to eliminate the optical-to-electrical signal bottleneck in high speed communication and data transfer. Recent efforts have concentrated on developing new optimized materials for integrated optics. Organic polymers have received much attention as potential candidates and are promising materials for integrated optics because they have low optical losses, low dielectric constants, and are compatible with existing silicon fabrication procedures.[1] Furthermore, polymers can be cast into multiple layer structures containing channels, ribs, and other routing structures. The ability to define multiple layer structures allows high densities of optical interconnects and facilitates high speed signal routing and multiplexing.

The successful incorporation of active polymeric materials into integrated optics depends on the degree to which the following parameters can be simultaneously optimized: optical loss, waveguide formation, optical nonlinearity, and thermal stability. This optimization will likely be based on a series of trade-offs. For example, large nonlinearities can be obtained by operating near resonance with a large chromophore concentration at the cost of severe optical absorption. The

[4]Current address: Motorola Corporation, Northbrook, IL 60062

0097–6156/95/0601–0381$12.00/0

first two parameters, optical loss and waveguide formation, characterize the ability of these materials to serve as passive optical interconnects. Since low optical losses in polymers are achievable, a natural step in the development of nonlinear optical polymers is to convert low optical loss polymers into nonlinear optical materials by mixing chromophores with large nonlinearities ($\mu\beta$ products) into the polymer matrix and inducing noncentrosymmetric ordering by electric field poling. However, the ordering induced by poling places the polymer in a state of non-equilibrium leading to time and temperature dependent decay of the optical nonlinearity. Initial nonlinear optical hosts such as polymethylmethacrylate, polystyrene and polycarbonate have low optical loss, but also low glass transition temperatures, which leads to rapid decay of poling induced orientation at near ambient temperatures. The nonlinearity of these materials is limited because increased dopant loading causes aggregation and plasticization.

Further attempts to optimize the $\mu\beta$ product and thermal stability included the use of main chain chromophores, side chain chromophores, and crosslinked systems.[2–4] These materials have higher glass transition temperatures and reduced mobility as compared to their guest-host analogs and consequently reduced chromophore relaxation. Guest-host polyimides have been employed to optimize thermal stability of induced orientation while maintaining functionality. Polyimides are well known in the electronics industry for their thermal integrity, resistance to organic solvents, and compatibility with silicon fabrication processes. Initial guest-host polyimides required imidization and densification at temperatures greater than 300° C to retain their induced ordering while exposed to a 200°C environment.[5,6] The high processing temperature of these materials requires dopant molecules to have high thermal decomposition temperatures. In studies where the chromophores did not thermally decompose, sublimation and plasticization occurred.[7,8] Alternatives to high temperature imidization have been developed and include chemical imidization and preimidized soluble polyimides.[9,10] In these cases, densification for extended periods of time was necessary to minimize the free volume around the chromophore and reduce thermal relaxation of induced orientation.[11] More recently, side-chain polyimides have been developed.[12–14] These materials imidize at temperatures below the thermal decomposition temperatures of most chromophores and have demonstrated excellent thermal stability of induced orientation for over 200 hours at elevated temperatures . The additional orientational stability is presumably due to the hindered rotational motion of the large polyimide chains. The optical losses and processability of these materials remain unreported.

In this chapter, we report on our approach to the development of device-quality electro-optic materials. We have sought to optimize the optical losses and processing properties first, and later to functionalize the materials having learned how to synthesize and process them for devices. We started with perfluorinated preimidized fully aromatic polyimides. Preimidization makes processing doped polymer systems more flexible, by making functionalizing easier while allowing spin coating of soluble fully-imidized polymers. Fluorination increases solubility while decreasing optical loss and refractive index. Fully aromatic polymers allow for the best high temperature properties. We found that by introducing alkylated aromatic crosslinking groups which can be photo- or thermally activated, optical losses were greatly reduced, and high quality waveguides could be fabricated. Another advantage of the crosslinking groups is that they open up new processing capabilities. For example, cross-linking allows simple multilayer formation, since the crosslinked polymers do not dissolve when subsequent layers are spin deposited. The photo-crosslinking also permits a simple liquid etch process to define waveguides, and provides a chemical hook to which chromophores may be covalently attached to these polymers. After optimizing the loss and processing using this crosslinking chemistry, we studied the functionalization of these polymers using poled guest chromophores. We have uncovered a number of issues relating to functionalization, such as increased optical loss due to long absorption tails in the chromophore and enhanced mobility of small chromophores. We have also studied poling issues related to multilayer films. We believe that our approach for material development, focused on the processing and operation of devices, has led to a flexible material system with exceptional promise for the development of electro-optic polymer devices.

1. Polymer Waveguides

We studied optical losses in polyimide waveguides in order to understand the mechanisms responsible for losses and to develop new polyimides with lower losses. Waveguide losses are caused by absorption and scattering which can be either extrinsic or intrinsic.[15] Intrinsic absorption at long wavelengths is due to (C-H) carbon-hydrogen bond combination and overtone bands, and at short wavelengths, from electronic absorptions. Extrinsic absorption and scattering are caused by waveguide nonuniformities and contaminants in the polymer. The first polyimide synthesized for optical applications, PMDA-ODA (Pyromellitic dianhydride oxydianiline),had high losses.[16] Losses were significantly reduced by attaching perfluoro (CF_3) groups to the polymer backbone. The addition of these groups decreased absorption in the visible wavelengths by reducing charge transfer and creating polymer deformation.[17–19] Losses were cure dependent. Scattering losses, which are attributed to waveguide refractive index fluctuations, are also dependent on cure and structure. The inability to separate scattering and absorptive losses has limited the development of optical polyimides.

The most direct method to measure absorption is UV-visible spectroscopy. Unfortunately, UV-visible spectroscopy is not sensitive at the level of 1 dB/cm in thin film transmission measurements. Uncertainties in material properties lead to large errors in absorption coefficients calculated from absorption spectra.[20] More sensitive absorption techniques such as photothermal deflection spectroscopy (PDS) have been used to measure pure absorptive losses in polymers.[21] Using PDS along with waveguide loss spectroscopy (WLS), which measures the sum of scattering and absorption losses, the mechanisms of optical loss can be isolated.

Waveguide losses were measured using the scattered streak method. Measurements were made by coupling polarized light into the lowest order guided mode and capturing a video image of the guided streak with a CCD camera. This method assumes that the intensity of guided light is proportional to the intensity of scattered light. The digitized image was converted to an intensity vs distance plot from which the optical loss was determined. The sensitivity of WLS allows losses between 0.3 and 50 dB/cm to be measured. Losses greater than 50 dB/cm produce a guided streak too short to be measured accurately, and losses lower than 0.3 dB/cm do not scatter enough light out of the waveguide.

The PDS measurements were carried out in the transverse mode using a highly stable, single mode, 632 nm He-Ne laser.[22] The heating beam was provided by a 19 Hz chopped beam from a short arc, high pressure Xenon lamp passing through a single pass grating monochrometer with a band pass of 12 nm. A partially masked Si photo-cell served as the probe beam detector, while the heating beam intensity was continuously monitored with a second Si photo-cell for normalization purposes. Spectrum normalization was accomplished by comparing the sample PDS response with that obtained with a reference sample fabricated by baking a flat black organic coating of low reflectivity (4%) on quartz substrates. Independent measurements of the reflectivity of this calibration standard were made by routine methods. With proper vibration isolation of the PDS apparatus, the noise level for absorbance was 3 ppm. Flourinert FC-75 was chosen as the deflection medium. In the data reduction it was assumed that the reflectivity of the polymers in FC-75 was not a function of photon energy. Because the optical indices of these polymers did not differ greatly from FC-75, the reflectivity at the polymer/liquid interface was small, and variations of this quantity with photon energy will not introduce serious errors in the calculation of the absorption coefficient. Sample thicknesses were measured with a Sloan Dektak in several locations in the region heated by the PDS pump beam.

Polyimides were synthesized with increasing amounts of fluorination as shown in Fig. 1 (in order of increasing fluorination). The details of synthesis have been reported elsewhere.[23] Refractive indices in Fig. 2 were measured at 632 nm using TM and TE polarized He-Ne laser light. Several trends are apparent from the refractive index data. First, the refractive index and birefringence decrease with increasing fluorination. Decreased refractive indices are attributed to the reduction of conjugation due to twisting of the polymer backbone. The steric hindrance of fluorine-substituted polyimides prevents in-plane packing which decreases the birefringence. The only exception to this trend is Ultradel 9020D, which has a birefringence greater than the unalkylated polyimides. The larger birefringence is a result of its rigid backbone. The low refractive index results from its low density due to the presence of alkylated aromatic crosslinking groups[24] and a high degree of fluorination. Second, birefringence increases with curing temperature for all polyimides due to densification even in highly fluorinated samples. The temperature dependence of optical properties requires that care be taken when processing optical polyimides.

PDS was used to determine the absorption loss in thin polyimide films. Increasing fluorination also decreased absorption as shown in the PDS data of Fig. 3. The decrease in absorption is caused by (CF_3) groups that inhibit charge transfer and interrupt conjugation between electron-deficient imide and electron-rich amine sections. Additional (CF_3) groups were attached to the polymer chain to

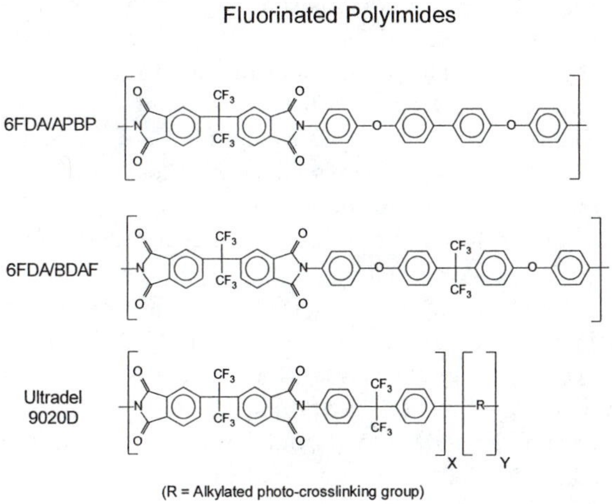

Fluorinated Polyimides

6FDA/APBP

6FDA/BDAF

Ultradel 9020D

(R = Alkylated photo-crosslinking group)

Fig. 1 Chemical structures of polyimides used in this study are listed in order of increasing fluorination. Ultradel 9020D is a crosslinkable polyimide. (R = alkylated aromatic crosslinking group as described in reference 24) Reproduced with permission from reference 29.

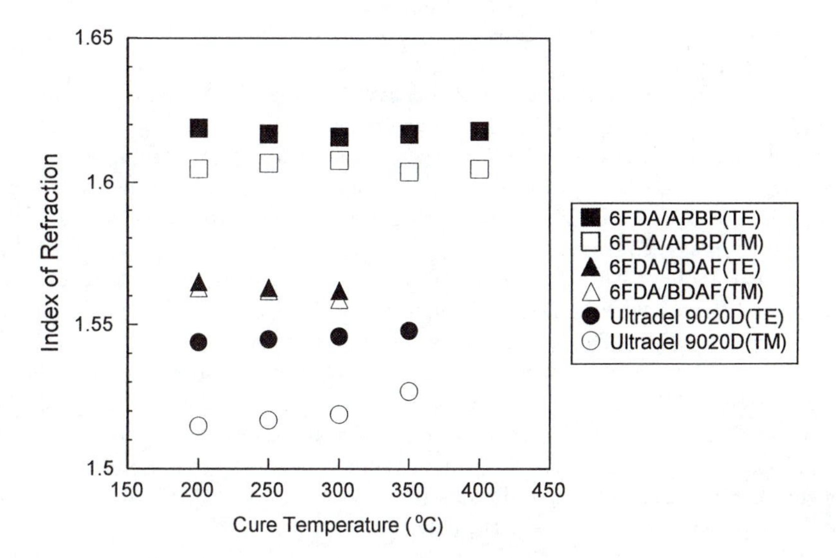

Fig. 2 TM and TE refractive index measurements at $\lambda = 632$ nm for polyimides shown in Fig. 1. Refractive indices and birefringences decrease with increasing fluorination. Reproduced with permission from reference 29.

further increase transparency as well as solubility. The additional fluorines that aid solubility, as in 6FDA/BDAF, also caused the fully imidized material to be susceptible to organic solvents. This problem was overcome by attaching alkylated aromatic crosslinking groups onto the polyimide backbone as in Ultradel 9020D.[24] Crosslinking can be accomplished by photo or thermal methods. The crosslinking polyimide showed the lowest absorptive loss, although it did not have the smallest birefringence. Waveguide losses are shown in Fig. 3 as data points along with PDS spectra(lines). For the polyimides in this study, fluorination reduced the waveguide losses considerably. Fig. 3 contains WLS data that is less than PDS data which is unphysical because WLS measures the sum of absorptive and scattering losses. This discrepancy between PDS and WLS data is a result of slight differences in thermal curing in the different samples used in each measurement as well as experimental uncertainty in both measurement techniques. The cure temperature dependent waveguide losses of Ultradel 9020D are plotted in Fig. 4 from 200° C to 400° C. At temperatures greater than 350° C, increased losses are due to polyimide decomposition. Waveguide loss measurements give information on loss mechanisms including both absorption and scattering. PDS measures the absorption of the material and is not susceptible to scattering effects present in WLS measurements and is more sensitive than ordinary UV-vis spectroscopy. Performing both experiments allows absorption losses and scattering losses to be separated by subtraction. Although both birefringence and absorption decrease with increasing fluorination, the agreement between PDS and WLS data for these polyimides indicates that losses are primarily due to absorption and not scattering.

Active devices must have low optical losses to allow efficient nonlinear interactions to take place over length scales appropriate for integrated optics. Nonlinear optically active devices are made by mixing nonlinear optical chromophores into the host material. The introduction of nonlinear chromophores may introduce additional losses in the form of scattering sites and absorption tails which can be further enhanced upon poling.[25] Fig. 5 depicts the structure of our nonlinear optical dopants, DCM[26,27] and DADC[28]. Fig. 6 shows the PDS spectra for undoped Ultradel 9020D, DCM/Ultradel 9020D, and DADC/Ultradel 9020D at 17% weight fraction. The spectra shows absorptive losses exceeding $\alpha = 11cm^{-1}$ (50.0 dB/cm) at 1.4ev (830nm) for DADC cured at 300° C and absorptive losses of $\alpha = 1.1cm^{-1}$ (5.0 dB/cm) for DCM cured at 175° C. In both cases significant losses are introduced by the chromophore, even at long wavelengths. The additional losses in DADC/Ultradel 9020D at high cure temperatures are attributed to chromophore degradation and possibly phase separation. It is important, then, to develop new chromophores with reduced absorption tails.

2. Electro-Optic Properties

The excellent linear optical properties, preimidized form, and crosslinking nature of Ultradel 9020D make it a suitable candidate for high temperature electro-optic

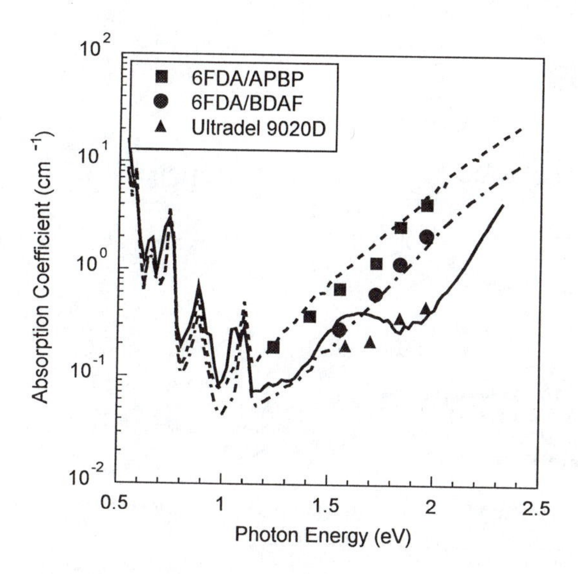

Fig. 3 Comparison between PDS (lines) and WLS (data points). The losses decrease with increasing fluorination. The agreement between PDS and WLS suggests that losses are primarily absorptive. Reproduced with permission from reference 29.

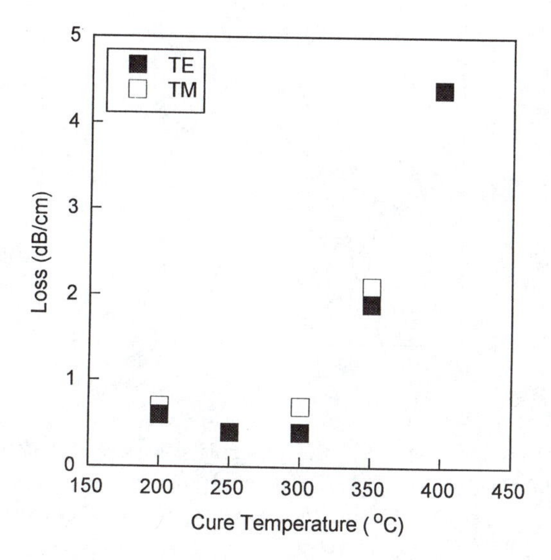

Fig. 4 WLS data at $\lambda = 800$ nm plotted as a function of curing temperature. Thermal crosslinking of Ultradel 9020D occurs near 200^0C. Processing variables significantly alter the optical properties of polyimides. Reproduced with permission from reference 29.

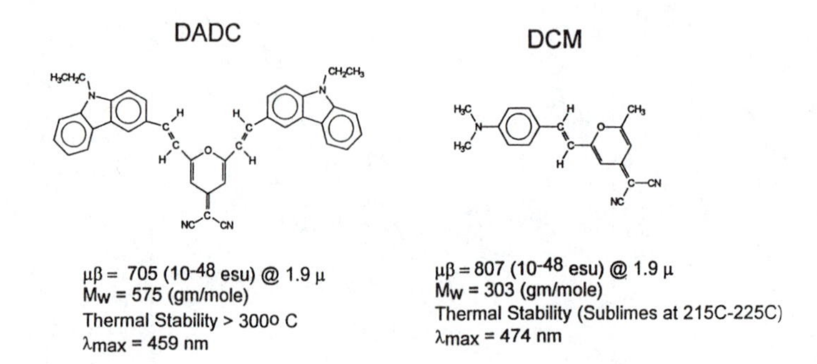

Fig. 5 Chemical structures and properties of thermally stable high temperature chromophores DADC and DCM.

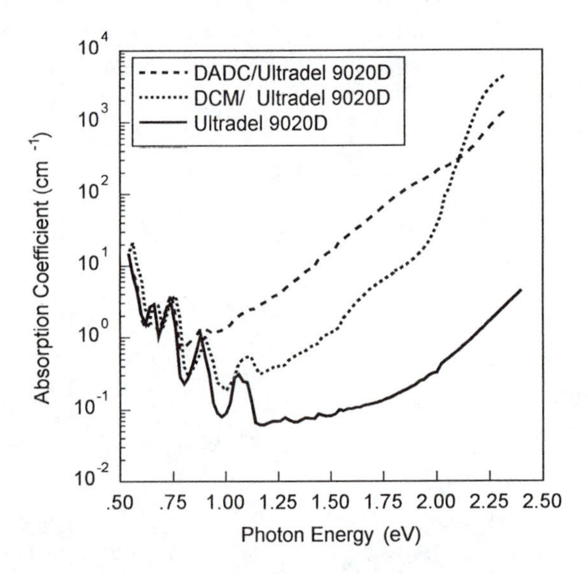

Fig. 6 PDS spectra showing the absorptive losses for Ultradel 9020D, DCM/Ultradel 9020D, and DADC/Ultradel 9020D. Chromophore loading is 17% by weight fraction. Curing temperatures for these samples were 300° C, 175° C, and 300° C, respectively.

applications.[23,29] Ultradel 9020D was doped with thermally stable chromophores, DCM and DADC, and electric field-poled to impart an electro-optic response. DCM is a highly nonlinear chromophore with low molecular weight that is photo-bleachable in the UV.[26] Using DCM as a dopant, we can obtain large nonlinearities because high number densities are possible. While DCM's decomposition temperature remains above the processing temperature of Ultradel 9020D, other researchers have reported sublimation at temperatures as low as 220° C.[8] DADC, a chromophore developed for high temperature electro-optic applications, has a high nonlinearity, high molecular weight, and is also photo-bleachable in the UV.[28] Using DADC we expect lower nonlinearities than DCM because of lower number densities at a given weight fraction. DADC's larger structure eliminates sublimation even at temperatures as high as 400° C. TGA (thermogravimetric analysis) experiments have shown only a 2% weight loss in nitrogen at temperatures as high as 400° C.[26]

The optimum conditions for poling during crosslinking were investigated. Ideally, the crosslinking reaction and chromophore alignment should proceed simultaneously. Poling at low temperatures can result in reduced alignment because of limited chromophore mobility, while poling the sample at too high temperature limits the ordering due to increased thermal effects and higher conductivity. The optimum poling temperature was determined by comparing samples poled with the same field at varying temperatures. Orientation and relaxation were probed by measuring the in-plane electro-optic coefficient at 800 nm with a Mach-Zehnder interferometer.[30] The electro-optic coefficient was measured immediately after poling for both sets of samples. Afterwards, the samples were placed into an oven at 125° C for approximately 50 hours, and the electro-optic coefficient was re-measured. Table I shows the results for both DADC and DCM. For DADC the electro-optic coefficient before aging is constant with temperature. After thermal aging, the sample poled below 200° C decayed while those above 200° C remained stable. This indicates that crosslinking occurs just above 200° C. The 200° C crosslinking is also observed in dielectric loss measurements as shown in Fig. 7, and in greatly decreased solubility when films are processed at that temperature.

Table I shows that the ratio, $r_{33}(t)/r_{33}(0)$, for DCM is fairly constant for temperatures above 200° C. At higher temperatures, $r_{33}(0)$ was lower. We attribute this to sublimation of the chromophore at elevated temperatures. Evidence for sublimation is shown in Fig. 8 where a decrease in optical absorption occurred when the doped polymer was processed at 225° C. Interestingly, the chromophore retention depends on the initial concentration which may be evidence of the chromophore plasticizing the polymer. The enhanced orientational decay of DCM at all temperatures may be attributed to plasticizing of the polyimide.[10]

The electric field dependence of the electro-optic coefficient for DADC-doped films is plotted in Fig. 9. Samples were poled using electric fields between 0.1 and 1.3 MV/cm at 200^0 C using a ramp, soak, and cool cycle for a total of 45 minutes. The in-plane electro-optic coefficient was measured as a function of applied field and compared to theoretical calculations using the two level oriented gas model with μ and β obtained from electric field-induced second harmonic generation

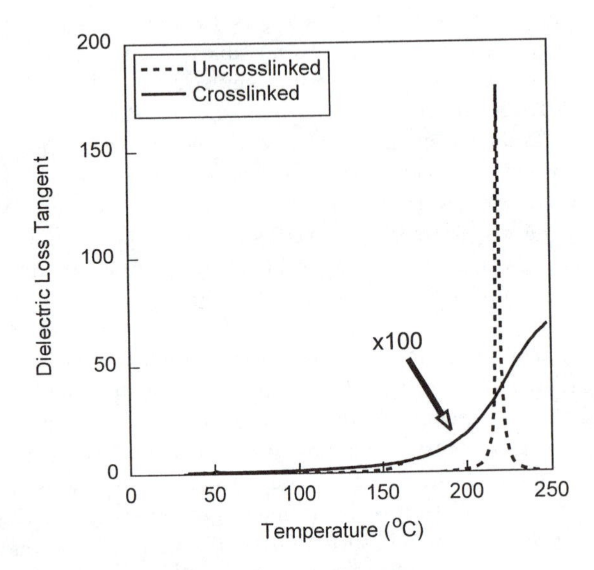

Fig. 7 Dielectric loss tangent of U9020D is plotted as a function of temperature for a sample processed at 175° C (dotted line). After cooling to room temperature the sample was remeasured (solid line). The solid line is multiplied by a factor of 100 to offset it from the dotted line for clarity.

Table I Dependence of electro-optic coefficient on poling temperature. Maximum electro-optic coefficient for DCM and DADC immediately after poling is scaled to unity for comparison purposes. Samples processed at temperatures above 200° C have larger ratios.

	DCM/Ultradel 9020D				DADC/Ultradel 9020D		
$T_{poling}(^oC)$	175	195	215	235	175	225	275
$r_{33}(0)$(a.u.)	0.74	1.0	0.97	0.61	1.0	1.0	0.96
$r_{33}(t)$(a.u.)	0.23	0.49	0.48	0.27	0.58	0.96	0.92
$r_{33}(t)/r_{33}(0)$	0.31	0.49	0.49	0.44	0.58	0.96	0.96

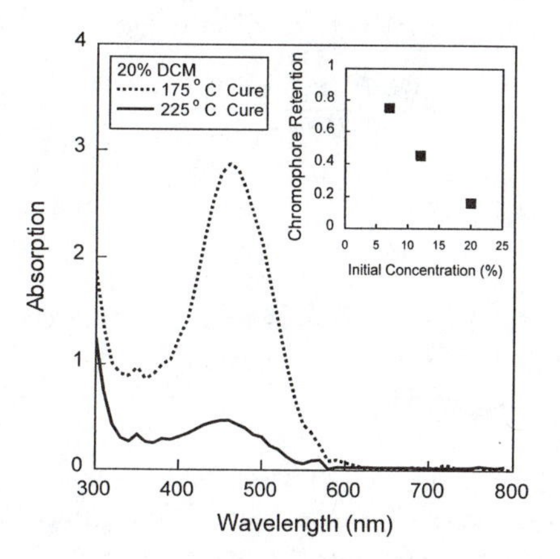

Fig. 8 Optical Absorption spectra of 20% DCM/U9020D at different curing temperatures. The inset shows the chromophore retention as measured by the change in optical absorption for samples of different initial weight fraction.

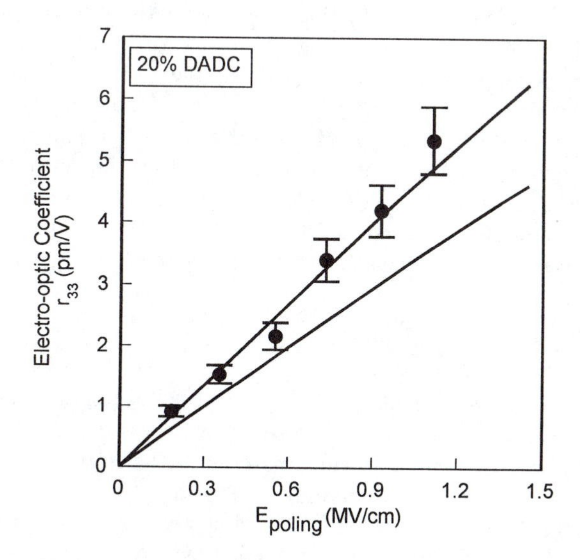

Fig. 9 Electric field dependence of electro-optic coefficient for 17% DADC by weight fraction. Solid lines represent (+/-)15% uncertainty in $\mu\beta$ values as input to the theory. Data points are from electro-optic experiment where $r_{33} = 3r_{13}$ was assumed.

(EFISH). The equation relating the electro-optic coefficient to its microscopic parameters at a particular wavelength is given by[31],

$$r_{ij,k}(-\omega;\omega,0) = \frac{-4N\beta_{zzz}L_3(p)}{n_\omega^4}\frac{f^\omega f^\omega f^o}{f^{2\omega'}f^{\omega'}f^{\omega'}}\frac{(3\omega_o^2-\omega^2)(\omega_o^2-\omega'^2)(\omega_o^2-4\omega'^2)}{3\omega_o^2(\omega_o^2-\omega^2)^2} \quad (1)$$

$$p = f^o\frac{\mu E_p}{k_B T_p} \quad (2)$$

where N is the chromophore number density, β_{zzz} is the second order microscopic susceptibility, $L_3(p)$ is the third-order Langevin function, μ is the dipole moment, E_p is the poling field, k_B is the Boltzmann constant, T_p is the poling temperature, f^o is the Onsager local field factor for the static poling field, $f^\omega f^{2\omega'} f^{\omega'}$ are the Lorenz-Lorentz local field factors at optical frequencies, ω' and $2\omega'$ are the frequencies used in the second harmonic generation experiment, ω_o and ω are the resonance and electro-optic experiment frequencies, respectively. Fig. 9 shows the agreement between Eq. 1 and experiment indicating full poling and verifies the electronic nature of the electro-optic effect. The solid lines define the (+/−)15% error in the measurement of $\mu\beta$. The slight offset of electro-optic coefficient at larger poling fields is probably due to increased birefringence with increasing poling fields. Long term thermal stability of induced orientation was measured for DCM- and DADC-doped films. In this experiment, the electro-optic coefficient was measured immediately after poling. Samples were then placed in an oven at 125° C, and samples were retested at subsequent times. The decay of the electro-optic coefficient is shown in Fig. 10. The data show the decay at elevated temperatures for both samples. DCM undergoes a more rapid decay of induced orientation at elevated temperatures indicating that the higher number densities may be plasticizing the host. After an initial rapid decay to 60%, DCM slowly decays to 40% after 200 hours at 125° C. DADC is more stable retaining 75% of its original electro-optic coefficient after 200 hours at 125° C.

3. Three-Layer Slab Waveguides

Nonlinear optical materials must be formed into channel, slab, or rib waveguides for integrated optical applications. The resistance of crosslinked Ultradel 9020D to organic solvents makes it a suitable material for layered devices. We formed three-layer slab waveguides by spinning and curing successive layers of Ultradel 9020D on ITO-coated glass. The layered structures consisted of DCM- and DADC-doped Ultradel 9020D cores surrounded by neat Ultradel 9020D cladding layers. To obtain noncentrosymmetric ordering as well as large poling field induced nonlinearities in three-layered devices, it is essential for the cladding layers to be more conductive than the core layer so that the voltage drop, and hence the poling field, across the active layer is as large as possible. Direct measurements of the poling field are not possible in three-layer samples because the voltage splitting between core and cladding layers is unknown. The effective poling field can be determined indirectly by measuring the macroscopic susceptibility's dependence on the externally applied field and comparing to theory. The macroscopic susceptibility was measured using the Rotational Maker Fringe (RMF) technique as a function of poling field for single- and three-layer samples.[32] Gold electrodes were sputtered onto the samples for contact poling. After contact poling, the gold was removed and the RMF experiment was performed. Measurements were made using the 2nd Stokes line ($\lambda = 1.367\mu$) of a hydrogen-filled Raman cell that was pumped by the output of a Nd:YAG pumped pulsed dye laser. For single-layers the poling field-induced nonlinearity is described by,

$$d_{33} = N f^{2\omega'} f^{\omega'} f^{\omega'} f^{o} \beta_{zzz} \frac{\mu E_p}{5 k_B T_p} \tag{3}$$

which relates the macroscopic susceptibility to microscopic parameters.[32] Fig. 11 shows the results for DCM and DADC single-layer samples. The solid line represents the predicted value for the nonlinear optical susceptibility, d_{33}, which includes the microscopic parameters obtained from EFISH experiments and the data points from RMF measurements. The agreement between theory and experiment shows that the oriented gas model adequately describes the poling induced ordering in polyimide materials.

The data for three-layer samples appear in Figs. 12a and 12b for DCM- and DADC-doped samples, respectively. The two sets of data points on the graphs represent different methods of calculating the poling field across the active layer. The open data points assume the resistivities of all three layers are identical and the resulting poling field is obtained by dividing the applied voltage by the total thickness. The shaded data points use a scaling factor to fit the experimental data to the theoretical prediction of Eq. 3 with the poling field as an adjustable parameter. If we assume the three-layer sample can be modeled by resistors in series, then the scaling factor can be related to the cladding and core resistances by the following equation,

$$V_{eff} = \frac{R_{core}}{R_{core} + 2R_{cladding}} V_{applied} \tag{4}$$

where R_{core} and $R_{cladding}$ are the resistances of the core and cladding layers, and V_{eff} and $V_{applied}$ are the effective and applied poling voltages.[33] The open data points show the inadequacy of assuming equal resistivities to determine the poling field across the active layer. For DCM triple-layers, a value of 0.4 is used for the best fit adjustable parameter, while for DADC triple-layers, a value of 0.11 is obtained. Conductivity of thin single-layer films was used to independently calculate the effective poling field. Conductivity measured as a function of temperature for DCM/Ultradel 9020D, DADC/Ultradel 9020D, and Ultradel 9020D is shown in Fig. 13. The conductivity and thickness of DADC triple-layers poled at 250° C predicts a correction factor of 0.1 in excellent agreement with the best fit. Thus, the voltage division model appears valid. For DCM triple-layers poled at 225° C Eq. 4 predicts an effective poling field an order of magnitude smaller than the

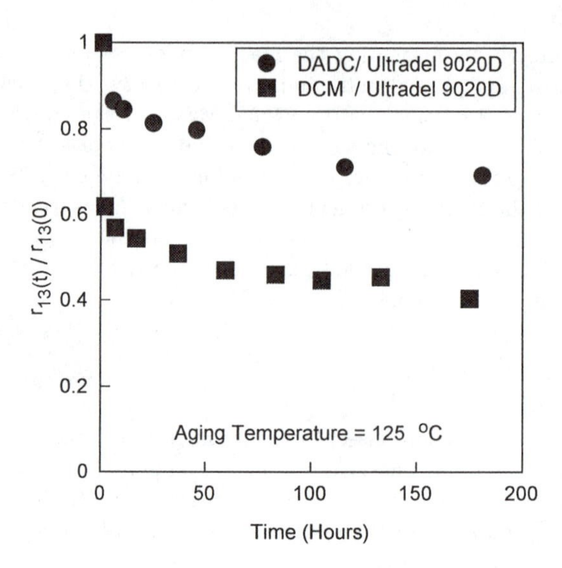

Fig. 10 Decay of electro-optic coefficient for DCM and DADC samples stored at 125° C. Electro-optic coefficients for DADC and DCM are normalized at $t = 0$.

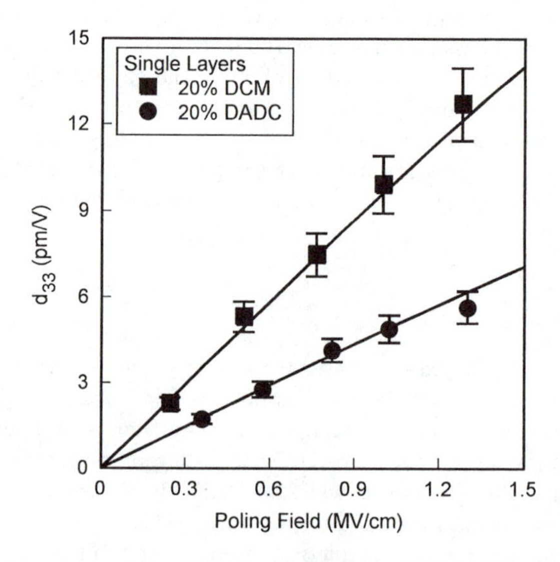

Fig. 11 d_{33} versus poling field for single layer DCM/Ultradel 9020D and DADC/Ultradel 9020D obtained from RMF experiments. Solid line represents calculation using Eq. 3 along with $\mu\beta$ values obtained from EFISH experiments. Agreement between theory and experiment shows the validity of using the oriented gas model to describe the poling process in guest-host polyimides.

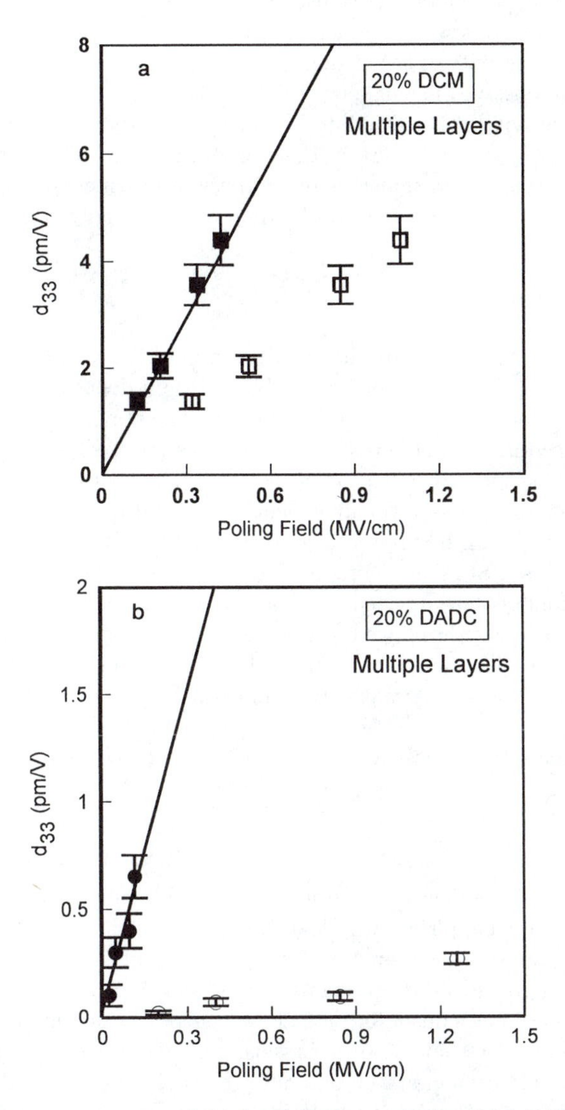

Fig. 12 d_{33} versus poling field for three layer stacks of (a) DCM/Ultradel 9020D and (b) DADC/Ultradel 9020D. Solid lines represents Eq. 3. Data points are measurements from three-layer samples using the poling field as an adjustable parameter. Best results (shaded data points) are obtained using 0.4 as the best fit parameter for DCM samples and 0.1 for DADC samples. (unshaded data points) assume that resistances of all layers are equal.

best fit adjustable parameter. This discrepancy may be due to migration of DCM into the cladding layers which lowers the conductivity of the core layer while increasing the conductivity of the cladding layers. This migration is consistent with the sublimation observed in single-layer films. The optimization of poling requires that the conductivity of the cladding be much larger than the core to maximize the electric field in the active layer. Finding methods to adjust conductivity are important. We have found that the conductivity of polymer compositions can be controlled by adjusting the degree of fluorination of the polymer.

4. Channel Device Fabrication

A processing scheme for creating channel devices has been developed. This process is shown in Fig. 14. The photosensitive polymer is spin-deposited onto a substrate, and soft-cured (below 200^oC) to remove solvent. The polymer is exposed through a photomask leading to photo-crosslinking, and then partially etched with organic solvent developers into a rib or channel pattern. The coating is then post-baked at $300 - 350^oC$ to remove residual solvent and to fully crosslink. Either rib or channel waveguides can be produced with this method.

Fig. 15 shows a SEM micrograph of a rib structure fabricated using photosensitive polyimides showing their high quality structure. Channels and ribs 4-5 microns wide have been produced. After post-cure, the films can withstand processes such as metallization, overcoating, thermal cycling, and solvent exposure. Waveguide loss spectroscopy was carried out on three-layer rib waveguides composed of two variations of Ultradel 9020D polymers with appropriate refractive indices. No excess optical loss was introduced by channel waveguide fabrication.

5. Conclusions

We have described the development of a crosslinkable polyimide material system for integrated optical applications. First, we optimized the optical and waveguide processing properties, and later introduced functionality by guest-host inclusion and electric field poling. From optical waveguide and PDS measurements we determined the primary mechanism for optical waveguide loss in polyimide waveguides is absorption from electronic absorption tails. The measured waveguide loss in Ultradel 9020D of 0.4 dB/cm at 800 nm and predicted losses of 0.3 dB/cm at 1300 nm make these materials excellent candidates for integrated optical applications. Introduction of the active chromophores, DCM and DADC, increased the optical losses as determined by PDS. We believe that these long absorption tails may be a general problem in the realization of electro-optic polymers.

We then studied the electric field poling conditions. The optimum poling temperature was found to be slightly below the sublimation temperature of DCM and well below the thermal decomposition temperature of most thermally stable chromophores. Both guest-host systems display excellent thermal stability of the electric field-induced orientation at room temperature. At elevated tempera-

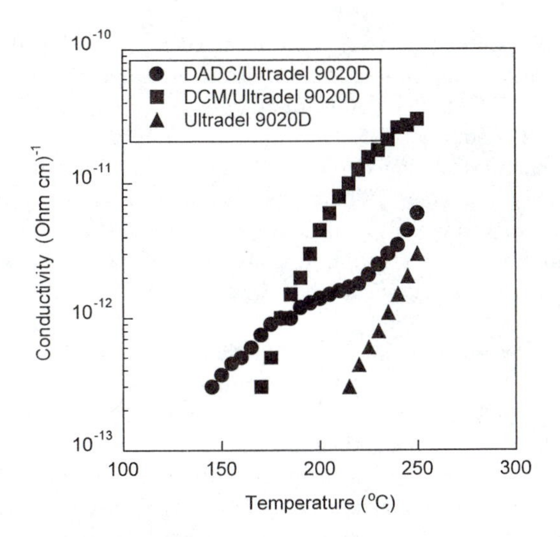

Fig. 13 Temperature dependent AC conductivity (0.1 Hz) for DCM/Ultradel 9020D, DADC/Ultradel 9020D, and neat Ultradel 9020D. An optimum three-layer stack would have cladding layers more conductive than the core.

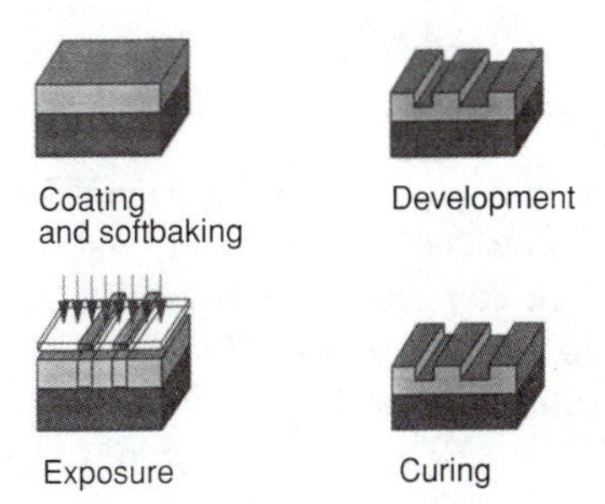

Fig. 14 Processing schematic for fabrication of single mode polyimide waveguides. Reproduced with permission from reference 23.

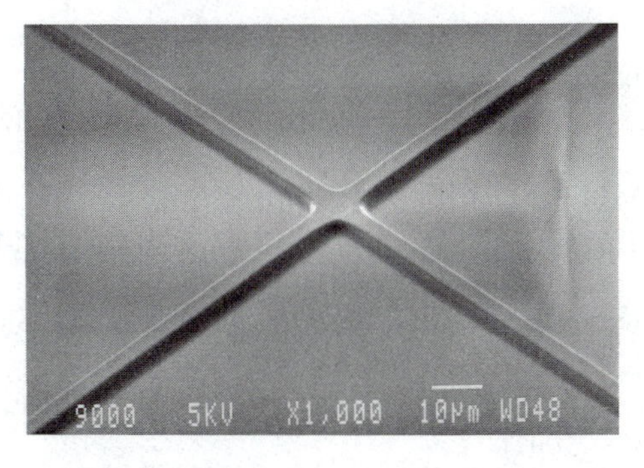

Fig. 15 SEM of 5.0 μm wide crossed rib waveguides etched in Ultradel 9020D.

tures, significant orientational relaxation occurred in DCM-doped samples. The enhanced decay may be attributed to plasticizing of the polyimide and greater rotational mobility of DCM. The DADC-doped polyimide retained 75% of its initial orientation after 200 hours at 125^0 C.

Three-layer samples of doped polyimide showed nonlinearities smaller than comparable single-layer films. The decrease in nonlinearity is attributed to lower poling fields due to voltage division across the three-layer sample and possibly chromophore migration into cladding regions. The voltage splitting can be eliminated by substituting existing cladding layers with more conductive cladding layers.

We have also described a simple waveguide processing procedure for forming channel and rib structures which takes advantage of the photocrosslinking nature of polyimide materials and does not introduce excess optical losses. We believe that this material system is an excellent candidate for active and passive integrated optical applications. Further refinement in processing and refractive index control are all that is necessary for passive applications. For active applications, the identification of suitable chromophores which do not introduce absorption is necessary. Covalent attachment of active chromophores to crosslinking sites should be straightforward, and would lead to vastly improved performance regarding optical nonlinearity and stability.

6. Acknowledgements

This work was performed, in part, at Sandia National Laboratories and was supported by the U.S. Department of Energy under contract DE-AC04-94AL85000. Partial support for this work was provided by AFOSR under grant # 49620-93-1-0202. The authors wish to thank S. Ermer (Lockheed)for chromophore synthesis and C. Allen (Amoco) for her technical assistance.

Literature Cited

1. G.I. Stegeman and W. Torruellas, MRS Proc., A.F. Garito, A. Jen, C. Lee, and L.R. Dalton, ed., Materials Research Society, Pittsburgh, **328**, 397 (1994).
2. P.M. Ranon, Y. Shi, H Steier, C. Xu, B. Wu, L.R. Dalton, Appl. Phys. Lett. **62**, 2605 (1993).
3. W. Sotoyama, S. Tatsuura, and T. Yoshimura, Appl. Phys. Lett. **64**, 2197 (1994).
4. B. Wu, C. Xu, L.R. Dalton, S. Kalluri, Y. Shi, and W. H. Steier, in MRS Proc., A.F. Garito, A. Jen, C. Lee, and L.R. Dalton, ed., Materials Research Society, Pittsburgh, **328**, 529 (1994).
5. J.W. Wu, J.F. Valley, S. Ermer, E.S. Binkley, J.T. Kenney, G.F. Lipscomb, and R. Lytel, Appl. Phys. Lett. **58**, 225 (1991).
6. J.W. Wu, E.S. Binkley, J.T. Kenney, R. Lytel, and A.F. Garito, J. Appl. Phys. **69**, 7366 (1991).

7. J.F. Valley, J.W. Wu, S. Ermer, M. Stiller, E.S. Binkley, J.T. Kenney, G.F. Lipscomb, and R. Lytel, Appl. Phys. Lett. **60**, 160 (1992).
8. H.H. Fujimoto, S. Das, J.F. Valley, M. Stiller, L. Dries, D. Girton, T. Van Eck, S. Ermer, E.S. Binkley, J.C. Nurse, and J.T. Kenney, in MRS Proc., A.F. Garito, A. Jen, C. Lee, and L.R. Dalton, ed., Materials Research Society, Pittsburgh, **328**, 553 (1994).
9. J.W. Wu, J.F. Valley, S. Ermer, E.S. Binkley, J.T. Kenney, and R. Lytel, Appl. Phys. Lett. **59**, 2213 (1991).
10. S.F. Hubbard, K.D. Singer, F. Li, S.Z.D. Cheng, and F.W. Harris, Appl. Phys. Lett. **65**, 265 (1994).
11. K.Y. Wong and A.K.Y. Jen, J. Appl. Phys. **75**, 3308 (1994).
12. B. Zysset, M. Ahlheim, M. Stahelin, F. Lehr, P. Pretre, P. Kaatz, and P. Gunter, Proc. SPIE **2025**, 70 (1993).
13. W. Sotoyama, S. Tatsuura, and T. Yoshimura, Appl. Phys. Lett. **64**, 2197 (1994).
14. D. Yu, A. Gharavi, and L. Yu, Appl. Phys. Lett. **66**, 1050 (1995).
15. T. Kaino in *Polymers for Lightwave and Integrated Optics*, L. A. Hornak, ed., Marcel Dekker, New York, 1 (1992).
16. T.P. Russel, H. Gugger, and J.D. Swalen, J. Polym. Phys. **21**, 1745 (1983).
17. H. Franke in *Polymers for Lightwave and Integrated Optics*, L. A. Hornak, ed., Marcel Dekker, New York, 207 (1992).
18. R. Reuter, H. Franke, and C. Feger, Appl. Optics **27**, 4565 (1988).
19. J.M. Salley, T. Miwa, and C.W. Frank, Proc. MRS **227**, 117 (1991).
20. R.A. Norwood, D.R. Holcomb, and F.F. So, Nonlinear Optics **6**, 193 (1993).
21. A. Skumanich, M. Jurich, and J.D. Swalen, Appl. Phys. Lett. **62**, 446 (1993).
22. A.C. Boccara, D. Fournier, and J. Badoz, Appl. Phys. Lett. **20**, 1333 (1981).
23. A.J. Beuhler, D.A. Wargowski, T.C. Kowalczyk, K.D. Singer, Proc. SPIE **1849**, 92 (1993).
24. A.J. Beuhler and D. A. Wargowski, "Photodefinable Optical Waveguides", U.S. Patent No. 5,317,082 (May, 1994)
25. C.C. Teng, M.A. Mortazavi, and G.K. Boudoughian, Appl. Phys. Lett. **66**, 667 (1995).
26. S. Ermer, J.F. Valley, R. Lytel, G.F. Lipscomb, T.E. Van Eck, D.G. Girton, D.S. Leung, and S.M. Lovejoy, Proc. SPIE **1853**, 183 (1993).
27. S. Ermer, J.F. Valley, R. Lytel, G.F. Lipscomb, T.E. Van Eck, and D.G. Girton, Appl. Phys. Lett. **61**, 2272 (1992).
28. S. Ermer, D. Leung, S. Lovejoy, J. Valley, and T.E. Van Eck, in *Organic Thin Films for Photonic Applications Technical Digest*, 1993 Vol. 17 (Optical Society of America, Washington, D.C., 1993) pp. 70-72.
29. T.C. Kowalczyk, T.Z. Kosc, K.D. Singer, P.A. Cahill, C.H. Seager, M.B. Meinhardt, A. Beuhler, and D.A. Wargowski, J. Appl. Phys. **76** 2505 (1994).

30. K.D. Singer, M.G. Kuzyk, W.R. Holland, J.E. Sohn, S.J. Lalama, R.B. Comizzoli, H.E. Katz, and M.L. Schilling, Appl. Phys. Lett. **53**, 1800 (1988).
31. K.D. Singer, M.G. Kuzyk, and J.E. Sohn, J. Opt. Soc. Am. B **4**, 968 (1987).
32. K.D. Singer, J.E. Sohn, S.J. Lalama, Appl. Phys. Lett. **49**, 248 (1986).
33. H.C. Ling, W.R. Holland, and H.M. Gordon, J. Appl. Phys. **70** , 6669 (1991).

RECEIVED April 19, 1995

Chapter 29

Bifunctional Dyes for Cross-linked Nonlinear Optical Polymers

Kenneth M. White[1], Elisa M. Cross, Cecil V. Francis, and Robert S. Moshrefzadeh

3M Company, Photonics Research Laboratory, Building 201–2N–19, St. Paul, MN 55144–1000

The incorporation of bifunctional, nonlinear optical dyes into isocyanate-based, cross-linked polymers via a two-step pole and cure process has produced materials that have significant potential for use in thin film electrooptic devices. Experimental results for two dyes that have been designed and synthesized for these polymer systems are presented and compared. Second-harmonic generation, electro-optic, and thermally stimulated current measurements have been employed to determine the magnitude of the nonlinear optical response and its temporal stability in these materials. Thermal stability of the dyes themselves was also investigated.

Many materials requirements must be met by nonlinear optical (NLO) polymers in order for these materials to be utilized in electrooptic (EO) devices. Some of these requirements, such as incorporation of a high concentration (N) of NLO dyes with large molecular hyperpolarizabilities (β) and processability of the resultant polymer into thin films, have been achieved in a wide variety of polymer systems (*1-5*). The need for a large EO coefficient, which increases with N, β, and the extent of polar order induced in the film by orienting the NLO dyes with a very high electric field (poling), has also been met with some success (*6,7*). Other demands, like producing EO films containing channel waveguides that exhibit low loss at diode laser wavelengths, have proven to be somewhat more difficult to satisfy (*8,9*). The ultimate challenge of course is to develop a single material that will simultaneously meet all the requirements.

A substantial amount of research has been devoted to the particular need for retaining a desired value of the EO coefficient over an extended period of time in the temperature range of interest. Decay of the NLO response results primarily from relaxation of the ordered dyes in the polymer toward a state of random orientation. However, thermal degradation or other alteration of dyes with a

[1]Current address: 3M Center, 209–BS–01, St. Paul, MN 55144–1000

0097–6156/95/0601–0401$12.00/0

concomitant decrease in EO activity has also been reported in NLO polymers (*10-12*). Consequently, we have considered both of these decay mechanisms in our design and development of polymer films with highly stable EO coefficients.

An increasing number of polymers have been developed with the intent of improving orientational stability of NLO dyes by means of cross-linking during poling (*13-17*). Two strategies have been employed; namely, one in which the NLO-active molecule is attached to the polymer network at one end and the polymer is cross-linked around it, and another in which the NLO-active molecule is attached at more than one point by making it part of the cross-link itself. Following the latter strategy, we have produced polymer films made by incorporating bifunctional NLO dyes into polyurea-polyurethane networks via reaction with an isocyanate cross-linker. The reaction involves amino and hydroxyl groups on the dyes that differ in reactivity toward isocyanate so that the polymer can be easily processed by curing in two separate steps. In the first step, the amino group reacts in solution slightly above room temperature to allow coating of a smooth, optically clear film of uniform thickness. In the second step, the hydroxyl group reacts at elevated temperature during poling to form a cross-linked network intended to restrict relaxation of the oriented NLO dyes when the poling field is removed from the cooled film.

Two dyes that we have synthesized, 3-amino-5-[4′-[*N*-ethyl-*N*-(2′′-hydroxyethyl)amino]benzylidene]rhodanine (RA) and 1-(2′-aminoethyl)-2-(4′′-[4′′′-[*N*-ethyl-*N*-(2′′′′-hydroxyethyl)amino]phenylazo]phenyl)-5-nitrobenzimidazole (BIN2), have been tested in this cross-linked system (*10,11,18*). Films made by reacting the dyes with a trifunctional isocyanate (see Figure 1) have been poled and cured and subsequently characterized for their NLO properties and thermal stability. While both types of polymer networks exhibited promising alignment stability at temperatures as high as 100 °C, significant differences in the stability of the two chromophores and in relaxation processes occurring in their respective films were observed. In this paper, we compare and contrast the pertinent properties observed in cross-linked films made from these two NLO dyes.

Film Processing

Films of RA-HDT and BIN2-HDT were made by reaction with the trifunctional isocyanurate Tolonate HDT (from Rhone-Poulenc). Typical preparation of samples was accomplished by dissolving the appropriate dye in dried, distilled pyridine along with 10% equivalent excess HDT and gently heating the solution to react 40-50% of the isocyanate, as determined by infrared (IR) spectroscopy. Solutions were filtered and spin-coated onto glass substrates covered with indium tin oxide electrodes to produce films with nominal thicknesses of 2-3 μm. After drying the films at 75 °C (2 h for RA-HDT, 1 h for BIN2-HDT) and evaporating gold electrodes on top, the samples were poled under an applied field of 100 V/μm and simultaneously cured. RA-HDT was cured for 2 h at 124 °C and BIN2-HDT was cured first at 109 °C for 35-50 min and then at 122 °C for a 2 h period. Multistage cure of BIN2-HDT was used since it yielded a larger EO coefficient. To complete the processing, the samples were cooled to room temperature (2-3 °C/min) and the poling field was removed. IR spectra of films coated onto salt plates and cured in this manner without poling revealed production of urea linkages in the first stage of the process, followed by generation of urethane linkages produced when the films were cured while being poled. By monitoring the IR band at 2270 cm^{-1}, it was determined that an equivalent amount of isocyanate had reacted with each dye as a result of this process.

Comparison of EO Coefficients

The poled and cross-linked RA-HDT films exhibited EO coefficients (r_{33}) near 3.5 pm/V, as determined by a transmission measurement at 1300 nm similar to the crossed polarizers technique reported previously (*19*). Using poling fields estimated at 2.5 MV/cm, an r_{33} of 7.8 pm/V was obtained. Mach-Zehnder devices with V_π values as low as 15 V and photobleached waveguide losses near 2 dB/cm have been fabricated using RA-HDT.

Values of r_{33} for BIN2-HDT were measured in the range of 6 to 8 pm/V. Larger values approaching 15 pm/V were obtained from films poled under fields exceeding 2.5 MV/cm. Mach-Zehnder devices made from these materials had V_π values down to 10 V, with losses in the photobleached waveguides as low as 2.4 dB/cm before poling, but significantly higher after poling. In one sample, the V_π was unchanged after the device had been temperature cycled 10 times from -40 to +70 °C (0.5 h ramp, 2 h soak).

Stability of the EO Response

EO stability in the cross-linked films was monitored via the second-order nonlinear optical response ($\chi^{(2)}$) of the materials, since a fractional change in $\chi^{(2)}$ produces a corresponding change in the EO coefficient. Measurements of $\chi^{(2)}$ by means of second-harmonic generation (SHG) were employed in the case of RA-HDT and EO measurements were utilized for BIN2-HDT. The SHG measurement employed a *p*-polarized laser beam having a fundamental wavelength of 1580 nm that was directed at the sample 45° from normal incidence. The experimental setup has been described previously (*10*).

The upper plot displayed in Figure 2 shows decay of $\chi^{(2)}$ (obtained as $\sqrt{I_{SHG}}$) in a film of RA-HDT that had been stored in air in an oven maintained at 75 °C. The data have been normalized to the $\chi^{(2)}$ value obtained for the film prior to the start of the stability test. Despite the fact that the NLO response decreased only 15% over the three month period at elevated temperature, it was unexpected that a cross-linked film such as this one would exhibit this much decay in $\chi^{(2)}$ at a relatively modest temperature. Measurements of UV-visible spectra from a film prepared without poling and aged in a similar manner provided an explanation, as shown in Figure 3. In this graph, absorbance by RA dye in the film was observed to decrease in the region of the peak at 474 nm while absorbance in the region around 375 nm increased. The change was not reversible. Furthermore, by plotting normalized values of the peak absorbance as a function of time (lower plot of Figure 2), we found that the decay rates of both $\chi^{(2)}$ and peak absorbance were identical. This implies that the same mechanism is responsible for each of the instabilities. As both $\chi^{(2)}$ and absorbance are linearly dependent on the concentra-

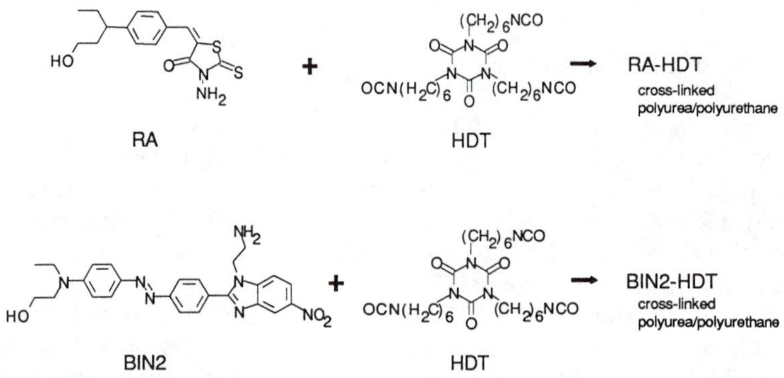

RA + HDT → RA-HDT cross-linked polyurea/polyurethane

BIN2 + HDT → BIN2-HDT cross-linked polyurea/polyurethane

Figure 1. Chemical structures of the RA and BIN2 dyes and the Tolonate HDT cross-linking agent that comprise the polymer films.

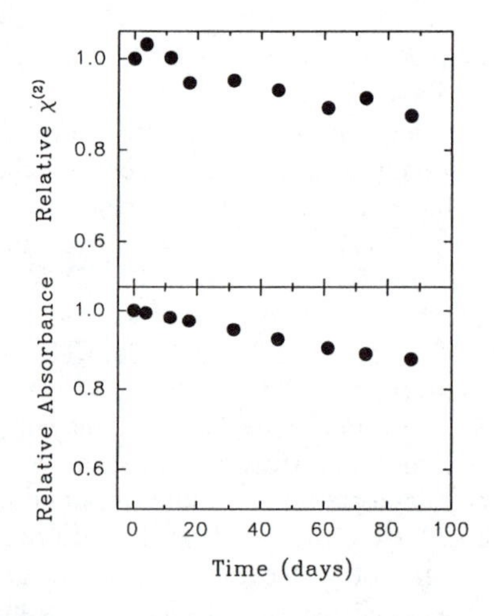

Figure 2. Decay of the second-order nonlinear optical response (upper plot) and the peak absorbance at 474 nm (lower plot) in cross-linked RA-HDT films maintained at 75 °C in air. The relative $\chi^{(2)}$ values were obtained from SHG measurements.

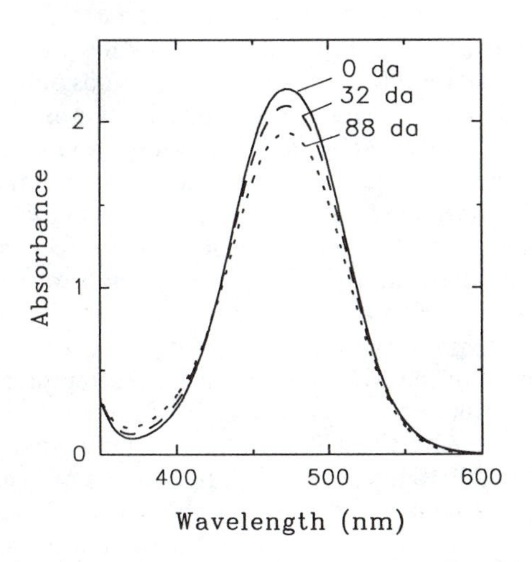

Figure 3. UV-visible spectra of a cured RA-HDT film stored at 75 °C, indicating alteration of the chromophore over a 3 month period.

tion of NLO dyes in the poled polymer, degradation or some other alteration of the RA chromophores is occurring when the films are stored at 75 °C in air. When similar tests were performed on RA-HDT films maintained at 100 °C, the $\chi^{(2)}$ and absorbance decay rates were also comparable during the time period following a more rapid drop-off of $\chi^{(2)}$ over the first several days (*10*).

Stability of the EO response in BIN2-HDT was monitored for films maintained in ovens under ambient atmosphere at 85, 100, and 110 °C. Figure 4 displays the normalized r_{33} values of these films measured for a period of over two months. In contrast to $\chi^{(2)}$ decay observed in RA-HDT, the decay in these films was due solely to orientational relaxation of the dye molecules. The thermal stability of BIN2 in films that were cured but not poled was investigated by means of UV-visible spectra measured from samples kept in the dark at 100 °C under ambient atmosphere. The peak absorbance of BIN2 dye at 445 nm decreased by only 1% after 27 days. Similar tests on a sample held at 110 °C revealed comparable stability over a few days. Indeed, the known resistance of benzimidazoles to thermal degradation was one of the motivations behind the design of this class of dyes for use in NLO polymers (*20*). Instability was observed, however, in the UV-visible spectra of BIN2-HDT films that were exposed to light, as shown in Figure 5. After leaving a sample in room light at ambient temperature for 45 min, the absorbance band decreased in intensity significantly, with an accompanying increase in absorbance near 375 nm. Reversal of the effect was accomplished by storing the film in the dark at 100 °C, giving complete recovery of the original spectrum. The cause of this light sensitivity is believed to be a *trans-cis* isomerization which has been demonstrated in a variety of azo dyes (*21*).

An examination of the EO data reveals that at temperatures up to 100 °C, the BIN2-HDT films retained greater than 80% of the $\chi^{(2)}$ value determined prior to aging for a period of more than 70 days, but at a temperature just 10 °C higher, the NLO response greatly diminished in less than one day. This not only suggests that the T_g of the BIN2-HDT films is not far above 110 °C, but also implies — in terms of an Arrhenius model for the kinetics of the decay — that the activation energy for relaxation of the poled dyes is extremely high. A large energy barrier to orientational relaxation is highly desirable and offers optimism for successful implementation of these cross-linked NLO polymer networks in electrooptic devices.

TSC Measurements

Because RA dye degrades while undergoing thermal stability testing, it was not possible to directly compare the relative $\chi^{(2)}$ stability provided by the cross-linked networks in RA-HDT and BIN2-HDT films through long-term stability measurements. Consequently, we employed concurrent measurements of thermally stimulated discharge current (TSC) and SHG from poled and cured films while applying a linear temperature ramp to observe the depoling behavior of these two polymer systems. The samples were heated at a rate of 2.5 °C/min and the current, which results from release of charge from the sample as it depolarizes, was detected in a short-circuit configuration with electrode area 0.8 cm^2. It was shown

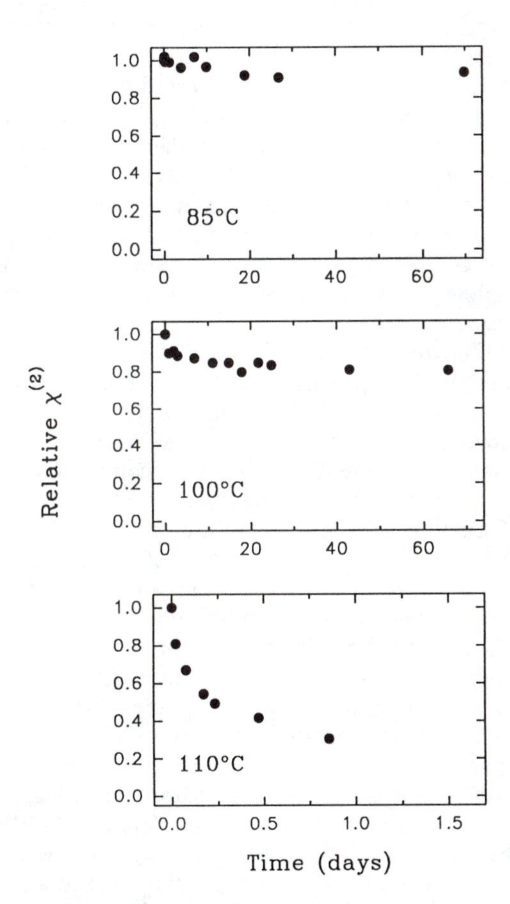

Figure 4. Decay of the second-order nonlinear optical response in cross-linked BIN2-HDT films maintained at various temperatures in air. The relative $\chi^{(2)}$ values were obtained from EO measurements.

by UV-visible spectroscopy that RA did not degrade under the conditions of this short-term experiment.

The data for RA-HDT are displayed in the upper graph in Figure 6. Orientational relaxation of the dyes is seen to occur at 142 °C, chosen as the temperature at which $\chi^{(2)}$ decline is steepest. This also corresponds to the location of the first peak in the TSC spectrum, identifying this as the α band which arises from flow of image charge as the molecular dipoles relax towards random orientation (*22*). The band at higher temperature, known as the ρ band, occurs due to relaxation of space charge (from impurities or other sources) that was polarized in the sample during poling (*23*). The narrow width of the α band is consistent with the steep decline observed in $\chi^{(2)}$ and together these suggest a high energy of activation for orientational relaxation in RA-HDT (*11,23*). Another indication of substantial stability imparted by the cross-linked network to the oriented RA dyes is the persistence of $\chi^{(2)}$ response above the 142 °C depoling temperature. Covalent attachment of the dyes at both ends of the molecule into the cross-linked polymer restricts their complete return to a state of random orientation until significantly higher temperatures.

For BIN2-HDT, the data are shown in the lower plot of Figure 6. The depoling temperature in this material is 122 °C and unlike RA-HDT, orientational relaxation of the dyes is essentially complete just above this temperature. Furthermore, instead of a single peak, multiple peaks that were found to be reproducible are observed in the TSC α band indicating that the BIN2 dyes were cross-linked into a range of different structures. Such structures may possibly be due to the use of multiple curing temperatures or the occurrence of secondary isocyanate reactions that form biurets and/or allophanates. The results are consistent with the long-term stability results (Figure 4) in at least two aspects: 1) the $\chi^{(2)}$ response does not begin to fall off until the film temperature has exceeded 100 °C and 2) above this temperature a sharp decrease in $\chi^{(2)}$ is observed. However, the drop in NLO response is slightly more gradual than is observed for RA-HDT and this is reflected in the breadth of the α band.

An examination of the α bands for the two polymers reveals that over twice as much charge was released (as determined by the area under the bands) by orientational relaxation in BIN2-HDT as compared to RA-HDT. This observation is likely the result of a larger ground-state dipole moment in BIN2 molecules (*24*) and incomplete depolarization of RA dyes, especially since dye concentration in the films is 20% larger in RA-HDT. The flow of space charge was also much greater in BIN2-HDT (current exceeded 40 nA above 150 °C), as indicated by comparison of the ρ bands.

Conclusion

Side-by-side comparison of RA-HDT and BIN2-HDT EO films has revealed that while each of these cross-linked materials possesses properties desired for use in electrooptic applications, neither of them is superior to the other in all aspects. Both show good long-term $\chi^{(2)}$ stability imparted by their respective cross-linked networks, indicating that the polymer system employed here is potentially useful. However, RA-HDT undergoes dye degradation that is accompanied by reduction

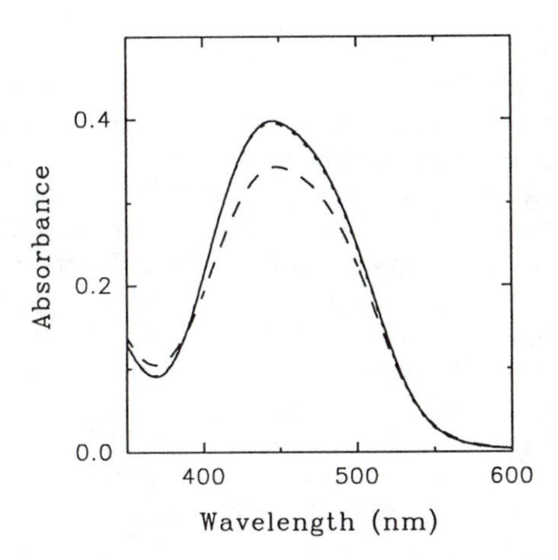

Figure 5. UV-visible spectra of a cured BIN2-HDT film before exposure to light (solid line), after exposure to room light for 45 min (dashed line), and after storage in the dark for 4 days at 100 °C (dotted line).

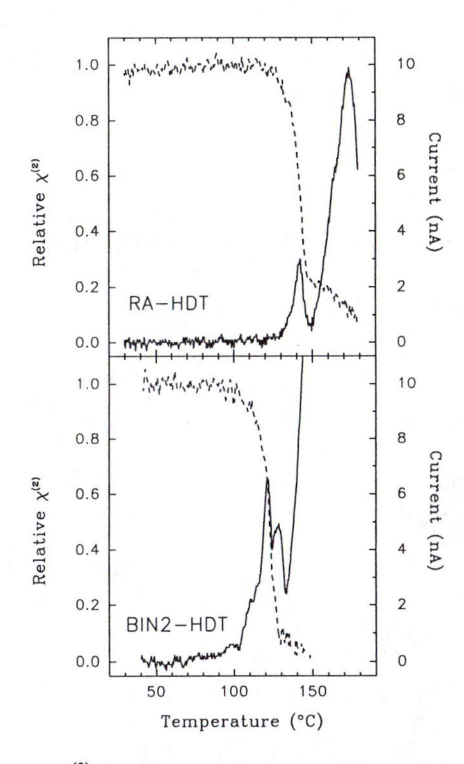

Figure 6. Relative $\chi^{(2)}$ (dashed line) from SHG measurements and TSC current (solid line) from RA-HDT and BIN2-HDT films during the application of a 2.5 °C/min temperature ramp.

of $\chi^{(2)}$ at only modest temperatures. BIN2 dyes are far more stable and remain poled in BIN2-HDT films held at 100 °C for over two months, yet they readily relax to a state of random orientation just 10 °C higher at 110 °C. Short-term SHG/TSC experiments suggest that RA-HDT exhibits greater orientational stability of its NLO dyes. It does not depole until 142 °C and displays residual $\chi^{(2)}$ response above this temperature. BIN2-HDT depoles at 122 °C to a completely random state. It is interesting to note that RA-HDT depoles at a temperature that is nearly 20 °C higher than the temperature at which it was cured, whereas BIN2-HDT depoles right at its curing temperature. The reasons for this are not fully understood, especially when considering that final cure temperatures and extent of cure were nearly identical for the two materials.

In terms of properties that pertain to EO devices, best values for r_{33} in BIN2-HDT (about 15 pm/V) were twice as large as those measured in RA-HDT (7.8 pm/V). However, in a Mach-Zehnder waveguide device, the BIN2-HDT material was unacceptably lossy. Losses in RA-HDT devices were lower, the best being 2 dB/cm.

Additional research to develop a thermally stable, bifunctional dye with large β is clearly warranted. Modifications to increase the depoling temperature of the resultant film and improve the long-term $\chi^{(2)}$ stability are also needed, as is a means to reduce the loss in waveguide devices. Nevertheless, as a material system that is easily processed and poled to produce stable EO coefficients for application in polymer waveguides, bifunctional dyes cross-linked into polymer networks comprise one of the primary candidates.

Acknowledgments

We thank Teng Lo for synthetic efforts, Alan Isackson for sample preparation and analytical measurements, Robert Williams and Dr. Marc Radcliffe for polymer processing, Mark Kleinschmit and Paul Pedersen for device fabrication, and Drs. Gary Boyd and Charles Walker for helpful discussions and support of this work.

Literature Cited

1. Rao, V. P.; Wong, K. Y.; Jen, A. K.-Y.; Mininni, R. M. *Proc. SPIE* **1993,** *2025,* 156.
2. Wong, K. Y.; Jen, A. K.-Y. *J. Appl. Phys.* **1994,** *75,* 3308.
3. Shi, Y.; Steier, W. H.; Chen, M.; Yu, L.; Dalton, L. R. *Appl. Phys. Lett.* **1992,** *60,* 2577.
4. Sotoyama, W.; Tatsuura, S.; Yoshimura, T. *Appl. Phys. Lett.* **1994,** *64,* 2197.
5. Dai, D.-R.; Hubbard, M. A.; Park, J.; Marks, T. J.; Wang, J.; Wong, G. K. *Mol. Cryst. Liq. Cryst.* **1990,** *189,* 93.
6. Möhlmann, G. R.; Horsthuis, W. H. G.; McDonach, A.; Copeland, M. J.; Duchet, C.; Fabre, P.; Diemeer, M. B. J.; Trommel, E. S.; Suyten, F. M. M.; Van Tomme, E.; Baquero, P.; Van Daele, P. *Proc. SPIE* **1990,** *1337,* 215.
7. Buckley, A. *Adv. Mater.* **1992,** *4,* 153.
8. Tumolillo, T. A., Jr.; Ashley, P. R. *IEEE Photon. Technol. Lett.* **1992,** *4,* 142.

9. Horsthuis, W. H. G.; Hofstraat, J. W.; Heideman, J.-L. P.; Möhlmann, G. R. *Proc. SPIE* **1993**, *2025*, 233.
10. Francis, C. V.; White, K. M.; Boyd, G. T.; Moshrefzadeh, R. S.; Mohapatra, S. K.; Radcliffe, M. D.; Trend, J. E.; Williams, R. C. *Chem. Mater.* **1993,** *5,* 506.
11. White, K. M.; Francis, C. V.; Isackson, A. J. *Macromolecules* **1994,** *27,* 3619.
12. Nahata, A.; Wu, C.; Knapp, C.; Lu, V.; Shan, J.; Yardley, J. T. *Appl. Phys. Lett.* **1994,** *64,* 3371.
13. Eich, M.; Reck, B.; Yoon, D. Y.; Willson, C. G.; Bjorklund, G. C. *J. Appl. Phys.* **1989,** *66,* 3241.
14. Jungbauer, D.; Reck, B.; Twieg, R.; Yoon, D. Y.; Willson, C. G.; Swalen, J. D. *Appl. Phys. Lett.* **1990,** *56,* 2610.
15. Marturunkakul, S.; Chen, J. I.; Li, L.; Jeng, R. J.; Kumar, J.; Tripathy, S. K. *Chem. Mater.* **1993,** *5,* 592.
16. Xu, C.; Wu, B.; Dalton, L. R.; Shi, Y.; Ranon, P. M.; Steier, W. H. *Macromolecules* **1992,** *25,* 6714.
17. Hubbard, M. A.; Marks, T. J.; Lin, W.; Wong, G. K. *Chem. Mater.* **1992,** *4,* 965.
18. Cross, E. M.; White, K. M.; Moshrefzadeh, R. S.; Francis, C. V. *Macromolecules*, in press.
19. van der Vorst, C. P. J. M.; van Weerdenburg, C. J. M. *Proc. SPIE* **1990,** *1337,* 246.
20. Cassidy, P. E. *Thermally Stable Polymers: Syntheses and Properties;* Marcel Dekker: New York, 1980.
21. Kumar, G. S.; Neckers, D. C. *Chem. Rev.* **1989,** *89,* 1915.
22. van Turnhout, J. In *Electrets;* Sessler, G. M., Ed.; Topics in Applied Physics; Springer-Verlag: Berlin, 1980, Vol. 33; pp. 81-215.
23. van Turnhout, J. *Thermally Stimulated Discharge of Polymer Electrets;* Elsevier: Amsterdam, 1975.
24. Leung, P. C., 3M Company, unpublished results.

RECEIVED January 30, 1995

Thin Film Devices

Chapter 30

Integration of Electrooptic Polymers in Optoelectronic Devices

R. Lytel, A. J. Ticknor, and G. F. Lipscomb

Akzo Nobel Electronic Products Inc., 250 C Twin Dolphin Drive, Redwood City, CA 94065

Organic electro-optic (EO) polymer materials offer some new opportunities in integrated optics. This paper reviews the application of EO polymer materials to highly integrated waveguide devices.

Organic electro-optic (EO) polymer materials offer potentially new opportunities in integrated optics for high-performance interconnections. The electronic *(1)* EO effect in organic materials yields moderate EO coefficients, low dispersion, and low dielectric constants *(2)*. EO polymer materials have been modulated flat to 40 GHz and exhibit few fundamental limits for ultrafast modulation and switching. Polymeric integrated optic materials also offer fabrication flexibility. The materials are spin-coatable into high quality, multilayer films, and can be patterned, metallized, and poled. Channel waveguides and integrated optic circuits can be defined by the poling process itself *(3)*, by photochemistry of the EO polymer *(4,5)*, or by a variety of well understood micro-machining techniques. Multi-layer integrated optic waveguide structures can be fabricated in much the same manner as Si-substrate, multilayer multichip modules. To date, EO polymer materials have been used to fabricate high-speed Mach-Zehnder modulators *(6)*, directional couplers *(7)*, Fabry-Perot etalons *(8)*, and even multitap devices *(9)*.

The potential impact of the application of EO polymer materials to highly integrated optical waveguide devices is particularly intriguing with respect to parallel processing systems. High-performance interconnections within optical multichip modules, backplanes, and optical connectors within massively parallel computers can greatly enhance the performance potential as such systems evolve to higher bisection bandwidths, wider data busses, and clock rates above 100 MHz. The incorporation of EO polymer materials into multilayered interconnection substrates built on wafers is of critical importance in the development of dense interconnection networks for modules, backplanes, and connectors. In these applications, EO polymer waveguide switches permit electronic devices to communicate via a waveguide network by driving ultralow-capacitance loads with logic levels signals.

To meet standard processing and packaging methodologies of standard electronic components, new EO polymer materials with extremely high thermal stability are required. Initial development proved promising*(10)* and today,

0097–6156/95/0601–0414$12.00/0

development continues throughout the world. These materials do not yet meet the EO performance of their lower temperature brethren but are expected to achieve comparable performance in the future. Meanwhile, there are numerous applications of EO polymers which do not require thermal stability at 350° C but instead demand stability and reliability in device performance at temperatures up to 80° C. Such applications include digital data communication links. Very good performance should be achievable with current materials, and even higher performance can be expected from materials that should be available in the near future.

Integrated optical transmitter arrays

Optical interconnections have traditionally been implemented as ultra-high-performance serial links. Serialization of all the communications in a massively parallel architecture would be foreboding to most system designers. A primary advantage of EO polymer integrated optics is the ability to take a reasonable parallel approach to optical interconnections. EO polymer waveguide switches and taps occupy only 0.02 mm^2 area and may be tiled to yield large scale integration. To an IC output pin, the switches load the output drivers primarily as only a small capacitive load (100 fF switch, plus via and tape, bump, or wire-bond capacitance). A single switch may be driven by 1-5 volts and be used to switch optical power for routing and distribution. Reflectors can turn light within a plane, providing high packing density and tiling of cellular elements. Finally, subtle modifications in waveguide dimensions, shapes, and polymer refractive indices can produce simple integrated lenses, which will be extremely useful as alignment features for fibers, lasers, and receivers.

Active optical transmitter arrays can be fabricated on planar surfaces such as silicon wafers using these methods. They are active because the energy required for the data stream is provided primarily by an external laser and the transducers appear as high-impedance loads to the electrical signals. Arrays could then be integrated into modules as could receiver arrays of similar configuration to provide optical I/O that could be coupled between modules using fiber-ribbon arrays with self-aligning terminals.

There are also good approaches being pursued in various projects to use directly-modulated laser arrays for similar applications. These have many of the advantages of parallel optical interconnection and can even be made to look to the signal source like high-impedance transducers through the use of active buffers and separate power supplies. The primary penalty of this approach is however power consumption and particularly the resultant on-module heat generation. As the data rates increase and the energy-per-bit stays steady (to maintain signal integrity), the laser drivers must switch greater and greater energy. This means they must switch more current more rapidly into the impedance of the laser circuit. The on-module power will increase approximately with a linear contribution from the frequency and a quadratic contribution from the peak power. To increase the power in the modulator array, one needs simply to increase the incoming laser power. Since the modulated current in the transmitter array starts out small, its linear increase with frequency to the power consumption has much less impact and the 'higher power' component only contributes a linear increase to power consumption instead of the quadratic increase in the equivalent direct-drive term. Furthermore, the continuous optical energy supplying the modulator array is supplied by fiber and the power-dissipating element (the laser) can be located remotely to the module to isolate its contribution to heating from the active circuitry. The figures in Table 1 show the power required for an eight channel, one-gigahertz/channel optical transmitter using direct modulation vs external modulation for three different power levels. The contributions are separated in terms of modulated (RF) current and low-frequency (DC) current so considerations of switching noise can also be made. External modulators are modeled as capacitative loads, while the directly modulated laser figures are an average of commercially available specs.

Table 1. Power Required for 1 GHz-by-8 Optical Transmitter Arrays

METHOD	- 3 dBm	0 dBm	3 dBm
DIRECT MODULATION			
• RF power, mW	20	80	320
• DC power, mW	128	128	128
EXTERNAL MODULATION			
• RF power, mW	11.2	11.2	11.2
• DC power, mW	53	105	210

Reducing modulator size

The length of device required for an EO modulator to fully switch or modulate an optical signal from logic-level electrical signals is of order 1 cm in any material system so far demonstrated. The width required is only a few tens of microns. This not only leads to significant space required for an EO transmitter array, but also means the electrodes necessary to drive the modulators must be transmission lines at frequencies around 0.3 GHz or above. There is incentive to decrease the active device length not only to occupy less area, but also to allow the drive electrodes to be designed as simple RF stubs for significantly higher impedance. One approach to this is to increase the EO sensitivity of the materials, and this has long been one of the great promises of polymer EO materials and is being pursued by many groups around the world.

Another approach is to make active complementary taps *(11)*. In this approach, a large amount of optical energy is fed into the modulator, and the modulator redirects a small fraction of that energy into one or the other of the two outputs. Since only a small portion of the light is actually being modulated, the active device length can be much shorter, of order 1 mm in current materials, for a given switching voltage. This lets the electrical load of the modulators be modeled as stub elements into the multi-GHz range, rather than requiring transmission lines as in the longer modulators, thereby reducing the power dissipated in both the modulator and the output driving it. The signal level is indicated by which of the two outputs that most of the output signal is directed towards. Since a relatively constant fraction of energy is removed from the supply beam, it can be used as the supply for subsequent complementary modulators with little additional noise. Since the signal is represented by the relative levels in two complementary channels, there can be greater tolerance to the additional noise that is introduced. Eventually, the supply beam will become too noisy and/or too weak for further tapping and must be dumped. This means that such a scheme makes much less efficient use of optical energy than other schemes and is only practical in systems where large amounts of single-mode light can be supplied economically, i.e. where wall-plug power is plentiful and power dissipation is the primary limitation on improved system performance.

EO polymer waveguide taps may be cascaded and tiled to provide an active optical tap array. This device permits efficient conversion of many electrical signals to optical data streams through the use of a single, CW laser. Tap nodes are small and compact and make very efficient use of the light. An immediate impact to the system designer is that the IC output drivers need only charge a 100 fF capacitor (plus via and TAB), for any interconnection length and at any frequency up to the link limit. This significantly reduces on-chip line driver power dissipation, and thereby reduces switching noise in the supply current and, consequently, the number of power and ground pins supplying a high-performance IC such as a CMOS microprocessor. An active-tap network can be used with existing chip designs since it requires no more amplification or power than is already present on the chip, but allows the chip designer to scale back output drivers, freeing up more chip real estate and power for logic functions, and reducing delay and noise. Active taps eliminate the problem of how to mount, align, and control hundreds of semiconductor lasers, and utilize instead a few fiber pigtails to CW sources that may be located elsewhere in the system.

Conclusion

Our analysis indicates that, for data rates of 50 MHz and above, EO polymer optical interconnects would require significantly less power than electrical interconnects for either CMOS or ECL devices. Other performance advantages to the system designer include simple fanout, impedance-matched lines at all frequencies, low noise, low power, low propagation delay, high density, and simple layout.

To verify these claims, good stable sources of EO polymers and an acceptable low-loss process are required. We anticipate that such devices will be built and demonstrated in the coming three years.

What about other applications of EO polymers? We feel compelled to comment on the potential for polymers to gain the commonplace utilization of optical semiconductors in optoelectronics.

Planar polymer waveguide technologies have the ultimate potential to gain widespread use in many electronic and fiber-optic system applications. Passive components could find use as splitters, couplers, multiplexers, and parallel array connectors in trunk, local loop, wide-area, and local-area networks. Applications of EO modulators beyond external modulation of lasers include fast network configuration switches, optical network units in Fiber-to-the-Home (FTTH), modulator arrays for data networks, filters, couplers, multiplexers, digital-analog and analog-digital converters, and pulse-shapers. The market potential for planar polymer waveguides is very large due to low wafer processing costs and potential to achieve low-cost single-mode fiber-attach and packaging. This means polymers may compete well with other technologies in conventional optoelectronic applications.

Polymer technologies offer some features that are also available with some (MQW semiconductor) but not other (crystalline) technologies. With polymers, high levels of integration have been demonstrated by using multiple levels of waveguides *(12)* as well as in-plane and out-of-plane mirrors *(13)*. The potential for low-cost manufacturing, packaging, and assembly arises from the demonstrated ability to perform hybrid integration of single-mode components using lithographically-defined registration techniques. Advanced products include processor multichip modules with high-bandwidth interfaces between CPU and second-level cache, optical mesh routers for massively parallel computers, and 8-12 bit, high-speed A-D's.

Cost, reliability, performance, and availability are the main drivers for obtaining and sustaining long-term interest in polymers by systems users. Polymer reliability is seen by customers as a major issue, particularly for EO poled polymers. Reliability needs to be proved with extensive test data of the packaged components, following the well-known standards for telecom and electronic components, in general. It is important to note that laser diodes have achieved success in the market, despite their propensity for drift, low-yields, limited lifetime, and failure. The market has accepted "correction" methods for laser diode performance, such as thermo-electric coolers, drift compensation circuitry, and elaborate packaging because the total cost of a laser transmitter has been reduced to acceptable levels in many cases. Similar techniques could be applied to polymer devices but will increase cost and reduce reliability. On the other hand, absolute stability of the EO coefficient and waveguide properties to thermal excursions and illumination is essential and not correctable with electronics. This may be an important issue for EO polymers.

What about competing technologies? For passive technologies, glass is the main competitor. $LiNbO_3$ and GaAs waveguides, and direct laser modulation provide competition for EO polymers. Underlying all of this is the inertia of electronic systems designers to change their solutions from wires to fiber-based systems: Whenever possible, electronic solutions will be thoroughly examined and selected, if economically feasible and practical. However, high-end communication in all markets is moving toward utilization of the bandwidth offered by optical fiber, and thus the growth of markets for all optoelectronic devices is inevitable.

With further development, electro-optic polymers have the potential to far-outdistance inorganic materials in figures-of-merit, and, in fact, already do in some key properties, such as length-bandwidth products. Polymers are not likely to ever exhibit insertion loss as low as glass for passive devices. However, intrinsic performance of polymers, measured against other materials, is not sufficient for judging the potential of the technology. Overall production costs, balanced against performance and reliability, will determine the utilization of polymer waveguide technologies.

References

1. S.J. Lalama and A.F. Garito, "Origin of the Nonlinear Second-order Optical Susceptibilities of Organic Systems", Phys. Rev. A 20, 1179 (1979).
2. K.D. Singer and A.F. Garito, "Measurements of Molecular Second-order Optical Susceptibilities Using DC Induced Second Harmonic Generation", J. Chem. Phys. 75, 3572 (1981).
3. J.I. Thackara, G.F. Lipscomb, M.A. Stiller, A.J. Ticknor and R. Lytel, "Poled Electro-optic Waveguide Formation in Thin-film Organic Media", Appl. Phys. Lett. 52, 1031 (1988).
4. G. R. Mohlmann, W.H. Horsthuis, C.P. van der Vorst, "Recent Developments in Optically Nonlinear Polymers and Related Electro-Optic Devices," Proc. SPIE 1177, 67 (1989).
5. M.B.J. Diemeer, F.M.M. Suyten, E.S. Trammel, A. McDonach, M.J. Copeland, L.J. Jenneskens and W.H.G. Horsthuis, Electronics Letters 26 (6) 379 (1990).
6. D.G. Girton, S. Kwiatkowski, G.F. Lipscomb, and R. Lytel, "20 GHz Electro-optic Polymer Mach-Zehnder Modulator", Appl. Phys. Lett. 58, 1730 (1991).
7. R. Lytel, G.F. Lipscomb, M. Stiller, J.I. Thackara, and A.J. Ticknor, "Organic Integrated Optical Devices", in Nonlinear Optical Effects in Polymers, J. Messier, F. Kajzar, P. Prasad, and D. Ulrich, eds., NATO ASI Series Vol. 162 (1989), p. 227.
8. C.A. Eldering, A. Knoesen, and S.T. Kowel, "Characterization of Polymeric Electro-optic Films Using Metal Mirror/Electrode Fabry-Perot Etalons", Proc. SPIE 1337, 348 (1990).
9. T.E. Van Eck, A.J. Ticknor, R. Lytel, and G.F. Lipscomb, "A Complementary Optical Tap Fabricated in an Electro-optic Polymer Waveguide", Appl. Phys. Lett. 58, 1558 (1991).
10. J.F. Valley, J.W. Wu, S. Ermer, M. Stiller, E.S. Binkley, J.T. Kenney, G.F. Lipscomb, and R. Lytel, "Thermoplasticity and Parallel-plate Poling of Electro-

optic Polyimide Host Thin Films", Appl. Phys. Lett. 60, 160 (1992); G.R. Mohlmann ed., Proc. SPIE 2025 (1993)
11. R. Lytel, G.F. Lipscomb, E.S. Binkley, J.T. Kenney, and A.J. Ticknor, "Electro-optic Polymers for Optical Interconnects", Proc. SPIE 1215, 252 (1990).
12. T.A. Tumolillo, Jr. and P.R. Ashley, "Multilevel Registered Polymeric Mach-Zehnder Intensity Modulator Array", Appl. Phys. Lett. 62, 3068 (1993).
13. B.L. Booth, "Optical Interconnection Polymers", in *Polymers for Lightwave and Integrated Optics"*, L.A. Hornak ed. (Marcel Dekker, New York), 1992, pp. 231-266.

RECEIVED April 24, 1995

Chapter 31

Polymer Electrooptic Waveguide Fabrication

M. Ziari[1], A. Chen[1], S. Kalluri[1], W. H. Steier[1], Y. Shi[2], W. Wang[3], D. Chen[3], and H. R. Fetterman[3]

[1]Department of Electrical Engineering, University of Southern California, Los Angeles, CA 90089–0483
[2]Tacan Corporation, 2330 Faraday Avenue, Carlsbad, CA 92008
[3]Department of Electrical Engineering, University of California, Los Angeles, CA 90024

Various methods for the fabrication of etched waveguide structures, environmentally stable electrooptic modulators and the integration of polymer waveguides with silicon circuitry is discussed. Dry etching of the active polymer layer by reactive ion etching was used to fabricate millimeter-wave modulators (dc-60 GHz) with long term stability. Further improvement in etch quality was achieved by employing the novel etch method of electron cyclotron resonance etching. Our polymer modulator was tested in a community access television optical-fiber link demonstration and general optical power handling requirements of such devices are outlined. We also discuss the fabrication of polymer waveguides on planarized silicon VLSI circuits. Multiple fiber attachment using v-grooves etched in Si/SiO_2 substrate is demonstrated.

Photonic applications of second-order optical nonlinear polymers require a well-developed and cost effective device fabrication process. The material advances made over the past few years that addressed problems with environmental and thermal stability of these active polymers have made this technology more attractive for commercial applications. Polymeric integrated optical devices, however, must still overcome the skepticism that exists regarding their performance and be able to compete favorably with mature technologies based on $LiNbO_3$ and III-V semiconductor materials. In this article, we will discuss various fabrications processes and demonstrate devices that are long-term environmentally stable, operate at very high frequencies (40 GHz) and perform well in an experimental community access television (CATV) optical link. In addition, we will discuss and demonstrate integration and packaging concepts that can eventually place polymeric devices one step ahead of the competing technologies in both functionality and production cost. This article is organized in the following manner. First we discuss general waveguide and modulator design and fabrication methods. We have extensively used dry etching processes and devote a section to present the details of reactive ion etching and electron cyclotron resonance etching. The next section provides the details of electrooptic device fabrication process, testing, and CATV applications.

0097–6156/95/0601–0420$12.00/0

The next section will discuss the subject of integration of polymer devices with electronic circuits. Finally, we present our efforts on device packaging and multiple fiber attachment.

Polymer Waveguides

A polymeric optical waveguide structure often consists of multiple polymer layers. A proper choice of the upper cladding, core, and lower cladding materials based on their thicknesses and refractive indices can provide single mode confinement in the vertical direction. In addition, the processing compatibility of these films and their linear and nonlinear optical properties such as loss and electrooptic, dielectric, electrical, and mechanical properties are all very important. Fabrication parameters like ease in spinning, poling, cutting and polishing also influence the choice of materials.

The availability of good cladding layer materials is critical for overall device performance and is the subject of current research. The cladding index of refraction affects the mode confinement and mode matching to optical fibers. For vertical poling through the cladding layer, the electrical resistivity of the cladding material must be less than that of the electrooptic layer to assure efficient poling.(*1*) As an alternative to polymer films, thermally grown SiO_2 on Si substrate can also be used for the lower cladding. The high optical quality of SiO_2 reduces scattering and absorption losses and provides a smooth surface for spinning the guiding layer. The waveguide must be designed however to prevent losses due to the high index Si substrate. Calculations show that an SiO_2 layer thickness of 3 to 4 μm is adequate to keep this loss below 0.1 dB/cm.(*2*). It is important, however, to note that the very high resistivity and the high breakdown voltage of SiO_2 lower claddings makes such structures impractical for corona discharge poling.

The lateral confinement of guided modes requires additional processing steps whose effectiveness, cost, and performance are the subject of many research efforts. Photobleaching(*3*), and physical etching or laser ablation have been the primary techniques used for this purpose. Photobleaching is a simple, low cost and highly accurate technique for patterning the refractive indices. The accuracy of UV-induced changes in the refractive indices is advantageous since it allows precise control of the guided mode profile which can improve fiber coupling efficiency. This method relies on photochemical processes such as isomerization, realignment of the optically active moieties, or material decomposition. We have used photobleaching with UV light for the fabrication of birefringent cladding waveguides and directional couplers(*4*). Although we have not observed any sign of degradation in these devices after many months in a dark environment, their long term stability under operation conditions requires further investigation.

Etched structures, on the other hand, are expected to offer greater stability because of their physically structured nature. Lateral confinement in these structures can be provided by one of several methods that involve the etching of either the lower cladding or the active layer. One approach that we have taken is to etch a trench in the polyurethane or SiO_2 lower cladding (see Fig. 1). In this case, the active layer is spun after the etching process and an inverted ridge structure is formed. Many of the process compatibility issues are relaxed because lateral confinement does not require any further processing. The other approach relies on forming a ridge structure on the active guiding layer by dry etching and requires compatible lithographic patterning. We have employed this technique in the fabrication of electrooptic modulators. The dry etching is performed by reactive ion etching (RIE) or electron cyclotron resonance (ECR) etching. The next section provides a more detailed description of the various etching methods.

One important requirement for successful polymer waveguide device fabrication is to have a reliable design tool that is consistent with the capabilities and tolerances of the fabrication process. The calculation of waveguide parameters such as the layer thicknesses, waveguide width, etch depth or photobleaching exposure dose, and electrode characteristics can be obtained using analytical methods for simple structures but require detailed computer simulations for more complex multilayer structures. We rely on the effective index method to simulate waveguide structures and calculate the propagation constant, the mode profile and, possibly the number of modes. One important parameter obtained from these calculations is the optimum etch depth for single mode operation. We have previously calculated the cut-off etch depth for multi-mode waveguides based on our active and cladding layer indices *(4)*. Recent experimental results and detailed simulations using the effective-index method *(5)* show that modal characteristics such as the mode shape and the number of modes can be effectively predicted and controlled. Figure 2 shows the out-coupled 1.3 mm pattern from a 4 μm wide and 1 μm thick PU-DR19 waveguide with a 4 μm lower and a 9 μm upper polyurethane cladding layers (n=1.55). The buried ridge height (etch depth) is 0.5 μm. The patterns show that two horizontal modes and one vertical mode can be excited in this structure which is in very good agreement with the predictions of the simulation. An etch depth of ~ 0.2 μm is requires for single mode operation of a 5 μm width ridge waveguides. This shallow etch depth is attainable with RIE and ECR dry etching methods.

Very high frequency (multi-GHz) electrooptic modulators are traveling wave devices both for the optical and the modulation frequencies and must be designed such that the whole structure including the polymer optical waveguides, the millimeter wave electrode structure, and the millimeter wave connectors meet both optical and millimeter wave requirements. These can often be conflicting requirements. For example the width and spacing of the metallic layers to achieve a required millimeter wave impedance may not be compatible with a low loss optical waveguide since the electrodes can introduce additional metallic losses for the optical waves. Fiber coupling of the optical beam and coaxial or metallic waveguide coupling of the millimeter wave energy all impose additional constraints on the device structure. It is important that all of these design issues be addressed from the beginning and the modulator be considered as a combination of an optical and a millimeter wave device. The complexity of the design is many-fold and the final structure is often a realistic compromise between optical and electrical requirements and the material and processing capabilities.

Etching

The large demand for semiconductor devices over the past decades has resulted in the development of various dry and wet etching technologies that can now be employed for the fabrication of optical polymeric devices. Dry plasma etching of organic polymers, polyimides, and photoresist layers is routinely performed in an oxygen plasma environment. We have used two forms of ion-assisted plasma etching methods, RIE and ECR, for patterning polymer films. The important issues here are again, the compatibility of the photolithography and the etch mask with the polymer, etch selectivity, the etch anisotropy required to achieve vertical etched walls, etch rate, and wall smoothness. The isotropic and purely chemical etch in a plasma discharge is the most common form of plasma etching *(6-8)*. Because of its isotropic nature, this method has limited use in polymer waveguide fabrication. It is rather the ion-assisted plasma etching methods such as RIE or ECR that are anisotropic and can result in vertical wall structures *(7, 8)*.

Reactive Ion Etching (RIE). This method is now widely used for creating polymer waveguide structures(*4, 9*). A radio frequency (RF) source (30 to 200 W) creates the plasma and the necessary bias to achieve vertical etching. The exact RIE etch rate is pressure, RF power, material and machine dependent.(*10*) We typically etch at a rate of 2200 ± 250 Å/min at 45 W power and 0.15 Torr pressure for PU-DR19 thermoset polymer. Higher etch rates result in wall roughness and depth nonuniformities. We also used RIE for etching thermally grown SiO_2 with an etch rate of 550 Å/min at 150 W and 0.1 Torr pressure.

In RIE, RF power is used to excite the plasma. The sample is placed on the electrode that is connected to the RF source and thus is unavoidably biased relative to the plasma(*6, 8*). The excitation of the plasma and the necessary bias are both created by the RF source and therefore it is not possible to generate the necessary high plasma density independent of the bias condition. This limits the control of the etch rate and quality of the sidewall structure.

Electron Cyclotron Resonance (ECR). In this novel method, a magnetic field is applied to the reaction chamber. A microwave source at approximately 2.54 GHz is then tuned to the electron cyclotron resonance frequency which creates a very high plasma density (*11*). This electrodeless mechanism of exciting the plasma relieves the biasing problems associated with the RIE method. Nevertheless, an optional RF power and bias is often used to independently energize the ions and achieve anisotropic etch. The etch rate, etch anisotropy, and smoothness can be controlled by adjusting the flow rate and pressure of gases, the microwave power, the RF power, and the bias. Very high selectivity and a broad range of etch characteristics are achieved with this technique. The smoothness of the vertical surfaces is fairly good and high quality mirrors for semiconductor lasers have been obtained by ECR etching of the facets.

We have etched polymer films with this novel method. We used a PlasmaQuest model 375 ECR machine to etch PU-DR19 polymer under a gas mixture of oxygen and argon. Our initial results indicate an etching rate of roughly 2000 Å/min for a microwave power of 400 W, an RF power of 153 W and –200 V bias, a cyclotron magnet current of 180 Amps at 18 ° C temperature, and O_2 and Ar flow rates of 17.6 and 49 sccm respectively. Further improvement in the etch quality, i.e. rate and smoothness, can be achieved by adjusting the above etching parameters.

Figure 3 shows a comparison of an RIE (top) and an ECR (bottom) etched PU-DR19 samples. The photographs, taken by a scanning electron microscope (SEM), demonstrate the improved smoothness of ECR-etched structures. The picture also illustrates the excellent vertical profile of the wall for the case of ECR etched sample. Note the ECR sample is still partially covered with patterned AZ 5214-E photoresist. The depth of the photoresist etch suggests that PU-DR19 and this resist have similar etch rates. A decrease in the etch rate of photoresist relative to the nonlinear polymer should be possible under different operating conditions.

Etching of SiO_2. Patterning of the SiO_2 lower cladding layer can be performed with dry etching methods such as RIE or with inorganic wet etchants such as buffered hydrofluoric acid (HF). The wet etching of SiO_2 is isotropic and causes a considerable undercut. However, it is still a useful method to define trench (or groove) patterns in the lower cladding because the required etch depths for single mode operation are typically only about 0.2 to 0.4 µm. The etch rate of buffered HF is typically 500 Å/min. We have also used the RIE and ECR instruments with a CF_4 gas source for etching SiO_2 and have achieved smooth vertical walls. The RIE etch rate is 500 Å/min. at 0.15 Torr pressure and 150 W of RF power. Higher pressures increase the etch rate but may also cause a problematic redeposition of SiO_2 on the sample. The availability of multiple gas sources and the improved control of ECR

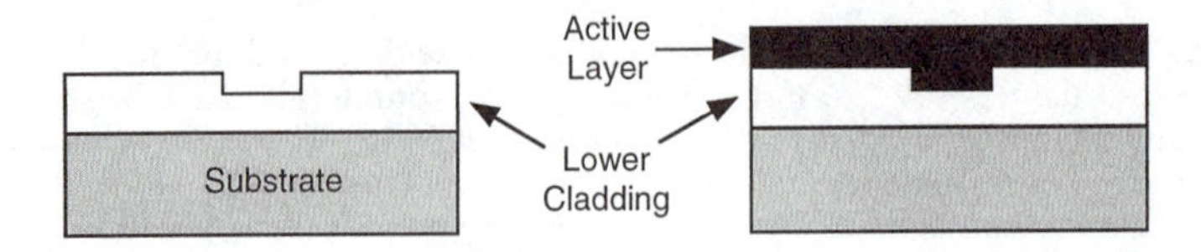

Figure 1. Schematic drawing of an etched trench in lower cladding (left) and an inverted ridge structure (right).

Figure 2. Out-coupled light from 4 μm wide waveguide showing the excitation of two horizontal modes. The 1.3 μm wavelength light was detected using an infrared camera.

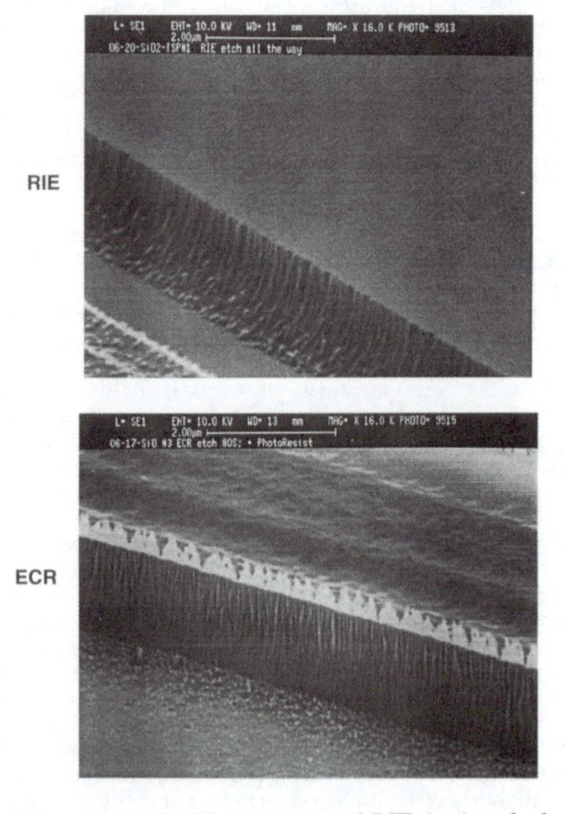

Figure 3. Comparison of ECR (bottom) and RIE (top) etched structures. (Scale is shown on the top).

also helps to accurately and smoothly etch the SiO_2 cladding layer. ECR etching can be considerably faster (~ 2000 Å/min) and more selective.

Electrooptic Waveguide modulator.

The fabrication and applications of electrooptic modulators are discussed in this section. It will also serve to present the materials currently being used and discuss device related issues in the fabrication process.

Material. We used the thermosetting polyurethane Disperse Red-19 (PU-DR19) for the active layer. This material is not a guest-host system and the nonlinear chromophore is covalently attached to the polymer backbone with a loading density of 32%. The nonlinear properties, poling characteristics and the thermal stability of this crosslinked polymer system have been reported earlier(*12*). In summary, the absorption resonance peak is at 470 nm and the indices of refraction at 850, 1.06 and 1.32 µm are 1.685, 1.673 and 1.667 respectively. This material has shown long term stability in its electrooptic coefficient for over two years at room temperature, longer than 3000 hours at 90° C with only 27% drop in nonlinearity, and for short term at 100° C (*4*). After curing the thermoset films are chemically resistant to the solvents used in the lithography process and subsequent film spinning. Using second harmonic generation at a fundamental wavelength of 1.064 µm and electrooptic measurements of r_{13} coefficient at 633 nm and 800 nm, we have extrapolated an r_{33} value of 12 pm/V at 1.064 nm and 10-11 pm/V at 1.3 µm wavelengths. Waveguide measurements at these wavelengths have now confirmed the r_{33}= 12 pm/V value at 1.064 µm wavelength. Commercially available polyurethane (Epoxilite 9651-1) with an index of 1.55 was used for the 9 µm thick upper and 4 µm thick lower cladding.

Fabrication. Electrooptic phase/polarization modulators were fabricated on gold coated quartz substrates. Figure 4 outlines the fabrication process. The Epoxilite cladding material was cured at 60° C for one hour or at room temperature for 24 hours. The active layer of PUR-DR19 was spin cast and left overnight at 60 °C vacuum for drying and precuring. Thermal setting and corona poling was done at 140 °C (Fig 4a). After spin coating with a photoresist layer (Fig 4b), and lithographic patterning (Fig 4c), the cured active layer was etched by RIE to form a ridge structure (Fig 4d). The lithographic patterning required a low dosage of UV illumination and two oven bakes of half hour duration at 90° C. Our waveguide pattern consisted of several sets of 2, 3, 4, 5, 8 and 10 µm wide stripes. After the application of the top cladding (Fig 4e), a gold microstrip line was patterned and gold-platted to a total thickness of 5 µm and a width of 36 µm (Fig 4f). The end surface preparation of many polymer waveguides can often be achieved by carefully cutting them with a dicing saw. Our triple-stack polymer structure required additional polishing which we believe is due to the mechanical properties of our cladding material. This particular cladding material stretches and tears during cleavage which is an indication of its rubber-like properties. This material was chosen mainly because it is easy to pole through and it is solvent and cure temperature compatible with our active layer.

The electrode dimensions and the total polymer layer thickness (15 µm) were chosen to achieve a 50 Ω impedance at microwave frequencies. This 50 Ω microstrip line electrode was used for high speed traveling wave operation and two Wiltron K connectors were utilized to launch and terminate the mm-wave signal through tapered contact pads. The connector to connector dc resistance of the modulator was 3 Ω. Our time domain microwave reflection measurements on the modulator structure suggest an effective mm-wave dielectric constant of 1.50 and a bandwidth-length product limitation of greater than 100 GHz.cm. A coplanar circuit structure

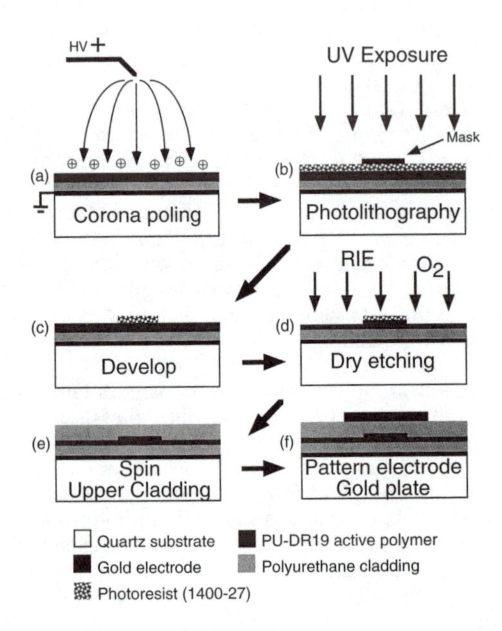

Figure 4. Fabrication steps of the inverted ridge electrooptic modulator.

that can benefit from this very high bandwidth is now under investigation for frequencies over 60 GHz.

Measurement. The electrooptic waveguide was used both as a phase or polarization (birefringence) modulator.(*13, 14*) We used a single frequency Nd:YAG laser with an input power level of 15 mW at 1.064 μm wavelength for dc to 18 GHz measurements(*13*). Measurements at higher frequencies utilized a Nd:YAG laser at 1.32 μm wavelength with an input power of 30 mW.

Full details of the 1.064 μm measurements are given in Reference 13. In summary, the half-wave voltage, V_{π}, of the amplitude modulator was determined to be 35 V at 1.064 μm, yielding an electrooptic coefficient r_{33} of 12 pm/V which is consistent with the extrapolations made with a two-level model from second harmonic generation measurements. The half-wave voltage is less than 24 V when the device is configured as a phase modulator. The waveguide loss was estimated to be ~2dB/cm from scattering measurements. This loss is mainly attributed to the surface roughness of the lower cladding layer due to fast drying during spin cast. We believe the etched waveguide walls also contribute to this scattering loss and that this loss can be reduced when the ECR etching process is employed.

For measurements under 20 GHz, we used an amplified sweep oscillator(*13*) with a driving microwave power of 24 dBm. The light output of the modulator was coupled to a fast photo detector via a single mode fiber. The detected signal was amplified and monitored by an electronic spectrum analyzer with 20 GHz bandwidth. Figure 5a shows the measured optical response. For higher frequencies, a combination of an HP8350B sweep oscillator and a Hughes Ka band (26.5-40 GHz) traveling wave tube (TWT) was used as the source(*14*). These measurements show a 3 dB bandwidth of 26 GHz for this device(*14*), however, optical modulation at frequencies up to 60 GHz has also been observed in this device. Due to the lack of a direct observation technique at these high frequencies, the modulation signal was down converted electronically using a microwave mixer and a 20 GHz local oscillator. Figure 5b shows the modulated signal at 37 GHz. Note that this spectrum analyzer plot shows an electronically down-converted signal (by 20 GHz) with a 12 dB loss at the mixer.

The bandwidth and sensitivity of the detection system was further improved by employing optical heterodyne detection and optically down-converting the mm-wave phase modulation. Using this method, we observed and characterized mm-wave modulation at frequencies up to 60 GHz. The details of the measurement system and 60 GHz results will be reported in future publication. In addition to our current efforts that utilize microstrip line electrodes, we are also evaluating coplanar electrode structures and better microwave coupling schemes for further extending the modulation bandwidth.

Our modulator device has so far operated for longer than 2 years under laboratory ambient conditions (room light, room temperature) and with measurements taken periodically at the above stated electrical and optical input power levels without any degradation in responsitivity. This is a strong confirmation of the stability of this polymer electrooptic modulator. Systematic long term measurements at higher input power levels are in progress.

Applications. Optical modulators have many potential applications in fiber telecommunication and optical interconnection systems (*15*). In addition to these, another possible application with a vast emerging market is the use of a polymer electrooptic modulator in the transmitter for the fiber-optic community access television (CATV) networks. Traditional fiber-optic CATV transmitters employ laser diodes for direct modulation or $LiNbO_3$ waveguide modulators for external modulation. To demonstrate the applications of polymer modulators in CATV

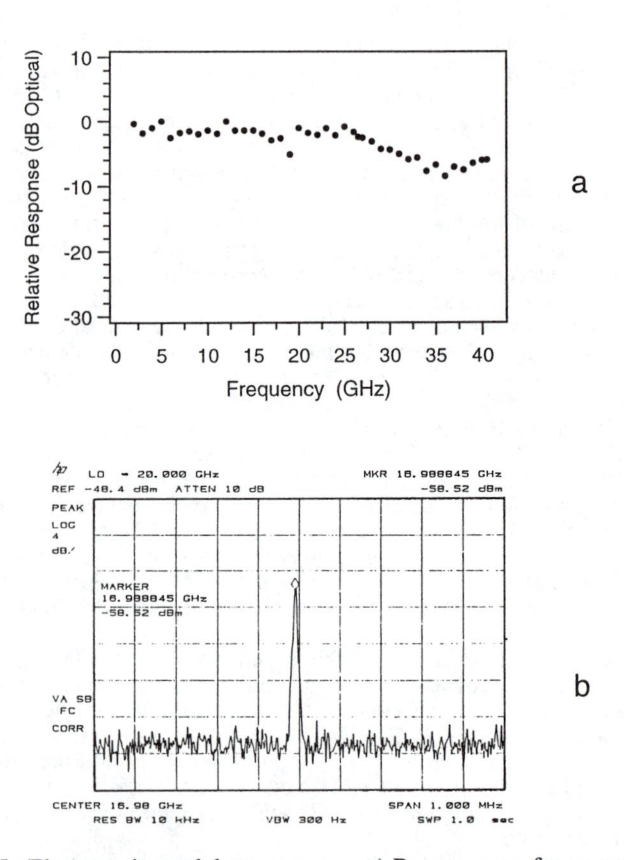

Figure 5. Electrooptic modulator response. a) Response vs. frequency from dc to 40 GHz; and b) electronic spectrum analyzer scan showing the detected 37 GHz optical modulation signal after electronic down conversion by 20 GHz.

transmission systems, we constructed an analog AM video link. The experimental setup consisted of a TV carrier generator, a video camera and a polymer modulator fabricated with the above process as the signal transmitter. The modulated optical signal was coupled into an optical fiber and received by a commercial CATV receiver. The detected signal was converted to an RF signal, and was monitored by a spectrum analyzer or a TV monitor. The preliminary results showed good signal transmission quality when adequate optical power reached the photo detector. A carrier-to-noise ratio of ~ 48 dB was measured in the system. Although only a single channel and large modulation depths were used in the preliminary tests, we obtained valuable information regarding the requirements of polymeric modulators for practical analog AM fiber-optic CATV transmission systems. The most prominent requirements are a low half-wave voltage, optical power handling capability and long term stability.

In order to obtain high quality signal transmission and to be competitive in RF power consumption with traditional laser direct modulation or $LiNbO_3$ systems, it was summarized *(15)* that V_π should be $\leq$ 5 V with an output optical power of $\geq$ 10 mW per fiber. The required V_π is based on ~4% per channel modulation depth and the RF circuit output level of 37-40 dBmV/channel used in direct current modulated laser systems or $LiNbO_3$ modulators. For a 1.3 μm wavelength fiber-optic link with a link loss of 10-12 dB and the receiving power of 0.6 mW (-2dBm) at the detector for sufficient carrier to noise ratio, the required optical power will be 7-10 mW per fiber link. For a fan-out of 4, the total modulator output power will be near 40 mW. The electrooptic polymer modulator used in commercial transmitters should have a life time of at least 5 years at an office-like environment.

The CATV optical power requirement is a challenging issue for the optical power handling capability and optical losses of active polymer waveguides. This requires both coupling of an intense laser into the modulator and a reduction in the device loss. We estimate a maximum power level of $\geq$ 100 mW for an output of 40 mW in a 3 cm long device with 1 dB/cm linear loss and 1 dB output coupling loss. The maximum power density in the waveguide can reach 100 kW/cm^2 to 1 MW/cm^2 for a waveguide cross section of 10 to 25 μm^2. We are currently studying the stability of nonlinear optical properties and possible optical damage at these high power densities in electrooptic modulators made of different polymer materials.

Integration of Polymer Waveguides with Silicon Circuitry

The capability to spin cast polymers onto any flat substrate is a major advantage over the competing $LiNbO_3$ and III-V semiconductor material systems. This makes possible the integration of polymer modulators with integrated Si drive and control electronics on the same Si substrate. This integration offers the possibility of increased functionality and possibly reduced cost compared to hybrid systems in which the modulator and electronics are separate interconnected packages. While it may be possible to flip chip bond $LiNbO_3$ or III-V based modulators onto Si substrates, this still remains a costly and difficult technology. On the other hand, the integration of polymers with Si seems relatively straightforward. This is a unique feature of polymeric systems and should be exploited for device development.

One approach to integrating EO polymer devices on VLSI circuits is to fabricate the active polymer device on top of a planarized VLSI Si circuit that provides the signal and control electronics. Planarization smoothes out the step heights inherent in the finished Si circuit and provides an optically flat substrate for the fabrication of polymer waveguides.

As a first step in this integration approach, we have fabricated passive slab polymer waveguides on planarized VLSI circuits and compared their optical loss to identical waveguides fabricated on high optical quality thermally grown SiO_2 on Si substrates. A planarizing polymer (PC3-6000, Futurex) was spun on the nonplanar VLSI circuit to a final thickness of approximately 4 to 6 µm. The substrate was then placed on a 200°C hot plate for 2 minutes. At this temperature small variations in height are smoothed out as the polymer reflows from the peaks into the troughs. Figure 6 is an SEM photograph of the cleaved edge of Si circuit and shows the planarized top surface on a non-planar Si circuit. The features on Si circuit are 1 to 2 µm high. There were also 6 µm deep by 4 µm wide vias on the surface that this process successfully planarized. For subsequent waveguide fabrication and processing, this polymer layer must be made chemically resistant to organic solvents. After additional curing in a 200°C vacuum oven for 30 minutes the polymer planarizing layer is resistant to solvents such as dioxane, tetrahydrafuran, acetone, etc. However, at this stage it is also difficult to wet the surface with either water or solvents which makes spinning of subsequent layers difficult. A very short and low power RIE etch was used to roughen the surface slightly for further spinning. A polyurethane (PU, Epoxilite) lower cladding of 5-6 µm thickness was spun on and cured. A photoresist polymer (AZ 5214E) was used as the guiding layer.

Prism coupling was used to couple a 1.06µm Nd:YAG laser in and out of the slab waveguide for waveguide loss measurements using the streak method. In this measurement technique, an IR camera is focused on the light scattered along the waveguide and line scans along the streak are fed into an oscilloscope. The measured waveguide loss is less than the minimum loss, 0.5 dB/cm, that can be resolved using our setup. Fig 7 shows a photograph of the waveguide and the underlying circuit with out coupled light on the right hand side. We also fabricated identical PU-phoresist slab waveguides on high optical quality polished SiO_2/Si substrates and the loss results of both kinds of waveguides are comparably low. However, for the planarized substrates, the loss for the waveguide over the VLSI circuit's bond pads is somewhat higher due to the extreme step heights. Thicker planarizing layers should solve this problem.

Planarized "vertical" integration is one approach to active polymer/VLSI device integration. The attractive feature of this approach is the conservation of chip area intrinsic in the technique since the VLSI circuit and the EO devices occupy different horizontal planes. Another approach is "side-by side" approach. In this approach we visualize dedicated circuit free areas on the Si wafer for polymer device fabrication. Though some design complexity is reduced it is at the expense of valuable real-estate. Moreover, aspects of the more important issues like circuit protection, etc. are similar for both approaches. A third approach we visualize is to integrate polymer devices and VLSI circuits on opposite sides of the same wafer and perform the interconnections through the wafer. The feedthrough interconnection technique itself is somewhat novel and therefore not a part of normal Si processing. The potential relative merits of this technique is therefore left for a future discussion.

With these encouraging results we have started the integration of phase and intensity (Mach-Zender) modulators on planarized VLSI circuits. Issues such as circuit protection from high static poling fields, circuit access through vias or bond pads, circuit and polymer device testing and packaging are all vital aspects that we are now investigating.

Fiber Attachment

Packaging and fiber attachment is a major obstacle in cost-effective fabrication of modulators. The sub-micron tolerances required for the positioning of the fiber prior to bonding is time consuming, labor intensive, and costly. The attachment of multiple fiber cables or ribbons, required for computer optical interconnect links, is more complex and demands much tighter positioning tolerances(*16*). Fiber attachment and packaging probably remains the costliest part of waveguide modulator fabrication for both polymer and competing technologies.

In an effort to develop reliable and cost effective solutions to the fiber attachment problem, we have investigated the use of self-aligning v-grooves. V-groove structures, such as the one schematically shown in Fig. 8, can be prepared using anisotropic etching of Si(*16*). Anisotropic etching is a very interesting property of semiconductors and has been used in micro machining and micro fabrication(*17*). For example, the etch rate of (100) oriented Si wafers in potassium hydroxide (KOH) solution along the <110> orientation is a few hundred times faster than the etch rate along the <111> orientation(*17, 18*). For a stripe opening oriented along the (110) direction on the etching mask, this etching results in a v-groove pattern. V-grooves can provide an ideal positioning platform for the fiber. In this scheme, the vertical height of the fiber core relative to any horizontal surface, depicted by t in Figure 8 , depends on the opening width of the V-groove and can easily be adjusted for any fiber cladding diameter(*18*). In the horizontal direction, the core position is always centered relative to the groove. Therefore a proper v-groove can be patterned relative to a waveguide structure such that the fiber and the waveguide modes have an optimal overlap. As a result, the mask design, lithography and etch process will provide a self-aligned platform for *all* the fibers and the tedious and costly alignment procedure need not repeat for every single fiber.

Though this technology has previously been considered for $LiNbO_3$ and GaAs devices (*15, 18*), its commercial use has been hindered by the fact that the self-aligned v-grooves and the modulators are on separate and different substrates, and the packaging involves the alignment and the attachment of the two substrates. Reliable alignment and attachment of devices and substrates from different material systems is itself a technological challenge and the costs of such hybrid devices are therefore high. Our approach is fundamentally different. We have etched the v-grooves on the same Si substrate that supports the waveguide devices. This approach is unique to polymer devices and assures a less complex self-alignment that is more economical to achieve. The other novelty of our approach is that we form the waveguide by spin casting *after* placing the fibers in the v-groove. This approach results in a direct bond between the fiber and the waveguide and avoids reflections from air gaps. As we will show later, the spinning forms a tapered coupling region that can reduce the coupling loss.

We used Si substrates with a 4 µm thick thermally grown SiO_2. The SiO_2 layer functioned as the lower cladding for waveguides and as the etch mask for the v-grooves. First, we used standard photolithography to pattern the SiO_2 and etched away the SiO_2 layer for the v-groove openings using RIE and wet etching. The v-groove opening is chosen such that the fiber core is properly placed slightly (~ 0.5-1.0 µm) above the 4 µm thick SiO_2 lower cladding layer. The v-grooves were separated by 250 µm which is the standard separation of fiber ribbons. We used KOH or other hydroxide based etchants for anisotropic etching of Si. One important etching artifact is the undercut of Si under the SiO_2 layer due to the limited anisotropy of the etchant. This undercut should be controlled or minimized for accurate positioning of the fiber. The etching process proceeds extremely slowly after the formation of the v-groove and timing is not critical.

After the formation of the v-grooves, the substrate is coated with photoresist again and the waveguide structures are accurately patterned on the SiO_2 using a matched set of waveguide stripes and alignment markers. The stripe patterns were etched on SiO_2 to form trenches that would laterally confine the guided modes in the electrooptic guiding layer. This step completed the Si substrate preparation. This prepared waveguide and fiber alignment v-groove can be mass produced by standard semiconductor processing facilities.

The next task is the positioning of the fiber in the v-groove. The 80 µm cladding diameter cleaved fibers is manually placed in the v-grooves and attached to the Si wafer using a UV curable epoxy. Figure 9 shows the finished product after spin coating the guiding layer and shows the fibers in the v-grooves. Although we

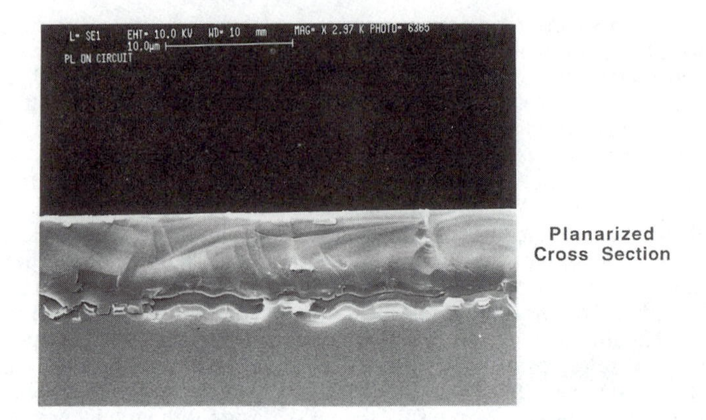

Figure 6. Cross-section scanning electron microscope view of planarized circuit. (Scale is shown on the top).

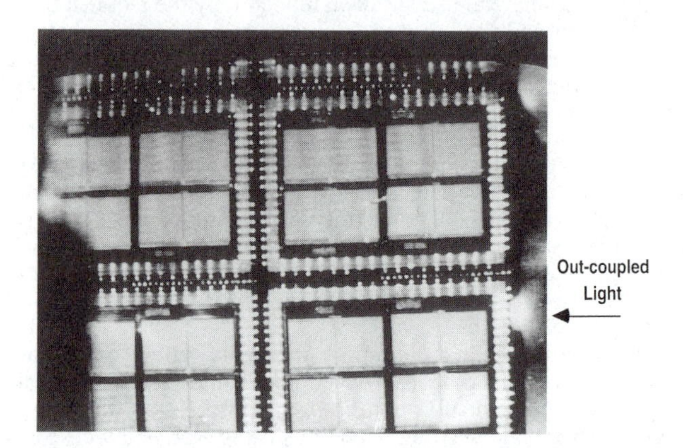

Figure 7. Infrared camera photograph of the slab waveguide on planarized VLSI silicon circuit. The light at 1.064 µm wavelength is prism coupled on the left side and out-coupled on the right (the bright spot). The very low scattered light from the guided mode, barely observable in this picture, indicates low scattering losses.

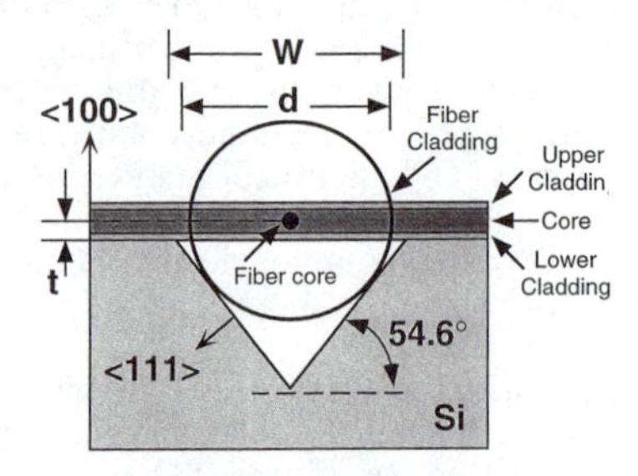

Figure 8. Schematic drawing of an optical fiber in an etched Si v-groove showing the proper orientation of Si substrate, the etch angle, the fiber position, and the waveguide cladding and core layers.

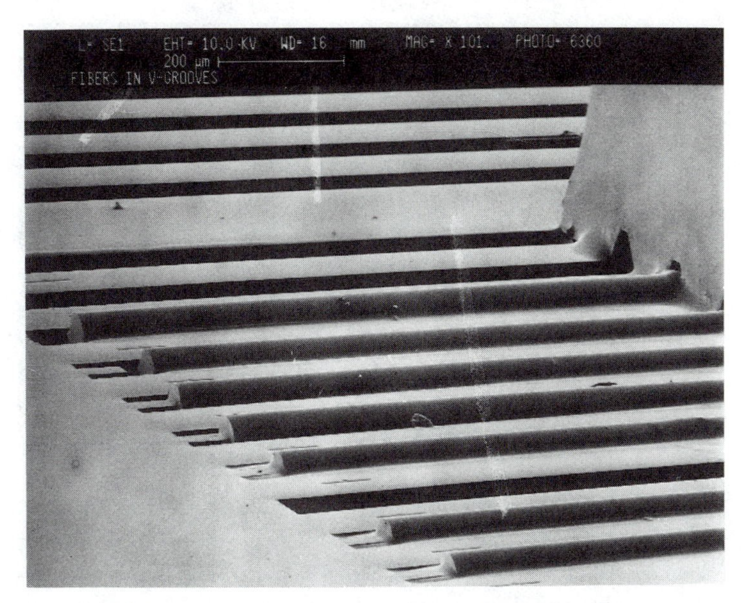

Figure 9. SEM photograph of multiple fibers aligned in silicon v-grooves after spinning of polymer waveguide.

find the adhesion property of UV curable epoxy adequate for most cases, we have also used a Si chip with matching v-grooves to hold the fibers in the grooves (see Figure 9). The next step is spin casting the high index core layer. We performed this task using a specially designed spin chuck that supports the fiber bundle which extends from the device. We examined the stability of the above structure during spinning at speeds up to 5000 rpm.

Lower V_π is beneficial for most applications and can be achieved by reducing the electrode spacing. However, this also results in a waveguide mode that is much smaller than the fiber mode which is 5 to 10 µm in diameter. One approach to reduce the mismatch is the use of tapered ends on the guide such that the guided mode diameter at the in coupling and out coupling points are larger than that in the rest of the guiding structure. The implementation of tapered ends in the horizontal direction is relatively simple and only requires a tapered stripe in the lithography mask. Vertical tapers for optical devices are, however, more difficult to fabricate(*19*). Figure 10 shows the tapered end of the polymer waveguide at the fiber contact point. The fiber core is at the lower right corner. This tapered structure, which is an artifact of spinning, is useful to match the often larger mode of the fiber to the smaller mode of the waveguide in the vertical direction. A complete characterization of the coupling losses and the effect of taper is under investigation. The study so far verifies the feasibility of our approach, namely patterning the waveguide and v-grooves on the same substrate; and spin coating the polymer waveguide layer with the fiber in place. Implementations of different poling methods and active devices on these patterned substrates are also under investigation.

Conclusion

The fabrication of polymer active devices is a complex task and involves many design and processing steps. We have addressed many of the fabrication issues and

Figure 10. SEM side view showing the tapered end of polymer waveguide at the fiber interface. The fiber core is at the lower right corner. (Scale is shown on the top).

have demonstrated an environmentally stable mm-wave modulator. The demands of the market and recent advances in material nonlinear properties have placed a greater emphasis on the device fabrication and design issues such as microwave design, fiber attachment, and packaging. A successful design and fabrication process should address all these often conflicting issues and provide a stable device with an acceptable performance and cost.

Acknowledgments

The authors would like to recognize the support of the funding agencies who made this work possible. They include AFOSR, ARPA (The National Center for Integrated Photonic Technology), and the Joint Services Electronics Program. The authors also acknowledge the support of Prof. L. R. Dalton of the Chemistry Department at USC.

References

1. Girton, D. G., et al., Polymer Preprints, **1994,** *35,* 219.
2. Beeson, K. W.; Horn, K. A.; McFarland, M.; Yardley, J. T., Appl. Phys. Lett., **1991,** *58,* 1955.
3. Steier, W. H., et al., Proc. SPIE, **1993,** *2025,* 535.
4. Surette, M. R.; Mikelson, A. R. *Slab and channel waveguide software* (Guided Wave Optics Laboratory, Dept. of Electrical Engineering, University of Colorado, Boulder, CO, 1990).
5. Melliar-Smith, C. M.; Mogab, C. J., *Plasma Assisted Etching Techniques for Pattern Delineation*, in Thin Film Processes, edited by J. L. Vossen and W. Kern, Academic Press: New York, NY, 1978, pp. 497.

6. d'Agostino, R., *Plasma depostion, treatment, and etching of polymers*, in Plasma material interactions, edited by O. Auciello and D. L. Flamm, Academic Press: San Diego, 1990
7. Wolf, S.; Tauber, R. N., *Silicon Procesing for the VLSI Era, Volume 1: Process Technology*, Lattice Press: Sunset Beach, CA, 1986.
8. Sullivan, C. T., SPIE Proc., **1988,** *994*, 92.
9. Hartney, M. A.; Greene, W. M.; Soane, D. S.; Hess, D. W., J. Vac. Sci. Technol. B, **1988,** *6*, 1892.
10. Asmussen, J., J. Vac. Sci. Technol. A., **1989,** *7*, 883.
11. Shi, Y., et al., Appl. Phys. Lett., **1992,** *60*, 2577.
12. Wang, W., et al., Appl. Phys. Lett., **1994,** *65*, 929.
13. Wang, W., et al., IEEE Photon. Techol. Lett., **1995,**
14. Shi, Y., et al., Polymer Preprints, **1994,** *35*, 223.
15. Murphy, E. J., J. Lightwave Technol., **1988,** *6*, 862.
16. Bean, K. E., IEEE Trans. Electron Device, **1978,** *ED-25*, 1185.
17. Petersen, K. E., Proc. IEEE, **1982,** *70*, 420.
18. Kaufmann, H., et al., Electron. Lett., **1986,** *22*, 642.
19. Koch, T. L., et al., IEEE Photon. Technol. Lett., **1990,** *2*, 88.

RECEIVED February 23, 1995

Chapter 32

Advances in Organic Polymer-Based Optoelectronics

R. Levenson, J. Liang, C. Rossier, R. Hierle, E. Toussaere, N. Bouadma, and J. Zyss

France Telecom, Centre National d'Etudes des Télécommunications, Paris B, BP 107, 196 Avenue Henri Ravera, 92225 Bagneux, France

Design and fabrication of low loss polymer waveguides with an attenuation of the order of 1 to 2 dB/cm are discussed. Passive functions such as optical combiners or splitters have been achieved with a polystyrene core waveguide. Electrooptic devices such as phase modulators or Mach Zehnder interferometers have been fabricated and demonstrated respectively with MAGLY, a new variety of crosslinked polymers and the classical methyl methacrylate-Disperse Red One (DR1-MMA) side chain polymer. The linear and nonlinear properties of MAGLY are characterized in comparison with DR1-MMA. Polymer films are shown to sustain, after combined curing and poling processes a high electrooptic coefficient of 12 pm/V at 1.32μm during several weeks at 85°C. The fabrication and characterization of Buried Ridge Structure (BRS) lasers monolithically integrated with a butt coupled polymer based buried strip waveguide are presented. The device exhibits a total waveguide insertion loss of less than 5dB.

The limitations of current semiconductor or lithium niobate based technologies in terms of efficiency, integrability, and cost can be surpassed by calling on the remarkable properties of functionalized polymers.[1]-[5] The major asset of this new family of materials is the unlimited flexibility of potentially available structures resulting from a predictive molecular engineering approach. Furthermore, adequately defined poling and processing technologies are shown to be compatible with hybrid polymer/semiconductor integration. In that perspective, the fabrication technology of unimodal low loss waveguides will be described as well as resulting passive and electrooptic applications. We will review the linear

0097–6156/95/0601–0436$12.00/0

and nonlinear properties of side-chain and crosslinked poled polymeric thin films. The wavelength dispersion of refractive index and absorption are obtained by spectroscopic ellipsometry. The second order susceptibility tensor $\chi^{(2)}(-\omega_3; \omega_1, \omega_2)$ is jointly estimated by transverse second harmonic generation ($\omega_1=\omega_2=\omega$, $\omega_3=2\omega$) and modulated reflection measurements ($\omega_1=\omega$, $\omega_2=\Omega$, $\omega_3=\omega+\Omega$ where Ω corresponds to a low frequency voltage). Comparison between the two approaches is in-keeping with a quantum two-level model of the molecular quadratic nonlinearities. Dynamical orientation and relaxation behaviour is inferred from second harmonic generation combined with in-situ corona poling. A waveguided phase modulator with 2-D confinement based on a crosslinked polymer strip waveguide over a doped silicon substrate has been demonstrated with a half wavelength voltage V_π of 30V at 1.06μm for an electrode length of 1.2cm, corresponding to an r_{33} coefficient of 4pm/V. In the rapidly developing field of photonic integrated circuits (PICs), monolithic integration of active and passive optoelectronic components is becoming increasingly important as a tool to produce low cost and high functionality optical modules with application in a wide range of systems. One of the key elements in such PICs is the connection of a laser diode to an external waveguide in a monolithical configuration.[6]-[10] We report here the monolithic integration of a laser diode with a polymeric based waveguide as a first step towards the development of monolithically integrated photonic devices.

I. STRUCTURE AND FABRICATION OF THE WAVE GUIDES:

The basic architecture of the waveguides comprises a strip of a high index polymer eventually endowed of nonlinear properties, depending on the applications in which it will be involved, buried between passive buffer polymers of lower refractive index and deposited on a semiconductor substrate (see Figure 1) [11]. The strip is designed by classical photolithography an dry etching techniques. The fabrication process is summarized in Figure 2. [12]. When the guiding core is made of a thermally stable electrooptic polymer, the electrical poling process occurs at step 2 under the optimal conditions that will be detailed in section III. In step 3, a classical photolithography process, currently used for microelectronics, is applied over the silicon nitride (Si_3N_4) layer. The photosensitive resin is spin-coated and insulated through a mask with UV light. The insulated resin is then dissolved with a developer leading to the structure shown in Step 3 of Figure 2 is then obtained. To achieve the configuration in step 4, the selective reactive ion etching technique is used : Oxygen and tetrafluorocarbon plasmas were chosen to etch respectively the organic layers (the photosensitive resin and the electrooptic polymer) and the Si_3N_4 layer. The sample is plasma etched three times: first based on the structure of step 3, a CF_4 plasma is used to etch the exposed Si_3N_4 section with no photosensitive resin on top. The O_2 plasma is then used to etch the organic layers through the Si_3N_4 mask. CF_4 is then used again to etch the Si_3N_4 on top of the electrooptic guide core.

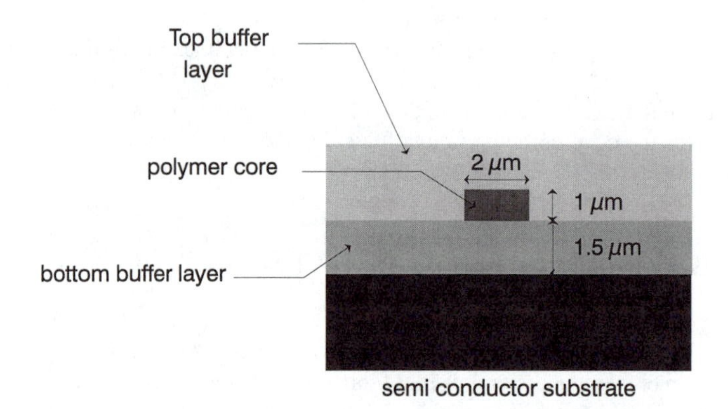

Figure 1: Transverse cross section of the waveguide.

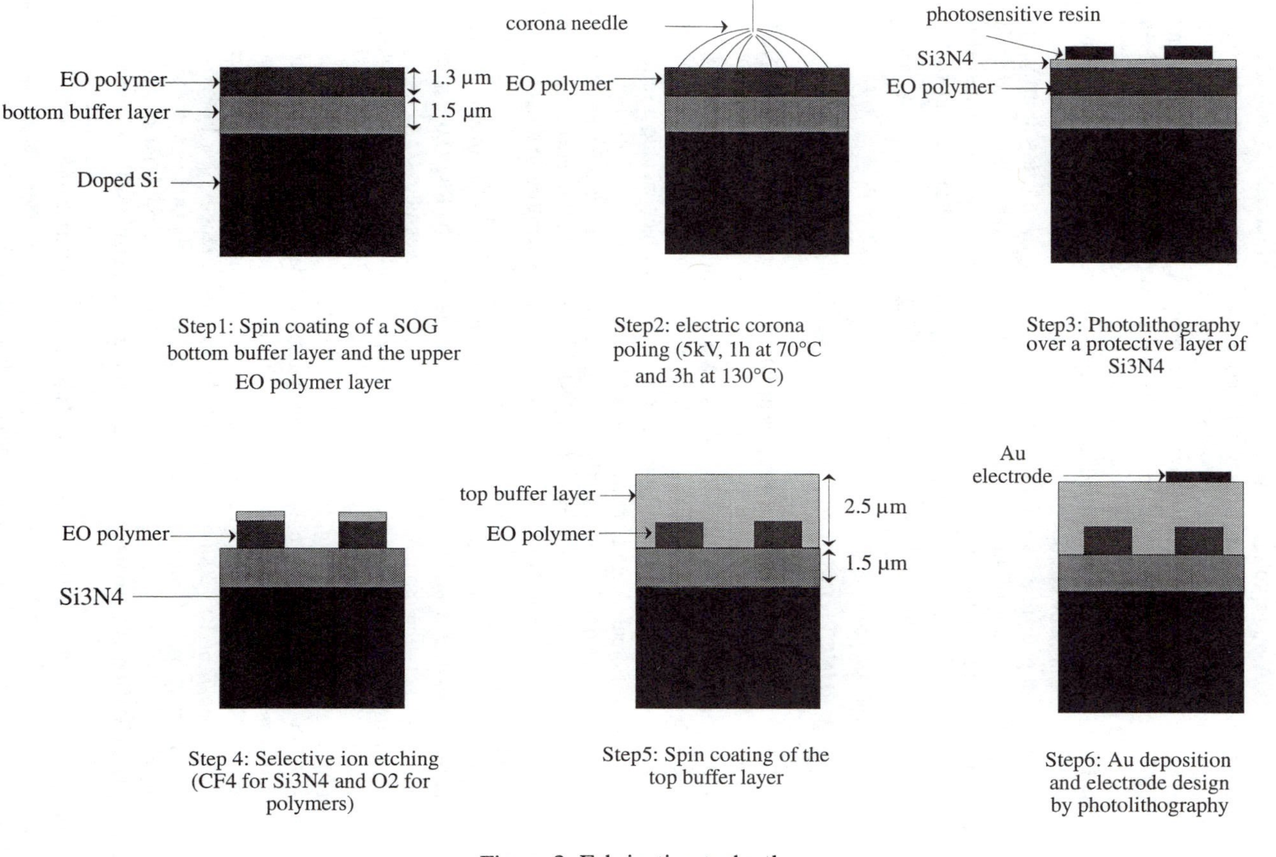

Figure 2: Fabrication technology .

A top buffer layer is spin-coated and cured at a temperature close to 100°C. In step 4, a passive polymer waveguide is obtained. For the electrooptic devices, a 1000Å-thick gold electrode is deposited by evaporation, with a final photolithography step whereby a thicker photosensitive resin is exposed through the electrode pattern mask. After photolithography, a potassium iodide solution removes the gold layer that is not covered with resin. A final UV insulation and application of a developer cleans the residual resin over the gold electrode. An electrooptic polymer modulator is finally obtained at step 6. A propagation loss of 2 dB/cm or less depending on the guiding material, has been measured by the cut-back method for such electrooptic waveguides.

These steps applied at 100°C for about 3 hours should not destroy the electrooptic properties of the film. This method is only valid for electrooptic polymers thermally stable at 100°C. For an electrooptic polymer unstable at 100°C, like MMA-DR1, electrical poling can only be applied at the end of the fabrication process, namely after Step 6.

II.PASSIVE FUNCTIONS

Light propagation in the polymer waveguides is simulated by BPM-CNET (ALCOR) [13]-[15] , software developed by CNET FRANCE TELECOM following the Beam Propagation Method (BPM). The propagation losses in a S-bend waveguide is shown in Figure 3. The propagation losses of 9 S-bend structures have been measured . Photolithography masks which include a series of S-bends with different angles (from 1° to 26°) and a series of Y junctions have been first fabricated. The loss is inferred from comparison between the output signal of a S-bend wave guide and a straight-line waveguide, the ratio of the two outgoing signals evidencing the losses due to the curvature radius of the waveguide with results presented in Figure 4.

Experimental results are in good agreement with the simulation . The curvature radius of S-bends is a function of the angle given by $R=e/(2\theta)^2$ where e is the distance between input and outputs of the S-bend waveguide.

Figure 4 shows that the propagation losses are negligible when the angle is less than 10°, corresponding to a curvature radius of 100µm. This value is one order of magnitude smaller than for LiNbO3 which limits typical S-bend angles to values of the order of 1°. A polymer based integrated device may thus be down scaled to a much smaller size. Firstly, the waveguide width is about 2µm instead of 10µm in LiNbO3 (5) and the gap between parallel waveguides is of the order of 10-20µm instead of 100-200µm. Secondly, the transition length of a low loss S-bend is shorter, 100-200µm instead of 1-2mm. The combination of these two advantages opens interesting perspectives for polymers in optronic devices. It permits the design of new integrated optics devices achievable with polymers but out of reach for LiNbO3, the latter technology being limited by the bulk crystal configuration and the small refractive index step as from titane in-diffusion.

Based on this technology, 1 to 4 junctions have been achieved with asymetric or symmetric outputs (see respectively Figures 5a and 5b).

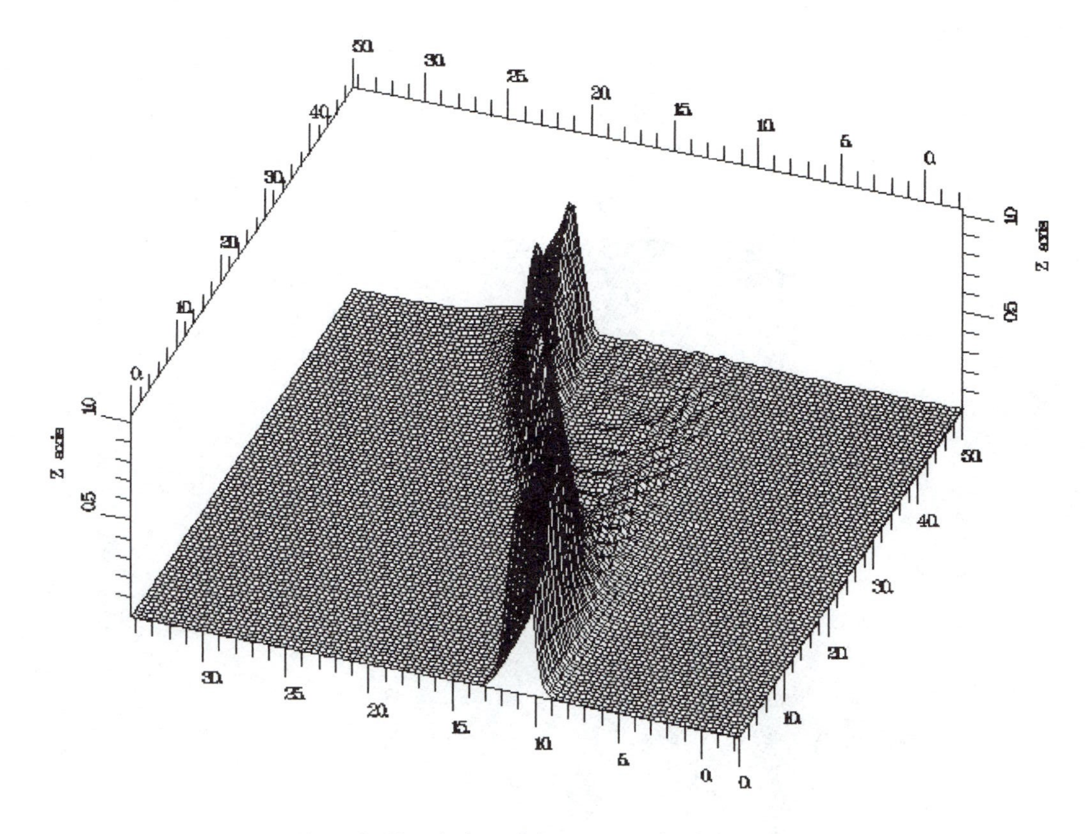

Figure 3: Simulation of the propagation in a S bend.

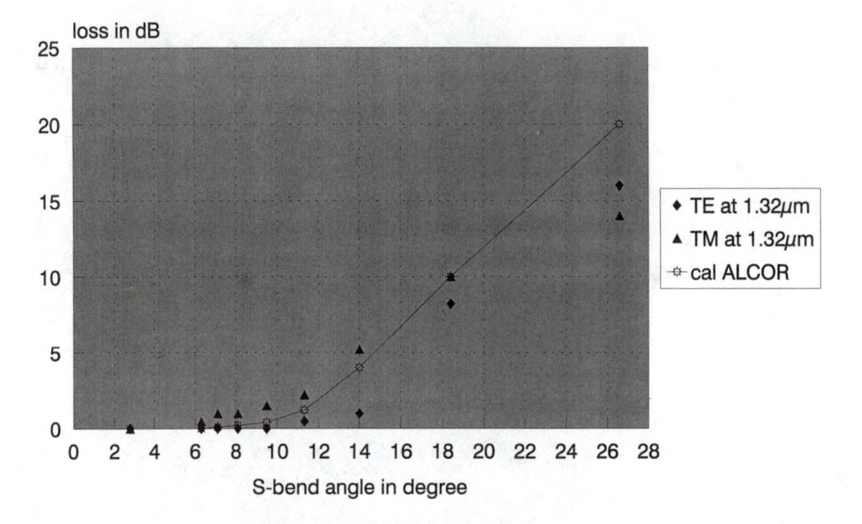

Figure 4: Losses in S-Bends of different angles.

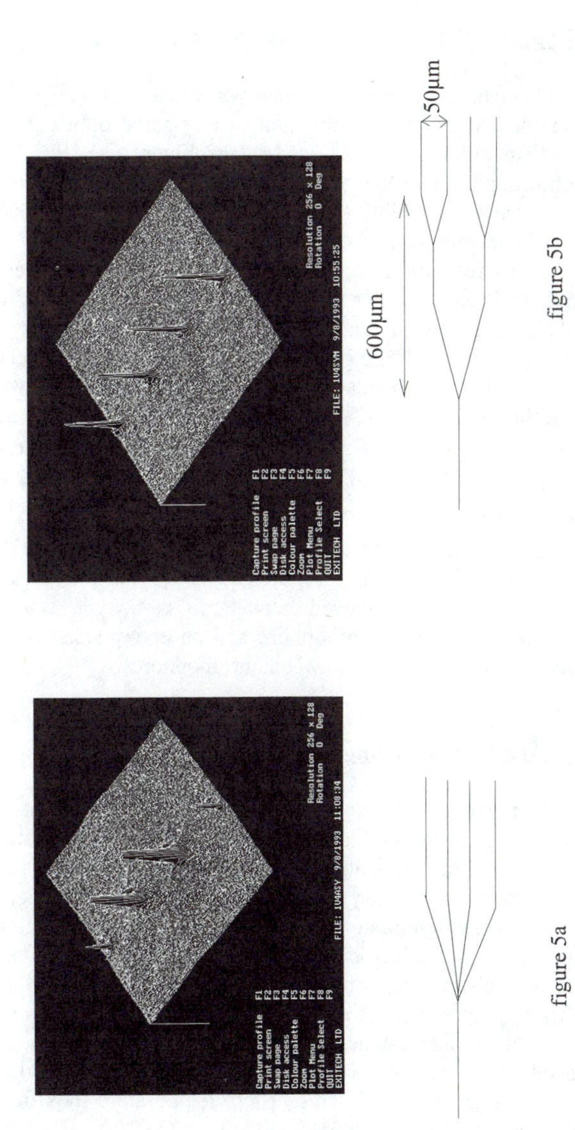

Figure 5: 3-D mapping of digitalized outgoing beams at 1.32µm of 1 to 4 passive junctions; 5a corresponds to an asymetric structure and 5b to a symetric one.

III. ELECTROOPTIC APPLICATIONS

III.1 MATERIAL:

III.1.a)Structure of the cross-linkable polymer

Nonlinear side-chain polymers have been widely studied over recent years towards applications in electrooptic modulation in integrated optics format. The most current side-chain polymer is DR1-MMA (see Figure 6a) [16]. The molar concentration of chromophores in the polymer matrix is given by x. This polymer presents a high quadratic susceptibility (d_{33}=56pm/V at 1.32μm), but suffers from a poor thermal stability leading to a strong decrease of the susceptibility at 70°C . This temperature is too low for telecommunication applications which require stability up to 85°C at least. In order to improve the thermal stability of DR1-MMA, a cross-linkable polymer known as "Red-acid Magly" has been synthesised and studied [CNET patent n° 9310572]. Pyroelectric relaxation and Thermally Stimulated Depolarisation experiments have permitted to evidence, within the framework of the Kohlrausch-Williams-Watt (KWW) model, the higher activation energy E_A of the Magly polymer. A βE_Avalue, where β is the stretching coefficient and E_A the activation energy for the relaxation lifetime, of 3.2 ± 0.6 eV has been measured by thermally stimulated depolarisation and is to be compared to 1.7 ± 0.06 eV for a DR1-styrene copolymer and 1.1 ± 0.4 eV for a DR1-PMMA guest host system[17]. Its structure is shown in Figure 6b.

Thermal crosslinking is achieved between a carboxylic acid function (COOH) located on the nonlinear chromophore and an epoxy side group. There can be no disruptive chemical reaction between chromophores.

III.1.b) Electrical poling of polymer films:

The polymer powder is dissolved in trichloro 1,1,2 ethane at 10% concentration in weight. One micron thick films are prepared by spin-coating (1000 to 2000 rpm). At this stage, the orientation distribution of the nonlinear polarizable units inside the polymer film is centrosymmetric. In order for these films to display electrooptic properties, centrosymmetry must be broken by electric poling. In the case of side-chain polymers, the electric poling procedure is achieved in two steps.

Firstly, the side-chain polymer film is heated near its glass transition temperature T_g, thereby enhancing the orientational mobility of the chromophores which are oriented through electrostatic interaction of their dipole moment with an electric field induced either by corona discharge or by planar electrodes. In the second step, the samples are cooled down to room temperature thus freezing the molecular orientation before the electric field is finally turned-off.

In the case of crosslinked polymers, the first step is modified since the polymer must simultaneously undergo a poling and a crosslinking process. Crosslinking will considerably improve the film stability due to the anchoring of chromophores to the polymer matrix. However this procedure also competes with the poling process since it will partly disorient the molecular dipoles as a result of chemical attachment, thus decreasing the nonlinearity of the film.

Since this competition depends on the temperature, we have further decomposed the poling and the crosslinking procedures into two steps. The first step takes place at low temperature, it favours the poling of the chromophores over the thermal crosslinking procedure which remains very slow. The second step takes place at high temperature, thus enabling complete crosslinking of the polymer while maintaining the same poling field. The optimal poling conditions are as follows:

-first the polymer is heated at 70°C during 1 hour while the corona voltage is maintained at 5kV. This step corresponds essentially to a poling procedure.

-secondly the temperature is increased to 130°C during 3 hours under the same 5kV corona discharge. The polymer is thus crosslinked with the molecular dipoles oriented perpendicular to the substrate.

After these two steps, the films are cooled down to room temperature in presence of the electric field.

After crosslinking, the polymer does not dissolve in any solvent. The transparency of the film is increased, and the colour turns from purple to red.

A decrease in the absorption peak and a blue shift of λ_{max} from 520nm to 488nm are observed in accordance with the observed increased transparency and colour change. The blue shift of λmax observed on the absorption spectrum shown in Figure 7 is due to the crosslinking process which modifies the electronic configuration of the chromophore. A possible mechanism may involve the influence of the acid group: before cross linking it acts as an electron acceptor acid group which adds its effect to that of the nitro group. Subsequent cross linking neutralizes or at least reduces its influence, hence the blue shift.

The efficiency of the thermal crosslinking procedure is a function of temperature. The higher the temperature, the faster the crosslinking proceeds. Table 1 shows the duration needed to reach complete crosslinking at different temperatures for Magly with a 30% chromophore molar concentration as observed by Differential Scanning Calorimetry.

Table 1: *Cross-link duration for Magly 30% at different temperatures.*

Temperature	Cross-link duration
130°C	170mn
180°C	30mn
230°C	2mn

After cross linking, the glass transition temperature of the polymer reaches 160°C. Its decomposition temperature is 250°C.

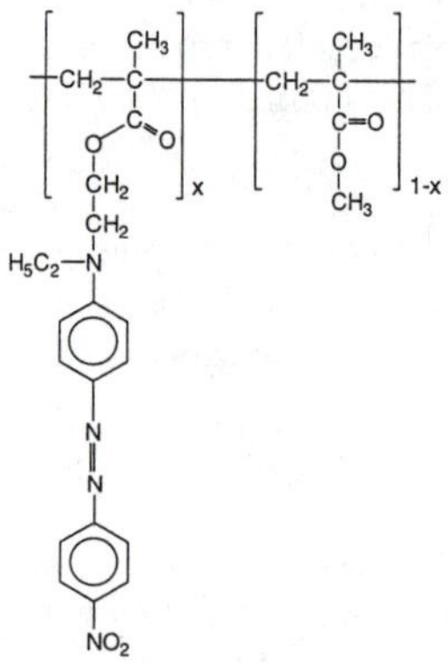

Figure 6a 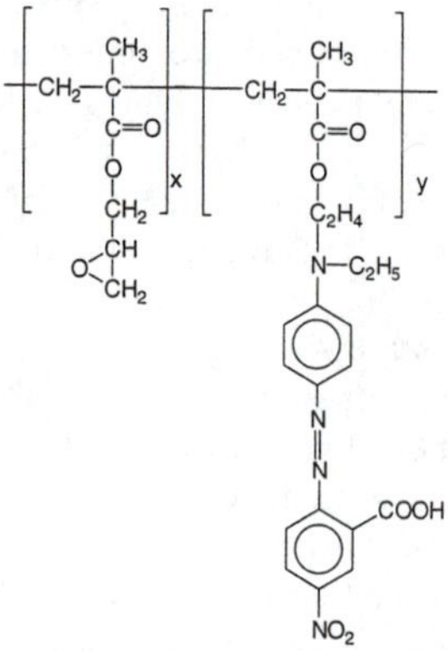 **Figure 6b**

Figure 6: Structure of the NLO chromophores.

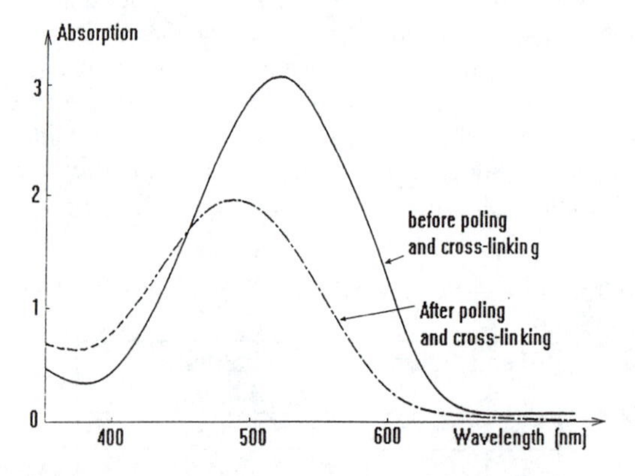

Figure 7: Absorption spectrum of Red Acid Magly before and after cross linking

III.1.c) Ellipsometric measurements:

Spectroscopic ellipsometry is an efficient method to determine refractive index, absorption, and thickness of thin birefringent polymer films.[18]-[19] In Table 2 we have listed the refractive indices at the four wavelengths used in our experiments:

Table 2: *Refractive index at different wavelengths for Magly at 8%, 17% and 30% molar concentrations.*

	670 nm	860 nm	1.06 μm	1.32μm
Magly 8% (COPO1)	1.59	1.55	1.545	1.54
Magly 17% (COPO8)	1.66	1.60	1.58	1.57
Magly 30% COPO9b	1.7	1.62	1.62	1.62

We observe an increase in the refractive index with the concentration. The spectroscopic dispersion curves of the refractive index n and the absorption coefficient k of Magly 30% are shown in Figure 8. Real and imaginary parts of the refractive index follow the Kramers-Krönig relation. The ellipsometric absorption curve is in good agreement with the direct absorption measurement shown in Fig.3.

III.1.d): Second Harmonic Generation measurements and thermal stability evaluation:

The stability of the polymer films was investigated by measuring the dependence of the quadratic susceptibility involved in the second harmonic generation of a 1.32μm YAG laser.

Second-harmonic generation is a reliable method to characterize the nonlinear properties of polymer films.[20]-[21] The signal detection is much more sensitive and is unambiguously of nonlinear origin as opposed to the modulated reflection method described in the next paragraph, which may contain other contributions such as of electrostrictive origin.

An important short term decay of the quadratic susceptibility d_{33} of a DR1-MMA film (29% molar concentration) was evidenced by the SHG method at 1.32μm .When comparing the d_{33} values just after poling and 24 hours after poling at room temperature, the chromophore relaxation brings the nonlinear efficiency from 56pm/V down to a stable value of 29pm/V. On the contrary, no significant decay is observed for Red-acid Magly in the same conditions, showing that short term relaxation in these crosslinked polymers is much weaker than in analogue side-chain polymers.

In order to assess the thermal stability of polymer films, the SHG response from a polymer film is measured following a constant in-situ heating rate. The

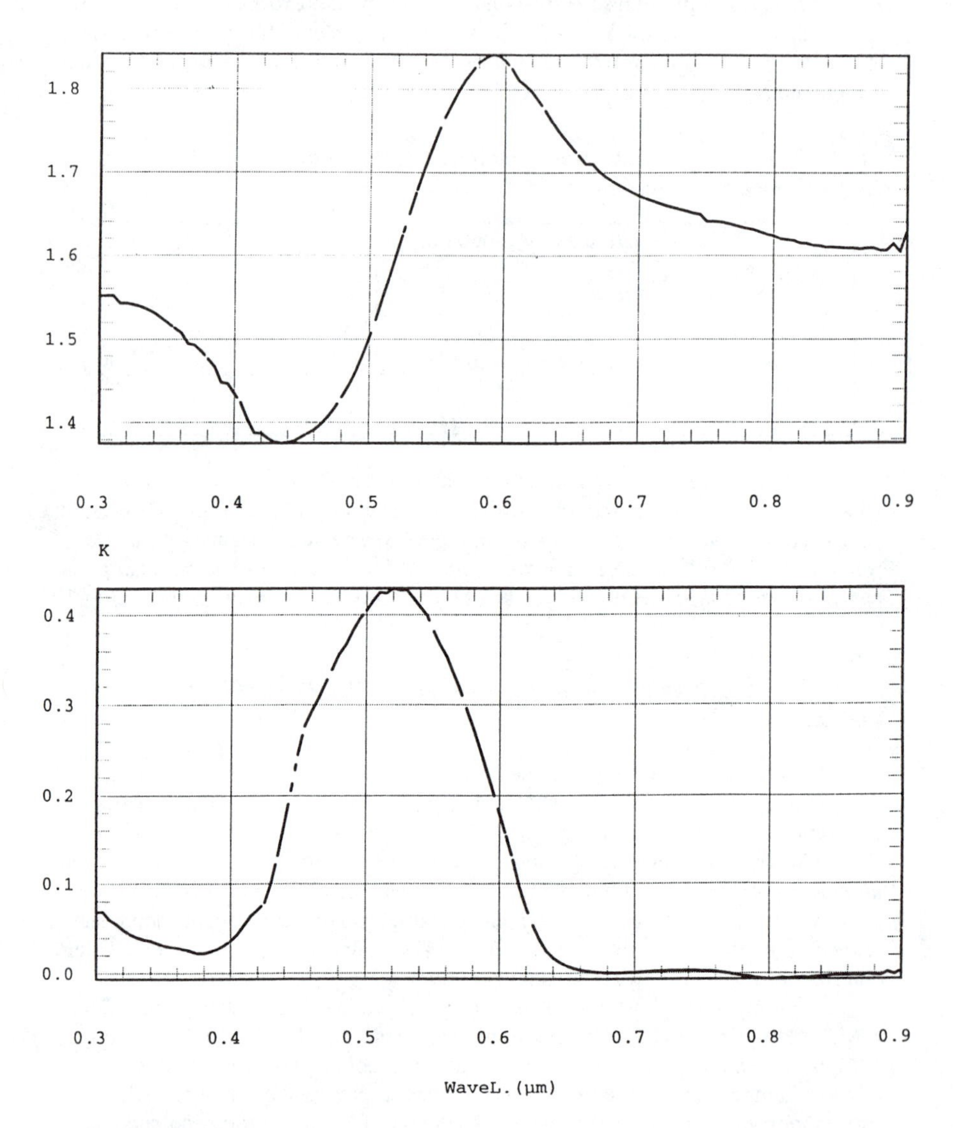

Figure 8: Ellipsometric measurements :wavelength dispersion of the real (n) and imaginary (k) parts of the refractive index.

temperature dependance of the SHG signal $I_{2\omega}$ for DR1-MMA 29% and Magly 30% are shown in Figure 9.

Thermal stability is increased in the case of Magly 30%: for the same amplitude of $I_{2\omega}$, a 30°C improvement can be noted.

Polar relaxation of the different polymers was also investigated at different temperatures. Figure 10 shows the decay of d_{33}(at 1.32μm) as a function of time at 55°C for DR1-MMA 29% and at 85°C for Magly 30%. Magly 30% displays a good stability at 85°C whereas DR1-MMA 29% exhibits comparable stability only up to 55°C. A 30°C improvement in thermal stability is observed here again in isothermal conditions when comparing the stability of the side-chain polymer to that of the crosslinked polymer.

We also measured the susceptibilities of different copolymers of the Red-acid Magly family prepared with different chromophore concentrations and compared them to DR1-MMA 29%.

Table 3: *d_{33} and d_{13} values of DR1-MMA 29% and Red-acid Magly of different molar concentrations at a fundamental wavelength of 1.32μm.*

polymer	d_{33}	d_{13}
DR1-MMA 29%(relaxed)	29pm/V	-
Magly 8%	10.2pm/V	-
Magly 17%	24pm/V	-
Magly 25%	24pm/V	-
Magly 30%	29pm/V	9.5pm/V

This table shows that d_{33} increases with the chromophore concentration, within the range of the studied concentrations.

Comparing d_{13} and d_{33} of Magly 30% shows that the d_{33}/d_{13} ratio coming close to 3 agrees well with the results of a one-dimensional molecular model[3] for this crosslinked polymer following the process described above.

III.1.e) Electrooptic measurements:

For the determination of the electrooptic r_{33} coefficient, we have used a modulated reflection method [22], whereby the electrooptic coefficients difference r_{33}-r_{13} is deduced from the modulation of the intensity of low power cw laser beams reflected by the samples under test. The r_{33} coefficient can be inferred assuming a r_{33}/r_{31} ratio of 3. This method remains a fast assessment technique which provides the electrooptic coefficient of polymer films to be compared to the susceptibilities obtained from SHG measurements.

The measurements, carried out at two wavelengths after eventual short term relaxation of the molecular orientation, for both DR1-MMA 29% and Magly 30%, yield similar results for both types of polymers : (Table 4):

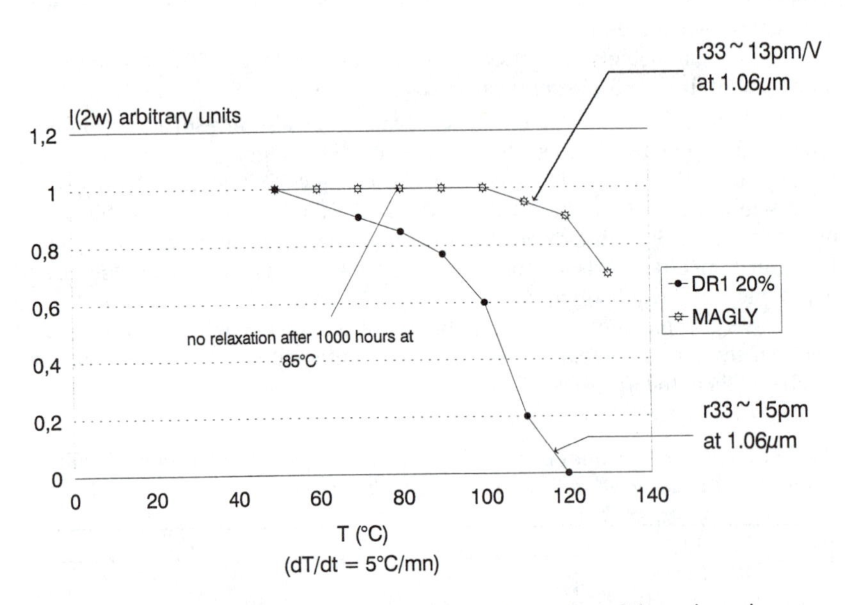

Figure 9: In situ SHG measurement :Thermal stability of the orientation as measured by second harmonic generation for Magly 30% and DR1-MMA.

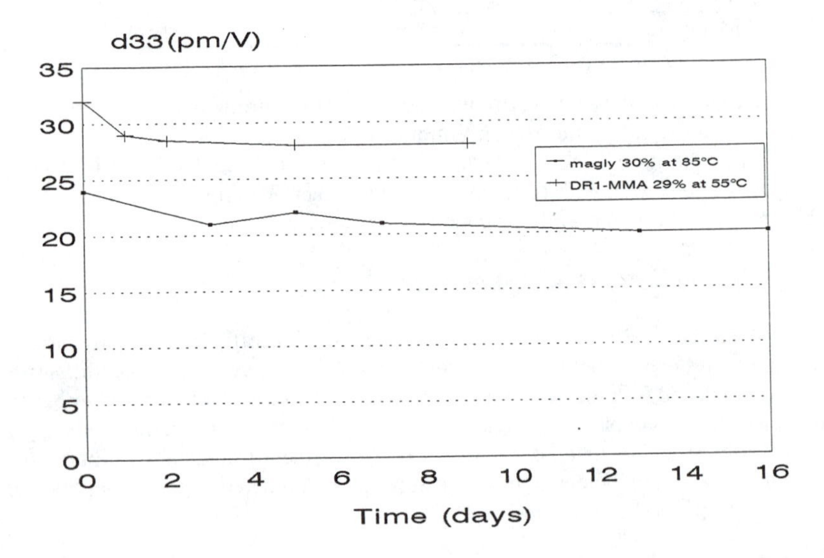

Figure 10: Stability of the $\chi(2)$ tensor at 85°C for Magly, and at 55°C for MMA-DR1.

Table 5: *Electrooptic coefficients r_{33} of DR1-MMA 29% and Magly 30%.*

	r_{33} at 1064nm	r_{33} at 1320nm
DR1-MMA 29%	13.5 pm/V	12.4 pm/V
Magly 30%	14.8 pm/V	12.6 pm/V

The quantum two-level model of the molecular quadratic nonlinearities [23]-[25] leads to the following relation between the second harmonic coefficient $\chi^{(2)}_{zzz}(-2\omega;\omega,\omega)$ and the electrooptic coefficient $r_{33}(-\omega;\omega,0)$.

$$r_{33}(-\omega;\omega,0) = -\frac{2}{n_\omega^4}\frac{f^0}{f^{2\omega}}\frac{(3\omega_0^2-\omega^2)(\omega_0^2-4\omega^2)}{3\omega_0^2(\omega_0^2-\omega^2)}\chi^{(2)}_{zzz}(-2\omega;\omega,\omega) \qquad \text{Eq.1}$$

where ω is the fundamental laser frequency, ω_0 is the absorption peak frequency of the polymer, n_ω the refractive index at frequency ω and $f^{2\omega}$ (resp.f^0) are Lorenz-Lorentz (resp. Onsager) local field factors respectively given by:

$$f^\omega = \frac{n_\omega^2+2}{3} \qquad f^0 = \frac{\varepsilon(0)(n_\infty^2+2)}{2\varepsilon(0)+n_\infty^2}$$

with $\varepsilon(0) \sim 4.5$ and $\chi^{(2)}_{zzz}(-2\omega;\omega,\omega) = 2d_{33}$

Experimental d_{33} and r_{33} values respectively equal to 29pm/V and 12.6pm/V are in good agreement with this model for the crosslinked polymer.

III.2: ELECTROOPTIC DEVICES

III.2.a) MMA-DR1 integrated Mach-Zehnder modulator

The nonlinearity of the side chain MMA-DR1 has been demonstrated in an amplitude modulator with integrated Mach-Zehnder geometry. For that application, the bottom buffer layer is SOG a commercial Allied Signal planarization resin (n~1.44) and the top buffer layer is pure PMMA (n~1.48). As MMA-DR1 exhibits poor stability above 60°C, the poling process is performed after the fabrication of the modulator.

The best figures obtained are $V\pi$=15V at 1.06μm with a modulation rate of 60% (the structure is not unimodal at that wavelength) and $V\pi$=18V with a modulation rate of 80% at 1.32μm for the Mach-Zehnder modulator operating in TM polarisation (figure 11).

These values should be decreased by an optimization of the poling conditions of the multilayer as it corresponds to a r_{33} coefficient of about 6.5 pm/V whereas the best r_{33} measured for the same polymer in thin film geometry is 13pm/V.

III.2.b) MAGLY phase modulator

Here the bottom buffer layer is unchanged and the top buffer layer is a fluorinated PMMA. (n=1.43)

The best figures obtained so far at 1.06μm are V_π = 30V for the phase modulator inserted between crossed polarizers. Here V_π represents the voltage necessary to induce by electrooptic effect a phase difference of π between the TE and TM modes. Electrooptic modulation was also observed at 1.32μm. The experimental set-up and the electrooptic modulation signal are presented in Figure 12. The modulation rates are the same as for the Mach-Zehnder configuration.

Here again, the r_{33} value in the phase modulator, as inferred from the relation $r_{33} = 3\pi\lambda/(2V_\pi L n^3)$, is of the order of 4pm/V. This value is weaker than in Table 4 for two reasons: thermal disorientation during the waveguide fabrication process (3 hours. at 100°C) and weaker poling efficiency for multilayers (Step 2 in Fig.2) than for single layers in thin film measurements.

VI . INTEGRATION OF A LASER DIODE WITH A POLYMER BASED WAVEGUIDE:

The monolithically integrated laser/waveguide device shown schematically on Figure 13 was prepared via two distinct processing steps : first the laser wafer with Buried Ridge Stripe structure (BRS) was fabricated using MOVPE and reactive ion beam etching (RIBE) [27]technique. After the p and n contact metallisation, the Fabry-Perrot laser mirrors have been made by $CH_4/H_2/Ar$ based RIBE.[28] Vertical and extremely smooth facets have been achieved. Single mode polymer waveguides were then fabricated using the process previously described. A bottom cladding layer of PMMA (n=1.48, 1.5μm thick and cured at 170°C) and a core layer of polystyrene (n=1.6, 1μm thick cured at 200°C) were spin coated onto the substrate. In order to improve the coupling efficiency, care has been taken to butt-joint couple the laser to the waveguide active layer, by photolithographically etching the bottom of the cladding layer to clear the laser facet.[29] Finally after the etching of the core ridge, a Teflon AF upper cladding layer was deposited. The light output power was then measured at the cleaved laser facet and compared to that emitted at the end of the waveguide. The laser cavity and the waveguide lengths were respectively 250μm and 600μm. The laser threshold current is 15mA and the optical power output from the laser and the waveguide facets were respectively 11 mW and 5mW at 100mA. Significant improvements may be expected by developing polymeric materials with low loss figures and by optimizing the processing steps. The integrated device shows high thermal stability against temperature after heating at 250°C for 1h, with no significant decrease in the waveguide power output.

V CONCLUSION

In conclusion, we have evidenced the improved stability of a thermally stable crosslinked nonlinear polymer and demonstrated a waveguide phase modulator based on this polymer. This polymer maintains its electrooptic coefficients over at least several weeks at 85°C. This efficiency-stability trade-off is believed to correspond to the current optimum for this class of materials and devices. A decrease of V_π by a factor of 3 to 4 can be expected from further

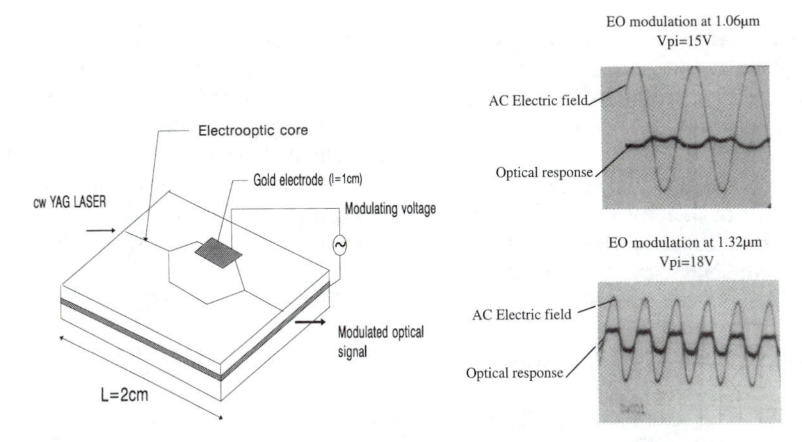

Figure 11: DR1-MMA based integrated Mach-Zehnder intensity modulator.

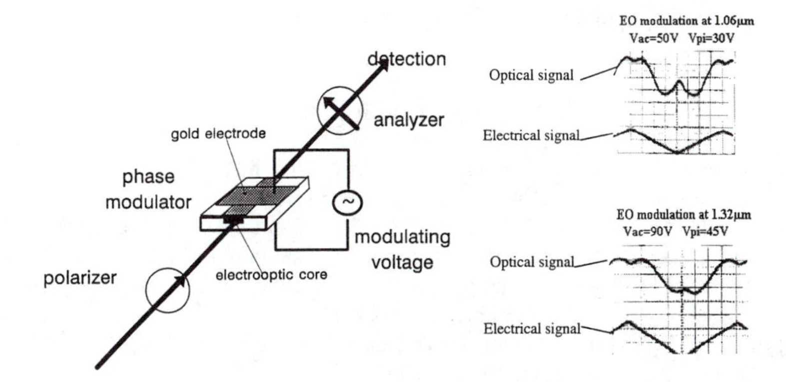

Figure 12: Magly based integrated phase modulator.

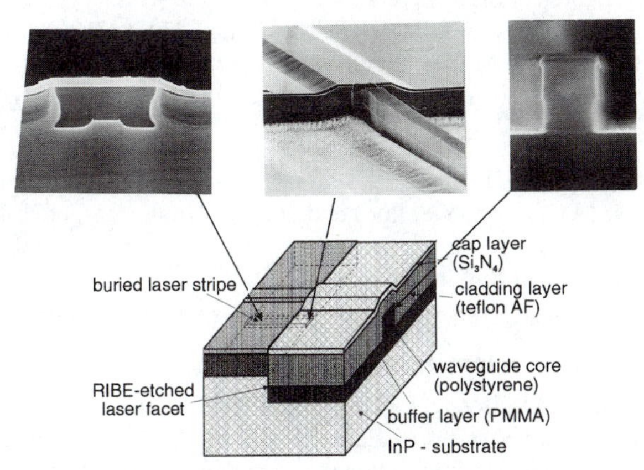

Figure 13: Monolithic integration of a laser diode with a polymer passive waveguide.

optimization such as a higher poling efficiency and better thermal stability of the crosslinked polymer. A waveguide fabrication process has been set-up and shown to lead to several passive or active applications. Furthermore, a 1.3μm BRS laser and polymer waveguide have been monolithically integrated using a high performance potentially low cost technology. The device exhibits a low threshold current and a total insertion loss smaller than 5dB. These results open the way to a variety of integrated III-V active functions together with passive or electrooptic polymer based functions such as optical combiners for WDM devices, splitters, switches and various other PICs.

VI ACKNOWLEDGEMENTS

The authors gratefully acknowledge P.Boulet, M.Carré, J.Charil, S.Grosmaire and F.Huet for technical assistance in the fabrication of the demonstrators and A.Rousseau and F.Foll from the Ecole de Chimie de Montpellier for supplying the electrooptic polymers.

Bibliography

[1] R.Lytel, G.F.Liscomb, J.L Thackara . Proc SPIE **824**, 152(1987).
[2] Donald R. Ulrich .Mol.Cryst. Liq. Cryst. **160**,1(1988).
[3] J.R Hill, P.Pantelis and G.J Davies. Inst. Phys. Conf. Ser. No 103: Section 2.5 .Conf. Materials for Nonlinear and Electrooptics. Cambridge, 1989.
[4] Emanuel Van Tomme , Peter p. Van Daele, Roel G. Baets, Paul P. Lagasse. IEE Journal of Quantum Electronics **27**, 3(1991).
[5] Emanuel Van Tomme , Peter p. Van Daele, Roel G. Baets, G.R Möhlmann , M.B Diemmer. J. Appl. Phys. **69** (9), (1991).
[6] P.J. Williams, P.M. Charles, I.Griffith, L.Considine, A.C. Carter. Electron.Lett. **26**, 142(1990).
[7] T.L.Koch, U.Koren. J.Lightwave Technol. **8**, 274(1990).
[8] T.Sasaki, I.Mito. Digest of OFC/IOOC'93, Comm. ThK1, 210(1993).
[9] A.Neyer, T.Knoche, L.Muller, Electron. Lett. **29**, 399(1993).
[10] C.Rompf, B.Hilmer, W.Kowalsky. Tech.Dig.ECOC'93 , Comm. WeP7.5 (1993).
[11] R. Pinsard-Levenson, J. Liang, E. Toussaere, N. Bouadma, A.Carenco, J.Zyss, G.Froyer, M.Guilbert, Y.Pellous and D.Bosc, Nonlinear Optics **4**, 233 (1993).
[12] CNET Patent n° 9114662
[13] M.Filoche,PhD Thesis, 6 March 1991, Université de Paris Sud.
[14] G.Hervé-Gruyer, Electron. Lett. **26**(17),1338(1992).
[15] G.Hervé-Gruyer, M.Filoche, F.Ghirardi. ECIO 93. Neuchâtel,(1993).
[16] K.D.Singer, J.E.Sohn, and S.J.Lalama, Appl.Phys.Lett. **49** ,5 (1986).
[17] S.Bauer et al., Proc. 8th International Symposium on the Electrets, Paris (1994).

[18] E.Toussaere, J.Zyss. Thin Solid Films, **234**, 454(1993).

[19] E.Toussaere, J.Zyss. Thin Solid Films, **234**,432(1993).

[20] J.Jerphagnon, S.K. Kurtz, J. Appl. Phys. **41**(4),(1970).

[21] D.Chemla, P.Kupecek. Revue de physique appliquée **6**, 31(1971).
[22] C.C. Teng, H.T. Man, App. Phys. Lett. **56,** 18 (1990).
[23] M.Sigelle and R.Hierle, J. Appl. Phys. **52,** 6 (1981).
[24] J.L.Oudar, D.S.Chemla. J. Chem. Phys., **66** (6) , 2264(1977).
[25] M.Sigelle, R.Hierle. J. Appl. Phys. 52 ,6(1981).
[26] K.D. Singer, M.G. Kuzyk, J.E. Sohn. J. Opt. Soc. Am. B.**4** (6), (1987).
[27] N.Bouadma, C.Kazmierski, J.Semo, Appl.Phys.Lett. **59**,1 (1991).
[28] N.Bouadma, J.Semo, J.Lightwave Technol. **12**(4) ,(1994).
[29] N.Bouadma, J.Liang, R.Levenson, S. Grosmaire, P.Boulet, S.Sainson. IEE Photonics Tech. Letters **6**,(1994).

RECEIVED March 27, 1995

Chapter 33

Electrooptic Polymer Mach–Zehnder Modulators

High-Speed Analog and Digital Considerations

D. G. Girton, W. W. Anderson, J. F. Valley[1], T. E. Van Eck, L. J. Dries, J. A. Marley, and S. Ermer

Lockheed Martin Palo Alto Research Laboratory, Research and Development Division, Organization 97–80, Building 202, 3251 Hanover Street, Palo Alto, CA 94304

Electro-Optic polymer Mach-Zehnder modulators are described with regard to design, fabrication, and testing at high operating speeds. The expected frequency response is discussed in terms of an electrical signal propagating co-linearly with an optical signal, as well as the conductivities of the electrodes. The effects of polymer conductivity and poling efficiency on modulator switching voltage are described. Modulators fabricated to have 5 V switching levels were tested using a bit error rate test system, and the results show excellent response times at a clock rate of 900 Mb/s. The response was limited by test equipment, and the actual performance is calculated to be significantly above the equipment limitations.

Electro-optic (EO) modulators based on certain polymer materials have significant advantages over inorganic devices, such as $LiNbO_3$, in four major areas: 1) higher operating speeds using simple electrode designs, 2) lower manufacturing costs using integrated circuit (IC) fabrication techniques, 3) ease of integration with underlying substrate circuitry, and 4) simpler optical fiber attachment using inherent characteristics of Si substrates. The high speed device operation using simple electrode designs, such as a true microstrip electrode, is due to a low effective dielectric constant, on the order of 2.5. This permits the electrical phase front of a microstrip electrode, which is directly over an optical waveguide, to travel at essentially the same velocity as the optical phase front in the waveguide. Thus, small changes in the polymer index of refraction caused by the electrode field will add cumulatively as the electrical and optical fronts travel together. Also significant to very high speed operation is the fact that the EO response in these polymers results from small changes in the system of electrons responsible for polarization, rather than a displacement of nuclear coordinates as in inorganic materials.[1]

The low cost of making polymer devices results from using small quantities of polymer materials fabricated on top of substrates such as Si or GaAs, while employing the same equipment that is used to build ICs. In addition, polymer waveguide structures can be easily fabricated directly on top of circuitry previously built into these substrates. This is much more difficult, or impossible, for inorganic materials such as $LiNbO_3$. Attachment of optical fibers to polymer devices can be assisted by using Si wafers with etched V-grooves that align precisely with the

[1]Current address: Applied Materials, Santa Clara, CA 95050

0097–6156/95/0601–0456$12.00/0

polymer waveguides. This approach will make it easier and less expensive to package polymer waveguide devices.

Polymer Materials for Electro-Optic Polymer Devices

Chemists have played a critical role in the development of polymers suitable for optical device fabrication. Organic chemists have made significant advances in the past decade by developing highly active chromophores that are compatible with optical grade polymers. These materials have been developed so that one polymer film can be stacked vertically on top of a previously cured film, making it possible to fabricate structures that form light waveguide channels. In addition these polymers are designed to be compatible with IC fabrication processes, including Au and Al metallization techniques. These materials have provided the basis for fabricating a variety of integrated optic polymer devices including multi-GHz modulators[2,3], logic level switches[4], active fanout devices[5], multilevel waveguide structures[6,7], and waveguide devices with selectively located passive and active EO polymers[8,9]. Today, there is an intense interest in developing high quality EO polymers that meet specific requirements, such as high EO activity, low optical loss, and good temporal and thermal stability.[10]

The polymer materials in this work utilize 4-(dicyanomethylene)-2-methyl-6-(p-dimethylaminostyryl)-4H-pyran (DCM) as a guest chromophore in Amoco Chemical Company's Ultradel 4212 polyimide as the host, as shown in Figure 1.

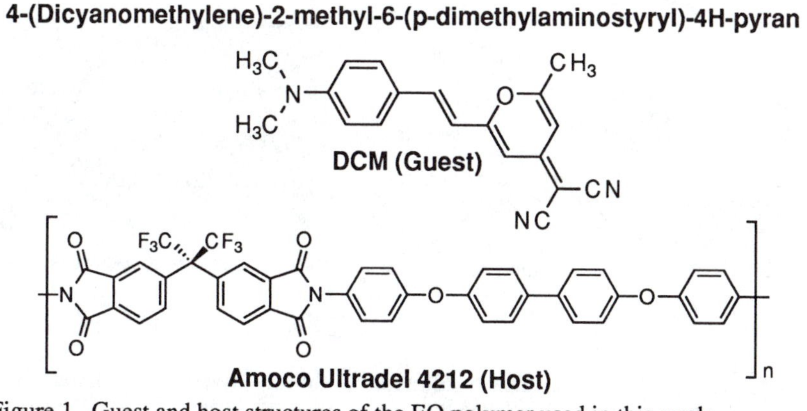

4-(Dicyanomethylene)-2-methyl-6-(p-dimethylaminostyryl)-4H-pyran

Figure 1. Guest and host structures of the EO polymer used in this work.

These materials were selected because they are commercially available and useful for building high speed EO devices. DCM is highly transparent near 830 nm and photobleaches to form waveguides. The thermal stability, EO activity, high transmittance, and commercial availability of DCM and Ultradel 4212 make this guest-host system a viable candidate for research on integrated optics.

EO Polymer Mach-Zehnder Modulators

A Mach-Zehnder modulator is an EO device that converts an electrical signal into an intensity modulated light signal. These modulators are significant devices with a wide variety of applications. An example of an integrated Mach-Zehnder modulator that is fabricated using EO polymers on top of a Si wafer is shown in Figure 2. The modulators were built up using three vertically stacked polymer layers: a bottom cladding layer, a core layer containing the EO polymer (DCM:4212), and a top cladding layer, as shown in Figure 2c. This structure is sometimes referred to as a polymer triple stack. Using IC fabrication equipment, each layer was spin coated and cured on an Al clad Si wafer, waveguides were photobleached, and gold electrodes deposited as previously described.[2] A completed modulator that was fabricated in this manner is shown in Figure 2d.

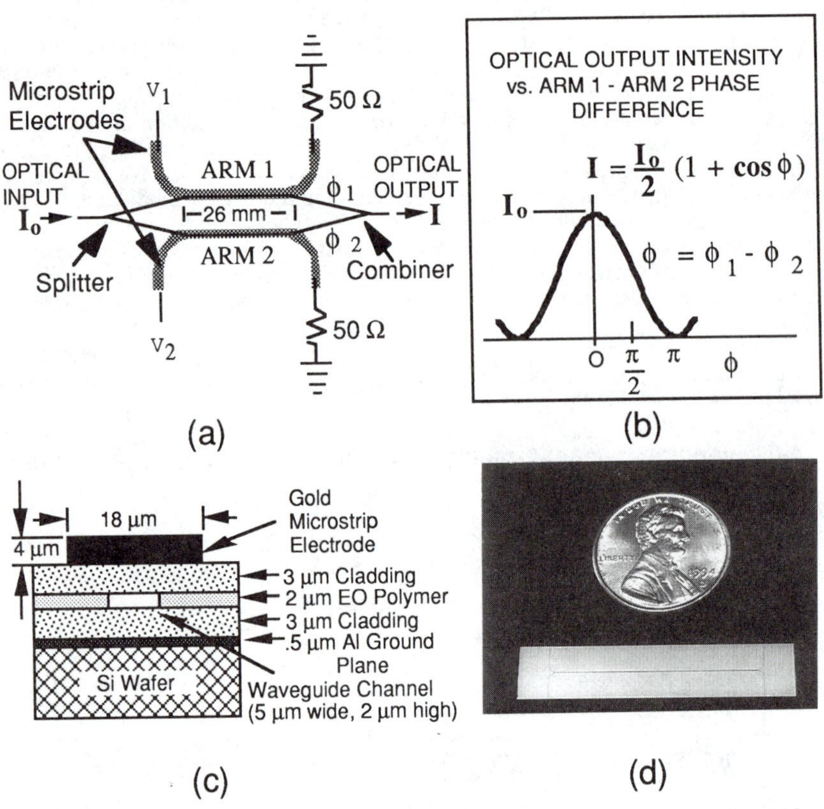

Figure 2. Integrated Mach-Zehnder modulator constructed using EO polymer materials. (a) Top view layout of modulator. (b) Ideal output intensity of modulator. (c) Cross-section through one arm showing each polymer layer and the waveguide channel. (d) Photograph of completed polymer modulator.

This polymer Mach-Zehnder modulator functions as follows: As voltage is applied between the electrode over one arm (e.g., arm 1) and the Al ground plane of Figure 2c, the index of refraction of the EO polymer changes, and the phase of the optical wave arriving at the combiner is changed relative to the other arm. The combiner sums the optical waves from each arm, and the phase change is converted to an amplitude change, resulting in optical intensity modulation. The phase change from an arm is proportional to the applied voltage, while the output intensity is related to the input voltage through the cosine function shown in Figure 2b. An important consideration for high speed performance is that the electrical phase front travels synchronously with the optical phase front. In Figure 2a both the electrical and optical phase fronts travel together, from left to right.

An important characteristic of a Mach-Zehnder modulator is its switching voltage, V_π. which is the voltage difference between the maximum and minimum output intensity. Using transverse magnetic (TM) light of wavelength λ as the input, the EO coefficient r_{33} of the polymer stack is related to V_π by :

$$V_\pi = \frac{\lambda h}{n^3 r_{33} L} \tag{1}$$

where λ is the wavelength of transverse magnetic (TM) light input to the device, h is the total thickness polymer stack, n is the polymer index of refraction, and L is the interaction length of an electrode over one arm of the modulator. For the work described hear L was 26 mm and λ was 830 nm. The index of refraction for both the core and cladding layers was 1.6. Another parameter used to characterize a Mach-Zehnder modulator is the ratio of maximum-to-minimum output intensity, called the extinction ratio.

The polymer Mach-Zehnder modulator shown in Figure 2 can be used as a stand-alone high speed device, or it can be integrated with other optical devices on the same substrate during fabrication. Another use of this modulator is in evaluating EO materials and device processing, including poling efficiency.

Frequency Response of Phase Modulators. The Mach-Zehnder modulator is a specific example of the general class of electro-optic modulators based on the modulation of the optical path length or phase of a propagating optical signal. Yariv has given an expression for the phase modulation of an optical signal due to an electrical signal propagating co-linearly with the optical signal.[11]

$$\Gamma(t) = a \int_t^{t+\tau_d} E(t', z(t'))dt' \tag{2}$$

where the definitions:

$z(t') = c(t'-t)/n$, coordinate of point on optical wave

$\tau_d \equiv nL/c$, transit time of point on optical wave

$E(t', z) = E_m \bullet e^{j(\omega_m t' - \gamma_m z)}$, modulating field

$\gamma_m \equiv k_m - ja_m$, complex propagation constant of modulating field

$k_m \equiv \omega_m \sqrt{\varepsilon_{eff}}/c$, phase factor of propagation constant

determine the solution for the integral of equation 2. In the solution equation 3, Γ_o is:

$$\Gamma(t) = aE_m\tau_d e^{j\omega_m t} \frac{e^{j\omega_m(n-\sqrt{\varepsilon_{eff}})L/c}e^{-a_m L} - 1}{j\omega_m(n-\sqrt{\varepsilon_{eff}})L/c - a_m L} = \Gamma_0 e^{j\omega_m t} \frac{e^{j\omega_m(n-\sqrt{\varepsilon_{eff}})L/c}e^{-a_m L} - 1}{j\omega_m(n-\sqrt{\varepsilon_{eff}})L/c - a_m L} \tag{3}$$

the low frequency (dc) phase retardation or modulation due to an electric field amplitude of E_m. The magnitude of the phase modulation as a function of frequency is thus given by:

$$|\Gamma(\omega)| = \Gamma_o \sqrt{\frac{1 + e^{-2y} - 2e^{-y}\cos x}{x^2 + y^2}} \tag{4}$$

where the phase walk-off and attenuation parameters are defined by:

$x \equiv (n_{eff} - \sqrt{\varepsilon_{eff}})\omega_m L / c$, phase walk-off parameter

$y \equiv a_m L = \alpha_m L / 8.69$, attenuation parameter.

In the above definitions, n_{eff} and $\sqrt{\varepsilon_{eff}}$ are the effective indices of refraction for the optical and modulating field propagation velocities respectively, a_m is the attenuation constant in the complex propagation constant definition above and α_m is the corresponding attenuation factor in dB/unit-length.

For microwave and millimeter wave planar transmission lines, the dominant contribution to the attenuation constant will be the skin effect resistance of the transmission line metal. Thus, a_m will scale as $a_m = a_{mo}\sqrt{f}$ at higher frequencies where the skin effect dominates. With an empirical determination of the skin effect attenuation, equation (3) was compared with the experimental frequency response measurements on $LiNbO_3$[12] and polymer[3] Mach Zehnder modulators with the results shown in Figure 3. In both cases, the measured results are shown as dots and theory, equation 4, is shown as the solid lines. The $LiNbO_3$ modulator was fabricated with a

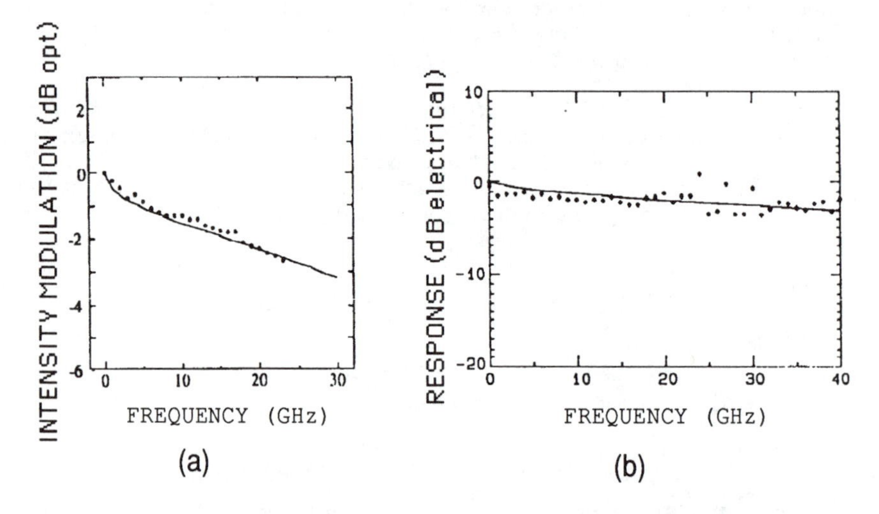

Figure 3. Theory and experimental response of Mach-Zehnder modulators for (a) $LiNbO_3$ device[12] and (b) polymer device[3].

thick, co-planar transmission line while the polymer modulator was driven by a microstrip transmission line. In both cases, equation 4 provided an accurate description of performance. Given the success of equation 4 in describing the performance of other modulators, we will use it to predict the performance of our polymer based modulator using the device dimensions of Figure 2c.

First, consider the phase walk-off parameter, x, when the attenuation parameter, y = 0. Equation 4 is down to 90% of its dc value when $x = \pi/2$ or $f_{max}L = c/(4|n_{eff} - \sqrt{\varepsilon_{eff}}|)$ Hz•cm, as shown in Figure 4.

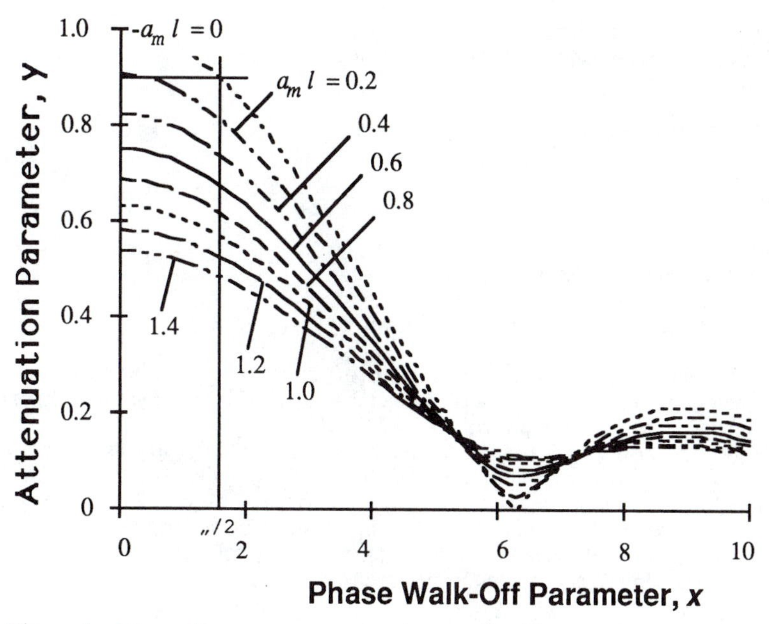

Figure 4. Attenuation parameter y vs. phase walk-off parameter x.

The microstrip effective dielectric constant to be used in equation 4 is given by[13]:

$$\varepsilon_{eff} = \frac{\varepsilon_r + 1}{2} + \frac{\varepsilon_r - 1}{2}\left(\frac{1}{\sqrt{1 + 12\frac{h}{W}}} - \frac{t}{2.3\sqrt{Wh}}\right) \approx 2.5 \tag{5}$$

with the parameters of equation 5 defined by:

W ≡ width of microstrip electrode (18 μm from Figure 2c)
h ≡ thickness of polymer dielectric (8 μm from Figure 2c)
t ≡ thickness of microstrip electrode (4 μm from Figure 2c)
$\varepsilon_r \approx 3.4$, relative dielectric constant of polymer materials used

Thus the maximum frequency of operation of the modulator due to phase walk-off is the order of $f_{max}L \approx 398$ GHz•cm when $n_{eff} \approx 1.6$, the effective dielectric constant of our polymer waveguide.

The attenuation factor due to microstrip conductor resistivity is conservatively[14] given by:

$$a_m = \frac{R_{sq_1} + R_{sq_2}}{2WZ} \tag{6}$$

where:

R_{sq} is the surface resistivity of the microstrip conductors (Ω/square)

W is the width of the top conductor (Figure 2)
Z is the characteristic impedance of the microstrip line.
The surface resistivity is given by:

$$R_{sq} = \frac{\rho}{t} \frac{t/_{\delta}[1 - e^{-2t/\delta} + 2e^{-t/\delta}\sin(t/_{\delta})]}{1 + e^{-2t/\delta} - 2e^{-t/\delta}\cos(t/_{\delta})} \tag{7}$$

where:
ρ is the conductor bulk resistivity
t is the conductor thickness
δ is the conductor skin effect depth given by $\delta = \sqrt{(\rho/\pi f \mu_o)}$

We note that R_{sq} varies smoothly from ρ/t at low frequencies to ρ/δ at high frequencies. In our device, the top conductor is gold with ρ_1 = 1.7 μΩ•cm and the bottom conductor is aluminum with ρ_2 = 2.7 μΩ•cm.

Figure 5 shows the calculated reduction in output intensity as a function of frequency for a polymer modulator having interaction lengths of L = 1 cm and L = 2 cm, and an Al ground plane thickness of 0.5 μm as used in our devices, as well as an Al ground plane thickness of 4 μm for comparison. Considering the dimensions of the microstrip electrode used in our device, the intensity roll-off evident in Figure 5 is due entirely to skin effect loss in the electrodes.

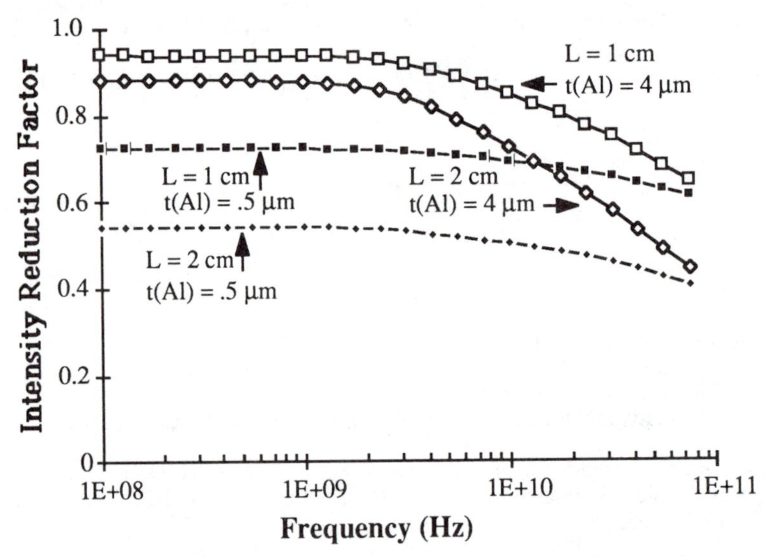

Figure 5. Calculated response of Mach-Zehnder modulator with structure and dimensions shown in Figure 2c.

Alignment of Chromophore Dipoles by Electric Field Poling. EO polymer films are amorphous when they are first deposited, and no second-order EO effects can be observed since the chromophore dipoles are randomly arranged. Electric-field poling is used to achieve a macroscopic alignment of the chromophores responsible for the EO effect.[15] The DCM:4212 material is heated above its glass transition temperature, a high electric field is applied by electrodes to align the nonlinear molecules in the direction of the field, and the material is cooled back to room temperature under the influence of the electric field. This process "freezes" in the dipole orientation and creates a macroscopic EO coefficient in the material, and the preservation of this molecular orientation is necessary to retain the EO effect. The dielectric response of a single polymer film at 1 kHz as a function of temperature for a concentration series of DCM in Ultradel 4212 is shown in Figure 6. This single film data at 1 kHz provides some guidance regarding usable poling temperatures. However, a three

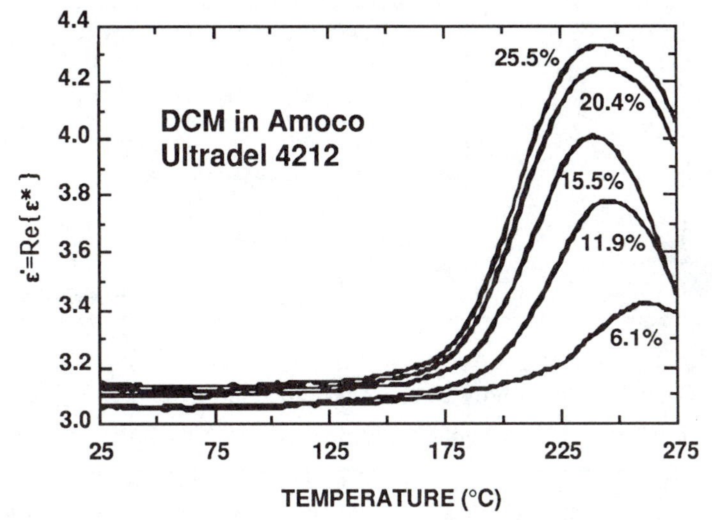

Figure 6. Dielectric spectroscopy of Amoco 4212 doped with various percentages of DCM.

layer polymer stack that is poled with a steady voltage is required to determine optimum poling conditions for devices such as Mach-Zehnder modulators. In a three layer stack the voltage that appears across the core layer depends on its resistivity in relation to the resistivities of the two cladding layers. Depending on this resistivity ratio, the actual poling field across the core layer can be very small or nearly the same as the total applied field, as shown in Figure 7.

To achieve efficient poling at reasonable voltages, a large portion of the applied field must appear across the core layer, with only a small field across the cladding layer. This means that the cladding layer should be at the very least equal in conductivity, and preferably much more conductive than the core layer at poling temperatures. When the three layers have the same resistivity, only one-third of the applied field appears across the core layer. Values of r_{33} obtained from triple stack device measurements may not agree well with those obtained from measurements on single polymer films since the core/cladding conductivity ratio will determine what portion of the applied field was applied to the core layer.

Although electric field poling has been used by workers to align chromophore dipoles in EO polymers, the dynamics of the process remain relatively unexplored. While poling is taking place, chemical rearrangements may occur during the temperature cycle, and effects such as chromophore diffusion or sublimation[16], residual solvent diffusion, and movement of ionic charges may occur. It is well known that small amounts of ions in polymers can have a large effect on the resistivity.[17,18] The variables of poling temperature and its duration, as well as the applied field strength and its duration, ionic content of each polymer layer, and

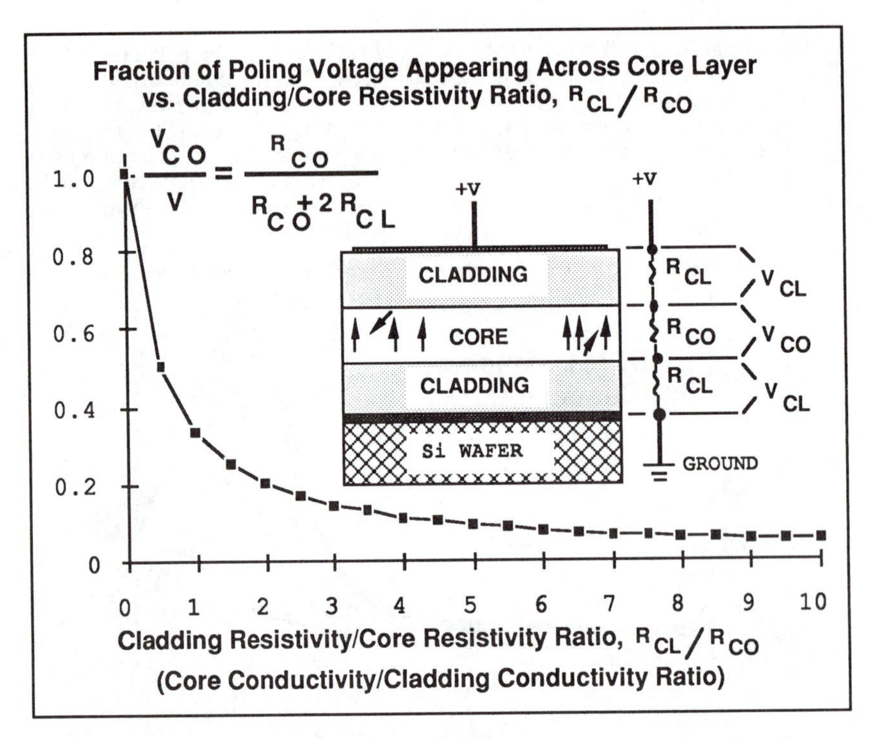

Figure 7. Dependency of core poling field on resistivity ratios in a three layer stack.

interface charges, must be explored for each polymer system in order to find the values that yield the optimum device characteristics.

Guest-host EO systems based on electronic grade polyimides pose exceptional challenges with regard to core-cladding conductivity matching. These systems are typically stacks of pure polyimide cladding layers and a core layer consisting of the same or related polyimide doped with a relatively small nonlinear chromophore. The electronic grade polyimides are engineered to be used as interlayer dielectrics and therefore have very low ionic content and conductivities. The core layer starts out with the same characteristics before doping, but the addition of the chromophore dopant raises the conductivity in two ways. First, NLO chromophores are rarely prepared or processed to microelectronic specifications. As such, they may be contaminated with ionic impurities which contaminate the core layer after doping. Second, most small molecules have a plasticizing effect on polymers, lowering the glass transition temperature T_g. This effect has been shown in polyimide systems containing Disperse Red 1 (DR1)[19] or DCM[20] as a guest. Ionic mobility increases as T_g is reached, so the core electrical conductivity increases at a faster rate than does the conductivity of the higher-T_g polyimide cladding.

Conductivity Measurements on Polymer Films. Electrical conductivity measurements were made on single polymer layers of Ultradel 4212, DCM:4212, and an in-house UV curable acrylate. Each single layer film had the same thickness and was prepared in the same manner as for that layer in a three layer stack used to build modulators. Figure 8 shows typical results for these materials.

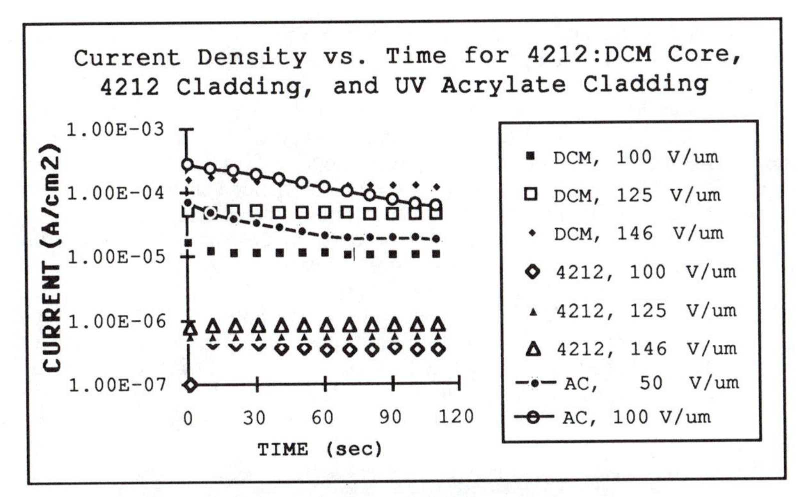

Figure 8. Conductivity of the core layer and two different cladding layers.

The data show that the 4212:DCM core layer is one to two orders of magnitude more conductive than the undoped 4212 cladding, and this should result in inefficient poling and high values of Vπ for devices. Mach-Zehnder modulators that were fabricated with these materials showed TM Vπ values near 30 V, corresponding to r_{33} near 2 pm/V. Conductivity values of the UV acrylate cladding in Figure 8 are in the same range as for the 4212:DCM core. Mach-Zehnder modulators fabricated with these materials yielded TM Vπ values of 5 V, corresponding to r_{33} = 13 pm/V. This six-fold improvement shows the importance of core/cladding conductivity ratios on poling efficiency and device characteristics.

Several Mach-Zehnder modulators were fabricated using UV acrylate claddings with a 4212:DCM core, and poled at various temperatures, times and poling fields. The results show that a total poling field of 130 V/μm at 190 °C for 1 min gave a minimum Vπ value for these materials.[21] Typical modulators made with this process required 5 V to switch from full on to off, and had extinction ratios of 15 dB or better. The modulator output mode pattern was examined using a 100x microscope objective and a Spiricon imaging system. The output pattern appeared as a single mode throughout a voltage sweep from full on to off. These mode patterns together with a 15 dB extinction ratio imply that these modulators are operating as single mode devices.

High Speed Bit Error Rate Testing of Polymer Mach-Zehnder Modulators

EO polymer modulators are expected to have a bandwidth as much as 10 times greater than current $LiNbO_3$ devices due to better matching of optical and electrical phase velocities.[22] Thus, a true microstrip electrode design can be utilized to achieve very wide bandwidths in EO polymer devices. Analog modulation of these devices has been reported up to 40 GHz.[3] We report here the digital testing of our polymer modulators at clock rates near 1 GHz.

Mach-Zehnder modulators were built as shown in Figure 1, and then tested at high speed using an Anritsu ME523 bit error rate tester (BERT) to evaluate their digital switching characteristics. The BERT system was used to drive one microstrip electrode using a pseudorandom bit pattern at a clock rate of 900 MHz, the maximum clock rate for this system. The modulated light output was detected by the receiver portion of the BERT system while a sampling oscilloscope monitored this output, resulting in an "eye-pattern". For these tests a laser diode with $\lambda = 980$ nm was used as the laser source since the optical detector in this system peaks at 1300 nm and was not sensitive at 830 nm.

The polymer Mach-Zehnder modulator was mounted in a special microwave test fixture. Contact to each end of the microstrip electrodes was made by precision micromanipulators, each containing an Omni Spectra 50 GHz connector (OS-50). Only one electrode was driven, and the end of this electrode was terminated with a 50 ohm resistor. The pins of the OS-50 connectors made direct mechanical contact with the ends of the microstrip line while contact to the Al ground plane was established nearby. The other electrode was not used for this test.

Bit Error Rate Test Results. A photograph of an eye-pattern from the oscilloscope at a clock rate of 900 Mb/s is shown in Figure 9. The device performance corresponded to a BER of 10^{-12}, with an eye-pattern showing clear transitions between on and off. These transitions were likely faster than observed by the sampling oscilloscope, which was limited by a 600 MHz bandwidth. The noise in the

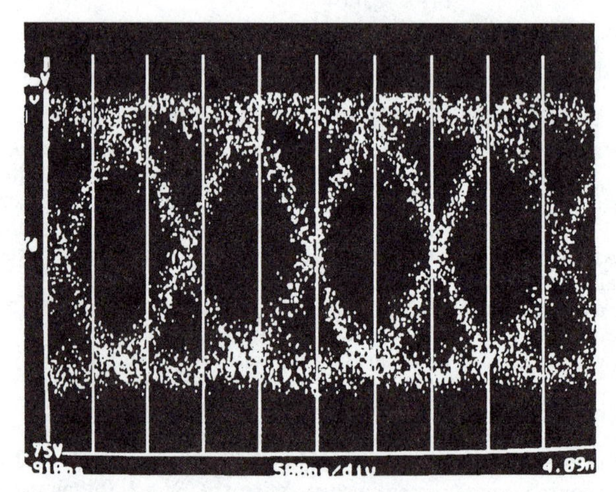

Figure 9. Eye-pattern of polymer Mach-Zehnder modulator at 900 MHz.

signal was caused by laser amplitude fluctuations, and this noise made it difficult to obtain long term BER data. The actual polymer modulator performance is expected to be significantly above the equipment limitations.

The high speed performance is attributed to driving the modulator with a microstrip electrode which was terminated with a 50 Ω resistor, and to the high speed response of the polymer as discussed above. The actual switching performance is expected to be faster than observed on the oscilloscope. Although the switching speed is good, a further improvement may be realized by increasing the thickness of the Al ground plane metallization to 4 μm, which would be the same as for the microstrip electrode.

Future Directions for High Speed Electro-Optic Devices

The polymer materials used in this work were chosen because of their availability. These materials were used to illustrate certain trends in EO polymer devices and should not be considered as the ultimate in polymer materials. As advances in EO polymer materials and processing are made, devices with lower switching voltages will be achieved. Since polymer devices can be integrated on top of high speed substrate circuitry, such as emitter coupled logic, very high speed EO modulation can be achieved. The speed at which these devices can be driven will be limited by the skin effect of the driving electrodes. However, as the EO coefficient of the polymer stack increases, the electrode interaction length can be shortened, and this will reduce the high frequency loss due to electrode skin effect. In recent years chemists have made significant progress in developing polymers for EO devices, and their future developments are expected to significantly improve the performance of modulators, switches and other EO polymer devices.

References

1. Williams, D. J. In *Nonlinear in Optical Properties of Organic and Polymeric Materials*, Williams, D. J., Ed.; ACS Symposium Series 223, American Chemical Society: Washington, DC 1983; preface.
2. Girton, D. G.; Kwiatkowski, S. L.; Lipscomb, G. F.; Lytel, R. S. *Appl. Phys. Lett.* **1991**, *58*, 1730-1732.
3. Teng, C. C. *Appl. Phys. Lett.* **1992**, *60*, 1538-1540.
4. Möhlmann, G. R.; Horsthuis, W. H. G.; McDonach, A.; Copeland, M. J.; Duchet, C.; Fabre, P.; Diemeer, M .B .J.; Tramel, E.S.; Suyten, F. M. M.; Van Tamme, E.; Baqero, P.; Van Daele, P. In *Nonlinear Optical Properties of Organic Materials III*, Khanarian, G., Ed.; Proc. SPIE 1337; Society of Photo-Optical Instrumentation Engineers: Bellingham, WA, 1990; pp 215-225.
5. Van Eck, T. E.; Ticknor, A. J.; Lytel, R. S.; Lipscomb, G. F. *Appl. Phys. Lett.* **1991**, *58*, 1558-1590.
6. Girton, D. G.; Lytel, R. S.; Lipscomb, G. F. *Conference on Lasers and Electro-optics*, 1992 Technical Digest Series; Optical Society of America: Washington DC, 1992; Vol. 12, p 580.
7. Tumolillo, T. A.; Ashley, P. R. *Appl. Phys. Lett*, **1993**, *62*, 3068-3070.
8. Kenney, J. T.; Binkley, E. S.; Nurse, J.; Jen, A. K.-Y.; Rao, V. P. *ACS Polymer Preprints* **1994**, *35*(2), 221-222.
9. Watanabe, T.; Amano, M.; Hikita, M.; Shuto, Y.; Tomaru, S.; *Appl. Phys. Lett.* **1994**, *65*, 1205-1207.
10. Ermer; S.; Lovejoy, S. M.; Leung, D. S., this volume.
11. Yariv, A. *Quantum Electronics*, John Wiley: New York, NY, 1975; 2nd ed., pp 349-351.
12. Rangaraj, M.; Hosoi, T.; Kondo, M. *Photon. Technol. Lett.* **1992**, *4*, 1020-1022.
13. Bahl, I. J. In *Handbook of Microwave and Optical Components*, K Chang, Ed.; John Wiley: New York, N. Y., 1989, Vol. 1, 1-59.
14. Schneider, M. V.; Glance, B.; Bodtmann, W. F. *Bell Sys. Tech. J.* **1969,** *48*, 1703-1726.
15. Singer, K. D.; Kuzyk, M. G.; Sohn, J. E. In *Nonlinear Optical and Electro-active Polymers*, P. N. Prasad and D. R. Ulrich, Eds., Plenum Press: New York 1988, pp 189-204.
16. Fujimoto, H. H.; Das, S.; Valley, J. F.; Stiller, M.; Dries, L.; Girton, D. G.; Van Eck, T. E.; Ermer, S.; Binkley, E. S.; Nurse, J. C.; Kenney, J. T. In *Electrical, Optical, and Magnetic Properties of Organic Solid State Materials*, Garito, A. F., Jen, A. K.-Y., Lee, C. Y.-C., and Dalton, L. R., Eds.; Materials Research Society Symposium Proceedings 328; Materials Research Society: Pittsburg, PA 1993, *27*, 1993; pp 553-564.
17. Neuhaus, H. J.; Day, D. R.; Senturia, S. D. *J. Electron. Mat.* **1985**, *14*, 379-404.
18. Smith, F. W.; Neuhaus, H. J.; Senturia, S. D. *J. Electron. Mat.* **1987**, *18*, 93-106.
19. Valley, J . F.; Wu, J. W.; Ermer, S.; Stiller, M.; Binkley, E. S.; Kenney, J. T.; Lipscomb, G. F.; Lytel, R. *Appl. Phys. Lett.* **1992**, *60*, 160-162.

20. Ermer, S.; Valley, J. F.; Lytel, R.; Lipscomb, G. F.; Van Eck, T. E.; Girton, D. G. *Appl. Phys. Lett.* **1992**, *61*, 2272-2274.
21. Girton, D. G.; Ermer, S.; Valley, J. F.; Van Eck, T. E.; Lovejoy, S. M.; Leung, D. S.; Marley, J. A. *Polymer Preprints* **1994,** 35(2), 219-220.
22. Lytel, R.; Binkley, E. S.; Girton, D. G.; Kenney, J. T.; Lipscomb, G. F.; Ticknor, A. J.; Van Eck, T. E. In *Advances in Interconnection and Packaging,* De Fonzo, A. P., Ed.; Proc. SPIE 1389; Society of Photo-Optical Instrumentation Engineers: Bellingham, WA, 1990; pp 547-558.

RECEIVED January 30, 1995

Chapter 34

Second Harmonic Generation by Counter-Directed Guided Waves in Poled Polymer Waveguides

A. Otomo[1], G. I. Stegeman[1,3], W. Horsthuis[2], and G. Möhlmann[2]

[1]Center for Research and Education in Optics and Lasers, University of Central Florida, Orlando, FL 32826
[2]Akzo Electronic Products, Arnhem, Holland

Doubling the frequency of laser light has been investigated since the early days of nonlinear optics. Until recently the goal has been to extend the frequency range of sources, mostly for scientific investigations. With the advent of inexpensive semiconductor lasers in the near infrared, interest has been strong in generating milliwatts of blue light for data storage and xerography. Still relatively unexplored are applications to signal processing, for example correlation, frequency demultiplexing etc. Again signal levels in the milliwatts range are required.

When dealing with sub-watt power inputs, either guided wave geometries or doubling in highly resonant cavities is required.*(1,2)* In fact most of the work has concentrated on interaction geometries in which the fundamental and second harmonic waves travel in essentially the same direction. In terms of waveguides, the best results to date have been obtained with ferroelectric materials such as $LiNbO_3$, KTP etc.*(1,3,4)* Organic materials, including poled polymers, have been shown to have large nonlinearities, but to date the loss in the blue region of the spectrum has been too large for these materials to be truly competitive with the oxide materials for blue light generation.*(5)* The problem has been that the harmonic has to travel multimillimeter distances to achieve large enough signal intensities.

We have been pursuing a different sum frequency generation geometry in waveguides.*(6)* Although the original goal was for signal processing, demultiplexing in our case, our approach may also prove useful for efficient blue light generation.*(7)* In our scheme, two oppositely propagating beams are mixed in a nonlinear ($\chi^{(2)}$-active) waveguide, and the resulting signal is radiated out of the waveguide, normal to the surface. Because the two input fields pass through (sample) each other, the time-dependence of the output signal contains information about the convolution of the input signals.*(8)* A nonlinear polarization source field polarized in the plane of the surface is required.

Counter-propagating SHG has been demonstrated previously in ferroelectrics and semiconductors.*(6,9)* The best conversion efficiencies have been obtained in modulated AlGaAs layered structures.*(10)* Recently we have been investigating poled

[3]Corresponding author

0097–6156/95/0601–0469$12.00/0

polymers and LB films for this purpose.*(11,12)* In this paper we discuss poled polymers for counter-propagating SHG.

The Counter-Propagating Geometry

The slab waveguide interaction geometry is shown in Figure 1. We write the incident guided wave fields as

$$E_i(r,t) = \frac{1}{2}\mathscr{E}_i(x)a_{\pm}(z)\exp^{i(\omega_{\pm}t \mp \beta_{\pm}z)} + c.c. \tag{1}$$

where the + and - signs refer to guided waves propagating along the + and -z axes respectively. For example, for TE polarized input fields, the case for poled polymer films, i ≡ y. The fields are normalized so that $|a_{\pm}(z)|^2$ is the guided wave power per unit distance along the y-axis. Note that we have chosen the case in which the input frequencies and propagation wavevectors, $\omega_{\pm}$ and $\beta_{\pm}$ respectively, are not necessarily equal. Assuming that the waveguide consists of a non-centrosymmetric material, the nonlinear polarization induced is

$$\begin{aligned} P_k(r,t) &= \frac{1}{2}[\wp_{k+}(x,z)e^{2i[\omega_+ t - \beta_+ z]} + c.c.] \\ &+ \frac{1}{2}[\wp_{k-}(x,z)e^{2i[\omega_- t + \beta_- z]} + c.c.] \\ &+ \frac{1}{2}[\wp_{k\pm}(x,z)e^{i[(\omega_+ + \omega_-)t - (\beta_+ - \beta_-)z]} + c.c.] \\ &+ \frac{1}{2}[\wp_{k\mp}(x,z)e^{i[(\omega_+ - \omega_-)t - (\beta_+ + \beta_-)z]} + c.c.] \end{aligned} \tag{2}$$

The first two terms correspond to co-propagating second harmonic generation (SHG), one for each of the input waves. The third term, $\wp_{\pm}$, is a polarization source oscillating at the frequency $\omega_+ + \omega_-$ with wavevector parallel to the surface of $\beta_+ - \beta_-$. It leads to the sum frequency signal of interest here. The fourth term at $\omega_+ - \omega_-$ oscillates at a low frequency and has too large a spatial wavevector $\beta_+ + \beta_-$ to lead to radiation terms for $\omega_+ - \omega_-$. For the case of interest

$$\wp_{i\pm}(x,z) = 2d^{(2)}_{ijk}(-[\omega_+ + \omega_-];\omega_+,\omega_-)\mathscr{E}_j(x)\mathscr{E}_k(x)a_+(z)a_-(z) \tag{3}$$

In the poled polymer case, i=j=k=y which implies in-plane poling.

The easiest case to understand is for $\omega_+ = \omega_-$, i.e. second harmonic generation (SHG). Because there is no spatial periodicity in the plane of the surface, the mixing results in a nonlinear polarization sheet, oscillating at 2ω, with variation along the x-axis given by the product $\mathcal{E}^2(x)$. The only direction into which this dipole sheet can radiate is orthogonal to the waveguide surface, along the x-axis. The variation along the x-axis of both the generated and source field is sketched in Figure 2. Because the SHG field oscillates, interference effects occur, limiting the magnitude of the output signal. Mathematically, the output signal intensity is given by

$$I(2\omega) = A^{NL} I_+(\omega) I_-(\omega) \tag{4}$$

where $I_\pm(\omega)$ ($\equiv |a_\pm|^2$) are the line intensities (W/m) of the input guided waves and A^{NL} is the nonlinear cross-section coefficient given by

$$A^{NL} = \frac{1}{2} c n_c(2\omega) \epsilon_0 |TS|^2 \tag{5}$$

Here $n_c(2\omega)$ is the cladding refractive index [x>0], T is an effective transmission coefficient which takes into account multi-reflections of the signal in the guiding film and S is given by

$$S = i \frac{\omega}{n_f(2\omega) c} \int_{-\infty}^{0} d^{(2)}_{yyy}(-2\omega;\omega,\omega) \mathcal{E}_y^2(x') e^{2in_f(2\omega)\frac{\omega}{c}x'} dx' \tag{6}$$

It proves more convenient to express the result of this nonlinear process in terms of input and output powers as

$$P(2\omega) = A^{NL} \frac{L}{W} P_+(\omega) P_-(\omega) \tag{7}$$

Here L is the length of the interaction region (along z) and W is the width of the guided wave beam (along y) in the plane of the surface. Clearly the narrower the beam the better, making channel waveguides the geometry of choice for optimizing signal power. For 1 cm long devices, the geometrical ratio L/W can be as large as 10^4. The disadvantage of small W is that this also represents the radiating aperture along the y-axis and the smaller the aperture, the larger the signal beam divergence. Thus a lens will be necessary to gather the signal and focus it onto a target in the y-dimension. Similarly the angular resolution along the z-axis λ/L is determined by L, the equivalent radiating aperture.

The case $\omega_+ \neq \omega_-$ has potential applications to real time wavelength demultiplexing. As indicated in Figure 1, the sum frequency is radiated at some angle away from the surface normal. Conservation of wavevector in the plane of the surface gives

$$\sin[\Delta\theta] = \frac{(\beta_+ - \beta_-)c}{(\omega_+ + \omega_-)n_c(2\omega)} \approx \frac{\omega_+ - \omega_-}{\omega_+ + \omega_-}\frac{n_f(\omega)}{n_c(2\omega)} \qquad (8)$$

In these equations, when we write ω as the argument for the refractive indices we mean $\omega \approx (\omega_+ + \omega_-)/2$. Note that if there is a cladding above the film, but the signal is detected in air, then from Snell's law the angle in air is obtained by replacing $n_c(2\omega)$ by 1. The key point is that the angle is effectively linear in the frequency (wavelength) difference between the two input signals. Therefore if one is a reference beam, then the different wavelength components of the input signal beam are separated spatially, with resolution $\Delta\lambda/\lambda \approx L/W$. This is one of the applications which we are pursuing.

Fabrication of DANS Poled Polymer Waveguides

The waveguide fabrication process consists of five distinct steps. First, gratings are milled into the substrate glass surface to facilitate coupling into guided modes. Second, parallel electrodes are deposited on the glass surface for poling of the polymer. Next, the polymer film is spun on to a predetermined thickness. Fourth, the waveguide structure is made by photobleaching through an appropriate mask. And finally, the film is poled at high temperatures.

Our technology did not initially allow us to couple high powers into channel waveguides by focusing the input light onto the end facet of the waveguide. Thus we adopted a grating coupler approach to excite the guided waves, coupled with a tapered section to decrease the beam width down to single mode channel dimensions. The grating couplers, the taper sections, and the single mode channel waveguide region are shown in Figure 3. The grating was fabricated in the glass substrate prior to spinning on the DANS film in the usual way. That is, photoresist was spun onto the substrate and exposed to the interference pattern of a He-Cd laser to produce a period of about 500 nm. The exposed photoresist was "developed" and the grating ion milled into the substrate to a depth of approximately 100 nm.

The particular interaction being used here requires large effective nonlinearities in the plane of the waveguide, i.e. $d_{yyy}^{(2)}(2\omega;\omega,\omega)$. This requires in-plane poling of polymers. Although transverse poling, primarily corona poling, has been by far the most frequently used approach, there are prior reports of successful in-plane poling.*(13-16)* Typically the problem has been arcing (electrical discharge) which has limited in many cases the maximum applied voltages to 50 to 100 V/μm.*(17)*

We used the side-chain polymer DANS which has been investigated extensively for its application to electro-optic devices.*(18,19)* The DANS molecule had about a 50% loading and was attached to the backbone polymer as a side-chain. Parallel aluminum electrodes were deposited with a spacing of approximately 20 μm. Sub-micron thick films were spun on. Using standard techniques we measured slab waveguide losses as low as 0.27 db/cm. Values of 0.7 db/cm were quite typical of high quality waveguides.

Channel-like waveguide structures were formed by photobleaching through an

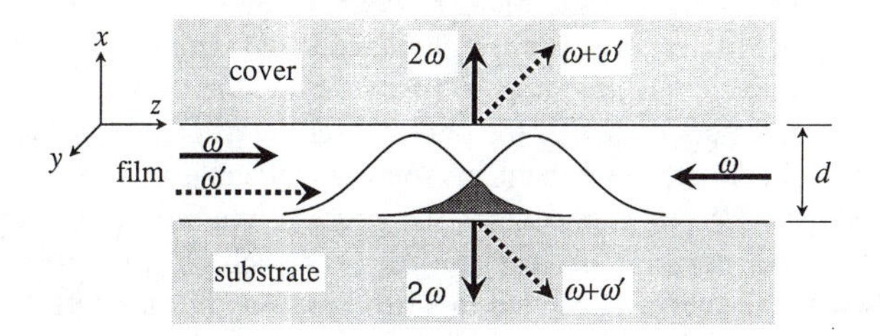

Figure 1. The waveguide counter-propagating SHG geometry for both equifrequency and unequal frequency inputs (ω'≠ ω).

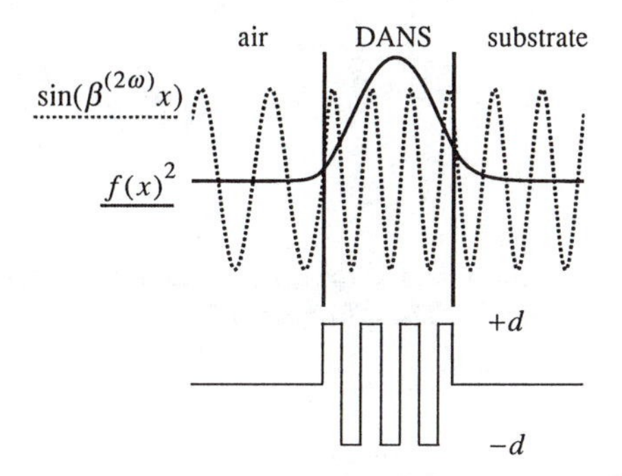

Figure 2. The variation in the polarization source field (solid line) and the generated second harmonic electric field (dashed line) along the transverse direction (along the normal to the surface).

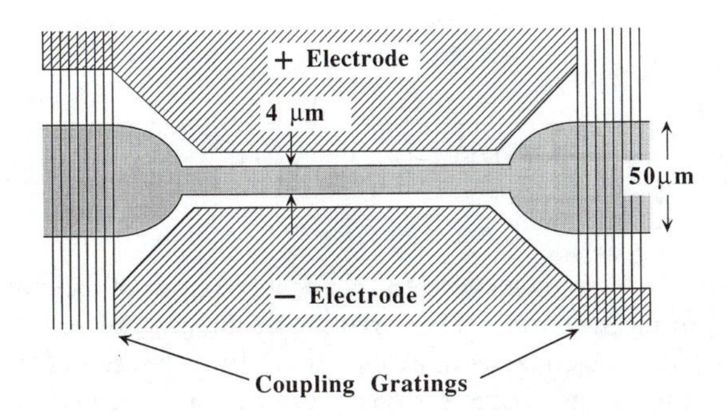

Figure 3. The SHG device structure viewed from the direction into which the SHG signal appears. Note the grating couplers at either end, the tapered transition region and the central narrow channel waveguide.

appropriate mask. Because photobleaching reduces the refractive index, it is necessary to block out the waveguiding region when illuminating the sample . In order to design the required structures it was necessary to first calibrate the photobleaching process. There is some debate about whether the photobleaching of DANS is due to cis-trans isomerization or due to other mechanisms such as oxygenation of one of the DANS bonds, or both.*(20-22)* Independent of the process, we begin our modelling by assuming that the number of unchanged (unbleached) molecules decreases exponentially with the energy deposited in a film small on the scale of the maximum inverse absorption coefficient. Thus the contribution to the linear susceptibility, $\Delta\chi$, due to the DANS molecules with an intact charge transfer state is given by*(23)*

$$\Delta\chi'(t) = \Delta\chi'_{max}\exp\left[-\beta_b\int_0^t I(t')dt'\right]$$

$$\Delta\chi''(t) = \Delta\chi''_{max}\exp\left[-\beta_b\int_0^t I(t')dt'\right] \quad (9)$$

where β_b is the bleaching constant and the single and double primes refer to the real and imaginary parts respectively. The refractive index, real [n(t)] and imaginary [κ(t)]

$$n(t) = \frac{1}{\sqrt{2}}\left[\sqrt{[\epsilon'_b+\Delta\chi'(t)]^2 + [\epsilon''_b + \Delta\chi''(t)]^2} + [\epsilon'_b+\Delta\chi'(t)]\right]^{\frac{1}{2}}$$

parts evolve as

$$\kappa(t) = \frac{1}{\sqrt{2}}\left[\sqrt{[\epsilon'_b+\Delta\chi'(t)]^2 + [\epsilon''_b + \Delta\chi''(t)]^2} - [\epsilon'_b+\Delta\chi'(t)]\right]^{\frac{1}{2}}$$

where ϵ_b refers to the background dielectric constant due to all of the other resonances in the UV part of the spectrum. The evolution of the photobleached species with distance into a film is obtained by modelling that film as a series of very thin slabs, thin enough so that the illumination intensity can be considered constant over each slab. The problem is then solved on a computer to calculate the index distribution in the film.

The transmission through a DANS film (absorption maximum at $\lambda = 430$ nm) was measured as a function of time. The source was the $\lambda = 442$ nm line from a He-Cd laser. The results are shown in Figure 4. The fit to the data is within the spread of the experimental data, yielding a value of $\beta_b = 7.5x10^{-7}$ m^2/J. This allowed us to calculate the refractive index as a function of depth into a thick film, as shown in Figure 5 for an illumination intensity of $40mW/cm^2$. This information is needed to obtain the photobleaching times etc. for the formation of channel waveguides.

The standard effective index method was then used to model the photobleached channel waveguides.*(24)* The conditions used to form effectively single mode waveguides in both dimensions are shown in Figure 6. Note that the in-plane mode structure allows a higher order mode, but it is difficult to excite because it is very near cut-off. Using grating couplers only one mode at a time is excited so that the higher order mode plays no significant role in our devices.

The taper section, shown in Figure 7, was used to increase the intensity in the channel waveguide by a factor of 12.5 over that of the coupling region, which increases the SHG signal by a factor of 150. The form of the taper is chosen to minimize radiation losses out of the waveguide, i.e. an adiabatic taper for which $\theta < \lambda/[N_{eff}(z)W(z)]$ where $N_{eff} = \beta(z)/k_0$ is the effective index.*(25,26)* It is given by

$$W(z) = \sqrt{2\frac{\lambda}{N_{eff}}z + W_{min}^2} \tag{12}$$

In our case L = 1.9mm, $W_{max} = 50\mu m$, $W_{min} = 4\mu m$, $\lambda = 1064nm$ and $N_{eff} = 1.62$. The taper was fabricated by photobleaching, leaving the region outside the taper with a lower refractive index. The additional propagation loss due to the taper and channel waveguide structure was measured to be only 24% (relative to a slab waveguide on the same substrate). The total throughput, including input coupling, taper transitions, channel propagation and output coupling was about 10%. The effective channel device length was 1 cm.

Over a period of one year we made various improvements to increase the maximum poling voltage from the frequently reported 50-100 V/μm.*(27)* Initially we found arcing to occur at 100 V/μm or less, typical of results reported in the literature. We were poling just below the glass transition temperature of 142°C, using glass slides as substrates with air above the film. The current flow into one of the electrodes was monitored and found to be tens of μA just before film damage occurred. This was improved by using high purity fused silica substrates, eliminating the charge injection and migration, and increasing the breakdown voltage to over 100 V/μm. Most of the experiments reported here were performed with these samples. At higher poling fields, we found that arcing occurred along the air-film interface. In order to eliminate this problem we added a cladding layer of lower refractive index, spun on to a thickness of about 15 μm. As a result, poling voltages up to 370 V/μm were recently achieved before breakdown occurred. The films were poled for about 30 minutes and then the temperature was decreased slowly (<1°C/min.) with the voltage maintained.

Standard Maker fringe techniques were used to measure the film nonlinearity. This was done at the wavelengths 1064nm and 1550nm for a sample poled at 200 V/μm, apparatus shown in Figure 8. The results are shown in Figure 9. It is clear that the $d^{(2)}$ coefficient for 1064nm inputs is enhanced due to the proximity of the

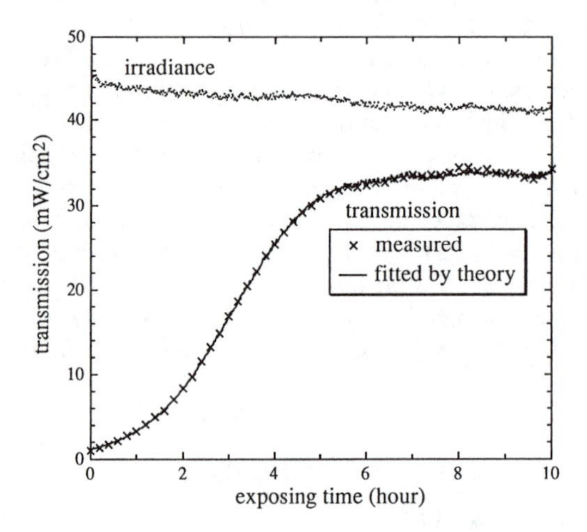

Figure 4. Measured transmission change during bleaching for a 320 nm thick DANS film.

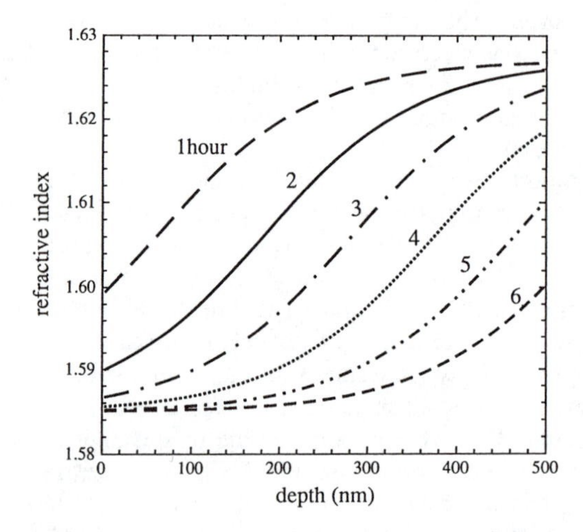

Figure 5. Calculated bleaching depth as a function of exposure time.

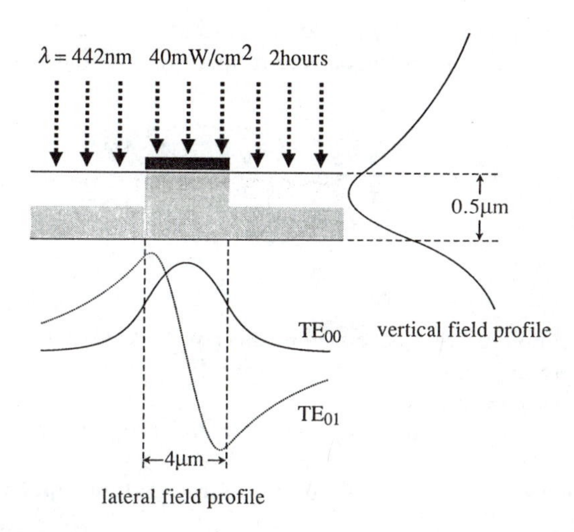

Figure 6. Photobleaching conditions and calculated electric field profiles in a channel waveguide.

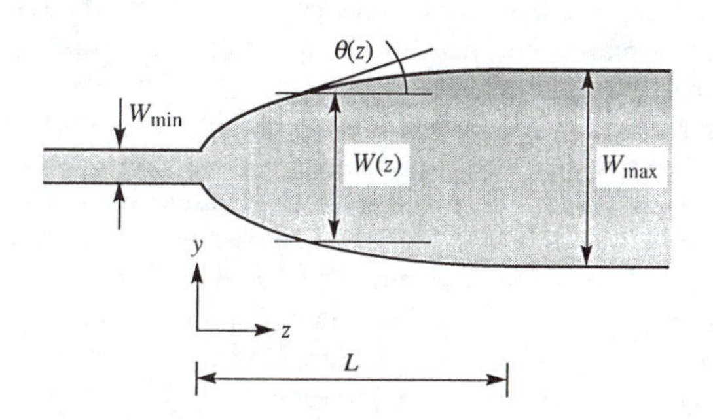

Figure 7. Parabolic shape tapered mode coupler.

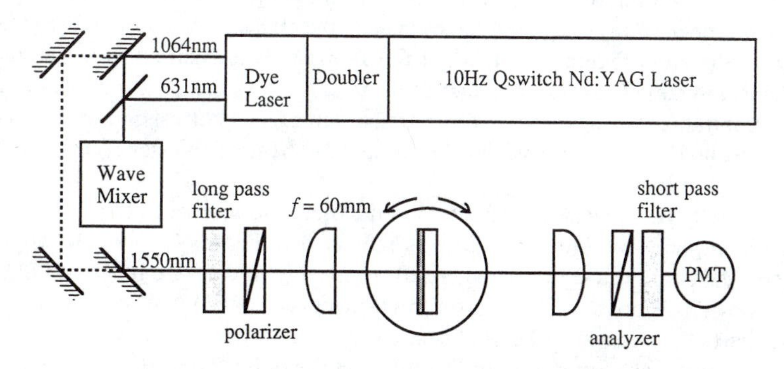

Figure 8. Apparatus used for Maker Fringe measurements of nonlinearity.

resonance at 430nm to the second harmonic signal at 532 nm. In fact we obtained good agreement with the dispersion derived from a two level model of the form*(28)*

$$d_{22}^{\omega} = d_{22}^{\omega'} \frac{f_2^{2\omega} f_2^{\omega} f_2^{\omega}}{f_2^{2\omega'} f_2^{\omega'} f_2^{\omega'}} \frac{(\omega_0^2 - \omega^2)(\omega_0^2 - 4\omega^2)}{(\omega_0^2 - \omega^2)(\omega_0^2 - 4\omega^2)} \tag{13}$$

The solid line in Figure 9 was generated by this equation. Over the range of poling fields used in our experiments, the nonlinearity $d^{(2)}$ was found to be linear in the applied poling field, see Figure 10.

An end view of the sample structure is reproduced in Figure 11.

Second Harmonic Generation by Oppositely Propagating Inputs: Experiments

We first examined numerically the trade-offs in S and hence the cross-section coefficient A^{NL} for typical DANS waveguide parameters. The magnitude of the overlap integral oscillates with increasing film thickness t as discussed above and shown in Figure 12. Since the guided mode profile $f_y(x')$ for the fundamental mode is broadened with increasing thickness, the envelope of this oscillation decreases with increasing thickness when normalized to input power. The important tendencies are clearly shown in Figure 12 which is based on typical DANS parameters ($n_f(\omega) = 1.63$, $n_f(2\omega) = 1.8$). The first and largest peak value is obtained for t = 0.47μm. A cover polymer which has $n_c(\omega) = 1.55$ was used in producing Figure 12. Based on these predictions, the film thickness of 0.47μm was chosen because it optimizes the cross-section and it was fabricated to an accuracy of ±0.04μm determined from calibrated data on the spinning rate and the solid concentration in the spinning solution.*(23)*

The surface emitted SHG power was implemented by coupling 100 picosecond Q-switched mode locked pulses at 1064 nm into both sides of the waveguide. The experimental layout is shown in Figure 13. A pulse was split into two identical halves and the overlap of these two pulses in the waveguide was optimized by changing the optical path length of one arm. Note that back coupling was used so that the guided waves are generated by the gratings in essentially the opposite direction to the incident beams. Shown in Figure 14 is the line signal observed emanating from the surface (horizontal streak). Close examination of the streak shows that it decays at both ends of the central (channel) region. These transition regions correspond to the tapered coupler section where the fundamental intensity (and hence SHG signal) is lower than in the channel region.

The SHG power was measured in the direction normal to the surfaces by a photo multiplier tube through a lens, as indicated in Figure 13. For the 50 V/μm devices, the only ones to be characterized completely to date, the measured SHG power was 17mW for a 32W input power which yields 0.64GW/cm^2 of peak intensity in the channel waveguide. The nonlinear cross- section was evaluated as $9\text{x}10^{-9}\text{W}^{-1}$ by assuming a Gaussian shape for the pulses in the pulse train. Note that this value agrees very well with our calculated value of $1.5\text{x}10^{-8}$ W^{-1}. Such an efficient single

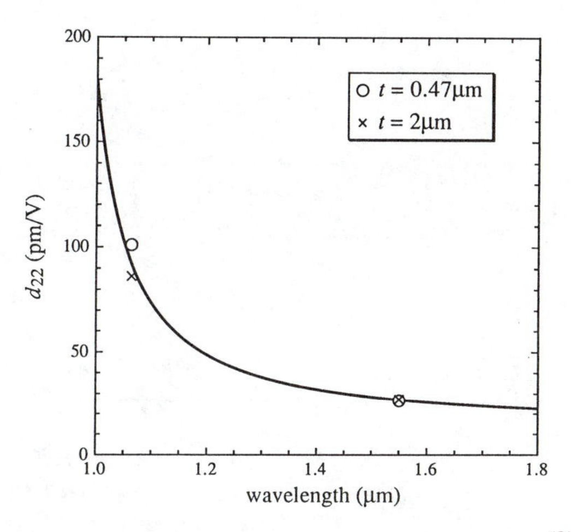

Figure 9. Wavelength variation of the nonlinear coefficient $d^{(2)}$ for DANS poled at 200 V/μm.

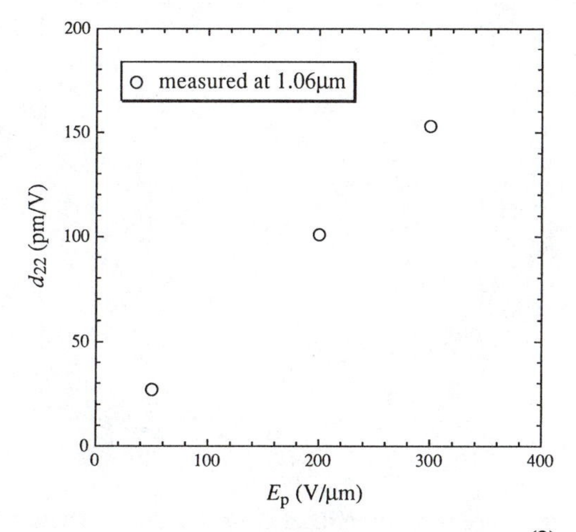

Figure 10. Variation of measured nonlinear coefficient $d_{yyy}^{(2)}(2\omega;\omega,\omega)$ for input at 1064 nm.

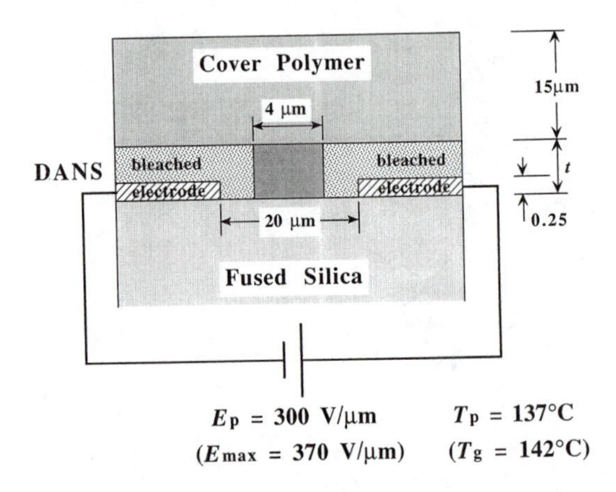

Figure 11. Schematic cross section of the waveguide.

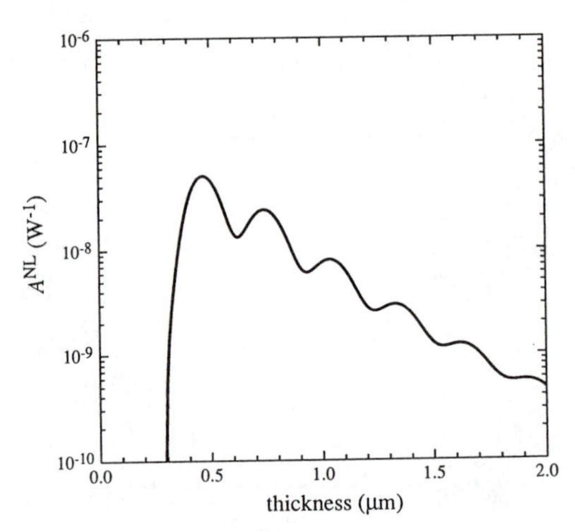

Figure 12. Dependence of the nonlinear cross-section coefficient A^{NL} on the DANS film thickness.

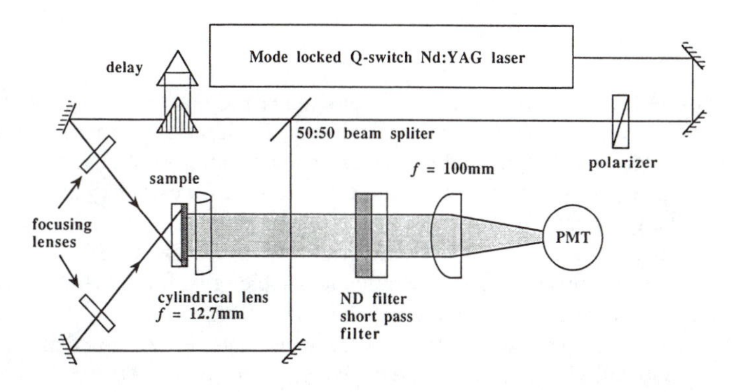

Figure 13. The apparatus used for measuring contrapropagating SHG.

Figure 14. The SH signal (horizontal line) emanating from the DANS waveguide, viewed normal to the waveguide surfaces.

layer device should make it possible to operate over a large frequency range, and to demonstrate new applications such as spectrometers etc. In principle, extending the device to multiple layer films with alternate films of zero activity should lead to even larger efficiencies.

Summary

In summary, SHG generated by the mixing of counter propagating guided waves was demonstrated with polymer channel waveguides. High efficiency was observed in poled DANS side chain polymer films. The reasonable agreement observed between the experimental values and theory supports the promise for even higher efficiencies with optimized devices. For example, by using our samples with 300 V/μm poling voltages the cross-section can be enhanced by a factor of 36. Additional enhancements could be obtained by operating at wavelengths for which the harmonic is closer to the resonance. Furthermore, by using advanced poling technology it should be possible to make multilayers with alternate layers poled in opposite directions. With such improvements we expect an increase with DANS to an $A^{NL} = 10^{-4} - 10^{-3}\ W^{-1}$. This implies that 100 mW inputs could produce as much as 10 mW of harmonic signal. We will pursue these and other approaches to optimizing the interaction cross-section in the future.

The fabrication technology and physical properties of the poled polymers make them very attractive for this particular nonlinear interaction. The projected polymer values for the cross-section coefficient are 4-5 order of magnitude better than the best previous results in $LiNbO_3$ waveguides, and 1-2 orders of magnitude better that the best existing semiconductor results. *(8-10)* Furthermore, the poling fabrication technology is more versatile and cheaper than the MBE methods used to fabricate semiconductor samples. However, semiconductors do offer the attractive possibility of combining the laser source and second harmonic convertor on the same chip.

The authors acknowledge the support of the Air Force Office of Scientific Research (AFOSR-91-0339). We also thank J. Ross for preparing the coupling gratings and the poling electrodes.

Literature Cited

1. Fejer,M.M.; Magel, G.A.; Jundt, D.H.; Byer, R.L., *IEEE J. Quant. Electron., QE-28*, 2631 (**1992**).
2. Dixon, G.J.; Tanner, C.E.; Wieman, C.E., *Opt. Lett., 14*, 731 (**1989**).
3. Yamamoto, K.; Mizuuchi, K., *IEEE Phot. Techn. Lett., 4*, 435 (**1992**).
4. van der Poel, C.J.; Bierlein, J.D.; Brown, J.B.; Colak, S., *Appl. Phys. Lett., 57*, 2074 (**1990**).
5. Shi, Y.; Ranon, P.M.; Steier, W.H.; Xu, C.; Wu, B.; Dalton, L.R., *Appl. Phys. Lett., 63*, 2168 (**1993**).
6. Normandin, R.; Stegeman, G. I., *Opt. Lett. 4*, 58 (**1979**).
7. Stegeman,G. I.; Burke, J. J.; Seaton, C. T., in *Optical Engineering: Integrated Optical Circuits and Components*, edited by Hutcheson,L. D., (Marcel Dekker, New York, **1987**), *Vol. 13.*
8. Normandin, R.; Stegeman, G. I., *Appl. Phys. Lett. 36*, 253 (**1980**).

9. Normandin, R.; Williams, R.L.; Chatenoud, F., *Electron. Lett., 26*, 2088 (**1990**).
10. Normandin, R.; Letourneau, S.; Chatenoud, F.; Williams, R. L., *J. Quant. Electron., QE-27*, 1520 (**1991**).
11. Otomo, A.; Mittler-Neher, S.; Bosshard, C.; Stegeman, G.I.; Horsthuis, W.H.G.; Möhlmann, G.R., *Appl. Phys. Lett., 63*, 3405 (**1993**).
12. Bosshard, Ch.; Otomo, A.; Stegeman,G.I.; Kupfer, M.; Florsheimer, M.; Gunter, P., *Appl. Phys. Let., 64*, 2076 (**1994**)
13. Tumolillo Jr, T.A.; Ashley, P.R., *Appl. Phys. Lett., 62*, 3068 (**1993**).
14. Tatsuura, S.; Sotoyama, W.; Yoshimura, T., *Appl. Phys. Lett., 60*, 1661 (**1992**).
15. Wu, J.W.; Valley, J.F.; Ermer, S.; Binkley, E.S.; Kenney, J.T.; Lipscomb, G.F.; Lytel, R., *Appl. Phys. Lett., 58*, 225 (**1991**).
16. Boyd, G.T.; Francis, C.V.; Trend, J.E.; Ender, D.A., *J. Opt. Soc. Am. B, 8*, 887 (**1991**).
17. Stahelin, M.; Walsh, C.A.; Burland, D.M.; Miller, R.D.; Twieg, R.J.; Volksen, W., *J. Appl. Phys., 73*, 8471 (**1993**).
18. Valley, J.F.; Wu, J.W.; Ermer, S.; Binkley, E.S.; Kenney, J.T.; Lipscomb, G.F.; Lytel, R., *Appl. Phys. Lett., 60*, 160 (**1992**).
19. Möhlmann, G.R.; Horsthuis, W.G.H.; Mertens, J.W.; Diemeer, M.B.J.; Suyten,F.M.M.; Hendriksen, B.; Duchet, C.; Fabre, P.; Brot, C.; Copeland, M.; Mellor, J.R.; Van Tomme, E.; Van Daele, P.; Baets, R., *SPIE 1560*, 426 (**1991**).
20. M. D. J. Diemeer, F. M. M. Suyten, E. S. Trommel, A. McDonach, J. M. Copeland, L. W. Jenneskens and W. H. G. Horsthuis, Electron. Lett. **26**, 379 (1990).
21. Van Eck, T.E.; Ticknor, A.J.; Lytel, R.S.; Lipscomb, G.F., *Appl. Phys. Lett., 58*, 1588 (**1991**).
22. Norwood, R.A.; Holcomb, D.R.; So, F.F., *Nonlin. Optics, 6*, 193 (**1993**).
23. Otomo, A., *Design and Fabrication of Channel Waveguides in a 4-dimethylamino-4'-nitrostilbene Side Chain Polymer*, MSc thesis, Un. Central Florida.
24. Kogelnik, H., "Theory of Dielectric Waveguides", in *Integrated Optics, Vol. 7 of Topics in Applied Physics*, edited by Tamir, T., (Springer-Verlag, Berlin, **1979**).
25. Milton, A.F.; Burns, W.K., *IEEE J. Quant. Electron., QE-13*, 828 (**1977**).
26. Otomo, A.; M.-Neher, S.; Stegeman, G. I.; Horsthuis, W. H. G.; Möhlmann, G. R., *Electron. Lett. 29*, 129 (**1993**).
27. Otomo, A.; Stegeman, G.I.; Horsthuis, W.H.G.; Mohlmann, G.R., *Appl. Phys. Lett., 65*, 2389 (**1994**).
28. Teng, C.C.; Garito, A.F., *Phys. Rev. B, 28*, 6766 (**1983**).

RECEIVED February 14, 1995

Chapter 35

Second Harmonic Femtosecond Pulse Generation in Nonlinear Polymer Thin-Film Structures

Andrew Dienes, Erkin Sidick, Richard A. Hill, and André Knoesen

Department of Electrical and Computer Engineering, University of California, Davis, CA 95616

Ultrashort-pulse second harmonic applications of nonlinear poled polymeric films (NPPF) and their optimization in terms of nonlinear chromophore concentration and thin film structure are discussed. The objective of such optimization is to maximize the conversion efficiency and, more importantly, minimize the pulse distortions in ultrashort-pulse second harmonic generation. It is shown that each NPPF has a specific optimum chromophore concentration density. It is also shown that the quasi-phase-matched structures of NPPFs not only enhance the efficiency of ultrashort-pulse second harmonic generation but also minimize pulse distortions.

During the last decade the potential of polymeric thin films for applications in nonlinear optics has received considerable attention. In particular, second-order nonlinear processes, which include second harmonic generation (SHG), two wave mixing and electro-optic modulation, have been extensively investigated. It has been demonstrated that polymeric thin films with large second-order nonlinearities can be created by permanently orienting molecular components with a large hyperpolarizability within a polymer host. The magnitudes of the nonlinearities can exceed those of inorganic second-order materials. In addition, the refractive index of organic polymers is low, a great advantage for electro-optic applications. Nonlinear poled polymeric films (NPPFs) have other attractive properties for SHG. Perhaps the most important of these is the flexibility which exists through both chemical synthesis and through the creation of multilayer structures for optimization of various physical properties for specific device requirements. One broad class of applications for NPPFs is guided wave nonlinear optics. The creation of nonlinear guided wave structures from these films for use in continuous-wave (cw) or quasi-cw SHG of optical waves has been explored recently (*1*). For these applications the main challenge lies in reducing the losses which severely limit the usable waveguide length.

NPPFs are uniquely suited for ultrashort-pulse nonlinear optical applications (2). Ultrashort optical pulses as short as 10 fs can now be generated directly from mode-locked lasers (*3*) and generation of new frequencies by nonlinear optical means is of great interest. In these applications the optical waves are not guided by the film but

0097–6156/95/0601–0484$12.00/0

must instead propagate *through* the film and the interaction length is on the order of microns. The short interaction length has the important advantage of reducing the pulse distortions caused by the frequency dispersion of linear refractive index. To avoid these distortions in inorganic nonlinear crystals, very thin materials must be used. This is particularly true if the center wavelength of the second harmonic (SH) pulse is in the UV. For pulses in the few tens of femtosecond range the required thickness is so small (on the order of 50 μm or less) that the crystals are difficult if not impossible to fabricate and polish (*4*). In contrast, single and multilayer NPPFs with excellent optical quality can be made in thicknesses ranging from 1 μm to 30 μm with relative ease, making NPPF very attractive for ultrashort pulse second-order nonlinear optics in general and for ultrashort pulse second harmonic generation (USP-SHG) in particular. Although a short interaction length limits the efficiency, the second order nonlinear coefficient (*d*-coefficient) of NPPFs can be large enough that this is not a severe restriction.

In this chapter we will first review and explain in simple terms the important issues in USP-SHG using nonlinear polymeric films, focusing on the effects that influence the generated pulse width and the SHG efficiency. Following that, experimental results using a single layer NPPF will be briefly reviewed. We will then present some recent theoretical results on the optimization of the pulse shape and the efficiency in single-layer and multilayer quasi-phase-matched (QPM) structures of NPPF. Limitations due to optical damage will also be briefly discussed. Although we will focus on SHG, it is obvious that the discussion and results are also relevant to the more general cases of sum and difference frequency generation as well.

Factors influencing pulse width and efficiency in USP-SHG

Although a complete theory of USP-SHG in NPPF is analytically complex, all of the important issues can be understood through simple physical arguments. The source term "driving" the generation of the electric field at the second harmonic (SH) frequency 2ω is the polarization $P_{2\omega}$. This polarization is proportional to an effective nonlinear *d*-coefficient and to the square of the fundamental electric field. Thus the SH intensity (power density) generated is proportional to the square of the effective *d*-coefficient. The basic geometry is illustrated in Figure 1. The *d*-coefficient is actually a tensor, but its dominant component coincides with the poling direction, which is in the present case perpendicular to the film plane. For SHG, the fundamental field must have component along these nonlinear dipoles and therefore the beam must be incident at an angle to the normal as shown and p polarized. Conversion efficiency is maximized near Brewster angle, which minimizes the Fresnel losses. The nonzero angle of propagation in the medium adds analytical complexity to the problem but does not alter the basic effects and their physical understanding. Therefore, in the subsequent discussions the nonzero propagation angle will be ignored.

Since the SH intensity is proportional to the fourth power of the fundamental field (i.e., the square of the fundamental power density), strong focusing is needed to obtain practical efficiency with short interaction lengths. One important simplification is that, since the interaction length is short, beam diffraction effects can be ignored. Clearly, SHG efficiency is maximized by achieving the largest possible *d*-coefficient and by using the tightest possible focusing. The first involves both "engineering" the active chromophore for highest hyperpolarizability and optimizing the alignment of the nonlinear dipoles (poling). Practical limitations to these processes exist but will not be discussed here. The minimum focused beam diameter is limited by optical fundamentals to the order of a few wavelengths but damage to the nonlinear material may pose a more severe limit. This issue will be addressed later in this chapter.

The various effects caused by the frequency dispersion of refractive index, as well as those caused by linear absorption losses, must be considered. These effects interact in a complicated fashion in a nonlinear material, but before trying to account for their complex interplay, it is important to obtain a basic physical picture for each of them separately. All the dispersive effects are due to the propagation of the optical electromagnetic fields in the nonlinear material. This propagation is described by $\exp[j(\omega_i t - k_i z)]$, where the propagation constant at frequency ω_i is $k_i = \omega_i n(\omega_i)/c$, the phase velocity is $v_i = c/n(\omega_i)$, and $i=1$ is for the fundamental and 2 for the SH. Because of material dispersion, normally $n(\omega_1) \neq n(\omega_2)$. A fundamental effect of this refractive index dispersion is the phase-mismatch which is important in both cw- and USP-SHG. Linear absorption results in a complex propagation constant, and its effects are also present in both cw- and USP-SHG.

Phase mismatch. The polarization of the nonlinear medium, $P_{2\omega}$, is proportional to the square of the fundamental field $E(\omega_1)$, so it propagates with the propagation constant $2k_1$. But the SH field generated by this polarization propagates with the propagation constant k_2. Normally $2k_1 \neq k_2$ because $n(\omega_1) \neq n(\omega_2)$, and thus a phase difference of $\Delta kz = (k_2 - 2k_1)z$ exists between the SH fields generated at planes separated by a distance z. If this phase difference equals π, exact cancellation occurs. The corresponding distance $L_C = \pi/|\Delta k|$ is called the nonlinear coherence length. Integration over z of the $\exp(j\Delta kz)$ phase mismatch gives the dependence of the total SH intensity on the nonlinear interaction length

$$F(L) = L^2 \left[\frac{\sin \Delta k L/2}{\Delta k L/2} \right]^2 . \quad (1)$$

If Δk is non-zero, only one coherence length effectively contributes to the SHG, and the SH intensity oscillates with length as shown with a solid-curve in Figure 2. If Δk can be made zero, a condition called phase matching, then the SH intensity builds up as L^2. Some inorganic crystals can be phase-matched by exploiting the anisotropy of their refractive indices. Unfortunately, NPPFs cannot be phase-matched, due to their particular symmetry properties, and phase mismatch will always be present in a single layer film. Typically, coherence lengths are on the order of a few microns, which are also the typical thicknesses of spin-coated NPPFs. Such short interaction lengths preclude the use of NPPFs for cw-SHG in non-guided configurations. Ultrashort pulses, however, have extremely high power densities and even such thin films can give practical efficiency for some applications. In fact, such a short nonlinear interaction length is another important advantage of NPPFs because ultra-thin films minimize pulse distortions in ultrashort-pulse nonlinear optical applications.

Absorption loss. Nonlinear organic materials often have absorption in the visible, near UV or near IR wavelengths. Such resonances have several effects on USP-SHG. One of these is the presence of linear absorption losses at either the fundamental or at the SH wavelength. Another is the modification of the refractive index through the well known Kramers–Kronig relations. (The real and imaginary parts of a resonant susceptibility contribute to the linear refractive index and to the extinction coefficient, respectively). This resonant index contribution can be exploited for a quasi-phase-matching effect by varying the active chromophore concentration. Changing the concentration, however, also changes the *d*-coefficient, the absorption loss, and other parameters. Here, we consider the effects of the linear loss. Because of the high fundamental powers involved, even a very small absorption at the fundamental wavelength inevitably results in damage to an NPPF. Therefore, only absorption at the SH wavelength can be tolerated, and the NPPF is useful only if this absorption is small.

Analysis of SHG in the presence of a linear loss at the SH wavelength introduces the physically obvious modification of replacing the real propagation constant k_2 by the complex equivalent $k_2 - j\alpha_2$, where α_2 is the absorption coefficient, assumed constant over the SH bandwidth. This $j\alpha_2$ term then appears in the phase mismatch function given in equation 1, which is modified to

$$F(L) = L^2 e^{-\alpha_2 L}\left[\frac{\sinh(\alpha_2 + j\Delta k)L/2}{(\alpha_2 + j\Delta k)L/2}\right]^2 . \tag{2}$$

The effect is to limit the efficiency and to damp out the oscillatory dependence with length as shown with a dashed-curve in Figure 2. If the value of α_2 is sufficiently small, the efficiency is not greatly reduced in a single layer of ~1 μm thickness. If α_2 is not constant over the SH bandwidth (loss dispersion), the effects are more complicated and cannot be fully described by equation 2. Increasing the absorption at the edges of the SH spectrum will act to broaden the pulses by decreasing the bandwidth. Clearly, a broad low absorption "window" at the SH wavelength is necessary for USP-SHG.

Group velocity effects. In addition to phase mismatch, the frequency dispersion of refractive index causes other more important effects specific to USP-SHG. To understand these we must consider the optical pulses in both the time domain and the frequency domain. It is well known from Fourier transform theory that a pulse modulated sinusoidal wave with a pulse width τ_p has a finite frequency bandwidth $\Delta\omega_p$ about the carrier ω_o, with $\Delta\omega_p \tau_p \approx \pi$ (5). While the sinusoidal carrier propagates with the phase velocity, the pulse envelope travels with a group velocity v_g. The group delay per unit length is $1/v_g = \partial k/\partial\omega$. In a dispersive material the group velocity is different from the phase velocity, and moreover, v_g itself is not constant with frequency. This is known as group velocity dispersion (GVD).

Intra-pulse GVD. It is well known that GVD within the frequency bandwidth of a pulse causes pulse broadening in a linear medium, due to the different group delays experienced by the different frequency components within the bandwidth. The physically obvious measure of the broadening, over the total propagation distance L, is the group delay difference between the opposite edges of the frequency band,

$$\Delta\tau_{ig} = L\frac{\partial}{\partial\omega}\left(\frac{1}{v_g}\right)\Delta\omega_p = L\frac{\partial^2 k}{\partial\omega^2}\Delta\omega_p . \tag{3}$$

This broadening is clearly present in any nonlinear medium as well, affecting the fundamental pulse as it propagates, and through it the SH (source) polarization $P_{2\omega}$. Additionally, the SH is similarly affected during its propagation. Owing to the short interaction length in USP-SHG, however, this intra-pulse GVD (IGVD) is negligible for all but the shortest pulses (~10 fs) that can presently be generated.

Group velocity mismatch. The frequency dependence of v_g also results in another more important effect, i.e., inter-pulse GVD or group velocity mismatch (GVM) between the fundamental and the SH pulses. Their envelopes propagate with different velocities. Since the $P_{2\omega}$ envelope also travels at the fundamental group velocity, a time delay of

$$\Delta\tau_g(z) = z\left(1/v_{g2} - 1/v_{g1}\right) = z\left(\dot{k}_2 - \dot{k}_1\right) \tag{4}$$

exists between the second harmonic pulses generated at points separated by distance z, as illustrated in Figure 3. Integration of the contributions over all z gives a stretched

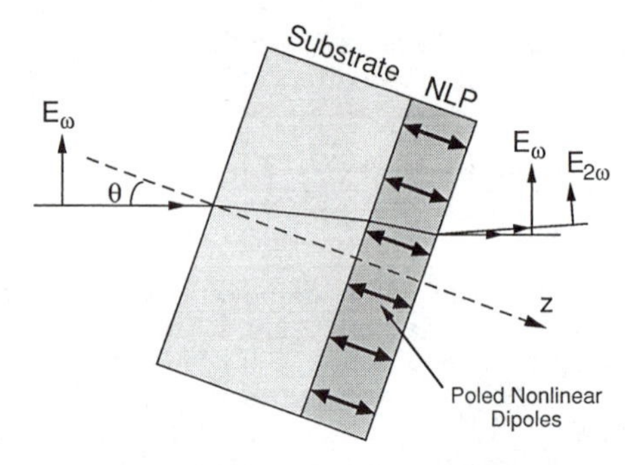

Figure 1. Schematic of a substrate-nonlinear polymer thin-film structure used in ultrashort-pulse second harmonic generation.

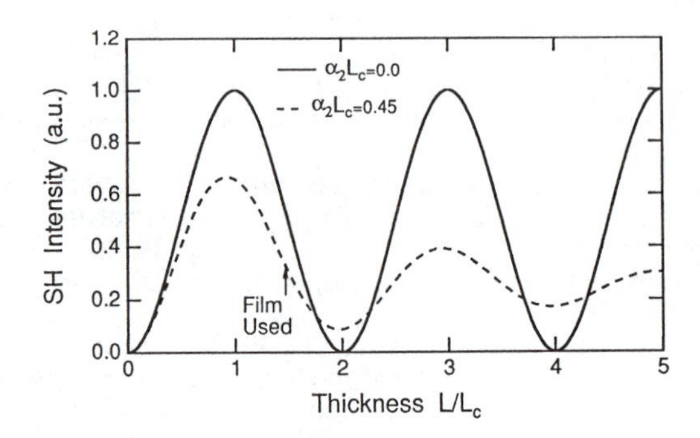

Figure 2. Normalized SH intensity versus film thickness L (in units of nonlinear coherence length L_C) calculated using equation 2.

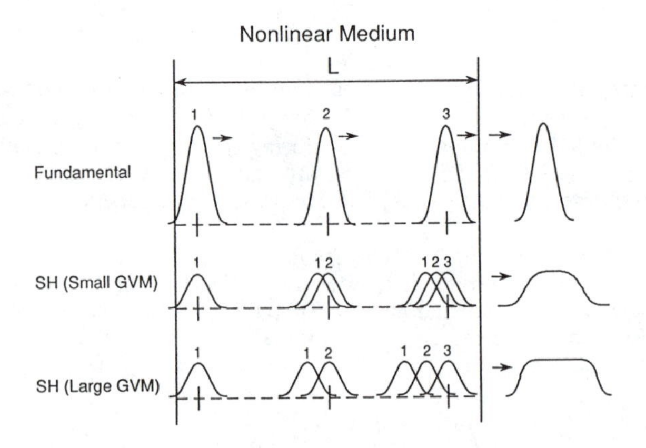

Figure 3. Broadening of SH pulses due to the group velocity mismatch. The SH pulses numbered 1, 2 and 3 represent those generated in the input side, middle, and output side of a nonlinear medium, respectively.

out, flat topped SH pulse if $\Delta\tau_g(L)$, the group delay difference over the interaction length, is large compared to the input fundamental pulsewidth as illustrated. On the other hand, if $\Delta\tau_g(L)$ is negligible, the generated SH pulse is shorter than the fundamental, due to the square law dependence of the SH power on the fundamental power. The value of the shortening factor is $\sqrt{2}$ for Gaussian shape pulses.

Further understanding of the effect of GVM may be gleaned from considering the process in the frequency domain. GVM arises from the different phase mismatches at different (corresponding) frequencies within the fundamental and the SH frequency bands. If we assume, for simplicity, that the center frequencies of the bands (ω_{o1}, ω_{o2}) are phase-matched, then it is obvious that phase mismatches must exist at other frequencies. Using the Taylor expansion $k(\omega)=k(\omega_o)+(\omega-\omega_o)\partial k/\partial\omega$, where $\partial k/\partial\omega$ is evaluated at $\omega=\omega_o$, at both the fundamental and the SH center frequencies we find that the phase mismatch at an arbitrary frequency within the fundamental frequency band is equal to $\Delta k(\omega)=z(\partial k_2/\partial\omega-\partial k_1/\partial\omega)(\omega-\omega_{o1})$ between points separated by distance z. Referring to the effects of a phase mismatch described above, it is not surprising that integration over z results in imposing a spectral filter over the SH band generated without GVM. The filter is

$$G(\omega)=\left(\frac{\sin[\Delta\tau_g(L)(\omega-\omega_{o1})]}{\Delta\tau_g(L)(\omega-\omega_{o1})}\right)^2. \tag{5}$$

The larger the value of $\Delta\tau_g(L)$, the narrower the filter, and the larger the spreading of the pulse in time. Quantitatively, multiplication by the filter function in the frequency domain corresponds to a convolution of the SH pulse shape with a rectangular pulse in the time domain. This is the obvious equivalent of the result derived above by the group velocity mismatch argument.

GVM is the most important limitation to USP-SHG with inorganic crystals, since typical values of the parameter $\Delta\tau_g(L)$ are on the order of 1 fs per μm. Thus, even the thinnest (50–100 μm) crystals result in broadening of the SH pulses when the fundamental pulses are shorter than ~50 fs. If the material is phase-matched at the carrier frequencies, then the energy conversion efficiency is not adversely affected since the carrier of the SH pulses originating from different points adds up coherently. However, the peak power conversion efficiency is obviously reduced by the same factor that the SH pulse is broadened. Thus, it is the peak power conversion efficiency that is a good measure of the performance of any medium for USP-SHG. Values of the GVM parameter, $\Delta\tau_g(L)$, in NPPFs are comparable to those in inorganic crystals. Therefore, in single layer films of a few microns thicknesses all effects of GVM are negligible, and even for fundamental pulses of around 10 fs duration the generated SH pulses are shorter than the fundamental (*6*). SHG efficiency of a single layer NPPF, however, is low even for materials with the highest nonlinear *d*-coefficients. As will be shown later, the use of multilayer quasi-phase-matched (QPM) structures can eliminate the limitations of the phase mismatch in NPPF and considerably increase the conversion efficiency. But GVM is still one of the important limiting effects in these structures. If *both* phase mismatch and GVM are present in a long interaction length, the combination of the destructive interference and the GVM time "slippage" can result in more severe pulse distortions, such as double peaks in the SH pulse profile (*7*). It will be shown later that in QPM structures of NPPFs such combined effects of phase mismatch and GVM are eliminated.

In the above, a simple physical description of the main factors influencing USP-SHG in a single layer NPPF has been presented, with emphasis on the two most important issues, namely preservation of the ultrashort pulse width and obtaining the best possible efficiency. A comprehensive analysis, in the form of coupled nonlinear differential equations, has been developed (*7*) that is able to account quantitatively for the sometimes rather complex interplay of the various factors. Additionally, it describes other phenomena such as the effects of non-transform limited fundamental pulses (i.e., pulses which contain frequency domain nonlinear phase, or equivalently have nonlinearly chirped carrier frequency), shaping of the generated SH pulses, etc. The analysis is also applicable to USP-SHG in various multilayer QPM structures and has been used to obtain the numerical results discussed later in this chapter.

Generation of 312 nm femtosecond pulses using a single layer NPPF

Before summarizing some of the results and predictions of the analysis, a brief summary of recent experimental results on the generation of ultrashort UV pulses using a single layer NPPF will be presented. The nonlinear polymer used in these experiments was a coumaromethacrylate copolymer [P(MMA-CMA)–11] (*8*). The spin-cast film had a thickness of 2.5 μm and a chromophore density of $9.8{\times}10^{20}$ mol/cm^3. The poling conditions were optimized, and the linear and nonlinear properties of the polymer were characterized (*9*). The absorption spectrum shown in Figure 4 has a low absorption window around λ=318 nm in fortunate coincidence with the second harmonic frequency of the well known R6G CPM dye laser. This window can pass the spectrum of pulses as short as ~5 fs ($\Delta\lambda$~30 nm). Shorter pulses will be broadened due to attenuation at their extreme wavelengths. Additionally, the absence of any absorption at 625 nm permits the doubling of pulses with large peak powers (*8*). The fundamental pulses were derived from a R6G CPM dye laser, amplified by an 8 KHz repetition rate Cu-vapor laser (*10*) to ~13 fs duration and ~50 nJ energy at 625 nm. The configuration was the same as shown in Figure 1, with the fundamental focused to a beam diameter of 26 μm in the film. The results are shown in Figure 4. This figure shows the measured SH pulse spectrum superimposed on the absorption spectrum of P(MMA-CMA). The full width half maximum is ~19 nm, which is consistent with a pulse length of ~11 fs. The GVM broadening parameter, $\Delta\tau_g(L)$, for this film was estimated to be around 1 fs. It was found that the interaction length was only about seven optical cycles which is approximately the FWHM of the fundamental pulse. These observations indicate that no broadening of the SH pulse occurred and that the SH pulsewidth is ~11 fs. The inset in Figure 4 shows the SHG efficiency vs average fundamental power (*8*). The dependence is linear, as expected. The highest efficiency observed was ~0.13% at an input energy of 51 nJ (peak power density of 740 GW/cm^2). This efficiency is high for such a thin film. However, at this power density the film was damaged and the SH signal gradually decayed. At the approximate damage threshold of 164 GW/cm^2 (11.4 nJ) the efficiency was 0.025% . By scaling the focal spot size and using the full 51 nJ fundamental, 13pJ UV pulses were obtained.

The film used in this experiment was not optimized for maximum efficiency. From the measured index at 625 nm, and a Kramers-Kronig prediction of the refractive index at 312 nm (see the next section), L_c was determined to be 1.7 μm. Using this and the measured α_2=$2.66{\times}10^5$ m^{-1} in equation 4, and accounting for the off-normal propagation direction, it was calculated that the 2.5 μm thickness of the film is longer than would be needed for maximum efficiency. The dashed-curve of SH intensity vs thickness shown on Figure 2 corresponds to the parameters of this film and the actual thickness is indicated. A factor of 2 improvement could be obtained by using a thickness of ~1.7 μm for this material at this particular chromophore concentration. The next issue that will be examined is how both efficiency and pulse duration can be optimized for a particular material by changing the chromophore concentration.

Optimization of a NPPF for the concentration of its active chromophore

In the experiments summarized in the preceding section, the fundamental and SH wavelengths were located at opposite sides of the main absorption band of the active chromophore of the NPPF. In such a case, the contribution from the real part of the resonant susceptibility causes anomalous dispersion and the index at the SH wavelength may be lower than that at the fundamental. The possibility of exploiting a trend toward anomalous-dispersion phase-matching (*11,12*) thus exists. By changing the concentration of the active chromophore, the coherence length can be increased and the absorption loss at the SH lowered as well. However, decreasing the concentration also lowers the *d*-coefficient and, unlike in a cw guided wave configuration, the increased interaction length may not be useful due to GVM and IGVD effects. An optimum concentration exists for a particular polymeric material which maximizes the SHG efficiency of a single layer film while still preserving the ultrashort pulse width. The optimum chromophore concentration has been calculated (7) for a representative NPPF, which is P(MMA-CMA) discussed in the previous section. In the method of Ref. 7, first the refractive index and the extinction coefficient were found from the measured absorption spectrum (Figure 4) at the original concentration using a Kramers-Kronig calculation. Next, an analytical fit to the complex permittivity of the material was obtained and related to the concentration through a scaling factor x_c. Finally, the analytical expression of the complex permittivity was used to calculate the coherence length L_c, the GVM and the IGVD parameters, and α_2 vs fundamental wavelength for various values of the x_c. The results for the calculation of L_c are shown in Figure 5(a). Parameters $\alpha_2 L_c$ and $d_{eff}L_c$ (d_{eff} is the effective *d*-coefficient) remained relatively constant for a given fundamental wavelength and are not shown. Following that, the comprehensive analysis was used to calculate the SH efficiency and pulse width, taking proper account of a number of factors not covered in this chapter, such as the optimization of the non-normal incidence on the film, etc. The results, which are given in Figure 5(b), show that there exists an optimum concentration (x_c=0.2) at which the efficiency is near maximum and the SH pulse is not broadened. It should be noted that this concentration is much higher than what would be needed for achieving exact anomalous-dispersion phase-matching. For the particular NPPF used in these calculations, the efficiency maximum is very broad, and there is approximately a factor of 10 range of nearly optimum concentrations. Nevertheless, the molar concentration of the chromophore cannot be arbitrarily increased. For other nonlinear polymers the results may be somewhat different, but the method used for optimization is very general.

USP-SHG in quasi-phase-matched structures

For NPPFs, which cannot be phase-matched by the conventional anisotropy method used in nonlinear inorganic crystals, the method of quasi-phase-matching (*11*) offers an effective way to increase the interaction length and thus the efficiency. In this method, the cancellation of the SH signal due to the π phase mismatch after one coherence length is eliminated by using multilayer structures as shown in Figure 6. Figure 6(a) is a so called Bipolar-QPM structure (B-QPM), in which the orientation of the nonlinear dipoles is reversed after each coherence length. This cancels the accumulated π phase mismatch after L_c, and the generated SH fields add in phase. The efficiency of each length, L_c, is still reduced by the phase mismatch, but the SH generated in successive layers now adds, enhancing the conversion efficiency. Such a structure using NPPFs can be made by mechanical assembly of multiple poled layers (*13*). Figure 6(b) shows a Unipolar-QPM structure (U-QPM) in which every second layer of length L_{c2} is a linear dielectric film, thus eliminating the out of phase contributions of these layers. U-QPM is considerably easier to make. Mechanical assembly is not required, therefore some unnecessary damage to the material can be

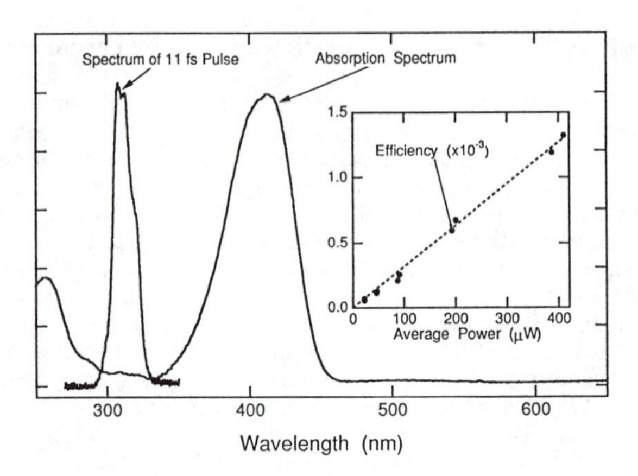

Figure 4. The measured spectrum of the SH pulse and the absorption spectrum of P(MMA-CMA). The inset shows the conversion efficiency as a function of the average power of the fundamental pulse. (Adapted from ref. 8).

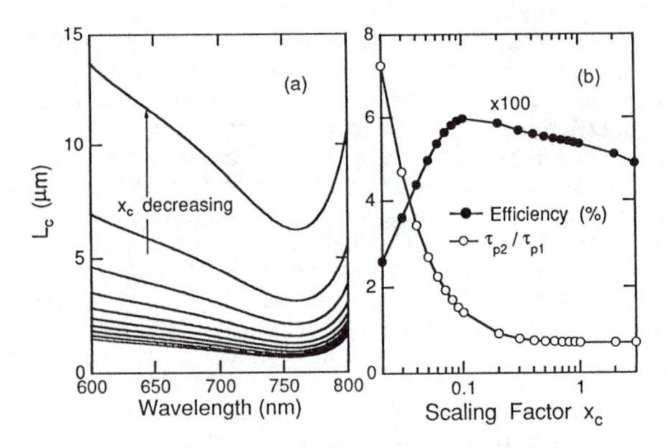

Figure 5. (a) Nonlinear coherence length L_C versus the fundamental wavelength for different chromophore concentrations. (b) Conversion efficiency and normalized SH pulsewidth τ_{p2}/τ_{p1} versus x_C for an initial fundamental peak intensity of 100 GW/cm^2, where the efficiency is magnified by the indicated factor.

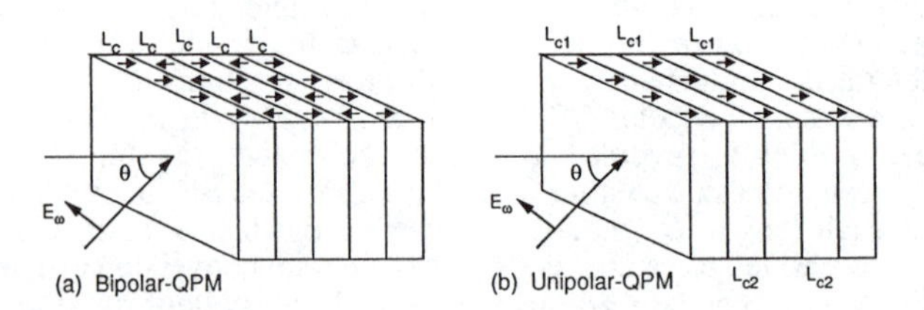

Figure 6. Schematic diagrams of Bipolar- and Unipolar-QPM structures. Each period of the Bipolar-QPM structure consists of two nonlinear layers oriented in the opposite directions. While each period of the Unipolar-QPM structure consists of one nonlinear layer and one linear layer, and all the nonlinear layers are oriented in the same direction.

prevented. In this case the multilayer structure can be fabricated by successively spin-coating the films and poling all the nonlinear layers together. Another advantage of this structure over B-QPM is that *in situ* poling (*14*) can be used to eliminate the orientational decay of the material. It is obvious, however, that a B-QPM structure is more efficient than a U-QPM structure consisting of same number of layers.

Beside optical damage, the effects of phase mismatch and absorption at the SH wavelength are the only practical limiting factors in single layer films, since IGVD is negligible and the effects of GVM can be avoided. In QPM structures, however, GVM, and even IGVD also become significant. Numerical analyses have been performed to clarify the effects of GVM and loss on both types of QPM structures, assuming $L_{c1}=L_{c2}=L_c$ in the U-QPM structure. The conversion efficiency and pulse distortion effects in the two types of structures were examined for a hyperbolic secant fundamental pulse shape and for some typical parameters, which correspond approximately to those of P(MMA-CMA) described previously with a concentration range $0.3 \leq x_c \leq 1.0$. Some quantities were normalized with the fundamental pulsewidth and the coherence length so they can be easily scaled. The parameters used correspond to a fundamental pulsewidth $\tau_{p1}=10$ fs, $L_c=2$ µm, $d_{eff}=6.4$ pm/V, and a normalized GVM parameter $\eta L_c=1.76\Delta\tau_g(L_c)/\tau_{p1}=0.2$. The normalized absorption coefficient $\alpha_2 L_c$ was varied between 0 and 0.3. IGVD was assumed to be negligible, and a representative value of 100 GW/cm^2 was used for the input power density. The results for the two QPM structures were compared to those for a thick single layer polymer, or equivalently, a multilayer polymer with identically oriented layers (with the same coherence length as the QPM layers).

The combined effects of phase mismatch, GVM, and linear loss can be best visualized in a 3-dimensional plot. The SH intensity profile produced by a single layer film is shown in Figure 7(a) for $\alpha_2 L_c=0$ and in Figure 7(b) for $\alpha_2 L_c=0.2$, respectively, as a function of film thickness. When only the phase mismatch and the GVM are present the SH pulses are stretched in time by the GVM. Additionally, if the nonlinear medium is very thick, the middle portion of the stretched SH pulse suffers destructive interference due to phase mismatch, resulting in double peaks in the SH intensity profile, or even complete splitting of the SH pulse. This effect has been observed experimentally in femtosecond pulse SHG in a thick KDP crystal (*15*). If linear loss is also present, the SH pulses generated earlier get attenuated, and the trailing peak of the SH becomes smaller than the leading one (*16*). For a thick medium, the trailing peak of the SH can completely disappear. Thus, the loss helps to reduce the broadening of the SH pulses. The conversion efficiency, however, is very low for either case as can be seen from the peak SH intensities. In fact, the highest value is reached after approximately one coherence length. Thus, nothing practical can be gained from using films of many coherence length thickness or multiple layer films that are identically oriented.

By contrast, as is shown in Figure 8, both QPM structures yield greatly improved efficiency with only the relatively small pulse distortions due to the GVM, which is not large even for the very short (10 fs) pulses considered. For the case shown ($\alpha_2 L_c=0.2$), the maximum SH peak intensities are ~18 and ~9 times larger, for the bipolar and unipolar structures, respectively, than that generated in the non-phase-matched thick films. The two QPM structures behave very similarly to an ideal (hypothetical) phase-matched single layer film of the same absorption and thickness, apart from a factor of ~2 and ~4 lower efficiency and small "dips" caused by the linear loss and "plateaus" caused by the phase mismatch and the linear layers. It should also be noted that for the cases shown, the peak SH intensity saturates after about 10 total layers due to the combined effects of fundamental pump depletion, absorption loss and GVM. For a lower loss, the saturation is somewhat slower. However, the pulse broadening effect of GVM always limits the conversion efficiency. In Figure 9(a) the pulse broadening versus the thickness of the bipolar structure is shown for various values of the absorption coefficient. The results for the unipolar structure are nearly

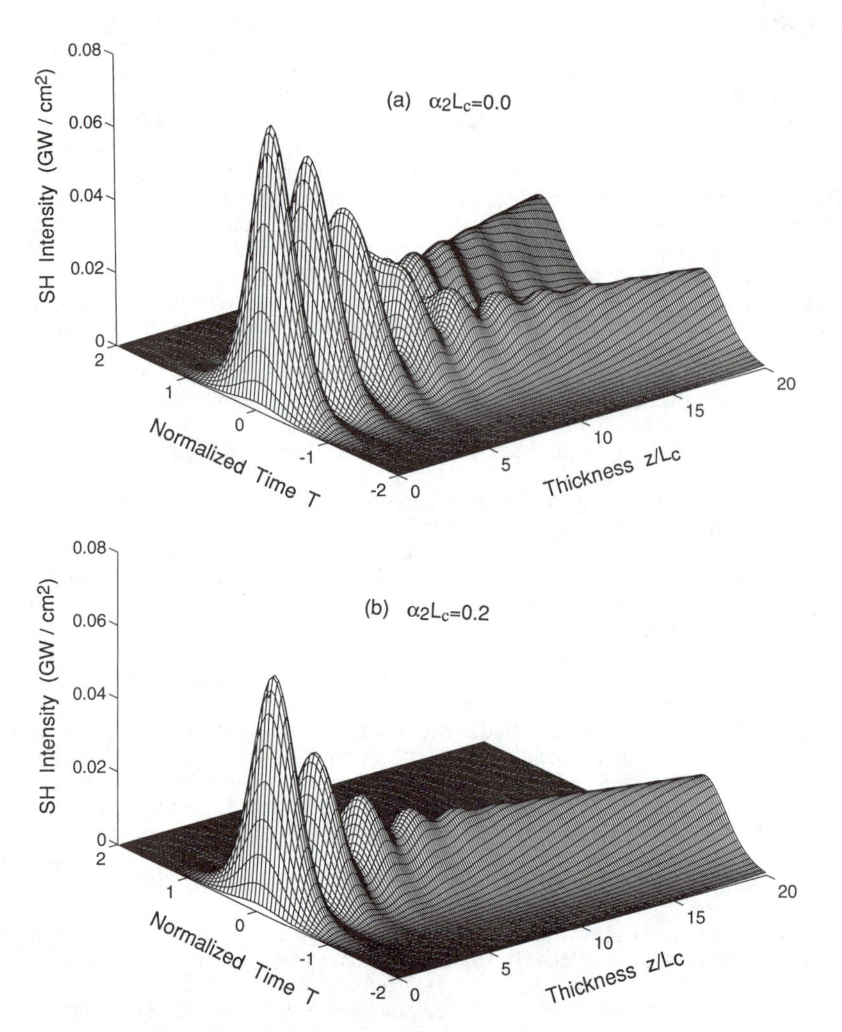

Figure 7. SH intensity profile produced in a single layer polymer film versus film thickness z/L_c for (a) $\alpha_2 L_c$=0 and (b) $\alpha_2 L_c$=0.2. Other parameters are τ_{p1}=10 fs, L_c=2 μm, d_{eff}=6.4 pm/V, ηL_c=0.2, and initial fundamental peak intensity 100 GW/cm^2. (Adapted from ref. 7).

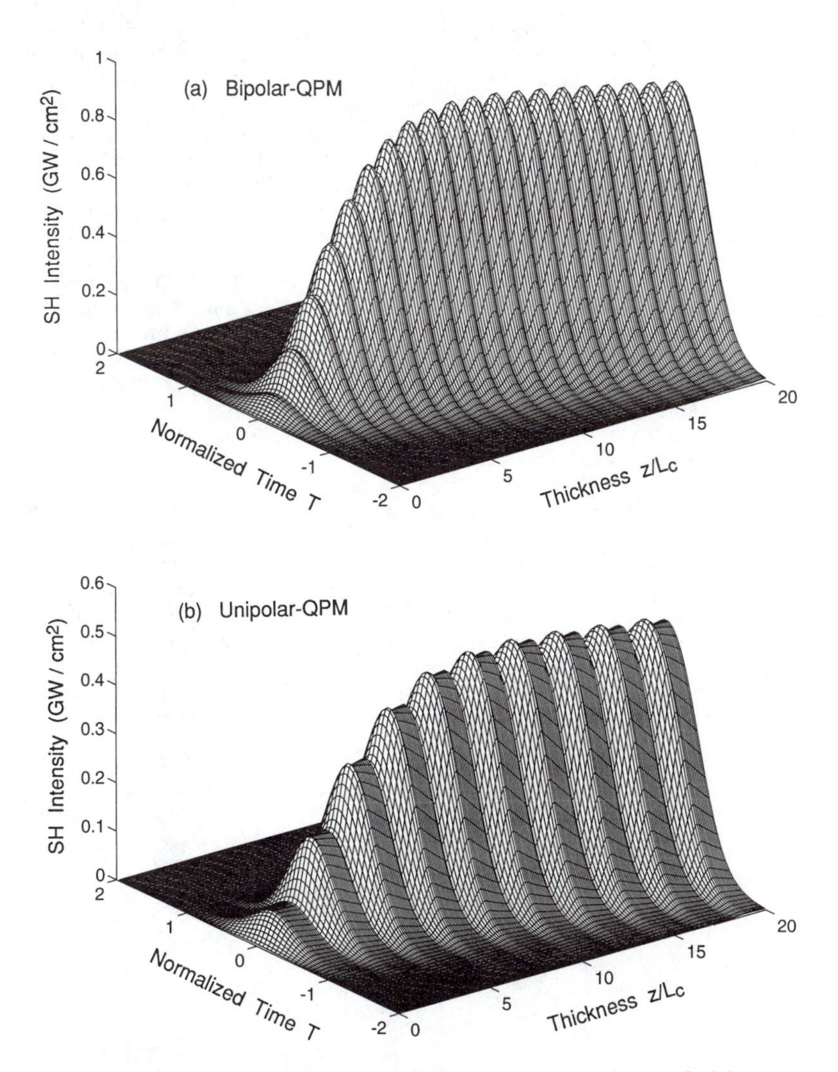

Figure 8. SH intensity profile produced in Bipolar- and Unipoler-QPM structures versus total thickness z/L_c for $\alpha_2 L_c$=0.2. Other parameters are the same as those in figure 7. (Adapted from ref. 7).

identical. It can be seen that a higher value of $\alpha_2 L_c$ prevents pulse broadening. However, as was seen in Figure 8, it also limits the buildup of the SH in additional layers. By contrast, if the loss is very low, it is the GVM induced pulse broadening which limits the optimum number of layers and the performance of the structure. In general, a SH pulse width no longer than the input fundamental pulse width is desired. In Figure 9(b), the maximum SH peak intensity that can be achieved under the conditions of $\tau_{p2}/\tau_{p1} \leq 1$ (τ_{p2} is the SH pulsewidth) and $z/L_c \leq 20$ is shown for the cases depicted in Figures 8. The result for the ideal single layer film was also included for comparison. Interestingly, both QPM structures require thicknesses of around $z/L_c=10$ to give the results shown in Figure 9(b). From these results it is evident that an optimum design exists for each particular multilayer NPPF, and a structure optimized for a particular fundamental pulse duration will not be optimum for a different length pulse, or for a different nonlinear material. Nevertheless, QPM structures will always improve the efficiency and reduce pulse distortions.

In this analysis it was assumed that the effects of IGVD are negligible. Calculations have also been performed with this effect included, with realistic parameters. The results show that when the number of layers is optimized to preserve the pulsewidth and maximize the efficiency, the detrimental effect of IGVD on both pulse broadening and efficiency is typically less than ~10% , even for the fundamental pulsewidth as short as 10 fs.

Optical damage limitations

It has been previously noted that optical damage to NPPF limits the fundamental power density that can be used for USP-SHG. This damage is typically detected as a gradual decay of the SH signal which cannot be attributed to orientational relaxation (*8*), (*17*). While a great deal of concern has been expressed regarding the long term orientational stability of these materials, little experimental work has been done to investigate the photo-chemical causes of the damage. *In situ* experiments that we have performed clearly indicate that the effect is irreversible, indicating permanent chemical changes have taken place rather than simply depoling. This damage limits the practical uses of NPPFs. Nevertheless, the experiments reported in Ref. 8 have shown that some

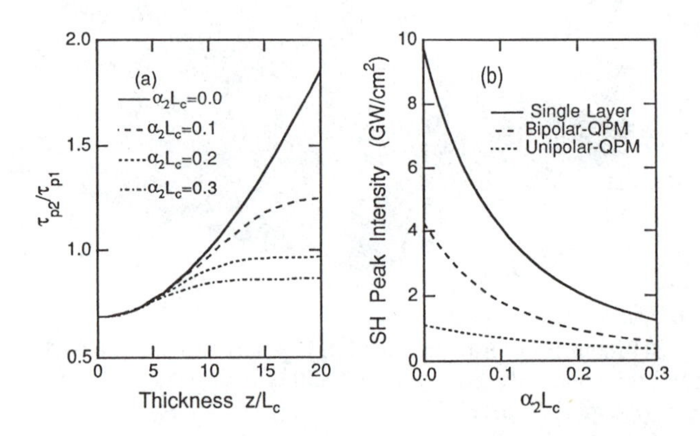

Figure 9. (a) Normalized SH pulsewidth τ_{p2}/τ_{p1} as a function of total thickness z/L_c in a Bipolar-QPM structure for various values of $\alpha_2 L_c$. (b) Maximum SH peak intensity that can be achieved under the conditions of $\tau_{p2}/\tau_{p1} \leq 1$ and $z/L_c \leq 20$ in phase-matched single layer, Bipolar-QPM structure and Unipolar-QPM structure versus $\alpha_2 L_c$. Other parameters are the same as those in figure 7.

materials are sufficiently rugged to withstand power densities near the damage threshold of inorganic crystals. Two issues that must be addressed are whether the damage is related to peak power or accumulated energy and whether a true damage threshold exists. To increase the SH efficiency, it is desirable to place the NPPF inside a laser cavity, thus taking advantage of the fundamental power which is typically 20-50 times higher than that available outside the cavity. Understanding the damage mechanisms is very important for this application because of the high peak power and rapidly accumulating energy that are present. Additionally, investigation of the damage threshold is important for long term stability. The same photo-chemical effects that influence USP-SHG applications will also most likely influence cw and quasi-cw applications that involve NPPFs. A great deal of work needs to be done to identify and quantify the damage mechanisms if rapid progress is to be made on the development of new and useful materials.

Acknowledgments: This research is supported by National Science Foundation grant ECS 9122168.

References

1. Khanarian, G; Norwood, R. A.; Haas, D.; Feuer, B.; Karim, D. *Appl. Phys. Lett.* **1990**, *57*, 977.
2. Knoesen, A.; Molau, N. E.; Yankelevich, D. R.; Mortazavi, M. A.; Dienes, A. *Intern. J. Nonlin. Optic. Phys.* **1992**, *1*, 73.
3. Curley, P. F.; Spielmann, Ch.; Brabec, T.; Krausz, F.; Wintner, E.;. Schmidt, J. *Opt. Lett.* **1993**, *18*, 54. Asaki, M. T.; Huang, C. P.; Garvey, D.; Zhou, J. P.; Kapteyn, H. C.; Murnane, M. M. *Opt. Lett.* **1993**, *18*, 977 A. Stingl, A.; Spielmann, C.; Krausz, F.; Szipöcs, R. *Opt. Lett.* **1994**, *19*, 204.
4. Edelstein, D. C.; Wachman, E. S.; Cheng, L. K.; Bosenberg, W. R.; C.. Tang, L. *Appl. Phys. Lett.* **1988**, *52*, 2211.
5. Siegman, A. E., *Lasers;* University Science, Mill Valley, CA, 1986, Chapter 9, pp 334-335.
6. Sidick, E.; Knoesen, A.; Dienes, A. *Opt. Lett.* **1994**, *19*, 266.
7. Sidick, E.; Dienes, A.; Knoesen, A. *Intern. J. Nonlin. Optic. Phys.* **1994**, *3*, to be published.
8. Yankelevich, D. R.; Dienes, A.; Knoesen, A.; Schoenlein, R. W.; Shank, C. V. *IEEE J. Quant. Elec.* **1992**, *QE-28*, 2398.
9. Mortavazi, M. A.; Knoesen, A.; Kowel, S. T.; Henry, R. A.; Hoover, J. M.; Lindsay, G. A. *Appl. Phys. B* **1991**, *53*, 287.
10. Knox, W. H.; Downer, M. C.; Fork, R. L.; Shank, C. V. *Opt. Lett.* **1984**, *9*, 552. Boyer, G.; Franco, M.; Schamberiet, J. P.; Migus, A.; Antonetti, A.; Georges, P.; Salin, F.; Brun, A. *Appl. Phys. Lett.* **1991**, *58*, 1119.
11. Armstrong, J. A.; Bloembergen, N.; Ducuing, J.; Peshan, P. S. *Phys. Rev.* **1962**, *27*, 1918.
12. Cahill, P. A.; Singer, K. D.; King, L. A. *Opt. Lett.* **1989**, *14*, 1137. Cahill, P. A.; Singer, K. D. In *Nonlinear Optical Materials*; Kuhn, H, and Robillard J., Ed.; CRC, Boca Raton, 1992.
13. Khanarian, G.; Mortazavi, M. A.; East, A. J. *App. Phys. Lett.* **1993**, *63*, 1462. Mortavazi, M. A.; Khanarian, G. *Opt. Lett.* **1994**, *19*, 1290.
14. Eich, M.; Sen, A.; Looser, H.; Bjorklund, G. C.; Swalen, J. D.; Twieg, R.; Yoon, D. Y. *J. Appl. Phys.* **1989**, *66*, 2559.
15. Noordam, L. D.; Bakker, H. J.; de Boer, M. P.; van Linden van den Heuvell, H. B. *Opt. Lett.* **1990**, *15*, 1464.
16. Ho, P. P.; Ji, D.; Wang, Q. Z.; Alfano, R. R. *J. Opt. Soc. Am. B* **1990**, *7*, 276.
17. Mortavazi, M. A.; Yoon, H. N.; Teng, C. C. *J. Appl. Phys.* **1993**, *74*, 4871.

RECEIVED February 14, 1995

Chapter 36

Periodic Domain Inversion in Thin Nonlinear Optical Polymer Films for Second Harmonic Generation

G. Khanarian and M. A. Mortazavi

Hoechst Celanese Research Division, 86 Morris Avenue, Summit, NJ 07901

This paper reviews recent progress in the fabrication of periodic structures in poled polymer films for quasi phase matched frequency doubling. In the case of waveguides a new two step poling process is demonstrated for periodic domain inversion, such that the sign of χ^2 changes every coherence length. Quasi phase matched frequency doubling is also demonstrated in periodic free standing films with picosecond lasers resulting in efficiencies of 10^{-2}%.

Methods for efficient second harmonics generation(SHG, frequency doubling) have been studied in nonlinear optical(NLO) poled polymers because of the ease of fabrication of thin film structures either as waveguides on silicon wafers or as free standing films, and also because of their potentially high NLO susceptibilities χ^2. A key element to obtaining high SHG efficiencies is phase matching the fundamental to the second harmonic wave(1). We have explored the use of the periodic modulation of the nonlinear susceptibility of poled polymers for quasiphase matching(QPM), and have reported efficiencies as high as 0.01%/W over a distance of 5 mm in a slab waveguide(2,3,4). However the periodic poling was unidirectional consisting of poled and unpoled regions and so only half the length of the waveguide was effectively phase matched. In this paper we describe a new poling technique for periodic bidirectional poling in a waveguide, i.e. periodic reversal of the sign of χ^2 which would utilize the whole NLO medium, and report QPM SHG from a slab waveguide.

0097–6156/95/0601–0498$12.00/0

We have also demonstrated efficient frequency doubling in free standing periodic NLO films(5,6). Such films would be useful for intracavity SHG devices and for frequency doubling femtosecond lasers. Single free standing poled NLO thin films with thicknesses of the order of 1 μm have been used in the frequency doubling of femtosecond lasers because they exhibit low dispersion(12). Recently, Knoessen et al.(13) have shown theoretically that the same is also true for periodic χ^2 films. We describe the fabrication of bidirectionally poled free standing films and demonstrate the enhancement of SHG using picosecond and nanosecond pulsed lasers.

Bidirectionally Poled Waveguides

The principles of QPM in polymer waveguides have been previously(2,3,4) described in detail. The key is the periodic modulation of the NLO susceptibility every coherence length L_c,

$$L_c = \frac{\lambda}{4[N(2\omega) - N(\omega)]} \quad (1)$$

where $N(\omega)$ and $N(2\omega)$ are the effective indices of the zeroth order modes of the fundamental and harmonic waves, respectively, and λ is the wavelength of the fundamental beam. Typically the coherence length in polymer waveguides is between 3 and 10 μm for $\lambda = 0.8$ μm. Previously we fabricated periodic unidirectionally poled waveguides via a one step poling process(2,3). The fabrication of the waveguide started with a silicon wafer coated with a metallic electrode. Then we spun coated cladding buffer layers and an NLO polymer guide layer. The NLO polymer MO3ONS/MMA 50/50 consisted of the dye 4-oxy,4'nitrostilbene attached as a sidechain to a polymethylmethacrylate(PMMA) backbone with a 50:50 mole percent composition .The buffer layers consisted either of PMMA or MO3ONS/MMA 30/70. Finally we deposited a top electrode which was patterned into a periodic electrode pattern using lithographic techniques. The whole structure was heated to near the glass transition temperature Tg of the polymer (115 C) and poled with an electric field(100 V/μm). The result was the periodic modulation of χ^2 i.e periodic regions of poled and unpoled NLO film. Phase matched SHG from such waveguides were reported(2-4) earlier .

New process for bidirectional poling. The most efficient structure for frequency doubling is one where there is a periodic reversal of the sign of the domains of χ^2 every coherence length(4). This is accomplished by a novel two step poling process where a waveguide is first poled uniformly, a periodic electrode structure is patterned at room temperature and then the waveguide is poled again with the reverse polarity at a lower temperature. This new approach(11) is based on our observation that the alignment of molecules in an electric field occurs at a faster rate than it's decay in the absence of an electric field, when the temperature of poling is below the Tg of the polymer. Thus the regions underneath the periodic electrodes undergo a reversal in the sign of χ^2 while the regions where there is no electrode undergo a slow decay of χ^2.

In order to verify this new approach to fabricating bidirectionally poled waveguides we characterized the dynamics of poling and the resultant modulation depth of periodic poling. Figure 1 shows the dynamics of poling and it's decay for the polymer MO_3ONS/MMA 50/50. We report the normalized χ^2 versus poling time when the electric field for poling was 140 V/µm and the temperature was 95 ^0C (20 ^0C below Tg of the polymer). The characteristic rise time for poling was only 20 minutes whereas the characteristic times for poling decay was 3 minutes(initial decay) and 110 minutes(long term decay) at the same temperature. The results of figure 1 verify that NLO polymers pole more rapidly than they decay when the temperature is below the Tg of the polymer. Recently, Kuzyk et al. (7,8) have reported similar observations.

Modulation depth of periodic poling. We have quantified the improvement in modulation depth by the two step poling process versus the one step poling process reported earlier (2-4). Let us suppose that the normalized χ^2 as a result of poling with positive polarity is 1. In the one step periodic poling process only 1/2 of the waveguide is periodically poled and so the modulation depth is 1/2. In the two step poling process, uniform poling with positive polarity is followed by periodic poling with the opposite negative polarity, and so the maximum theoretical modulation depth is [1 - (-1)]/2 = 1. In practice, the modulation depth as a result of the two step poling process is a function of the rise time of the second poling step and the decay time of the originally poled region. Figure 2 shows the modulation depth obtained as a function of poling time for the polymer MO_3ONS/MMA 50/50 at

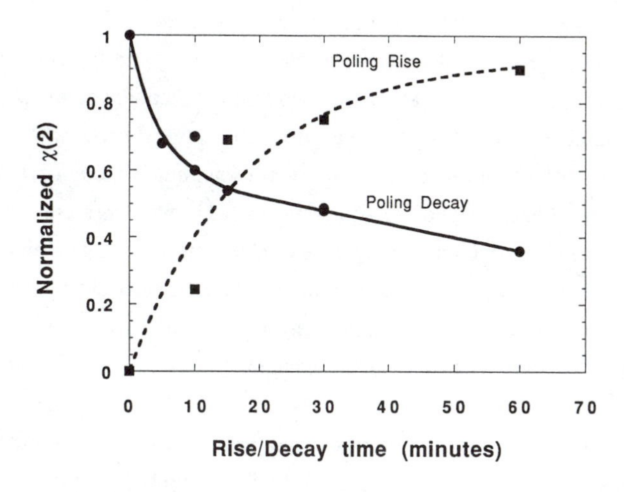

Figure 1. Rise and decay of normalized $\chi 2$ versus poling time.

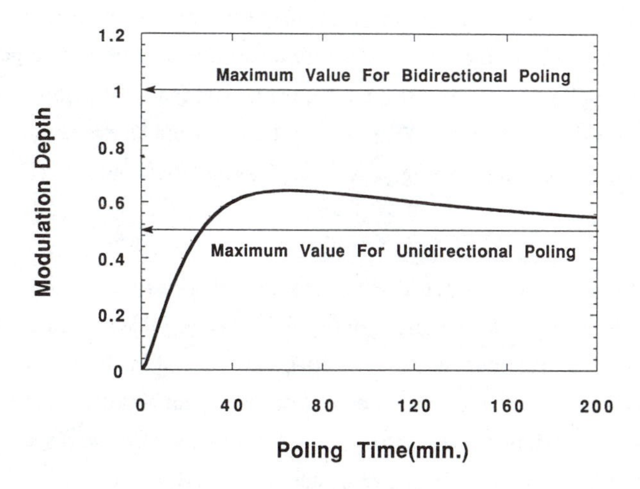

Figure 2. Modulation depth of periodic poling via the two step poling described in the text.

95 ^{0}C and 140 V/μm. Initially the modulation depth is zero because there is a uniformly poled substrate as a result of the first poling step. Then the modulation depth rapidly increases with time as the second poling step takes effect. Interestingly, the maximum modulation depth is reached 50 minutes after the beginning of the second poling step and has a value of 0.65 for this polymer and these poling conditions. It is clear that the two step bidirectional poling process gives greater modulation depth than the unidirectional poling process(0.5) described earlier(2-4) but does not reach the theoretically maximum value of 1. One could optimize the modulation depth by adjusting the poling temperature which in turn influences the poling rise and decay times. When the two step poling process is carried out for very long times the modulation depth approaches 0.5 because the originally poled regions are completely decayed to zero and only those regions underneath the periodic electrodes are now poled.

Yilmaz et al .(9,10) have proposed and demonstrated an alternative approach to creating periodic bidirectionally poled NLO films. The NLO film was sandwiched between two electrodes with the top electrode transparent to visible radiation. A laser beam was focussed tightly onto the film to locally heat it to near Tg while an electric field was applied. They demonstrated that poled films of lateral dimensions of about 7 μm could be written. By moving the film under the laser beam and periodically reversing the poling field, they claimed that they could fabricate a film with periodic domain reversal. Both our and Yilmaz's approach should be evaluated critically in terms modulation depth of poling, waveguide loss and SHG output efficiency from waveguides.

Fabrication of waveguides. The fabrication of a bidirectionally poled SHG waveguide is shown schematically in figure 3. A waveguide structure is fabricated as described above but with one difference. A thin layer of polyimide polymer(SIXEFF, Hoechst) is spun coated on top of the upper cladding layer. The purpose of the polyimide layer is to stop any deformation of the waveguide/cladding layer during the periodic poling process due to electrostriction. The polyimide has a Tg=250 C and so it does not deform at the poling temperatures of 95 ^{0}C used in this work. Then the two poling steps are carried out. The waveguide is first uniformly poled near Tg(115 ^{0}C) for a time t = 1-5 minutes at 140 V/μm and cooled down with the electric field on. The electrodes are periodically etched using a room temperature lithographic process so that no poling decay occurs. Then the second poling process takes place. The polarity is reversed and the waveguide structure is heated up to a temperature 20 ^{0}C below Tg. The electric field (-140 V/μm) is applied for t = 30 minutes to the periodic electrodes and then cooled down with the field on. According to the results of figure 2 the modulation depth should be near 0.5 for that time period of poling. We did not measure the optical waveguide losses.

The periodic waveguides were tested for phase matched SHG using a tunable Q switched nanosecond laser with a wavelength near 1.3 μm.The laser beam was coupled into the slab waveguide with cylindrical lenses and the output SHG signal measured as a function of wavelength. SHG versus fundamental wavelength is shown in figure 4. One notes the narrow tuning bandwidth showing that phase matching has occured over a distance greater than 3 mm. These results verify that a two step poling process is a viable approach to creating periodic domain reversal in polymer films. More work needs to be done to find the optimum temperature of poling, the optimum time for poling and characterizing the modulation depth and waveguide losses in these bidirectionally poled films. The principle of the two step poling process described here may also have applications wherever χ^2 regions of opposite sign are needed, for example, in a push pull Mach Zehnder modulator.

Bidirectionally Poled Free Standing Films

NLO polymers have also been studied for through the plane applications such as modulators for optical interconnects(14), photorefractive elements(15) for image processing and frequency doubling(5,6). The advantage of polymers is that large area mechanically robust NLO elements can be fabricated either by spin coating or by a film extrusion process. In contrast inorganic and organic crystals are difficult to grow and generally small in size. In the case of frequency doubling of femtosecond lasers there is another advantage in so far that very thin NLO films can be fabricated which have low group velocity dispersion(12). Recently we proposed and demonstrated quasi phase matched SHG from a periodic free standing film(5,6) depicted in figure 5. The thickness of each film equaled a coherence length at the optimum phase matching angle of 50 ^{0}. Knoessen et al.(13) analysed theoretically the transmission of a femtosecond pulse and the buildup of a femtosecond SHG pulse through the periodic structure depicted in figure 5 and concluded that under certain conditions, no pulse broadening should occur.

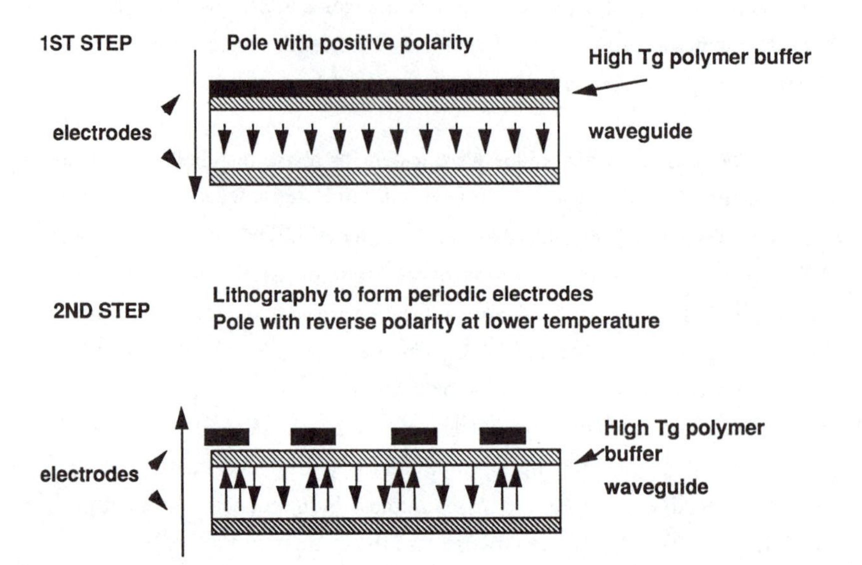

Figure 3. Schematic of fabrication steps of bidirectionally poled SHG waveguides .

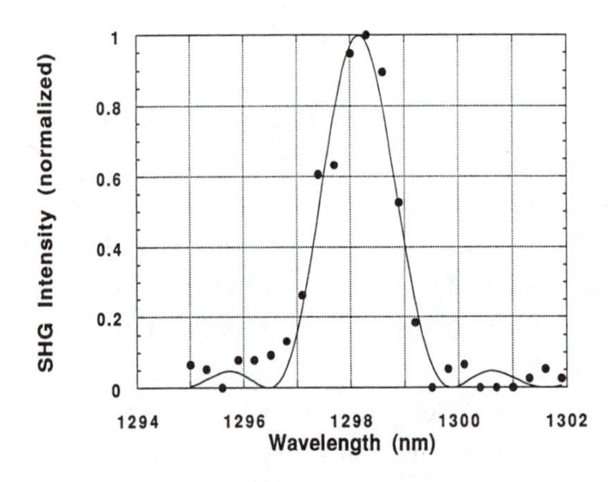

Figure 4. SHG versus fundamental wavelength from bidirectionally poled waveguide

Fabrication of free standing films. The fabrication of these periodic structures was as follows. The poled polymer used was MO3ONS/MMA with molar composition of 10/90. A silicon wafer was first vacuum deposited with 500 A° of gold(Au). This metallic layer served as the poling electrode and as the lifting-off layer for making free standing films. The polymer was spun coated and dried to give a thickness of 5.9 μm. The entire film surface was deposited with 1000 A° of Au and poled at a temperature of 110 °C and at a field of 70 V/μm. After the Au upper electrode was etched off, small squares (0.7x0.7cm) were cut into the film using an ablation process with a LPX 100 Excimer laser (λ = 193 nm). The wafer was then immersed in Au etchant for a few minutes until the Au dissolved and the square pieces floated to the surface. The films were picked up and placed on a Teflon sheet making certain that the poling direction was alternating between adjacent layers, resulting in a structure depicted in figure 5. Samples were prepared with N= 1, 5, 10,20, 30, and 43 layers.

By reversing the direction of poling one obtains a periodic structure whose χ^2 changes sign. When fundamental radiation is incident at angle θ, the mismatch in propagation vectors is

$$\Delta K(\theta) = K^{2\omega} - 2K^{\omega} \approx \frac{\pi}{Lc\sqrt{1 - (\sin\theta/n^{\omega})^2}} \tag{2}$$

where $K^{\omega} = 2\pi n^{\omega}\cos\theta^{\omega}/\lambda$ ($\omega=2\pi c/\lambda$) is the wave vector and θ^{ω} is the internal angle of incidence, L_c is the coherence length of the nonlinear optical material, $n^{\omega,2\omega}$ are the refractive indices at the fundamental and harmonic frequencies, λ is the fundamental wavelength and c is the speed of light, respectively. When the film thickness is $\Lambda = \pi/\Delta K(\theta_m)$ then the condition for quasi phase matching at θ_m is satisfied and the second harmonic adds up coherently. The optimum angle θ_m for obtaining maximum SHG is determined by a tradeoff between the projection of the $\chi 2$ tensor on the incident optical field and the linear t^{ω} and nonlinear $T^{2\omega}$ Fresnel transmission factors, and is near 50-60 degrees. A pulsed YAG laser (λ=1.06 μm) was incident on a periodic stack and the SHG was measured as a function of incident angle. The results are shown in figure 6. One notes that as the film thickness increases, the angular dependence of the SHG signal becomes sharper indicating a longer phase matching length. Figure 7 depicts the relative efficiency

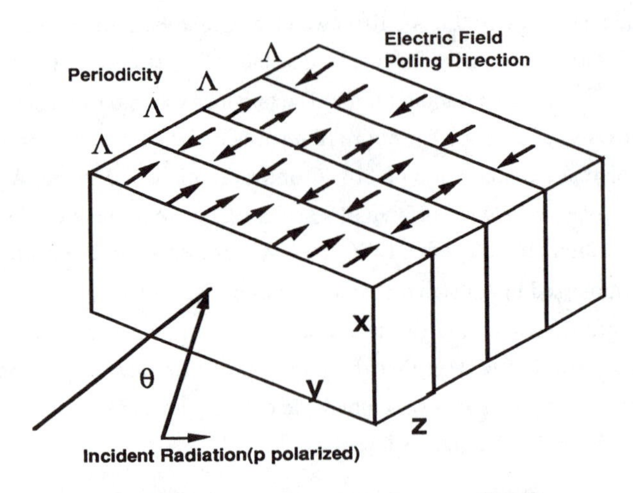

Figure 5. Schematic of periodic free standing film

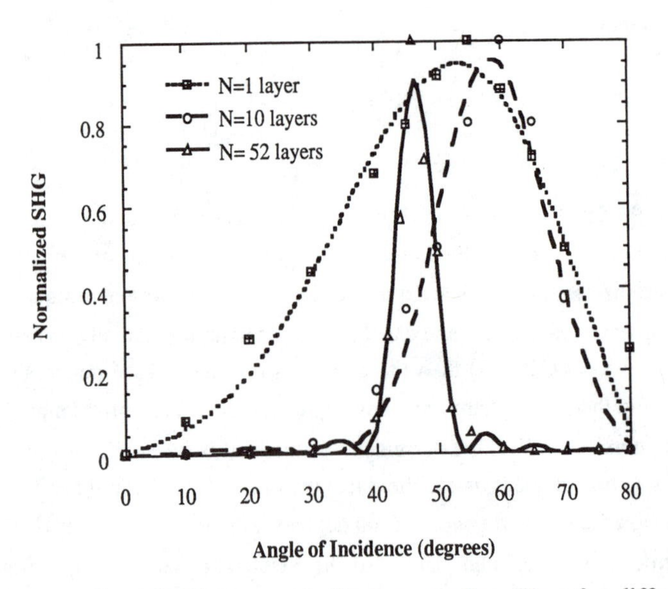

Figure 6. Normalized SHG from periodic free standing films for different film thicknesses.

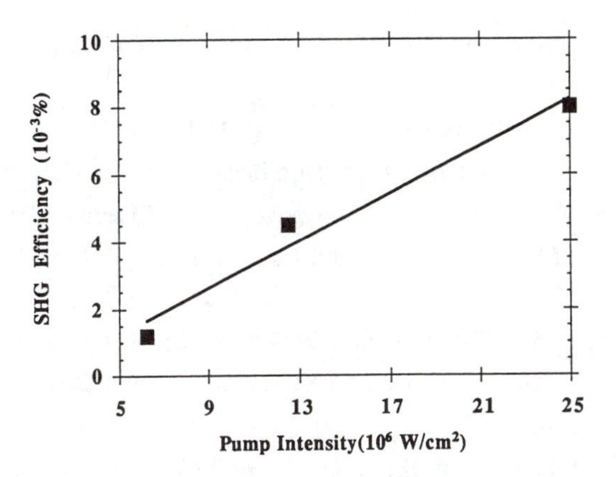

Figure 7. Relative efficiency of SHG versus fundamental power for a periodic free standing film of thickness 160 μm.

of SHG from a periodic stack of thickness 160 μm as a function of fundamental intensity. One observes bright green light from the polymer stack with an efficiency of 10^{-2} %. Further improvements will come from using nonlinear optical polymers with larger χ^2, more efficient poling and combining it with mechanical stretching(5). Therefore it is possible to fabricate polymer "synthetic crystals" for efficient frequency doubling and parametric processes.

Conclusions

We have demonstrated experimentally in this paper that periodic structures with domain reversal are possible in both waveguides and free standing films, and that efficient quasi phase matched SHG is observed. In the case of waveguides we describe a novel two step poling process with a reversal in the sign of the poling field resulting in periodic domain reversal along the length of the waveguide. In the case of the free standing films, we pole a film and then physically reverse the direction of the film during the fabrication of the multilayer film.

References

1. Zyss, J. and Chemla, D.S. in " Nonlinear Optical Properties of Organic Molecules and Crystals", Zyss, J. and Chemla, D.S.(eds.)(Academic Press, Orlando, 1987) chapter II-1,

2. Khanarian, G.; Norwood,R.A.; Haas, D.; Feuer, B. and Karim, D. *Appl. Phys. Lett.* **1990** ,57, 977

3.Norwood,R. A. and Khanarian,G. *Elec. Lett.* **1990**, 26, 2105

4.Khanarian,G. and Norwood,R. A. in "Nonlinear Optics: Fundamentals,Materials and Devices" S. Miyata (ed.)(Proc. Fifth Toyota Conf.,Elsevier, 1992) p.461

5. Khanarian, G.; Mortazavi, M. A. and East, A. J. *Appl. Phys. Lett.* **1993**, 63, 1462

6.Mortazavi, M.A. and Khanarian,G. , *Optics Lett.* **1994**, 19, 1290

7. Zimmerman, K.; Ghebremichael, F.; Kuzyk, M. G. and Dirk, C. W. , *J. Appl. Phys.* **1994**, 75,1267

8. Ghebremichael, F. and Kuzyk, M. G., *J. Appl. Phys.* (in press, 1995)

9. Yilmaz, S.; Bauer, S.;Wirges, W. and Gerhard-Multhaupt, R., *Appl. Phys. Lett.* **1993**, 63, 1724

10. Yilmaz, S.; Bauer, S. and Gerhard-Multhaupt, R., *Appl. Phys. Lett.* **1994**, 64, 2770

11.Khanarian, G., to be published

12.Mortazavi, M. A.; Yankelvich, D. ;Dienes, A.;Knoesen, A.; Kowel, S. T. and Dijali,S., *Appl. Opt.* **1989**, 28, 3278

13.Sidick, E.; Knoesen, A. and Dienes, A., *Opt. Lett.* **1994**, 19, 266

14. Yankelvich,D.R.; Hill, R. A.; Knoesen, A.; Mortazavi, M.A.;Yoon, H. N. and Kowel, S.T., *Photonics Lett.* **1994** , 6

15.Ducharme, S.; Scott, J. C.; Tweig, R. J. and W. E. Moerner, *Phys. Rev. Lett.*, **1991**, 66, 1846

RECEIVED March 10, 1995

Chapter 37

Cascading of Second-Order Nonlinearities

Concepts, Materials, and Devices

William E. Torruellas[1], Dug Y. Kim[1], Matthias Jaeger[1], Gijs Krijnen[1], Roland Schiek[1], George I. Stegeman[1], Petar Vidakovic[2], and Joseph Zyss[2]

[1]Center for Research and Education in Optics and Lasers, University of Central Florida, 12424 Research Parkway, Orlando, FL 32826
[2]France Telecom, Centre National d'Etudes des Télécommunications, Paris B, BP 107, 196 Avenue Henri Ravera, 92225 Bagneux, France

The concept of utilizing second order nonlinear optical processes to mimic third order nonlinearities, in particular the optical Kerr effect, is discussed. Nonlinear organic materials with their large second order nonlinearities offer definite advantages for the implementation of such concepts. As a demonstration, we present both experimental and numerical results in the two important telecommunication windows on a DAN single crystal core fiber originally designed for efficient second harmonic emission in the Čerenkov geometry at 820 nm.

When third order nonlinear phenomena are investigated in non-centrosymmetric materials, two successive (cascaded) second order processes can contribute to the intrinsic third order response for some effects.*(1)* As early as 1967, Ostrowskij predicted that cascading could play a prime role in self-action processes or any of the geometries involving the mixing of four degenerate waves.*(2)* The two contributions which lead to intensity-dependent effects for a single input beam are shown schematically in Figure 1. The presence of the so-called cascaded nonlinearities can interfere with and even mask the direct or intrinsic third order effects. In the early days of nonlinear optics such interference effects were reported in studies of nonlinear phenomena in GaAs.*(3,4)* Cascading has also been used to calibrate the third order nonlinear susceptibility of reference materials such as α-quartz whose nonlinearities are in turn are used as references for Third Harmonic Generation evaluations of nonlinearities in fused silica etc.*(5)* In general cascading in bulk materials has not led to a large intensity-dependent refractive index.*(6-9)* However, recent advances have produced both new organic second order materials with large nonlinearities as well as new methods of phase-matching existing materials in waveguides utilizing the large nonlinear coefficients normally not available for phase-matching bulk crystals.*(10-13)* This has led to new experiments in bulk organic materials as well as in waveguides.*(8,14-16)* The low powers expected in waveguides have led to a number

0097–6156/95/0601–0509$12.00/0

of predicted applications to signal processing, spatial solitons etc.*(17-22)* In fact, a novel form of all-optical switching has already been demonstrated.*(23)*

Experimental evidence for surprisingly large nonlinearities which were subsequently identified to be due to cascading was available already in 1990.*(24)* Yamashita and coworkers reported self-phase modulation in DAN crystal core fibers which corresponded to record high, instantaneous, intensity dependent refractive indices.*(24)* They used the effect to pulse compress the output of a femtosecond CPM dye laser.*(25)* These fibers were initially fabricated for efficient Cerenkov doubling of near infrared light into the blue.*(26)* By operating far from any molecular resonance, we have now clearly identified both experimentally and numerically the origin of Yamashita's SPM result as being due to cascading via the presence of a Cerenkov SHG in those fibers.*(15,27)* Additional confirmation of cascading was obtained from Z-scan measurements in a bulk single crystal of the same material used in the crystal core fibers.*(15)*

The latter results open, in our opinion, a very interesting area of research. Although SHG in organic crystal core fibers may be considered to be an attractive approach to direct doubling of diode lasers into the blue region of the spectrum, this approach has been slowed down because of photo-induced damage in the blue as well as the recent improvements in short wavelength diode lasers.*(28,29)* Phase matchable organic materials are, however, still extremely attractive because of their large figure of merit for SHG in devices using cascading as the source of nonlinear phase distortion, in particular within the two telecommunication windows demonstrated here. We show that such an effect gives rise to effective nonlinear coefficients of the same order of magnitude as the largest intrinsic optical Kerr coefficients reported at 1.3 and 1.6 μm.*(30,31)* Extrapolating these results to other materials and geometries, the possibility of exceeding these latter values by several orders of magnitude makes this approach in organic polymers and crystals probably the only viable one for reducing the switching power of a lossless all-optical switch to below the magical 1 Watt level.

Concepts:

As previously mentioned, cascaded nonlinearities play an important role in non-centrosymmetric materials with phase-matchable nonlinearities. In the case of SHG and for plane waves in the CW regime we can describe this effect by invoking a simple coupled mode analysis. This leads to the following two coupled partial differential equations describing the propagation of the two fields at the fundamental and second harmonic frequencies*(18)*:

$$\frac{\partial E^{\omega}}{\partial z'} = i\kappa E^{2\omega} E^{\omega *} e^{i\Delta k z'}$$
$$\frac{\partial E^{2\omega}}{\partial z'} = i\kappa (E^{\omega})^2 e^{-i\Delta k z'}$$

Here κ defines a coupling coefficient which is proportional to the effective second order nonlinear susceptibility $d_{eff}^{(2)}$ for the particular SHG interaction (propagation direction, light polarizations etc.), and $\Delta kz'$ is the linear phase mismatch between the two fields. Integrating these equations in the low depletion approximation is instructive for the understanding of the physics involved in the cascading process:

$$\frac{\partial I^{\omega}}{\partial z'} = \kappa^2 \frac{\sin(\Delta k z')}{\Delta k} (I^{\omega})^2$$

$$n_2^{eff} = \frac{\kappa^2}{k_{vac}} \frac{1 - \cos(\Delta k z')}{\Delta k}$$

Here I^{ω} refers to the fundamental light intensity, and n_2^{eff} (henceforth written as just n_2) is the effective cascaded nonlinearity. κ^2 is proportional to $[d_{eff}^{(2)}]^2/n^3$ which has to be as large as possible. Hence the same figure of merit occurs for cascading and SHG. An additional prerequisite for large nonlinearities is that the SHG interaction should be almost phase-matched. To first approximation, one can see that the "space averaged" nonlinearity will be proportional to $1/\Delta k$, for large ΔkL. Finally a consequence of the previous derivation is that the sign of the effective intensity dependent nonlinearity changes when the phase mismatch changes sign, a clear signature of the cascading process being present and dominant.

Note that the above analysis is only valid very far from phase matching. Because cascading is a consequence of a strong coupling between the two fields present, the low depletion approximation is primarily useful for pedagogical reasons and to gain some insight into the problem. Examining the above equations, some analogy with a two photon absorber is apparent. The phase-matching condition has in fact created an electromagnetic resonance for the fundamental field. However, when solved more completely, the above equations show significant differences relative to two-photon absorption.*(18,32)* However the most important one is that in the cascading case no photonic energy is lost to the material system as would be the case in two photon absorption. The energy is only transferred to other electromagnetic frequencies (second harmonic in this case). Therefore it is possible to recover the energy in the fundamental field. From this perspective the cascading approach to enhancing the effective Kerr nonlinearity is lossless in nature and hence does not induce thermo-optic effects if linear absorption is absent. If no second-harmonic is desired it is possible to design the "electromagnetic resonance" in order to induce large nonlinear phase shifts and no or small amounts of fundamental depletion as demonstrated in the following discussion.

Materials:

Table I clearly indicates that organic materials are far more promising than existing inorganics for cascading. For operation within one of the telecommunication windows at 1300nm or 1550nm these materials do not suffer from the detrimental absorption and damage associated with blue-green generation in organics.*(28)* Table I also shows that this approach could lead to an enhancement of several orders of magnitude over the existing non-resonant third order nonlinearities of semiconductors (~10^{-13} cm^2/W) and 1D π-electron conjugated polymers(~$4x10^{-12}$ cm^2/W).*(30,31,33)* As a result we believe that from a materials perspective, cascading in organics is extremely promising.

Table I:

λ_c refers to the estimated cut-off wavelength for linear and nonlinear absorption. d_{ii} and d_{ij} are respectively the largest diagonal and off-diagonal second order nonlinear coefficients. n_2 is the intensity dependent refractive index coefficient. P_s is the estimated switching power for a 2 cm long device with a 5 μm^2 channel waveguide area producing a 2π nonlinear phase shift at 1500 nm. It is approximately equal to $4x10^{-12}$ $/n_2$ (Watt).

	λ_c nm	d_{ij} (pm/V)	d_{ii} (pm/V)	n_2^{eff} cm^2/W	P_s Watt
AlGaAs[33]	1450 nm			$1x10^{-13}$	37.5
PTS[31]	1300 nm			$2x10^{-12}$	2
KTP[7]	700 nm	6	18	$1x10^{-13}$	40
LiNbO3[14]	700 nm	6	25	$2x10^{-13}$	25
DAN[15]	900 nm	50		$1x10^{-12}$	3.75
NPP[10]	900 nm	98		$5x10^{-12}$	0.8
MMONS[38]	900 nm	85		$4x10^{-12}$	1
Poled Polymer[39]	900 nm		200 (DANS)	$1x10^{-11}$	0.4
DAST[40]	1400 nm		600	$1x10^{-10}$	0.04

However at this time inorganic materials are more generally available and phase-matching techniques are much better developed in inorganics than for organic materials.*(12)* Hence one can expect a time lag in the implementation of organic cascading devices. Nevertheless we believe that because of their flexibility and integrability, organic polymers will ultimately be the material class of choice when all-optical devices and architectures are finally implemented. One should also note the extremely high d_{33} coefficient of DAST single crystals, in which the implementation of phase-matching involving other than a birefringence technique, quasi-phase matching or Čerenkov phase matching for example, could result in a very large enhancement.*(11)*

Nonlinearity Measurements:

Single crystal fibers of organic molecules of the paranitroaniline family have previously been grown for efficient SHG in the blue-green spectral region.*(26,34,35)* The single mode fiber geometry for DAN crystal core is shown in Figure 2, where the angle between the Z axis and the fiber axis has been determined to be 36 degrees.*(35)* Such fibers are designed to be phase matched under the Cerenkov condition for blue light generation.*(26)* As shown in Figure 3, while the fundamental wavelength at 1320 nm can be guided in the case of an SF1 cladding, the second harmonic polarized along the Y axis has a refractive index in the core smaller than that of the cladding. Efficient SHG for a 1320 nm (and previously 820 nm) input was observed. At the Cerenkov radiation angle in the cladding, the projection of the SHG wavevector along the fiber axis is twice that of the guided wave at the fundamental frequency. Figure 4 shows this condition. Although the radiated SHG is not strongly confined in the core of the fiber, efficient interaction between the guided mode at the fundamental frequency and the portion of the SHG trapped in the core can be realized. Such an interaction gives rise, in fact, to a phase matching condition similar to that observed in the more classical case of copropagating fundamental and harmonic.

We have studied the second harmonic and nonlinear phase shift processes numerically by using a standard BPM (Beam Propagation Method).*(36)* Shown in Figure 5 is the variation in nonlinear phase shift with the difference between the refractive index of the cladding at the second harmonic frequency and the effective refractive index of the guided fundamental. Similar to the bulk case, a change in sign occurs when going through such a phase matching resonance, allowing both negative and positive nonlinearities to be obtained in a Cerenkov geometry. Our fiber was designed with a change in refractive index, abscissa of Figure 5, close to 8×10^{-3}, implying as noted in the previous section that a phase shift of approximately -50° is reached with no considerable depletion being observed, less than 5%. Additional modifications of the phase matching conditions could improve in principle such a loss figure. Note that even though Figure 5 resembles the solution of the standard plane wave, infinite medium model, in our case the abscissa is an arbitrary parameter that we decided to vary in our model and is not the phase mismatch between the guided fundamental mode and the radiated Čerenkov wave.

A number of experiments were performed on single crystal DAN, both in fiber and bulk crystal form at 1064, 1320 and 1550 nm.*(15,27)* The most sophisticated utilized the DAN fiber in one arm of a Mach-Zehnder interferometer and propagation in air for the reference arm.*(15,37)* A typical fringe pattern observed when the length of the reference arm is changed (scanned) is shown in Figure 6. As the intensity of the radiation at 1320 nm in the fiber is increased, there is an additional phase shift due to cascading and the interferometer fringe pattern shifts relative to its low power case, as shown in Figure 6. Purposely moving one of the mirrors of the interferometer allowed us to calibrate the interferometer for a positive or negative nonlinear phase shift. The case shown in Figure 6 corresponds to a reduction of optical path length equivalent to a phase shift of $-\pi/4$. A value for the effective n_2 of -8×10^{-13} cm^2/W was obtained for a DAN core fiber with a SF1 glass cladding, in reasonable agreement with the calculated value.

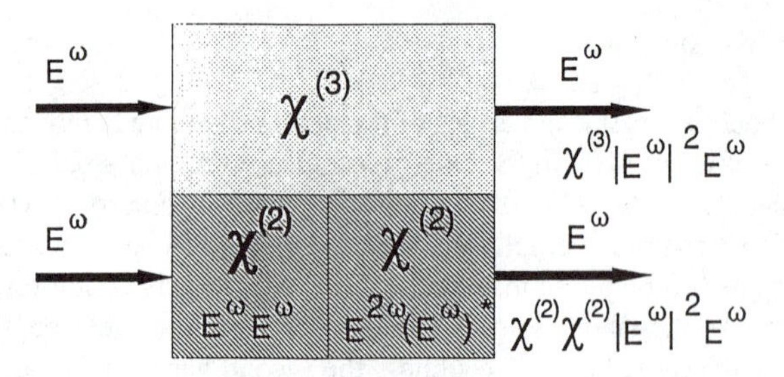

Figure 1. Schematics of the cascaded (small depletion limit) and direct Kerr processes. Both lead to effective nonlinearities proportional to $|E^{\omega}|^2$.

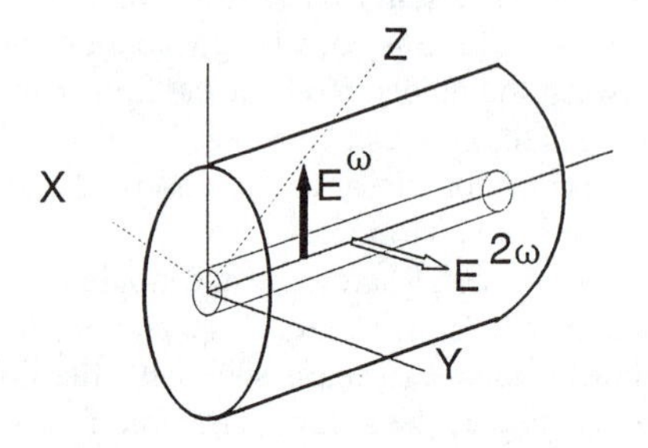

Figure 2. Single crystal DAN fiber geometry and the polarizations for the input fundamental and second harmonic generated by the Cerenkov process.

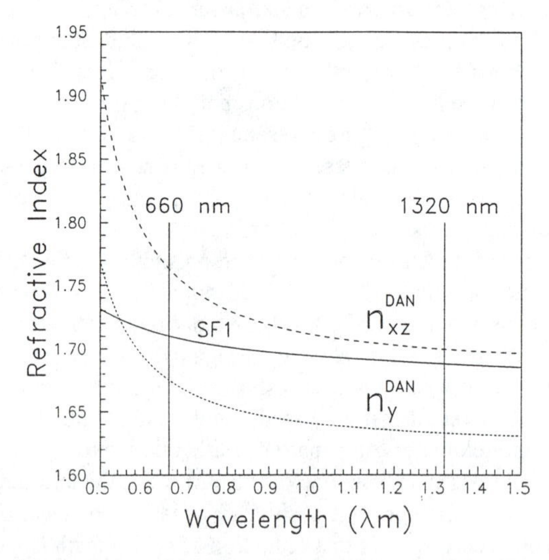

Figure 3. Refractive index dispersion with wavelength of the SF1 glass and the two normal mode polarizations of the DAN fiber. At 1320 nm n_{xz}(DAN) > n_{clad} showing that the fundamental is guided, where-as n_{clad}(660nm) > n_{xz}(DAN,1320nm) indicating the possibility of a Cerenkov phase matching condition.

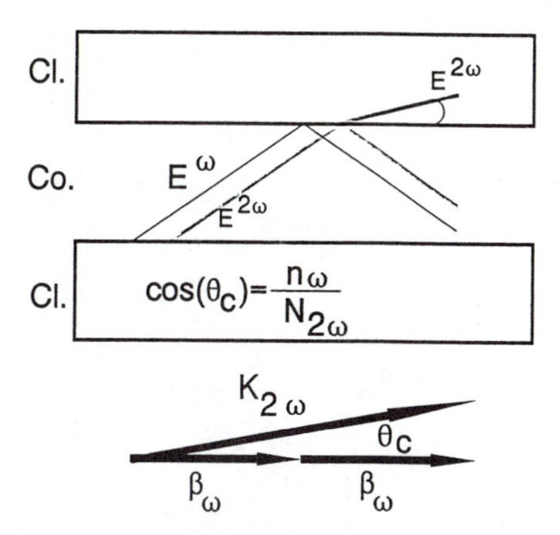

Figure 4. The Cerenkov phase matching condition with SHG generation at an angle θ_c into the cladding. Note that the SH field in the core is reflected at the core-cladding interface. For small angles θ_c the reflection coefficient of the SH field approaches unity, overlaps well the fundamental and can copropagate with it for fairly long distances. N_ω is the effective index of the fundamental and $n_{2\omega}$ is the cladding index at the harmonic frequency with $\cos\theta_c = N_\omega/n_{2\omega}$. The schematic Čerenkov phase matching condition is also represented.

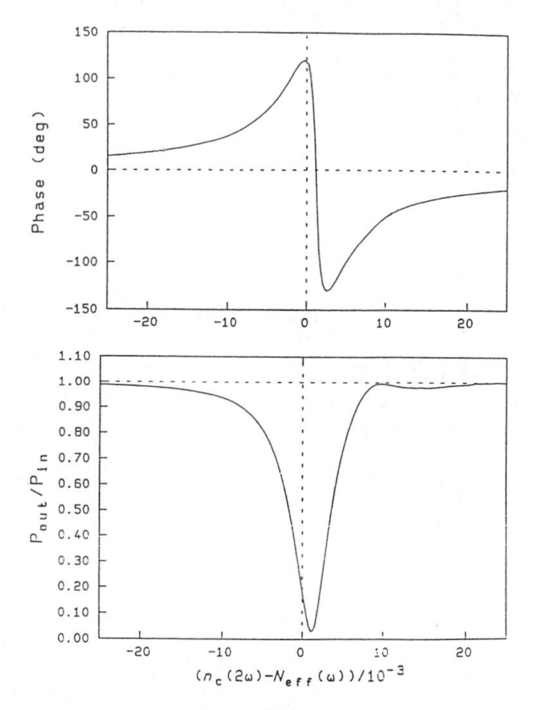

Figure 5. Results of a numerical calculation for the fundamental power and nonlinear phase shift at the output of a single mode DAN fiber under the Cerenkov condition. Here $n_c(2\omega)$ and $N_{eff}(\omega)$ are the cladding refractive index and guided wave effective index respectively.

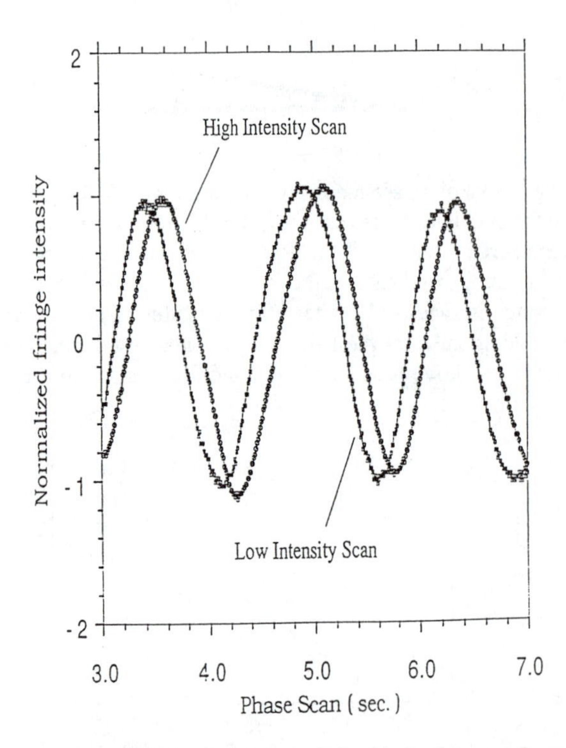

Figure 6. Low and high intensity scans of the fringe patterns from the Mach-Zehnder interferometer measured at 1320 nm corresponding to a negative intensity dependent refractive index.

Broadening of the frequency spectrum of a short pulse due to Self-Phase Modulation (SPM) was also used to estimate the effective nonlinearity at 1550 nm. Modulation of the carrier frequency could be observed when femtosecond pulses were launched into the DAN fiber. In this case the self-phase modulation signal (Figure 7) corresponds to an intensity dependent refractive index $|n_2|$ of $4x10^{-13}$ cm^2/W. Although only the magnitude of the effective nonlinearity can be deduced with this technique, reasonable agreement is obtained with the more accurate interferometric measurement at 1320 nm. (In fact the same value is not expected since the Cerenkov cross-section will change with wavelength.)

Both experiments had an important feature in common. In both cases the observation of an effective Kerr nonlinearity was correlated with the presence of a strong SHG field in the core. When different cladding glasses were used, the SHG intensity in the core was considerably reduced making the observation of a phase shift or a broadening due to SPM impossible at similar fundamental input powers. The correlation is that second harmonic conversion is necessary for large nonlinear phase shifts.

In the fibers the orientation of the crystal relative to the fiber axis is fixed by the growth technique used and the dependence of the nonlinearity on phase mismatch cannot be investigated conveniently. Therefore in order to verify the change in the sign of the nonlinearity at phase-matching, the nonlinear phase shift and effective nonlinearity n_2 were measured using Z-scan in a DAN bulk crystal at 1064 nm near the phase matching condition. The crystal was generously donated by the IBM group headed by Dr. G. Bjorklund. Figures 8 and 9 show the results for the fundamental and second harmonic powers, and the effective nonlinearity versus incidence angle near phase-matching. Integrating Eq.1 for a gaussian pulse in time and space under the

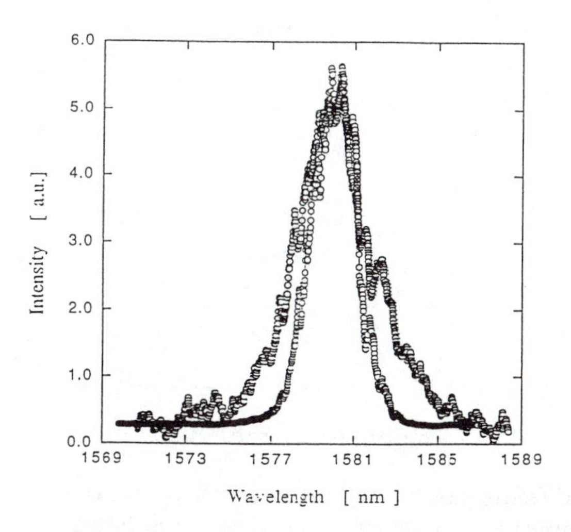

Figure 7. High power spectral broadening of 0.5 psec pulses output from a DAN crystal core fiber relative to the low input power case. Self-phase-modulation occurs due to cascading at 1550 nm, showing the ultrafast response of the nonlinearity.

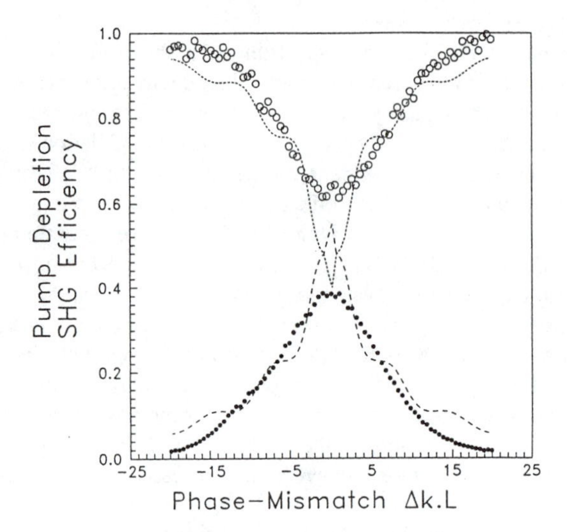

Figure 8. Phase matching resonance for SHG in a DAN single crystal. Note the large depletion of the fundamental beam (open circles) and the strong conversion to second harmonic (solid circles). ΔkL is the cumulative phase mismatch in the crystal L long.

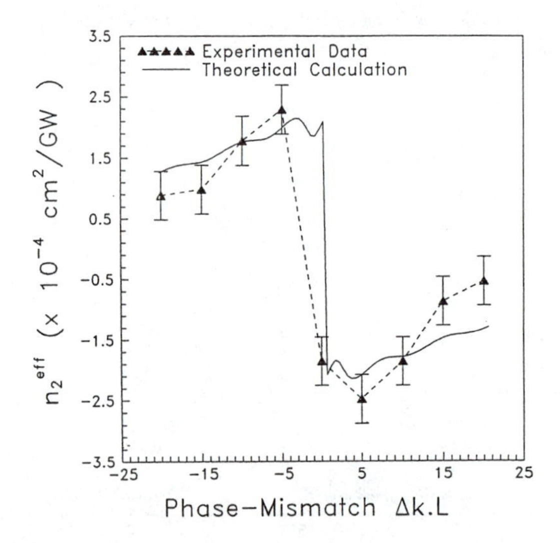

Figure 9. Effective intensity dependent refractive index coefficient measured at 1064nm with the Z-scan technique near the phase matching resonance versus cumulative phase mismatch ΔkL. The solid line represents the results of a numerical model which integrates Eq.1 for no walk-off and negligible diffraction.

assumptions of no diffraction and small spatial walk-off gives good agreement with the experimental results. The change in sign in n_2 clearly occurs at phase-match. Therefore we conclude that cascading of two second order processes is clearly the dominant mechanism for self action in DAN crystals when operating near a phase matching resonance.

In conclusion we have shown that within both telecommunication windows, cascaded nonlinearities in organic single crystal core fibers can mimic very large third order effects. Comparison with other organic crystals and polymers shows that this approach could lead to effects several orders of magnitude larger than those observed in the currently most effective third order materials.

The research at CREOL, was supported by a grant from the AFOSR. While at CNET, WT was a NSF/NATO postdoctoral fellow. The authors are grateful to the IBM group, for allowing them access to a DAN single crystal.

Literature Cited:

1. see for example Flytzanis,C., in *Quantum Electronics*; ed. Rabin, H.; Tang, C.L. (Academic, NY, **1975**), Vol 1, Part A.
2. Ostrowskij, L. A. *JETP Lett.*, *10*, 281 (**1967**).
3. Yablonovitch, E.; Flytzanis, C.; Bloembergen, N. *Phys. Rev. Lett.*, *29*, 865 (**1972**).
4. Gustafson, T.K.; Taran, J.-P.E.; Kelley, P.L.; Chiao, R.Y. *Opt. Comm.*, *2*, 17 (**1970**).
5. Meredith, G.R. *Phys. Rev. B*, *24*, 5522 (**1981**); Buchalter, B.; Meredith, G.R. *Appl. Opt.*, *21*, 3221 (**1982**).
6. Belashenkov, N. R.; Gagarskii, S. V.; Inochkin, M. V. *Opt. Spectrosc.*, *66*, 1383-6 (**1989**).
7. DeSalvo, R.; Hagan, D.J.; Sheik-Bahae, M.; Stegeman,G.I.; Vanherzeele, H.; Van Stryland, E.W. *Opt. Let.*, *17*, 28 (**1992**).
8. Danielius,D.; Di Trapani, P.; Dubietis, A.; Piskaskas, A.; Podenas, D.; Banfi, G.P. *Opt. Lett.*, *18*, (**1993**) 574.
9. Nitti, S.; Tan, H.M.; Banfi, G.P. and Degiorgio, V. *Opt. Comm.*, *106*, 263 (**1993**).
10. Zyss, J.; Chemla, D.S., "Quadratic Nonlinear optics and Optimization of the Second-Order Nonlinear Optical Response of Molecular Crystals", in *Nonlinear and Optical Properties of Organic Molecules and Crystals*; eds. Chemla, D.S.; Zyss, J., Vol 1 (Academic Press, **1987**) pp23-192
11. Marder,S.R.; Perry, J.W.; Schaefer, W.P., *Science 245*, 626 (**1989**).
12. Fejer, M. M.; Magel, G. A.; Jundt, D. H.; Byer, R. L., *IEEE J. Quantum Electron. 28*, 2631 (**1992**).
13. van der Poel, C. J.; Bierlein, J. D.; Brown, J. B.; Colak, S., *Appl. Phys. Lett. 57*, 2074 (**1990**).
14. Sundheimer, M.L.; Bosshard, Ch.; VanStryland, E.V.; Stegeman, G.I.; Bierlein, J.D., *Opt. Lett.*, *18*, 1397 (**1993**).

15. Kim, D.Y.; Torruellas, W.E.; Kang, J.; Bosshard, C.; Vidakovic, P.; Zyss, J.; Moerner, W.; Twieg, R.; Bjorklund, G., *Opt. Lett., 19*, 868 (**1994**).
16. Schiek, R.; Sundheimer, M.L.; Kim, D.Y.; Baek, Y.; Stegeman, G.I.; Suche, H.; Sohler, W., "Direct Measurement of Cascaded Nonlinearity in Lithium Niobate Channel Waveguides", *Opt. Lett.*, in press
17. Schiek, R., *J. Opt. Quant. Electron., 26*, 415 (**1994**).
18. Stegeman, G.I.; Sheik-Bahae, M.; VanStryland, E.; Assanto, G., *Opt. Lett., 18*, 13 (**1993**).
19. Assanto, G.; Stegeman, G.I.; Sheik-Bahae, M.; VanStryland, E., *Appl. Phys. Lett., 62*, 1323 (**1993**); Assanto, G.; Stegeman, G.I.; Sheik-Bahae, M.; VanStryland, E., "Coherent Interactions for All-Optical Signal Processing via Quadratic Nonlinearities", *J. Quant. Electron.*, submitted
20. St. Jean Russell, P., *Electron. Lett., 29*, 1228 (**1993**).
21. Werner, M.J.; Drummond, P.D., *Opt.Lett., 10*, (**1993**) 2390.
22. Torner, L.; Menyuk, C.R.; Stegeman, G.I., "Excitation of Soliton-like Waves with Cascaded Nonlinearities", *Opt. Lett.*, in press
23. Hagan, D.J.; Sheik-Bahae, M.; Wang, Z.; Stegeman, G.I.; VanStryland, E.W., "Phase Controlled Transistor Action by Cascading of Second-Order Nonlinearities in KTP", *Opt. Lett.*, in press
24. Yamashita, M.; Torizuka, K.; Uemiya, T., *Appl. Phys. Lett., 57*, (**1990**) 1301.
25. Yamashita, M.; Torizuka, K.; Uemiya, T.; Shimada, J., *Appl. Phys. Lett., 58*, 2727 (**1991**)
26. see for example, Chikuma, K.; Umegaki, S., *J. Opt. Soc. Am. B, 9*, 1083 (**1992**).
27. Torruellas, W.E.; Schiek, R.; Kim, D.Y.; Krijnen, G.; Stegeman, G.I.; Vidakovic, P.; Zyss, J., "Cascading Nonlinearities in an Organic Single Crystal Core Fiber: The Čerenkov Regime", *Opt. Comm*, in press
28. Heard, D.; Yano, K.; Inoue, Y.; Kijima, T.; Ide, T.; Arai, H.; Yano, S., *Technical digest of CLEO'94* (Opt. Soc. Am., Washington, **1994**), paper CTuK13 p90.
29. for example, Ishihara, T.; Brunthaler, G.; Walecki, W.; Hagerot, M.; Nurmiko, A.V.; Samarth, N.; Luo, H.; Furdyna, J., *Appl. Phys. Lett., 62*, 2460 (**1992**).
30. Kim, D.Y.; Lawrence, B,L.; Torruellas, W.E.; Baker, G.; Meth, J., "Assessment of Single Crystal PTS as an All-optical Switching Material at 1.3 µm", *Appl. Phys. Lett.*, in press
31. Lawrence, B.; Cha, M.; Kang, J.U.; Torruellas, W.; Stegeman, G.I.; Baker, G.; Meth, J.; Etemad, S., *Electron. Lett., 30*, 447 (**1994**).
32. Eckardt, R. C.; Reintjes, J. , *IEEE J. Quant. Electron., QE-20*, 1178 (**1984**).
33. Villeneuve, A. ; Yang, C.C.; Stegeman, G.I.; Lin, C.-H.; Lin, H.-H., *Appl. Phys. Lett, 62*, 2465 (**1993**).
34. Vidakovic, P.V.; Coquillay, M.; Salin, F., *J. Opt. Soc. Am., B, 4*, 998 (**1987**).

35. Kerkoc, P.; Zgonik, M.; Sutter, K.; Bosshard, Ch.; Gunter, P., *J. Opt. Soc. Am. B, 7*, **(1990)**
36. Hoekstra,H.J.W.M.; Krijnen, G.J.M.; Lambeck, P.V., *Opt. Comm., 97*, 301 (**1993**).
37. Kim, D.Y.; Sundheimer, M.; Otomo, A.; Stegeman, G.I.; Horsthuis, W.G.H.; Mohlmann, G.R., *Appl. Phys. Lett., 63*, 290 (**1993**).
38. Bierlein, J.D. ; Cheng, L.K.; Wang, Y.; Tam, W., *Appl. Phys. Lett., 56*, 423 (**1990**).
39. Otomo, A.; Stegeman, G.I.; Horsthuis, W.H.G.; Mohlmann, G.R., *Appl. Phys. Lett., 65*, 2389 (**1994**).
40. Lawrence, B.L., M.S. Dissertation, M.I.T., May **1992**.

RECEIVED February 14, 1995

Author Index

Affiliation Index

Subject Index

F

H

N

Production: Amie Jackowski
Indexing: Deborah H. Steiner
Acquisition: Anne Wilson & Barbara E. Pralle
Cover design: Amy Hayes

Printed and bound by Maple Press, York, PA

YORK COLLEGE OF PENNSYLVANIA 17403

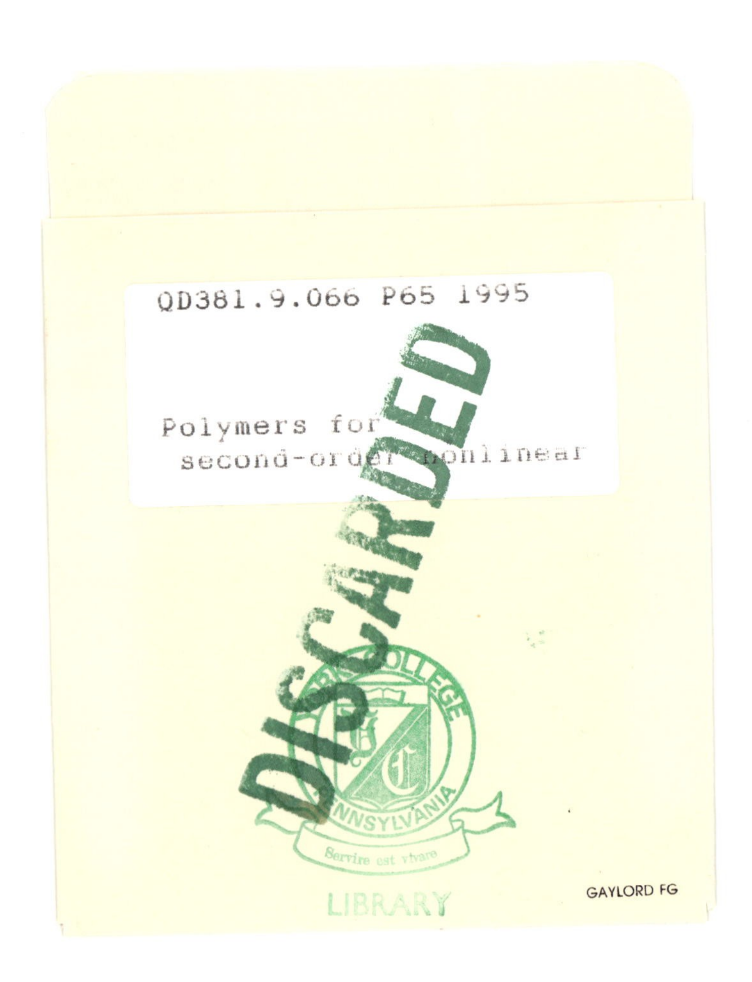
QD381.9.066 P65 1995
Polymers for
second-order nonlinear
DISCARDED
LIBRARY
GAYLORD FG